INTERNATIONAL MINING FORUM 2010

T0172584

19-21 SEPTEMBER 2010 – HUAINAN, CHINA

# Mine Safety and Efficient Exploitation Facing Challenges of the 21st Century

## International Mining Forum 2010

*Edited by*

Liu Zegong, Hua Xinzhu, Yuan Shujie, Dai Guanglong
*Anhui University of Science and Technology, Huainan, China*

Jerzy Kicki
*AGH University of Science and Technology, Cracow, Poland*
*Polish Academy of Sciences, Mineral and Energy Economy Research Institute, Cracow, Poland*

Eugeniusz J. Sobczyk
*Polish Academy of Sciences, Mineral and Energy Economy Research Institute, Cracow, Poland*

CRC Press
Taylor & Francis Group
Boca Raton London New York Leiden

CRC Press is an imprint of the
Taylor & Francis Group, an **informa** business
A BALKEMA BOOK

*CRC Press/Balkema is an imprint of the Taylor & Francis Group, an informa business*

© 2010 Taylor & Francis Group, London, UK

Typeset by Jacek Jarosz
Printed and bound in China

All rights reserved. No part of this publication or the information contained herein may be reproduced, stored in a retrieval system, or transmitted in any form or by any means, electronic, mechanical, by photocopying, recording or otherwise, without written prior permission from the publishers.

Although all care is taken to ensure integrity and the quality of this publication and the information herein, no responsibility is assumed by the publishers nor the author for any damage to the property or persons as a result of operation or use of this publication and/or the information contained herein.

Published by: CRC Press/Balkema
P.O. Box 447, 2300 AK Leiden, The Netherlands
e-mail: Pub.NL@taylorandfrancis.com
www.crcpress.com – www.taylorandfrancis.co.uk – www.balkema.nl

ISBN: 978-0-415-59896-5

International Mining Forum 2010, Liu et al. (eds) © 2010 Taylor & Francis Group, London, UK. ISBN 978-0-415-59896-5

# Table of Contents

# Preface

Coal is one of the world's basic energy resources, in 2009, global coal production was 3.409 billion tons of oil equivalent (equivalent to approximately 6.941 billion tons raw coal). Among them, China's coal output in the proportion of total global coal production was 45.6%.

In 2009 there were 10 countries whose coal production is more than 100 million tons. In the 10 countries total coal output was 6.212 billion tons, accounting for 89.5% of global coal production. China and Poland are the world's coal production powers. In the 80ths of the 20$^{th}$ century the average exploitation depth in Polish coal mines reached 1000 m, a lot of successful experience in safety and efficient exploitation at great depth has been obtained. Since the beginning of the 21$^{st}$ century, China is also faced with technical problems caused by exploitation at great depth in coal mines. Anhui University of Science and Technology in China and AGH University of Science and Technology in Poland are world-famous due to their effective contribution to coal exploitation and mine safety technology. Many experts and professors of both universities engaged in coal exploitation and safety research and teaching, obtained great achievements. The "International Mining Forum 2010 - Mine Safety and Efficient Exploitation Facing Challenge of the 21$^{st}$ Century" jointly organized by Anhui University of Science and Technology and AGH University of Science and Technology has great significance, providing an opportunity and platform of mutual technical exchanges to experts and scholars engaged in coal exploitation and safety from China, Poland and other countries. The purpose of the forum is, by exchanges and seminars, to promote worldwide research and development of mine safety and efficient exploitation theory and technology, providing theoretical and technical support for improvement of mine safety level.

In recent years, important progress in theoretical and technical research on coal mine safety and efficient exploitation in China has been made. In which theory and technology of integrated coal exploitation and gas extraction originated by Chinese researchers under leadership of professor Yuan Liang, an academician of the Chinese Academy of Engineering, has been successfully applied in the Huainan mining area, and promoted in other coal mines in China. Especially by using key technologies of pilarless integrated coal exploitation and gas extraction in coal seams group of low permeability, technical problems of integrated coal exploitation and gas extraction in complex geological conditions of high gas, low permeability and high strata stress was successfully resolved.

Key technologies of safety and efficient exploitation at great depth of coal mines are: strata control, gas and heat hazard control and ventilation, rock burst control, roadway layout and so on. Key problems needed to study further are: rock status and stress field and its distribution characteristics; gas disaster prevention technologies and equipments; rapid excavation and support technology of roadways (especially in soft rocks) and related equipments; techniques of rock burst control and technology of its monitoring and control; techniques of productive and efficient exploitation; technology and equipments of thermal hazard control. This is also the content wanted to be covered by the theme of this forum.

More than 100 articles from China, Poland and Australia were received, edited into two separate volumes, including: theory and technology of mine gas control, technology of mine gas utilization, integrated coal exploitation and gas extraction, theory and technology of fire prevention and control in mines, theory and technology of heat hazard prevention and control in mines, theory and technology of rock burst prevention and control in deep mines, theory and technology of roadway support in deep mines, technology of roadway rapid excavation and other related areas of research.

I would like to thank the authors and counterparts experts and scholars home and abroad for their support of the forum. I believe the forum will be successful.

I would like to express my thanks to the Materials and Engineering Division of the National Natural Science Foundation of China, National Engineering Research Center of Coal Mine Gas Control, Anhui Coal Institute, Mineral and Energy Economy Research Institute of Polish Academy of Sciences and Key Laboratory of Coal Mine Safety and Efficient Exploitation of Ministry of Education of China for support to the forum.

Chairman of Organizing Committee, IMF 2010 (Huainan, China)

Professor Liu Zegong

August 2010

# Organization

*Honorary Chairs*
Yan Shilong (China), Antoni Tajdus (Poland)

*Chairs*
Liu Zegong (China), Jerzy Lis (Poland)

*Vice-Chairs*
Hua Xinzhu (China), Piotr Czaja (Poland)

*Secretary General*
Yuan Shujie

*Members*
Yuan Shujie, Dai Guanglong, Liang Xing, Shi Biming, Lv Pin, Tu Min, Yang Ke, Qin Ruxiang, Li Tiefeng, Jerzy Kicki (Poland), Stanislaw Nawrat (Poland), Aneta Napieraj (Poland)

*Advisory Committee*
*(in alphabetical order)*
Prof. Andrzej Olajossy, AGH-UST, Poland
Prof. Andrzej Zorychta, AGH-UST, Poland
Prof. Bernard Nowak, AGH-UST, Poland
Prof. Bronisław Barchański, AGH-UST, Poland
Prof. Dai Guanglong, AUST, China
Prof. Gao Mingzhong, AUST, China
Prof. He Qilin, AUST, China
Prof. Hua Xinzhu, AUST, China
Prof. Ireneusz Soliński, AGH-UST, Poland
Prof. Jan Walaszczyk, AGH-UST, Poland
Prof. Jan Winzer, AGH-UST, Poland
Prof. Jerzy Lis, AGH-UST, Poland
Prof. Jolanta Marciniak-Kowalska, AGH-UST, Poland
Prof. Kazimierz Trybalski, AGH-UST, Poland
Prof. Krzysztof Broda, AGH-UST, Poland
Prof. Liu Zegong, AUST, China
Prof. Lv Pin, AUST, China
Prof. Maciej Mazurkiewicz, AGH-UST, Poland
Prof. Marek Cała, AGH-UST, Poland
Prof. Marian Branny, AGH-UST, Poland
Prof. Meng Xiangrui, AUST, China
Prof. Nikodem Szlązak, AGH-UST, Poland
Prof. Piotr Czaja, AGH-UST, Poland
Prof. Roman Magda, AGH-UST, Poland
Prof. Sheng Xue, CSIRO, Australia
Prof. Shi Biming, AUST, China
Prof. Stanisław Nawrat, AGH-UST, Poland
Prof. Stanisław Piechota, AGH-UST, Poland
Prof. Stanisław Wasilewski, AGH-UST, Poland
Prof. Tadeusz Majcherczyk, AGH-UST, Poland
Prof. Tadeusz Tumidajski, AGH-UST, Poland
Prof. Tu Min, AUST, China
Prof. Wacław Dziurzyński, AGH-UST, Poland

Prof. Waldemar Korzeniowski, AGH-UST, Poland
Prof. Wang Baishun, AUST, China
Prof. Wiesław Kozioł, AGH-UST, Poland
Prof. Xie Guangxiang, AUST, China
Academician Yuan Liang, Chinese Academy of Engineering, China
Prof. Yuan Shujie, AUST, China
Prof. Zan Liansheng, AUST, China
Prof. Zhang Guoshu, AUST, China
Prof. Zhao Guangming, AUST, China

*International Mining Forum 2010 is sponsored by:*
National Natural Science Foundation of China
Key Laboratory of Coal Mine Safety and Efficient Exploitation, Ministry of Education, China

International Mining Forum 2010, Liu et al. (eds) © 2010 Taylor & Francis Group, London, UK. ISBN 978-0-415-59896-5

# Fire extinguishing mechanism and application research of nitrogen injection displacement

Guoshu Zhang
*Key Laboratory of Coal Mine Safety and Efficient Exploitation of Ministry of Education; Anhui University of Science and Technology, Huainan, Anhui Province, China*

Shujie Yuan
*Key Laboratory of Integrated Coal Exploitation and Gas Extraction, Anhui University of Science and Technology, Huainan, Anhui Province, China*

Ming Deng
*Anhui University of Science & Technology, Huainan, Anhui Province, China*

ABSTRACT: Spontaneous combustion in sealed goaf has characteristics such as large scope, uncertain fire source and difficult extinguishment. Based on the analysis of disadvantages of traditional method of nitrogen injection for fire-prevention, a new fire-extinguishing technology, nitrogen injection displacement, was presented. The mechanism of nitrogen injection displacement, factors influencing effectiveness of fire extinguishing, system design and key technology were studied. The technology was successfully applied in longwall 1115(3) of Zhangji Coal Mine, Huainan.

KEYWORDS: Mine fire, nitrogen injection displacement, fire extinguishing

## 1. INTRODUCTION

At present, hybrid injection nitrogen is widely adopted to extinguish the fire in sealed goaf. The theory of hybrid injection nitrogen is to dilute the oxygen concentration in goaf. Nitrogen is injected to the sealed goaf with buried pipes and mixed turbulently with original air. Therefore, inert gas concentration is increased in goaf, and oxygen content is reduced less than 5% (critical concentration) relatively. Thus the targets of stifling fire source and inhibiting the development of spontaneous combustion are achieved. This technique requires the goaf be sealed strictly and air leakage never happen.

According to the characteristics of hybrid injection nitrogen flow field, high concentration nitrogen continuously distribute around the intakes in a certain range. With nitrogen injection time going on and gas emitting from goaf, air pressure increases a lot in goaf. As a result, in the region far away from nitrogen intakes, the turbulent mixed function decreases and nitrogen injected can not mix fully with original air. So nitrogen distributes unevenly in goaf. Nitrogen concentration in some place is high and some place low. The effect of hybrid injection nitrogen is determined by whether the region full of higher nitrogen covers the fire resource. If it can, the method takes effect; otherwise it doesn't.

In order to overcome the uncertainty of technology mentioned above, a new fire-extinguishing idea, nitrogen injection displacement, was elaborated and presented.

## 2. NITROGEN INJECTION DISPLACEMENT MECHANISM

The mechanism of nitrogen injection displacement is that nitrogen is poured into goaf artificially and displaces original air. A few nitrogen intakes and outlets are arranged in airflow inlet side and return side respectively, which make nitrogen flowed along the anticipated direction. By optimizing the nitrogen injection rate, nitrogen injected may not mix with air or mix lightly. Therefore, a moving nitrogen level or nitrogen wall will be formed in the form of laminar and diffusion. And the high oxygen atmosphere is released from the goaf from outer to inner (from stop line to the interior goaf) and from bottom to up (from floor to upper part). As a result, a nitrogen-lake is formatted in the lower part of goaf, which drowns the residual coal in lower level or fire resource. Therefore, nitrogen injection displacement is realized, and the full-standing inerting effect is achieved in goaf.

In additional, the atmosphere in goaf is made up of air, coal gas and carbon monoxide, etc. And the mixture is lighter than nitrogen. So nitrogen injected to goaf will easily load in the low level of goaf and form the nitrogen lake.

The mechanism of nitrogen injection displacement is different from that of hybrid injection nitrogen. The former can thoroughly displace the orient air in goaf with less nitrogen in less time. For example, if nitrogen mixes with atmosphere fully, it will take 15 times volume of nitrogen to make oxygen concentration reduce from 15% to 5%. But when nitrogen injection displacement is adopted, one times volume of nitrogen is enough to make oxygen concentration reduce to less than 1%. Therefore, if the volume of goaf reaches a few thousand cubic meters, the efficiency will be promoted dozens of times. So, the advantage of nitrogen injection displacement is self-evident.

## 3. DESIGN OF NITROGEN INJECTION DISPLACEMENT FIRE-EXTINGUISHING SYSTEM

The nitrogen injection displacement is made up of injection pipe, intake ports, extraction pipe, outlet ports, flow meter and flow measurement device. Figure1 shows a nitrogen injection displacement system. The system mainly includes two parts. One part is nitrogen injection system. The base of which is nitrogen injection pipe arranged in the airflow inlet side or embedded in goaf. It is connected to goaf by drills. Drills or embedded pipes are used as nitrogen injection ports. The other one is air extraction system. It consisted of pipes embedded in airflow return side or drills, which are used as exhaust ports.

Generally, a nitrogen injection displacement system is made up of several nitrogen injection ports and exhaust ports.

In order to detect the nitrogen injection quantum, the flow measurement device must be installed in injection pipes. In the air extraction pipe, sampling ports and pressure text devices are necessary, with which the component of gas exhausted from goaf is analyzed and the pressure difference between inside and outside of pipes is detected.

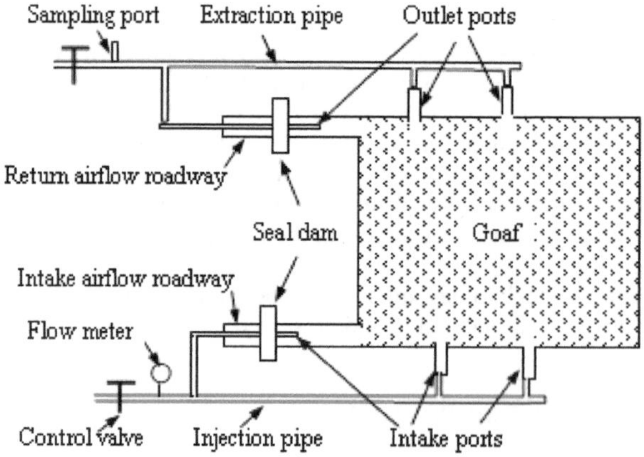


Figure 1. Nitrogen injection displacement system in goaf

## 4. FACTORS INFLUENCING ON NITROGEN INJECTION AND KEY TECHNOLOGIES

To achieve good effect in practice, nitrogen flow field got in goaf should shape as Figure 2 shows. The flow area can be divided into three parts: one part is high nitrogen concentration zone (above 95%), one part is laminar and diffusion zone and the other part is original air zone.

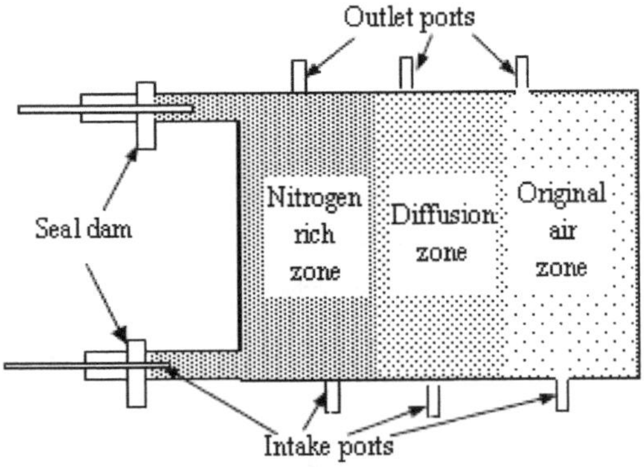


Figure 2. Flow field of nitrogen injection system

The nitrogen flow field formed in goaf is mainly determined by the locations of intakes and outlets and pipe discharge pressure.

Because of large scale of goaf, gas emission and air leakage, how to make nitrogen injected to the sealed goaf displace original air instead of mix with it is key to the technology. The general requirement is as following:

1) The locations of intakes and outlets are reasonably designed.

2) The intakes and outlets are rightly selected when nitrogen injected. If intakes are located in high nitrogen zone, outlets must be arranged in original air zone, and air exhausted from goaf can't contain nitrogen injected.

3) Nitrogen in goaf must flow along the anticipated direction in the form of laminar and format a nitrogen level or a nitrogen-lake, which can push original air out. Meanwhile, turbulent airflow and untouched corner are required to avoid.

4) In order to prevent air leakage, positive pressure difference must be kept in goaf.

5) The nitrogen injection rate is reasonable.

## 5. APPLICATION OF NITROGEN INJECTION DISPLACEMENT

The technology of nitrogen injection displacement was carried on in 1115(3) working face of Zhangji Coal Mine. Figure3 shows the circumstance of 1115(3) sealed goaf. The roadway above goaf is about 30 meters away from the coal seam roof.

1115(3) working face was stopped and sealed in July, 2005. And in October, carbon monoxide was found inside and outside of the seal dam at stopping line. The highest CO concentration reached 10000 ppm or so, which indicated that the self-heating or self-ignite of coal was happened. The comprehensive measures such as grouting, strengthening or spraying the seal dams with sealing material were taken. However, until March of 2006, the concentration of sign gas of coal spontaneous combustion couldn't reach the standard of extinguishment. So, the technology of nitrogen injection displacement was carried on in April, 2006.

The drills were constructed in roadway above goaf of 1116(3) working face, which were used as nitrogen intakes. Outlet ports were arranged in the seal dam in return airflow roadway and roadway above goaf of 1115(3). Gas sample was collected from outlets, and gas component was gotten.

Through analyzing, the nitrogen concentration of gas sample got from the seal dam in return airflow roadway reached 98%.

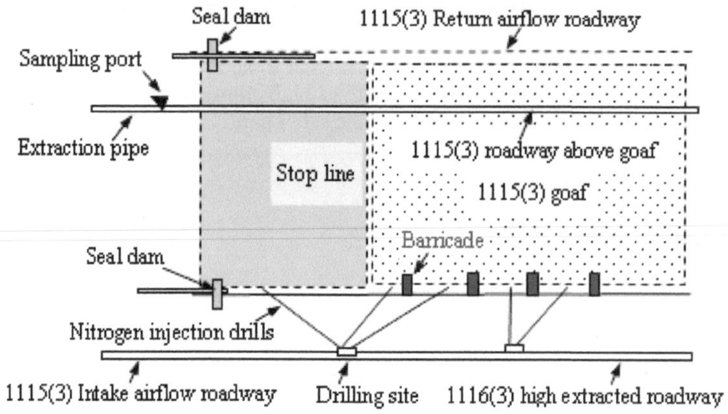


Figure 3. Isolation Project between working region and fire region

Figure 4 and Figure 5 give the analyzing result of gas sample from the roadway above goaf.

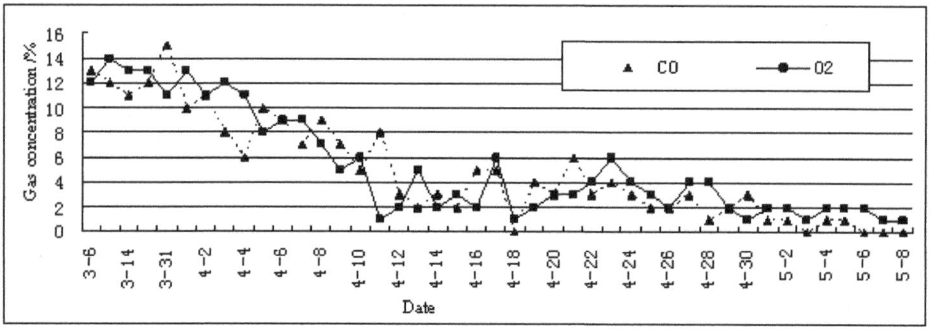


Figure 4. Relationship between concentration of $O_2$ or CO and time in the sampling port of roadway above goaf

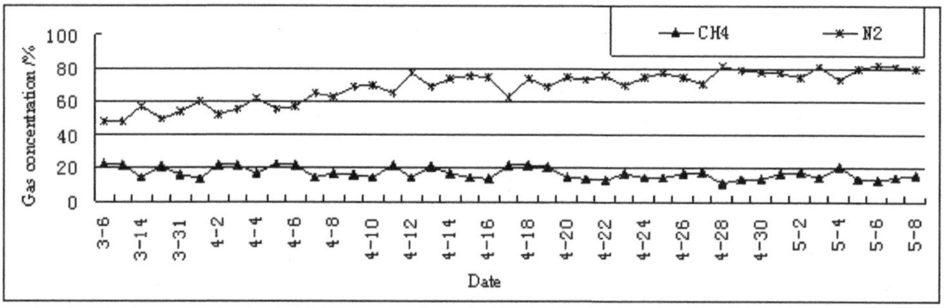


Figure 5. Relationship between concentration of $CH_4$ or $N_2$ and time in the sampling port of roadway above goaf

The sampling port was located in the extraction pipe embedded in the roadway above goaf. Because of high location, the nitrogen concentration didn't reach above 95%, the expected concentration.

## 6. CONCLUSIONS

The mechanism of nitrogen injection displacement for extinguishing fire in sealed goaf is that original air in goaf is displaced by nitrogen injected into goaf.

The reasonable location of intakes and outlets is the key factor for good effect of nitrogen injection. The position of nitrogen intakes must match that of outlets. And nitrogen can flow along the anticipated direction in goaf.

The theoretical analysis and experiment proved that nitrogen injection displacement is a new technology to extinguish fire in sealed goaf. Comparing with traditional method of nitrogen injection, it has the characteristics of much less nitrogen-using, more time-saving and better effect.

## ACKNOWLEDGMENT

The research is sponsored by Natural Science Funds of Anhui Province (050450404)

## REFERENCES

LI Qiang-min. 2000. Displacement ventilation: principles, design and applications [J]. Heating Ventilation and Air Conditioning 30(5): 41-46

YU Yan-ling, YOU Shi-jun, WANG Rong-guang. 2005. The application and research progress of displacement ventilation system [J]. China Construction Heating and Refrigeration (6): 61-65

ZHANG Guo-shu. 2000. Theory of Ventilation and Safety [M]. Xuzhou: Publishing House of China University of Mining & Technology

International Mining Forum 2010, Liu et al. (eds) © 2010 Taylor & Francis Group, London, UK. ISBN 978-0-415-59896-5

# Study on characteristics and control technology of mining induced stress shell in thick coal seam

Guangxiang Xie
*Key Laboratory of Coal Mine Safety and Efficient Exploitation of Ministry of Education; Anhui University of Science and Technology, Huainan, Anhui Province, China*

*Key Laboratory of Integrated Coal Exploitation and Gas Extraction, Anhui University of Science and Technology, Huainan, Anhui Province, China*

Ke Yang
*Key Laboratory of Integrated Coal Exploitation and Gas Extraction, Anhui University of Science and Technology, Huainan, Anhui Province, China*

ABSTRACT: In order to understand and master distributing and evolving patterns of stress induced by mining profoundly, and to control and utilize the action of rock pressure effectively, according to the geological and technological conditions of thick coal seam with fully mechanized top-coal caving (FMTC) of Huainan and Huaibei coal mines, in-situ observation, numerical and physical modeling tests has been carried investigate into the distribution patterns of stress field in surrounding rock of fully mechanized top-coal caving (FMTC) face in the large-scale and three-dimensional space. The results showed that a macro stress shell composed of high stress exists in surrounding rock of an FMTC face. The mining induced stress shell (MISS), which bears and transfers the loads of overlying strata, acts as the primary supporting system of forces, and is the corpus of characterizing three-dimensional and macro rock pressure distribution of mining face. Its external and internal shape changes with the variations in the working face structure as the face advances. Some factors affecting on MISS development have been synthetically considered and analyzed, which including mining-height, persevered pillar width, advancing rate, mining method and parameters. And their acting mechanisms and mechanical effects are opened up. The results based on discovery and analysis of MISS and its' development mechanism affected by geological and mining parameters are of guiding significance for engineering practice, i.e. control and utilization of rock pressure, design and choice of supports, supporting design and dynamic maintenance of gates, precaution of rock burst, gas drainage and safety monitoring etc.

KEYWORDS: Mining induced stress shell, fully mechanized top-coal caving, ground control, advancing velocity, pillar width

## 1. INTRODUCTION

Because of the high output and high efficiency compared with the slicing mining method, the fully mechanized top-coal caving (FMTC) and large mining-height (LMH) method are, at present, popular for thick coal seams in China. With regard to top-coal caving technology, abundant and fruitful researches have been done in the domain of rock pressure control in recent years (Jin 2001, Lu et al. 2002, Qian et al. 2003, Xie et al. 2006, Xie 2007, Yang 2007, Peng 2008, Yuan 2008). The Huainan and Huaibei coal mine is one of Chinese energy bases and plays an important role in economy development of Anhui province. With mining technology development, especially FMTC and LMH are applied and popular in mining thick coal seam, and that not only results in safety and high efficiency mining, but also needs foundational theory research to solve mining induced problems, such as strata behavior, support invalidation, rock burst, methane explosion and so on.

Therefore, basic theories are also the major problems of the high-speed development of FMTC and LMH technology. Engineering practice starves for being sustained by basic theories of mechanical characteristics distribution, in order to extend and apply thick seam mining technology.

In-situ measurement combined with large-scale and non-linear three-dimensional numerical simulation and equivalent material simulation are carried out integrate into MISS and its' development mechanism affected by geological and mining parameters.

## 2. MECHANICAL AND DEVELOPMENT CHARACTERISTICS OF MISS

### 2.1. *Mechanical characteristics of MISS*

There is a macro stress shell composed of high stress bundles in the surrounding rock of an FMTC or LMH face (Fig.1).

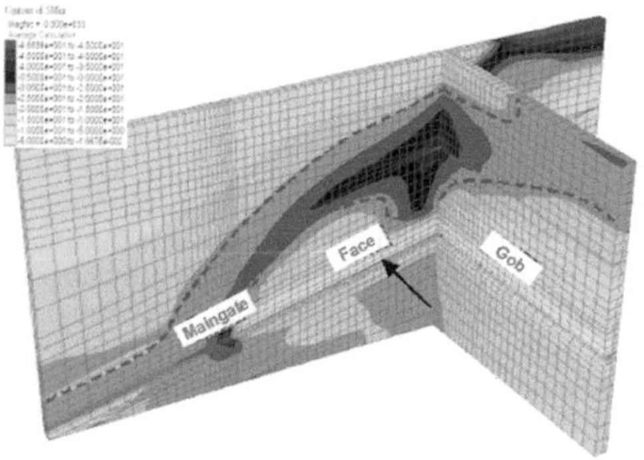


a. view of maingate side

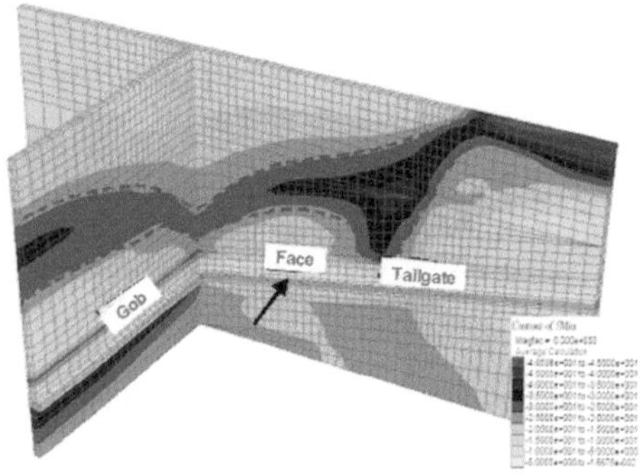


b. view of tailgate side

Figure 1. Three dimensional shape of MSS surrounding rocks (MPa)

The stress shell is mechanically characteristic of the following:

- the stress shell, made up of high stress bundles, is not an objective entity;
- the principal stress of the stress shell is larger than that of the rock mass inside and outside;
- the stress shell lies in the virgin rock and coal mass in the vicinity of the working face and in the sagging zone of the overlying strata. The shell skewback forms abutment pressure behind and ahead of the face and on the edge of the virgin seam on the sides. The stress of the shell skewback is the bearing stress of the face and its vicinity;
- the shape of the stress shell changes with the advancement of the working face. Throughout the process of mining, the stress shell changes all along. When a new face is prepared, the skewback of the former stress shell formed in the neighboring face above becomes the inner skewback, which goes higher and higher as the working face approaches and gradually disappears behind the face;
- the stress shell is geometrically asymmetrical and non-uniform as far as stress distribution is concerned. Its span does not agree with its thick. Generally speaking, stress is greatest at the

skewback, second at the vault and still less at the shoulders. The stress shell is not fully enclosed.

The voussoir beam lies in the stress-decrease zone below the macro stress shell. The inner profile formed by the caving zone and the fractured zone of the surrounding rock is the negative camber of the stress shell.

### 2.2. *Main mechanical characteristics of FMTC face*

An FMTC face is mechanically characteristic of its location in the low-stress zone protected by the stress shell of the overlying surrounding rock. The overlying rock mass passes its loads and pressures through the stress shell onto the working face and its vicinity. Stress peak value of coal seam transfer to the downside coal mass, and coal seam of the face is unloading. Hence the stress shell is the primary supporting body and the voussoir, which only bears partial loads of the strata under the stress shell. Breakage and instability of the voussoir, being situated in the low-stress zone under the shell, can give rise to periodic pressures, instead of great dynamic pressures, onto the face. Only off-balance of the stress shell can result in violent strata behaviors, such as shock bump or rock burst. Since an FMTC face is always situated in the low stress zone below the stress shell, strata behaviors tends to be eased.

The strata behaviors of the face and its neighboring gates are under control of the stress shell. Reasonably adjusting structure parameters of working face can improve dynamic balance of surrounding rock stress shell. It plays a positive role in protecting working face and reducing rock pressure influence. The discovery and analysis of the stress shell has revealed the mechanical nature of the top coal of an FMTC face acting as a "cushion".

### 2.3. *Abutment pressure patterns in asymmetrical layout*

The concentration extent and variation law of FMTC face are different obviously in different space positions under the condition of unsymmetrical mining. There are distinct differences between FMTC face and fully mechanized mining (FMM) face. Abutment pressure peak value and advanced distance on the top of face are larger than those in the centre and at the bottom. The maximum principal stress lies in the tailentry corner of face. Prescribed limit front of wall and top-coal lies in the low stress area. Abutment pressure peak value in coal pillar commonly lies in front of face, and lies in low stress area in the neighborhood of face. There isn't stress peak value in the lateral coal mass in the front of face on the strike. The peak value of abutment pressure appears the rear of face. There is peak value of abutment pressure, and isn't abutment pressure in the gob.

### 2.4. *Mining-height effect in MISS*

Based on synthetically analyzing, the stress shell, made up of high stress bundles bearing stress of the face and its vicinity, is not an objective entity. MSS development characteristics of different mining thicknesses are as follows:

- with mining thick increasing, the height, span, body ply, and skewback width of MISS increase;
- the unsymmetrical geometry shape, heterogeneous stress redistribution, and non-uniform span of MSS are basically invariable in different mining thicknesses but the MISS may be out-of-balance at the MSS shoulder of rear face.

The mechanism of MISS forming gradually reinforces with mining thick increasing that is the essence of dynamic load coefficient decreasing, strata behavior mitigating, lower peak value of abutment pressure and mining induced and influenced scope expanding in FMTC. Lots of engineering practices have proved FMTC's strata behaviors are gentler than FMM that it is acted by differences of MISS mechanical characteristics mining thick effects on. FMTC's MISS height is obviously bigger than FMM (Fig. 3) and some strata behaviors, i.e., shield support load and wall caving etc., will be gradually gentle because of more mining induced extension and energy of stress building down during MISS dynamical evolving or unbalancing. Therefore, MISS development effected by mining thick increasing results in MISS composed of high stress bundle is still apart from rear face and its mechanical configuration is more stable that face is more effectively protected by MISS bearing most loads of upper strata with FMTC.

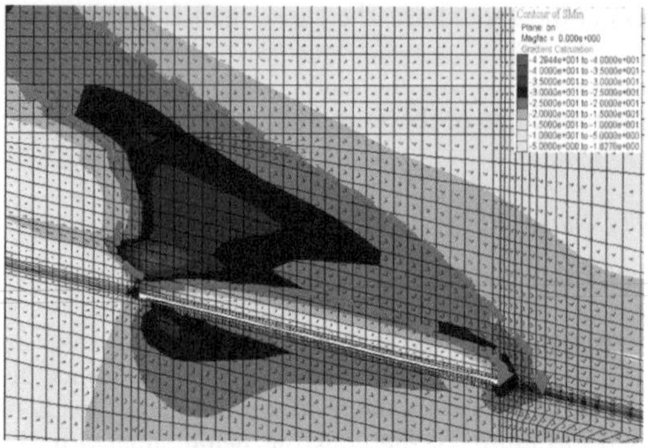

a. 2 m

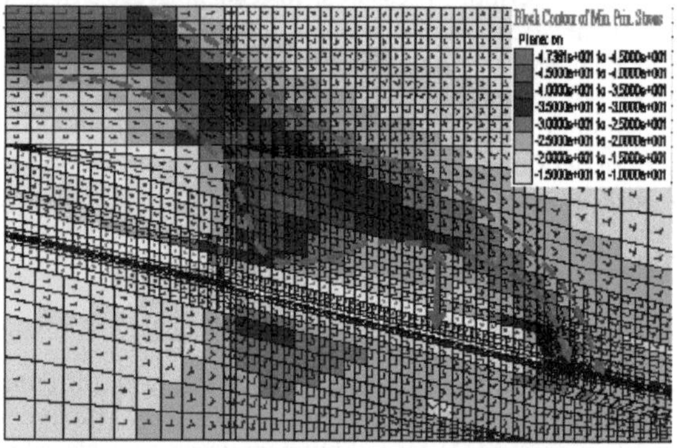

b. 3 m

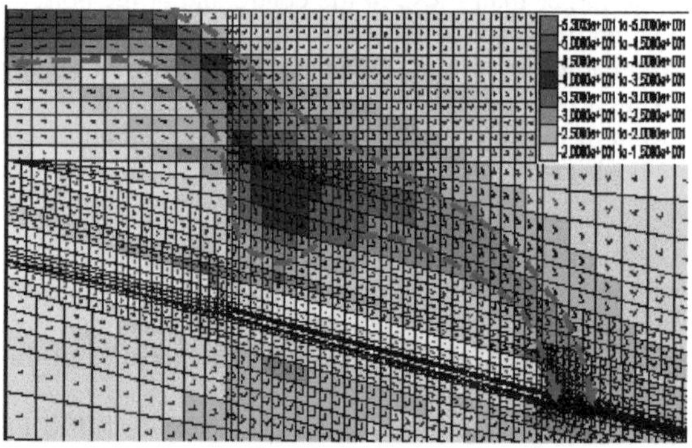

c. 5.4 m

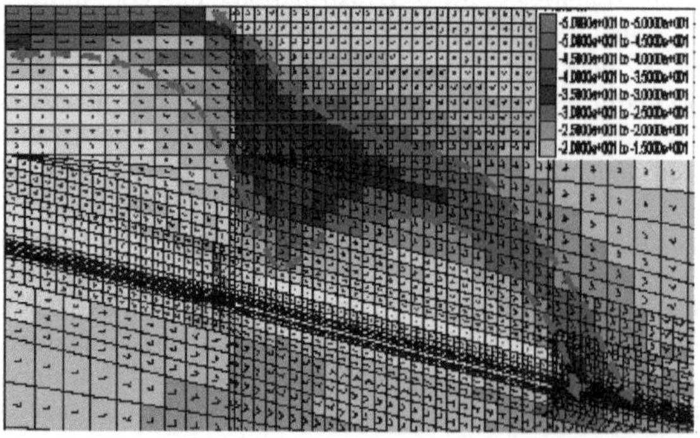

d. 8 m

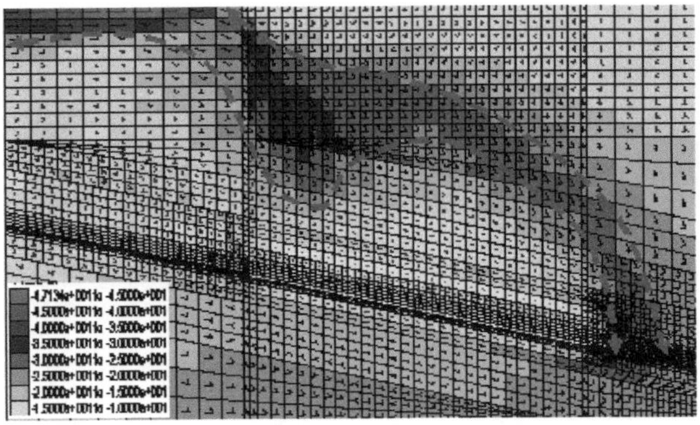

e. 12 m

Figure 2. The maximum principal stress field in working face to the dip with different thick coal seams (MPa)

Because of different MISS skewback locations and stress values affected by caving different mining thicknesses, vertical stress peak value gradually decreases and its location apart from face wall gradually increases while mining thick increases from 2 m to 12 m (Fig. 3).

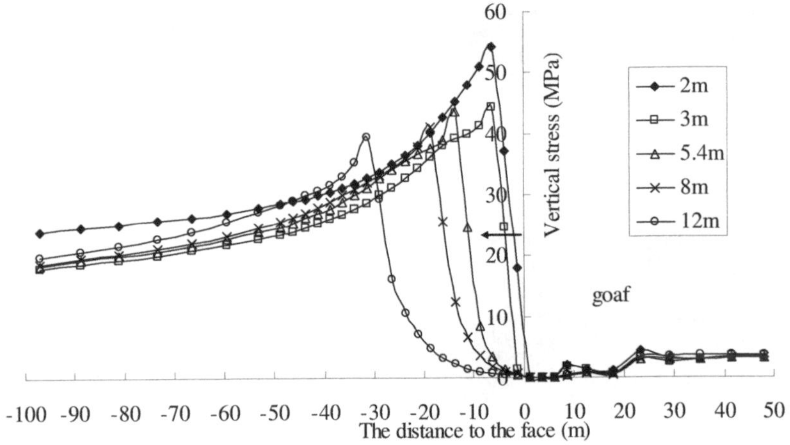


Figure 3. Abutment pressure distribution with different thick coal seams

## 2.5. *Longwall-length effect of MISS*

The effect of longwall-length on mechanical characteristics of MISS was discovered. There is marked influence of mechanical characteristics of MISS with the length of face changed (Fig. 4). With the increase of mining face length, the level of concentration of MISS located in front face and surrounding rock of roadway are amplified, the height and rate of flat are all enlarged, and the three dimensional stress shell is focused in the working face. The damage extent lies in the head entry corner of face and the vertical displacement are reduced, but horizontal displacement is enlarged. The dynamic balance of surrounding rock stress shell is improved with rational adjustment of face length, there have active effects for protection working face and reducing strata behaviors.

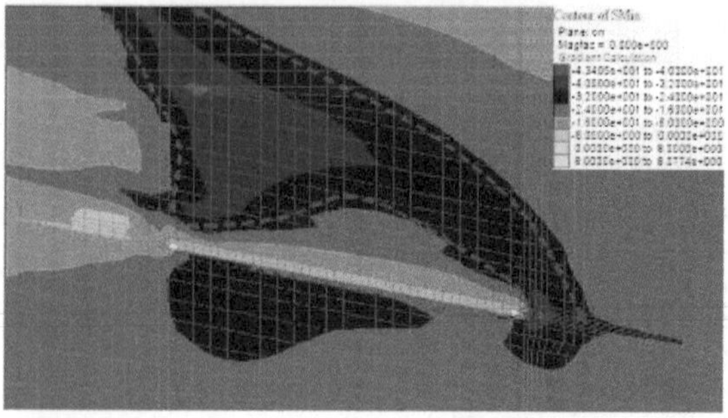


a. 180 m

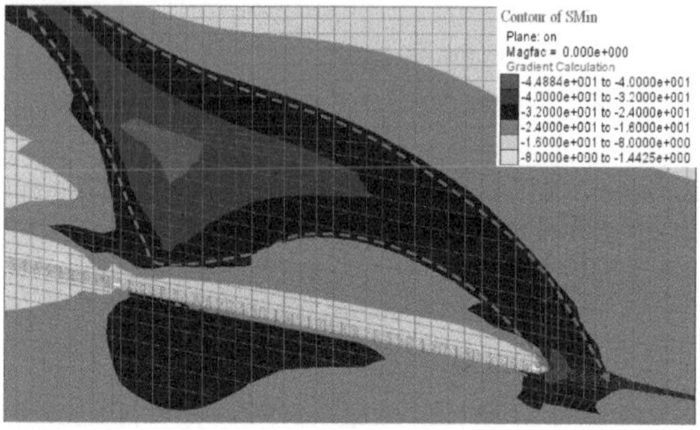


b. 231.8 m

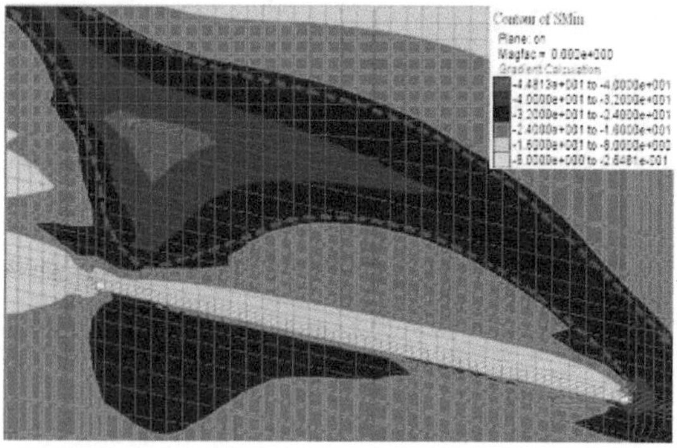


c. 300 m

Figure 4. Principal stress field at 5 m distance ahead of FMTC to the dip with different lengths

## 2.6. *Pillar-width effect of MISS*

With the variation of entry protection coal pillar width, the mechanical field characteristics not only in coal pillar but also in coal mass of adjacent face are different. The both distributions of mechanical field will transform with the variation of entry protection coal pillar width. The maintenance of entry depends on the both mechanical fields. When the width of coal pillar changes, not only stress distribution in coal pillar will change, but also stress distribution, displacement deformation and fractured area of face and surrounding rock are different. The variation of width of hasn't influence on stress state both in face coal mass and in coal pillar (Fig. 5).

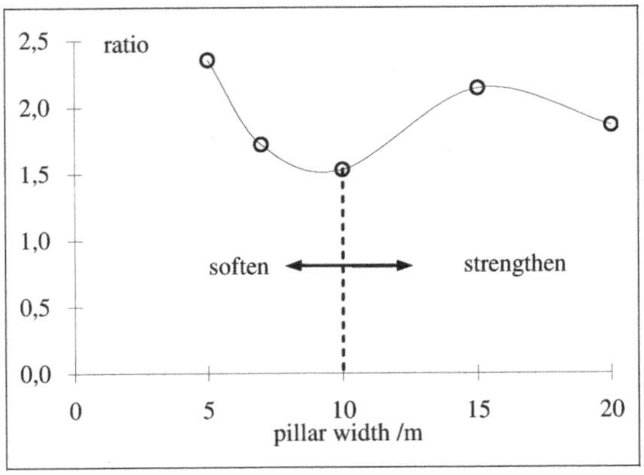


Figure 5. Ratio of vertical and horizontal stress peaks

## 2.7. *Advancing-rate effect of MISS*

The advancing rate affects the results of stress distribution, deformation and failure process in surrounding rocks in mining thick coal seam. The rate effect includes the completion of stress transition in surrounding rock under different mining rates, the variation of loading and unloading rate in process of different cutting depths in unit period of time, and the development of deformation and stress distribution in rock due to creep behavior. Experimental investigation are carried out to reveal the fact that the reduced stress zone around the mining face decreases with the increase advancing rate. Consequently, the failure zone decreases and the peak stress locates nearer the mining face while the stress value increases. Since the deformation period becomes shorter, as a result, the displacement reduces.

# 3. ENGINEERING APPLICATIONS AND DISCUSSIONS

1) Reasonable design and choice of support must be consistent with the characteristic of the distribution of rock pressure of FMTC face, and improve the capability of support. It has been seen from above that the pressure of overlying strata absorbed by the top coal of an FMTC face acting as a "cushion" is limited. It is important that surrounding rock near the support is lower-stress, and it is protected by a MISS at all. The loading of support is from deformation pressure due to deformation movement of the top-coal, the immediate roof and the main roof. Design idea of support in FMTC face, which uses supporting resistance for the roof to make the support stable, is not comprehensive following the idea of medium-thick seam. To be consistent with the characteristic of low stress and large deformation of the roof, stable design idea of support structure is put forward in FMTC face. In addition, the mechanical characteristics of an FMTC face are distinctly different along the dip, traditional and integrative structure design doesn't accord with nonuniform essential of the stress, failure and deformation in an FMTC face. So Reasonable design and choice of support must be consistent with the characteristics that FMTC face lies in lower stress zone below the stress shell and distributing nonuniformly along the dip. It makes support match movement of surrounding rock, and plays the capability of support at large.
2) Supporting design of gates should be changed from load control to deformation control. The displacement, deformation and failure zone of the gate beside original coal mass are smaller than those of the gate beside goaf. Not only is the reason relate to movement of the roof due to mining of the adjacent face beside goaf, but also it is the most important that the gates beside the face lies in different stress field. The gate beside original coal is the lower-stress zone of stress shell skewback all along. However, the gate beside goaf develops bigger displacement and deformation when the stress transfers and releases due to rising of the skewback

continuously with advancing of the face. The theory of supporting design based on controlling the load is not consistent with the gates within low-stress zone near face. The essential way of controlling gate surrounding rock is to carry out supporting design based on controlling deformation, and change deformation and load-bearing characteristics from mechanism, and adopt necessary dynamic maintaining means at the same time.

3) The action of rock pressure is controlled and utilized effectively. MISS bears and transfers the loads of overlying strata, acts as the primary supporting system of forces. The strata behaviors of the face and its neighboring gates are under control of the stress shell. Off-balance of the stress shell or transfer of the forces system of the stress shell can result in violent strata behaviors, such as shock bump or rock burst. So reasonably adjusting structure parameters of working face can change the dynamic balance of stress shell, and avoid the ill-effects of rock pressure, and reduce the probability of rock burst. On the contrary, high-stress circumstance of stress shell is utilized to presplit the hard top coal and roof, and to reduce dynamic disasters, and to improve the mining efficiency.

4) Effective position of methane gas drainage and safety monitoring should lie in the strata below the stress shell. Surrounding rock of stress shell and its above is not destroyed because FMTC face is protected by stress shell all the time. There are not fissures below the stress shell and gas permeability of rock mass is low. It is adverse to drainage of methane gas. The rock mass below stress shell will be failure with advancing of the face, and fissures are generated. So, effective position of methane drainage should be laid in the strata below MISS. The stress shell is of compaction and sealing for overlying strata above the goaf because the stress shell is high stress. It composes of the covering of methane gas and harmful gas accumulating. So the inferior border of stress shell is the important position of safety monitoring.

## 4. ACKNOWLEDGEMENTS

This study was financially supported by the National Basic Research Program of China under Grant No. 2005cb221503 and No. 2010CB226806, the Science and Technological Fund of Anhui Province for Outstanding Youth under Grant No. 08040106839, the Natural Science Research Project of Anhui Province for Colleges and Universities under Grant KJ2009A139, and the Outstanding Innovation Group Program of Anhui University of Science and Technology.

## REFERENCES

Jin Zhongming. 2001. Theory and technology of top-coal caving method. Beijing: Coal Industry Publishing House

Lu Mingxin, Hao Haijin, Wu Jian. 2002. The balance structure of main roof and its action on top coal in longwall top coal caving workface. Journal of China Coal Society 27(6): 591-595

Peng Syd S. 2008. Coal mine ground control. Third Edition, Published by Syd S. Peng

Qian Minggao, Mia Xiexing, Xu Jialin. 2003. Theory of key strata control. Xuzhou: China University of Mining and Technology Press

Xie Guangxiang. 2007. Three-dimensional mechanical characteristics of rocks surrounding the fully mechanized top-coal caving mining faces. Beijing: China Coal Industry Publishing Press

Xie Heping, Peng Suping, He Manchao. 2006. Basic theory and engineering practice of deep mining. Beijing: Science Press

Yang Ke. 2007. Study on the evolving characteristics and the dynamic affecting of surrounding rock macro stress shell and mining induced fracture. Huainan: Anhui University of Science and Technology

Yuan Liang. 2008. Theory and Practice of Integrated Pillarless Coal Production and Methane Extraction in Multiseams of Low Permeability. Beijing: China Coal Industry Publishing Press

*International Mining Forum 2010, Liu et al. (eds) © 2010 Taylor & Francis Group, London, UK. ISBN 978-0-415-59896-5*

# Thermal hazard in Chinese coal mines and measures of its control

Shujie Yuan
*Key Laboratory of Integrated Coal Exploitation and Gas Extraction, Anhui University of Science and Technology, Huainan, Anhui Province, China*

*Key Laboratory of Coal Mine Safety and Efficient Exploitation of Ministry of Education, Huainan, Anhui Province, China*

Zhenbin Zhang, Kai Deng, Chaomei Li
*School of Energy and Safety, Anhui University of Science and Technology, Huainan, Anhui Province, China*

ABSTRACT: With increase of mining depth of coal mines in China, thermal hazard problem in mines is getting more and more serious, affecting safety and productivity. In the paper thermal hazard situation in Chinese coal mines was presented, including geographical distribution of high-temperature mines, and diseases caused by high temperature. The reasons of thermal hazard formation and its influencing factors were analyzed. The fact was pointed out, that in deep mines heat release of rocks is the major reason of high temperature. Thermal conditions standards in mines of China and main countermeasures of high-temperature control were presented.

KEYWORDS: Mine disasters, thermal hazard, air conditioning

## 1. INTRODUCTION

Coal is China's primary energy sources. With mining depth growing, rock temperature in coal mines is increasing. In 1980, the average mining depth of Chinese coal mines was 288 m, in 1995 reached 428 m, and at present mining depth is increasing at an average rate of 8-12 m annually. There are dozens of coal mines with mining depth over 1000 m. According to survey data from around the world, the global average geothermal gradient is about 3°C/100 m. According to China's statistics, geothermal gradient in Chinese coal fields is 2-4°C/100 m. According to incomplete statistics, among China's state-owned key coal mines there are more than 50 coal mines with mining depth greater than 700 m, the deepest coal mines have depth of more than 1000 m. More than 80 coal mines suffers from different degrees of thermal hazard, of which 38 coal mines have working face of air temperature over 30°C (Wang, Yang 2008). China is a country with the most coal mines of thermal hazard. 73.2% of the predicted reserves of coal is underground deeper than 1000 m. With increase of mining depth, heat hazard is increasingly prominent, and became one of major disasters in coal mines.

## 2. THERMAL HAZARD IN CHINESE COAL MINES

In China, form 60ths of $20^{th}$ century, heat hazard occurred in coal mines of Huainan Mining Area. Rock temperature in Huainan Mining Area is listed in Table 1. Predicted rock temperature at different depths in some mines in Huainan Mining Area is shown in Table 2. (Li et al. 2007). According to survey in 1999-2001, there were 88 coal mines (not including coal mines of annual output below 300,000 Mg/year), in which air temperature in working faces is over 30°C. Among the 88 coal mines, there were 31 coal mines with air temperature of 30-32°C in working faces, 37 coal mines with air temperature of 32-35°C in working faces, and 20 coal mines with air temperature over 35°C. According to administrative division in China: 18 coal mines in Central China, in which Pingdingshan, Fengcheng and Xuchang coal mining areas suffer from the most serious thermal hazard; 39 coal mines in East China, in which Huainan, Huaibei, Yanzhou, Xinwen, Xuzhou (including Datun) and Juye coal mining areas suffer from the most serious thermal hazard; 26 coal mines in North China and Northeast China, in which Fengfeng (Handan), Xingtai, Datong, Kailuan, Tiefa, Beipiao, Fushun, Liaoyuan and Jixi coal mining areas suffer from the most serious thermal hazard; 5 coal mines in other regions (2 coal mines in Hunan, 1 in Gansu, 1 in Guangxi and 1 in Fujian Province) suffer from the most serious thermal hazard. According to recent survey, in Xinjiang region, the northwest China, thermal hazard occurs in coal mines as well. This showed a wide distribution of coal mines with high temperature, and most of them are large state-owned coal mines. Output of the 88 coal mines accounted for more than 60% of the total output of all state-owned coal mines. In the 88 coal mines, the total number of underground production miners was 0.1976 million.

Table 1. Rock temperature in Huainan Mining Area

| Coal mine | Mine capacity, $MMg \cdot a^{-1}$ | Depth, m | Rock temperature, °C | Geothermal gradient, $°C \cdot hm^{-1}$ |
|---|---|---|---|---|
| Jiulonggang | 0.60 | 830 | 31.6 | 1.82 |
| Xieli Shenbu | 3.00 | 990 | 27-32.5 | 1.1-1.6 |
| Xinzhuangzi | 1.80 | 574 | 26.8-27.3 | 1.4-1.6 |
| Panyi | 3.00 | 550 | 35.0 | 3.6-3.8 |
| Pansan | 3.00 | 680 | 32.3-37.6 | 2.6-3.8 |
| Xieqiao | 4.00 | 637 | 34.0-36.0 | 2.8-3.0 |
| Zhangji | 4.00 | 625 | 31.0-36.6 | 3.50 |
| Guqiao | 5.00 | 681 | 34.7-40.0 | 3.10 |

Table 2. Predicted rock temperature at different depth in some mines in Huainan Mining Area (°C)

| Depth, m | | | -500 | -600 | -700 | -800 | -900 | -1000 |
|---|---|---|---|---|---|---|---|---|
| Coal mine | Panyi | North part | 33.5 | 37.0 | 40.6 | 44.1 | 47.7 | 51.2 |
| | | South part | 27.1 | 29.3 | 31.5 | 33.7 | 35.9 | 38.1 |
| | Pansan | East part | 31.6 | 34.7 | 37.8 | 41.0 | 44.1 | 47.3 |
| | | West part | 32.9 | 36.3 | 39.7 | 43.1 | 46.6 | 50.0 |
| | Xieqiao | | 30.4 | 33.3 | 36.2 | 39.1 | 42.0 | 44.9 |
| | Zhangji | | 33.3 | 36.7 | 40.2 | 43.7 | 47.2 | 50.7 |
| | Guqiao | | 31.4 | 34.5 | 37.6 | 40.7 | 43.8 | 46.9 |

At present in China the most serious thermal hazard occurs in No.8 Coal Mine of Pingdingshan Mining Area, Xiezhuang Coal Mine of Xinwen Mining Area, Jianxin Coal Mine of Feng Mining Area, Sanhejian Coal Mine of Xuzhou Mining Area, Zengjiashan Coal Mine of Yongrong Mining Area and the newly developed Juye Coal Mining Area and so on.

With extension of deep-level mining, the number of coal mines with high air temperature, thermal hazard degree, and accidents affecting safety will increase year by year. In the future the bottleneck restricting development of coal industry in deep mining technology will be thermal hazard and high strata pressure (Hu 2008, Liao et al. 2009).

Most of explosive accidents in coal mines are closely related to high air temperature, which leads to coal spontaneous combustion, and coal fire makes gas rapid burning, resulting in gas explosion or produce large amounts of carbon monoxide. On March 29, 2002, gas explosion occurred in Roadway 201 at depth of 2000 m underground in No.2 Coal Mine of Xinfeng Mining Bureau, Hung-Chang Town, Yuzhou City, Henan Province. The accident was because air temperature in the roadway is too high, and air can not flow into the roadway, resulting in high gas concentration.

Thermal hazard not only brings many obstacles to safe exploitation at great depth in mines, but also serious threat to health of underground miners. Thermal Hazard Control in deep mines is difficult key point of occupational hazard prevention.

Xuzhou Coal Mining Area is a 125-year-old mining area. Shallow coal seams are gradually mined out, most of mines begin exploit coal seams at great depth (average mining depth is close to -900 m), in some coal mines mining depth reached -1,200 m. Thermal hazard with high temperature and humidity, etc. and occupational hazards become increasingly serious. In 2006, in Jiahe Coal Mine of Xuzhou Coal Mining Group 36 miners suffered heatstroke caused by high air temperature, in which two miners died.

Mining depth in coal mines of Pingdingshan Coal Mining Group is close to 1000 m. Ambient air temperature in working faces is getting higher and higher, usually 30-34°C, and in some working faces up to 35-37°C, and relative humidity close to 100%, additionally with about 43°C hot water leaching out from the roof, resulting in mugginess and dampness in working faces. Miners working in the working faces frequently suffer eczema, dizziness, heatstroke and other thermal illnesses.

In the months of June, July and August of every year, local atmospheric temperature is high in Xuzhou, Jiangsu province. Underground production miners in Baiji Coal Mine are susceptible to suffer skin disease, its pathological features are: red skin rash-like, distribution in the limbs, chest, abdomen, etc., unbearable sting and itching, affecting rest of miners, whole body discomfort, and lack of concentration, resulting in misoperation increase and impacting production safety. The statistical results of miners suffering from thermal illnesses caused by high temperature are shown in Table 3 (Liu, Zhao 2009).

Table 3. The statistics of miners suffering from thermal illnesses caused by high temperature in Baiji Coal Mine

| Year | Eczema, athlete's foot, persons | Heatstroke, persons |
|---|---|---|
| 2005 | 78 | 5 |
| 2004 | 85 | 5 |
| 2003 | 100 | 7 |
| 2002 | 100 | 5 |
| 2001 | 69 | 1 |

Note: The above is statistics in the months of June to September

Domestic and foreign research statistics show that on the basis of current climate standards, when air temperature increases per 1°C, mine productivity decrease 6-8%. When air temperature increases per 1°C, labor insurance for medical expenses for miners increase 8-10%. According to the latest statistics of South Africa, when miners works in the wet bulb temperature of 32.8-33.8°C, death rate of heat stroke per thousand miners is 0.57. When air temperature increases per 1°C, failure rate of underground electrical and mechanical equipment increases 1 times.

## 3. THE CAUSES OF HIGH AIR TEMPERATURE IN COAL MINES

Factors resulting in high air temperature in coal mines are: high temperature air flowing into coal mines, geothermal release, heat release of electrical and mechanical equipments, heat release of exothermic oxidation of coal or sulfide ores and other heat source, etc. Geothermal release generally includes heat release of rock, geothermal water and water vapor.

Influence of different underground heat sources on thermal conditions in a mine is showed in Table 4 (Liao et al. 2009).

Different geological conditions and different production conditions result in different proportion of different major heat sources causing thermal hazard in different mining areas. But overall the major heat source is heat release of rock.

High temperature of inlet air is main reason of high temperature in small shallow mines and large deep mines in construction during summer period. In large areas of southern China in July average maximum temperature exceeds 33°C, and sometimes as high as above 40°C.

Table 4. Influence of different underground heat sources on thermal conditions

| Heat source | Influence degree, % |
|---|---|
| Changes of atmospheric state in the surface | 18 |
| Self-compression of air | 13 |
| Heat release of rock | 38 |
| Heat release of mechanical and electrical equipments | 11 |
| Heat release of ore oxidation | 10 |
| Heat release of groundwater | 4 |
| Other heat sources | 6 |
| Total | 100 |

Geothermal heat is the main reason of high air temperature in deep mines. The deeper from the surface, the higher the rock temperature is. When underground water conducts with heat source in deep strata through faults and fractures, underground hot water activities can form local geothermal anomalies.

Heat release of mechanical and electrical equipments is an important heat source in high mechanized mines. About 80% power consumption of coal cutting machinery transfers into heat. In fully mechanized longwall, power of mechanical and electrical equipments is very big, in some of loangwalls the power is more than 1000 kW, resulting in air temperature rise a lot.

Heat release of oxidation of sulfide ores or coal is another reason of high air temperature in working face. Sometimes the heat release can account for more than 20% heat in airflow out of a working face.

Other heat sources, such as heat release of human body and blasting, filling materials, production water and so on. Their impact on thermal conditions is different due to different specific conditions. Generally, influence of these heat resources on air temperature in mines is not big.

## 4. THERMAL CONDITIONS STANDARDS IN MINES

China's "Provisions of Geothermal Measurement in Coal Resources Geological Exploration" states: regions of average geothermal gradient not more than 3°C/100 m belong to normal temperature zone; above 3°C/100 m belong to high temperature anomalies. Regions of original rock temperature above 31°C belong to thermal hazard zone of first category, the original rock temperature above 37°C belong to thermal hazard zone of second category.

In the world there are 3 kinds of temperature used as thermal conditions standards in mines. They are air dry bulb temperature, air equivalent temperature and air wet bulb temperature. In China air dry bulb temperature is used. In foreign countries thermal conditions indicators in mines are listed in table 5 (Hu 2008).

Table 5. Thermal conditions indicators in mines in foreign countries

| Country | Maximum allowable temperature, °C | Remarks |
|---|---|---|
| South Africa | $t_{wb}$=31.5 | When airflow>1.5m/s, in gold mines |
| | $t_{db} \leq 26$ | Φ<90%, in coal mines |
| Russia | $t_{db} \leq 25$ | Φ>90%, in coal mines |
| | $t_{db} \leq 25$ | In chemical ore mines, metal mines |
| United States | $t_{eff} \leq 32$ | In coal mines |
| | $t_{eff} > 32$ | In coal mines, operation is prohibited |
| Germany | $25 < t_{eff} < 29$ | Limited operating time to 6h |
| | $29 < t_{eff} < 30$ | Limited operating time to 5h, 10min break per hour |
| | $30 < t_{eff} < 32$ | Limited operating time to 5h, 20min break per hour |
| | $t_{eff} > 32$ | operation is prohibited |
| United Kingdom | $t_{wb} < 27.8$, $t_{eff} \leq 29.4$ | |
| Poland | $t_{db} \leq 26$ | In coal mines |
| | $t_{db} > 26$ | Labor ration can be reduced 4% |
| | $28 < t_{db} < 33$ | Limited operating time to 6h |
| New Zealand | $t_{wb} < 23.3$ | |
| | $t_{wb}$=23.3 | Limited operating time to 7h |
| | $t_{wb} \leq 23.8$ | Limited operating time to 6h |
| Czech Republic | $t_{db} \leq 28$ | Φ<90% |
| | $t_{db} \leq 30$ | Φ<80% |
| Zambia | $t_{wb} \leq 31$ | In copper mines |
| Belgium | $t_{eff} \leq 31$ | Operation is prohibited, when the temperature is exceeded |
| France | $t_{eff} \leq 31$ | Operation is prohibited, when the temperature is exceeded |

Note:
1. Data in this table is taken from foreign laws, regulations, procedures and standards; 2. $t_{db}$ - air dry bulb temperature, $t_{eff}$ - air equivalent temperature, $t_{wb}$ - air wet bulb temperature, Φ - humidity.

Article 102 of "Coal Mine Safety Regulations" implemented in China from March 1, 2010 requires: inlet air temperature below the mouth of shaft (dry bulb temperature, the same below) must be more than 2°C. Air temperature in exploitation or excavation working faces of mines in production must not exceed 26°C, in mechanical and electrical equipment chambers air temperature must not exceed 30°C; when the air temperature is exceeded, staff working hours in over-temperature location must be reduced, and care for high-temperature treatment paid. When air temperature in exploitation or excavation working faces exceeds 30°C, in mechanical and electrical equipment chambers air temperature exceeds 34°C, operation must be ceased. In time of design for newly built or expanded mines, air temperature in mines must be predicted by calculation, for over-temperature locations refrigeration cooling design must be done, and cooling facilities must be installed.

## 5. COUNTERMEASURES OF HIGH TEMPERATURE CONTROL IN MINES

Mine air cooling measures can be divided into two general categories:

- The first one is, cooling air flow by adjustment of original ventilation system and full use of thermal-regulation circle of old roadways, keeping away from heat release of mechanical and electrical equipments and roadways of large heat release of hot mine water, as much as possible isolation of high temperature heat sources in inlet airflow system, increasing air flow and speed to improve thermal conditions.
- The second one is, by use of air-conditioning system to force inlet airflow in high temperature locations cooling. Cost of this measure is big, only used in case of other methods invalid.

1) Cooling by ventilation

(1) Cooling by increasing air flow. A large number of in-situ experiments showed that good cooling effect can be achieved by increasing air flow, by which airflow temperature in working faces can be reduced 1-4°C. Experiments of Japanese researcher showed that when ventilation air flow increases, air temperature decreases dramatically, and temperature drop has sharply accelerated trend when ventilation airflow reaches a certain amount. If airflow continuously increases, then rate of air temperature drop gradually slows down. The most economical ventilation air flow is the amount equivalent to 0.56-0.84 times of roadway length (Liu et al. 2005). According to the experience of air cooling by increasing airflow in China, in longwall of high air temperature in case of net cross-sectional area of 6-8 $m^2$, the upper limit of reasonable amount of airflow for air cooling is 800-1000 $m^3$/min (Wang et al. 2009).

(2) Cooling air by improved ventilation methods in working face. Using reasonable ventilation system to shorten ventilation route as much as possible, and increase air flow and speed. In case of conditions permitted by "Coal Mine Safety Regulations", in exploitation working face area downward airflow ventilation, or W-type ventilation system in which inlet air flows in upper and lower roadways and outlet air flows out of the middle roadway, can significantly reduce air temperature in working faces. In high air temperature locations of relative concentration of workers in working faces, using various ejectors or small fans to increase air flow speed can improve conditions of heat release of human body.

2) Cooling air by exploitation with goaf filling

Application of exploitation with goaf filling is propitious to air cooling, because heat release of rocks in goaf is reduced, and air leakage in goaf is greatly reduced, meanwhile filling materials can absorb a large number of heat to cool air in mines.

3) Cooling air by refrigeration

Cooling air by refrigeration is an effective measure of high air temperature control in mines. The method is used usually only in case of cooling by ventilation invalid or not economical, because of its high cost. Air-cooling equipment consists in refrigeration system and cooling system. The former supplies cold water and the later cools air by cold water.

(1) fixed air cooling equipments

Fixed air cooling equipments layout is in three ways:

- Refrigeration and cooling systems located in the surface, cooling main inlet air flow. The construction, maintenance and cooling water treatment are convenient, but cold air on its way to the underground absorbs heat, and the cooling effect is reduced.

– Refrigeration system located in the surface, cooling system located in the underground. The investment is large, but easy to deal with cooling water, cooling effect is good. High-quality insulation materials are available to prevent cold water to absorb heat in its transportation process. The method is now widely used.
– Refrigeration and cooling systems located in the underground. This approach requires solving the problem of proper recycling of cooling water.

(2) Mobile air-cooling equipments

Mobile air-cooling equipments commonly are used in working faces and other locations for cooling air, using with local fan delivers cooled air to working face. Its refrigeration system and cooling system are assembled into one complex, can move with mining face advance. Another mobile air-cooling equipment has fixed refrigeration system and moving cooling system with working face advance.

4) Personal protection from thermal hazard

Personal protective measures can be used in high air temperature mines with dispersed operation locations and small number of miners. Some countries are developing different kinds of cooling helmets or cooling clothes to cool the air inhaled, and reduce surface temperature of human head or body.

5) Reduction of heat release within mines

Mechanical and electrical equipment chamber of large heat release should be ventilated independently, to avoid heat delivering to working face; the walls of water drenching roadways and ditches should be covered to reduce heat transfer and evaporation of hot water; prevent compressed air pipelines to heat inlet air; deliver hot water to the surface through air return roadways and shafts; ahead mine water drainage measures can be used in high temperature mines of hot water type; cover or spray walls of roadways with heat isolation materials to reduce heat release of rocks in the walls.

## 6. CONCLUSION

On the basis of analysis of thermal hazard situation in Chinese coal mines, the following conclusions were obtained:

China is a country with the most coal mines of thermal hazard. Geothermal gradient in Chinese coal fields is 2~4°C/100 m, and 73.2% of the predicted reserves of coal are underground deeper than 1000 m. With increase of mining depth, thermal hazard is increasingly serious, and became one of the major disasters in coal mines.

China's high temperature coal mines mainly distribute in Central China, East China, North China and Northeast China. Distribution of coal mines with high air temperature is wide, and most of them are large state-owned coal mines. Output of the coal mines accounted for more than 60% of the total output of all state-owned coal mines. In the coal mines, the total number of underground production miners is more than 0.1976 million.

At present among China's state-owned key coal mines there are more than 50 with mining depth greater than 700 m, the deepest coal mines have depth of more than 1000 m. More than 80 coal mines suffers different degrees of thermal hazard, of which 38 coal mines have working face of air temperature over 30°C. High temperature results in spontaneous combustion of coal, gas explosions and occupational diseases, affecting safety and productivity. In most of high-temperature coal mines, mechanical cooling measures must be used to solve thermal hazard problem.

## ACKNOWLEDGEMENTS

The research was performed in the range of China-Poland Bilateral Inter-Governmental S&T Cooperation Project "Study of effective cooling system in deep coal mines" (project No. 33-27) and supported by Doctoral Fund of Anhui University of Science and Technology.

## REFERENCES


Hu Chun-sheng. 2008. Strengthening heat hazard monitoring and control to improve safety level of coal mine production. www.minecooling.com/news/Detail.aspx?cid=38&id=255. (In Chinese)

Li Hong-yang, Zhu Yao-wu, Yi Ji-cheng. 2007. Ground temperature change law and abnormal factors analysis of Huainan Mining Area. Safety in Coal Mines (11): 68-71. (In Chinese)

Liao Bo, Jing Liu-jie, Tian Qiu-hojng. 2009. The situation of heat hazard in mine in our country and the discussion of using geothermal energy. Shanxi Architecture 35(8): 193-195. (In Chinese)

Liu Heqing, Wu Chao, Wang Weijun et al. 2005. Review of mine temperature drop technology. Metal Mine (6): 43-46. (In Chinese)

Liu Zeng-pin, Wang Jian-zhi, Sun Jing-kai. 2009. Measurement of thermal environment parameters in coal mines and cooling technology. www.sdcoal.org.cn/AddHtml/2009010104.htm. (In Chinese)

Liu Zhongchao, Zhao Wen. 2009. Analysis of Deep-seated Heat Damage and its Prevention in Baiji Colliery. Energy Technology and Management (6): 61-62. (In Chinese)

Wang Chang-yuan, Zhang Xi-Jun, Ji Jian-hu 2009. On technology of thermal hazard control in coal mines. Mining Safety and Environmental Protection 36(4): 62-64. (In Chinese)

Wang Cheng, Yang Sheng-qiang. 2008. Summary of Cooling Measures in Coal Mines. Energy Technology and Management (1): 15-17. (In Chinese)

Wang Wen, Gui Xiang-you,Wang Guo-jun. 2003. The Emergence and Control for Heat Harm in Mines. Industrial Safety and Environmental Protection 29(4): 33-35. (In Chinese)

International Mining Forum 2010, Liu et al. (eds) © 2010 Taylor & Francis Group, London, UK. ISBN 978-0-415-59896-5

# Research on surrounding rock control techniques of the second gob-side entry retaining for Y-type ventilation

Xinzhu Hua
*Laboratory of Coal Mine Safety and Efficient Exploitation of Ministry of Education; Anhui University of Science and Technology, Huainan, Anhui Province, China*

Yingfu Li, Denglong Zhang
*Anhui University of Science and Technology, Huainan, Anhui Province, China*

ABSTRACT: Traditional U-type ventilation is difficult to better dilute gushing gas from the upper-corner of goaf, which causes gas accumulation in the upper-corner. Y-type ventilation can change the direction of air flow, so it is able to change the laws of gas movement in the goaf, and if supplemented by other measures to control gas, the problem of gas accumulation and gas over-flow can be fundamentally solved in the upper-corner of high gas coal face. Taking No.512(5) working face of Xieyi mine in Huainan as project background, it is analyzed that the purpose of the second gob-side entry retaining and deformation characteristics of surrounding rock in different periods, and some suggestions are offered to control surrounding rock deformation of the second gob-side entry retaining. The industrial experiments show that the technique of the second gob-side entry retaining achieves good results in No.512(5) working face of Xieyi mine.

KEYWORDS: Y-type ventilation, the second gob-side entry retaining, surrounding rock control technique

## 1. INTRODUCTION

Most coal mines are high gas mine in our country, for the sake of safety mining, many scholars at home and abroad did a lot of experiments on gas control of high gas coal face, after years of efforts and quests, a series of research results have been gained, and the system of mine safety technology is initially established, but traditional U-type ventilation is easy to form eddy in the upper-corner of return airway, and difficult to better dilute gushing gas from the goaf, thus causes gas accumulation in the upper-corner, as shown in the Figure 1(a), additionally coal seam of some mines is of soft low-permeability, currently the problem of gas overflow is not yet foundationally solved in the upper-corner, and it is recently one of the prime reasons causing gas accidents in the coal mines of our country, with the increase of mining depth and mining intensity, the frequency of gas disaster increases on the daily basis.

Numerous studies show that U-type ventilation is a traditional and backward mode of ventilation in high gas coal face, Y-type ventilation can change the direction of air flow in the goaf, thus is able to change the laws of gas movement in the goaf, as shown in the Figure 1(b), and if supplemented by other measures to control gas, the problem of gas accumulation and gas overflow can be fundamentally solved in the upper-corner of high gas coal face.

To use Y-type ventilation needs to adopt the technique of the gob-side entry retaining, traditional technique of the gob-side entry retaining is to retain haulage entry of original working face to provide working face of the next section as return airway (here, one side of retained roadway is goaf, the other side is coal body) by strengthening support or adopting other effective methods after the working face of the previous section is mined, the purpose is to make a roadway be used twice Thus this way is also known as "one roadway being of two application", which the author calls as "the first gob-side entry retaining".

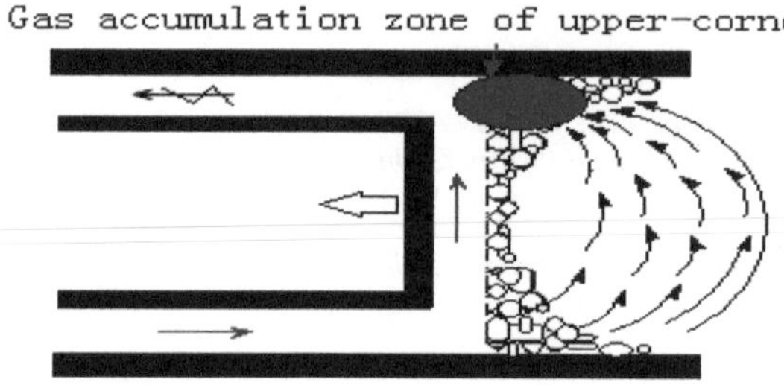


(a) U-type ventilation

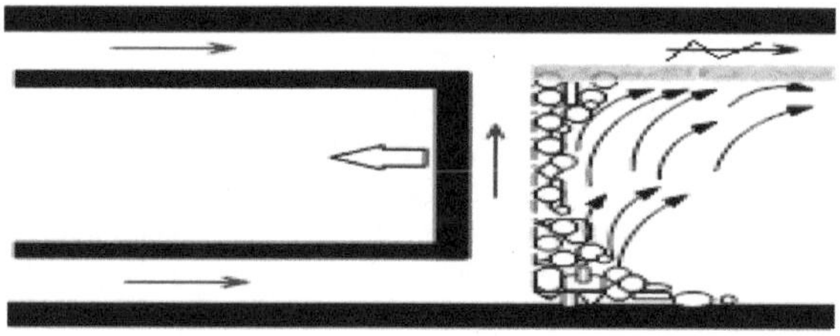

(b) Y-type ventilation

Figure 1. Gas flow direction of different type of ventilation in the goaf

The second gob-side entry retaining is to retain roadway again on the basis of the first gob-side entry retaining, finally two sides of retained roadway are both goaf, the middle is retained roadway in order to meet the needs of Y -type ventilation, as shown in Figure 2. The second gob-side entry retaining has to bear superimposed stress caused by tunneling and mining of two or more times. Since the service time of roadway is very long, strata behavior is severe, and cumulative deformation of surrounding rock is large, so roadway-in support, roadway-side support and reinforcing support must be of higher requirements.

The technique of the second gob-side entry retaining is a new way of gas control in high gas working face, especially when it is difficult to drain gas in the high gas working face, provides technical support for canceling coal pillar of section and saving coal resource. It is of important theoretical significance and practical value for reducing the cost of roadway excavation, for relieving the contradiction between mining and tunneling, and for canceling island face.

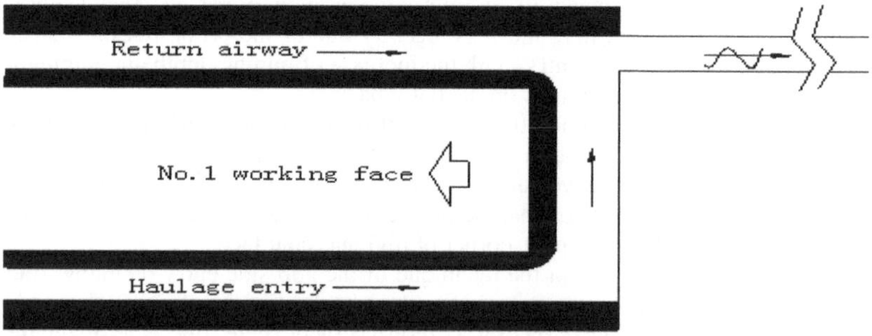


(a) The condition before haulage entry of No.1 working face retained
Figure 2. Process of the second gob-side entry retaining for Y-type ventilation

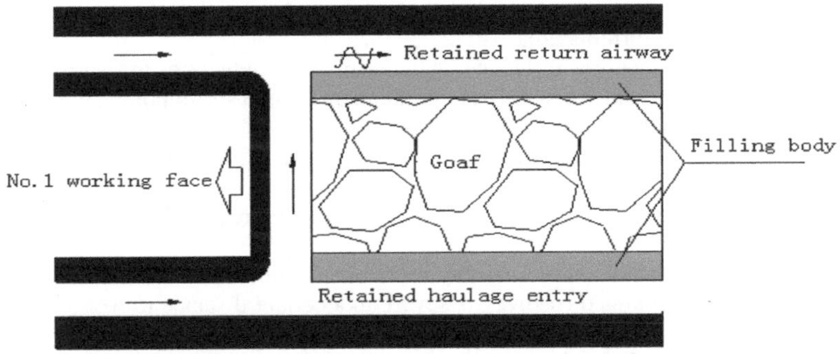


(b) The process of retaining haulage entry of No.1 working face for the first time

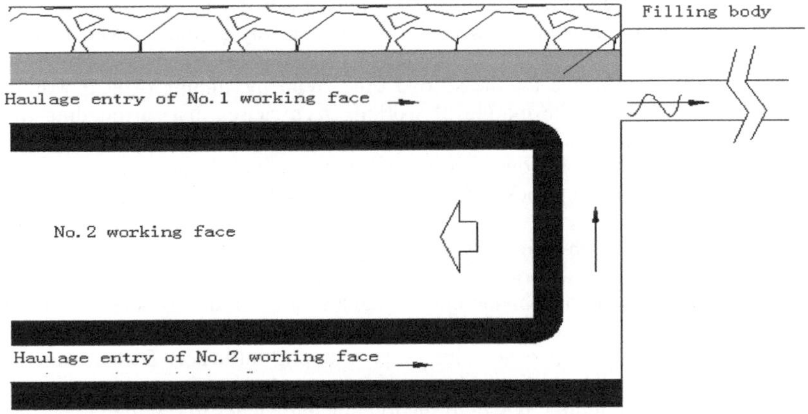


(c) Retaining haulage entry of No.1 working face

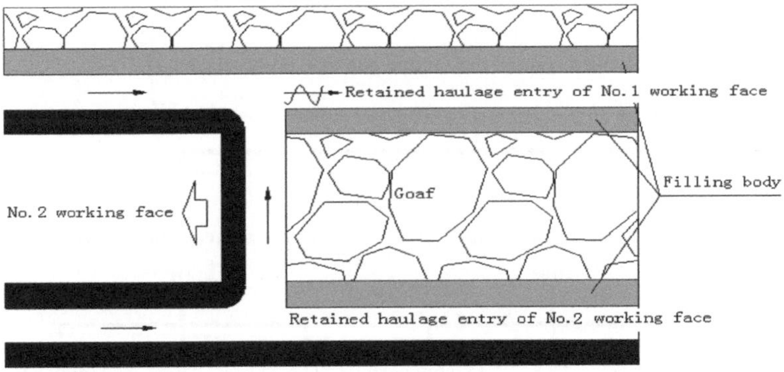


(d) The process of the second gob-side entry retaining

Figure 2. Process of the second gob-side entry retaining for Y-type ventilation

## 2. ANALYSIS ON DEFORMATION CHARACTERISTICS OF SURROUNDING ROCK OF THE SECOND GOB-SIDE ENTRY RETAINING IN DIFFERENT PERIODS

According to deformation characteristics of surrounding rock of the second gob-side entry retaining, the process from extraction to rejection can be divided into eight different periods, as shown in the Figure 3.

(1) The period of tunneling influence. After tunneling destroys original rock stress, stress concentration of surrounding rock arises, which causes surrounding rock deformation in the process of forming plastically deforming area.

(2) Stabilization period of tunneling influence. After a time of tunneling, the surrounding rock stress caused by tunneling tends to stabilize; simultaneously, the surrounding rock deformation also tends basically to stability.

(3) The period of mining influence for the first time. Due to the influence of abutment pressure caused by the mining of working face, plastic zone of surrounding rock in a certain distance to working face has a significant increase, and the surrounding rock deformation also increases dramatically.

(4) The period of the first gob-side entry retaining influence. With the advancement of working face, the strata behavior of retained roadway is severe in a certain distance to working face, and the surrounding rock deformation increases significantly.

(5) Stabilization period of the first gob-side entry retaining influence. With the advancement of the working face, waste rock far from working face tends gradually to stabilize by compaction, broken key block stops rotation and sinking, the activity of overlying rock tends to stabilize; de-formation velocity of surrounding rock of retained roadway gradually decreases and tends to stabilize.

(6) The period of mining influence for the second time. With the mining of next working face, abutment pressure caused by the mining of working face causes that surrounding rock stress of roadway redistributes again. Due to the simultaneous, superposition of abutment pressure caused by the mining of the previous section and next section plastic zone of retained roadway further enlarges, surrounding rock deformation is larger than that in the period of previous section mining influence.

(7) The period of the second gob-side entry retaining influence. For both sides of retained roadway are goaf, surrounding rock deformation velocity of the second gob-side entry retaining is much severer than that of the first gob-side entry retaining in a certain distance to working face.

(8) Stabilization period of the second gob-side entry retaining influence. Distant from the working face, surrounding rock deformation velocity of the second gob-side entry retaining decreases significantly again, surrounding rock deformation gradually tends to stabilize.

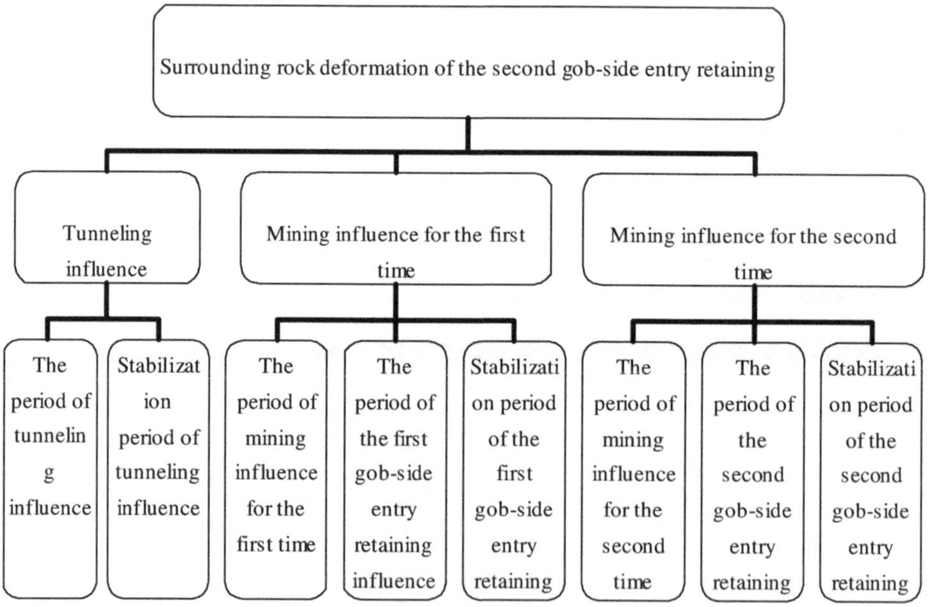


Figure 3. Eight different periods of surrounding rock deformation

## 3. CONTROL TECHNIQUE OF SURROUNDING ROCK DEFORMATION ABOUT THE SECOND GOB-SIDE ENTRY RETAINING

### 3.1. *Basic roadway-in support during the period of tunneling*

The second gob-side entry retaining has to bear supercoated stress caused by tunneling and twice or several times of mining, the service time of roadway is very long, strata behavior is severe, cumulative deformation of surrounding rock is very large, so roadway-in support, roadway-side support and reinforcing support must be of higher requirements, roadway-in support has an enormous influence on the surrounding rock deformation of retained roadway. It is the key to the success of gob-side entry retaining, if original roadway-in support adopts traditional passive support such as steel timber. It is very difficult to retain roadway, high-strength or super-high-strength combined active roadway-in support such as beam and bolting with wire mesh should be a preferred choice.

### 3.2. *Roadway-in reinforcing support before and after the first gob-side entry retaining*

(1) According to the regional characteristics of abutment pressure in front of working face (which is divided into severe influence zone and slight influence zone), advanced support measures of different strengths are adopted.

(2) With the advancement of working face, in order to ensure the integrity of roof on top of filling body, it is necessary to open gap in the face end in advance, and to use combined support such as bolting, cable, metal strip, beam, and wire netting to support the roof of gap.

(3) After roadway is retained for the first time and based on the characteristics of strata behavior of retained roadway, the regional reinforcing support of different strengths is adopted in a certain distance to working face.

### 3.3. *Roadway-in reinforcing support before and after the second gob-side entry retaining*

Roadway-in reinforcing support of the second gob-side entry retaining is basically the same as that of the first gob-side entry retaining, but reinforcing support needs to upgrade a rank. The measures such as shotcreting, grouting, and reinforcing floor can be adopted to reinforce the roadway globally if desired.

### 3.4. *Roadway-side support of the first gob-side entry retaining and the second gob-side entry retaining*

Appropriate roadway-side support is the key to ensuring the stability of retained roadway. Roadway-side support must use roadway-side filling body of early strength, high strength and certain compressibility, width and strength of filling body must meet the needs of support and breaking-off.

## 4. ENGINEERING PRACTICE OF THE GOB-SIDE ENTRY RETAINING

No.C13 Coal seam is the main commercial seam of Huainan mining area, but it is loose outburst coal seam of low permeability, the technique with which mines protective layer to eliminate gas outburst is the key to safe and efficient mining in outburst coal seam of high gas. In recent years, through mining protective layer to relieve pressure in line with gas drainage in Huainan mining area, the accident of coal and gas outburst is basically eliminated in strong outburst coal seam. Interlamellar spacing from No.C15 Coal seam to No.C13 Coal seam is 16 m in Xieyi mining of Huainan, No.C15 coal seam is mined firstly, the floor of its goaf occurs to dilatational deformation after mining, the permeability of No.C13 Coal seam below can increase significantly. Mining coal seam group of high gas, low permeability and gas outburst must change traditional U-type ventilation to Y-type ventilation, the key to realizing coal mining without pillars is the gob-side entry retaining, the paper plans to apply the technique of the gob-side entry retaining to haulage entry of No.512(5) working face in Xieyi mining of Huainan.

## 4.1. *Geological condition of experimental working face*

Ground elevation of No.512(5) working face in Xieyi mining of Huainan is from +19.0 m to +25.5 m, according to the distribution of coal seam in the section, the length is 1340 m along the strike, the length of working face is from 150 to 195 m; The level of return airway is -706 m, the level of haulage entry is -78 0 m, average thickness of No.C15 coal seam is 1.0 m, dip angle of coal seam is from 19°to 22°, average value is 20°, immediate roof is mudstone, average thickness is 2.3 m, brittle, broken; main roof is fine sandstone, average thickness is 2.5 m, crack is growing, hardness is very large; immediate floor is shale, average thickness is 1.5 m. In order to adopt Y-type ventilation in No. 512(5) working face, apply the technique of the gob-side entry retaining to haulage entry.

## 4.2. *Basic roadway-in support design for haulage entry of No.512(5) working face*

Haulage entry *is of* vertical wall and a half arch cross-section, clear width of 5.0 m, and clear height of 4.0 m; the form of support with anchor, cable, metal strip and wire netting is adopted to support roof and sidewall in regular zone, combined support with U-type yielding mental arch is adopted in especial zone (such as faultage, shuttered zone which is difficult to support by anchor, cable, wire netting etc.)

The diameter of anchor is 22 mm, its length is 2500 mm, steel type of anchor is 20MnSi, the anchor in line with metal strip and No.10 wire netting supports the roadway. 15 anchors are assigned to support roof and sidewall, line-row spacing of anchors in segmental arc is 800×800 mm, line-row spacing of anchors in vertical wall is 730×800 mm. Line-row spacing of anchors in the lap-joint of vertical wall and a half arch is 600×800 mm. All anchors are vertical to the surface of roadway except two anchors which is at an angle of 30°. All anchors adopt lengthened anchoring, one K2360 Resin Anchoring Agent and one Z2380 Resin Anchoring Agent are used to each anchors. Pretension moment of every anchor is not less than 150 Nm, pretension is not less than 50 KN, and anchored force is not less than 120 KN.

The diameter of cable is 17.8 mm, and its length is 6800 mm; Two cables are installed every two row anchors; arc length from the installation location of cable to roadway centerline is 1960mm, cable is at angle of 45° to horizontal plane, row spacing is 1600 mm; The length of channel steel attached to cable is 300 mm, stranded wire gets through the hole of No.16 channel steel to insert surrounding rock, outer end of stranded wire is directly connected to lockset.

After all anchors are installed, cable should be installed timely, its pretension is not less than 100 KN, anchorage force is not less than 250 KN; The diameter of drilling is 28 mm, every cable consumes one K2360 resin Anchoring Agent and three Z2380 resin Anchoring Agent, anchorage length is not less than 3000 mm.

Metal strip of KT-M5 type is used to support the roof and sidewall, of which two metal strips of 4100 mm length are used to support arc segment of across-section. Metal strip of KT-M3 type is used to support vertical wall, and its length is 1600 mm. The overlapping part of metal strip adds one hole in order to be beneficial to connecting metal strip, as shown in Figure 4.

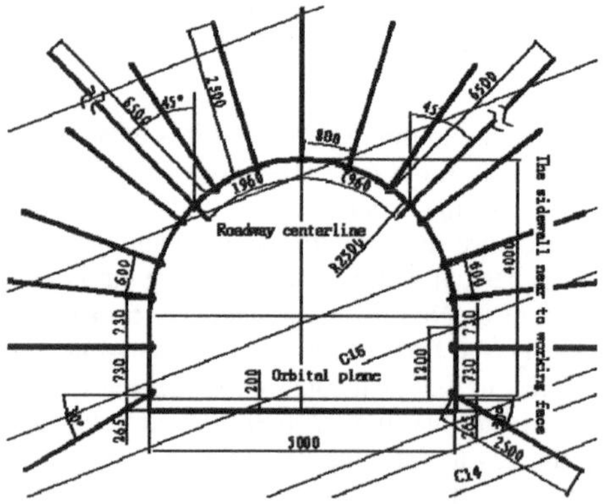


Figure 4. Support parameters of haulage entry

### 4.3. *Roadway-in reinforcing support before and after roadway is retained for the first time*

Experimental roadway belongs to traditional large across-section roadway of deep mining, practices have been shown that strata behavior of roadway is severe, so positive and effective measures should be adopted to support roadway.

(1) Beyond the range of 150 m to the rib of working face, advancing cable reinforcement is used in the segment which retains roadway. Combined reinforcement is adopted to support roadway, which the diameter of cable is 22 mm, and its length is 8000 mm. The type of metal strip is M5–280×3600 mm, and the type of cable pallet is 16–300×300 mm, row spacing of cable is 1600 mm, every row is 3 cables, and overlapping spacing of metal strip is 200 mm, as is shown in the Figure 3.

(2) Within the range from 20 m to 80 m to the rip of working face, individual hydraulic prop of DZ series and metal articulated roof beam of HDJA-1000 series, which are attached to one row of metal beam, are used to support the roof, row spacing of individual hydraulic prop is 1 m.

(3) Within the range of 20 m to the rip of working face, individual hydraulic prop of DZ series and metal articulated roof beam of HDJA-1000 series, which are attached to three rows of metal beam, are adopted to support the roof of roadway, two rows of metal beam are used in the underside of transfer conveyor, line-row spacing of metal beam is not less than 1m, one row metal beam is used in the upside of transfer conveyor, row spacing of individual hydraulic prop is 1 m.

(4) Two rows of metal beam are used within the range of 30 m to the rip of working face, the distance from metal beam to upper-sidewall and below-sidewall is both 1.5 m, row spacing of individual hydraulic prop is 1 m; in addition, if deformation and pressure of roadway are both very large, one row of metal beam should be added at a distance of 0.3 m to upper-sidewall.

(5) Two rows of wooden beam are used along the axis of roadway within the range from 30 m to 100 m to the rip of working face. Namely, it is to replace original metal articulated roof beam and individual hydraulic prop with wooden beam and wooden prop, every wooden beam is attached to three wooden prop, the diameter of wooden prop is 22 cm, its length is 3.2 m; the diameter of wooden prop is 22 cm, its length is from 2 m to 3 m, the distance from wooden beam to upper-sidewall and below-sidewall is both 1.5 m, row spacing of wooden prop is 1 m.

Reinforcing support design for haulage entry of No.512(5) working face is shown in Figure 5.

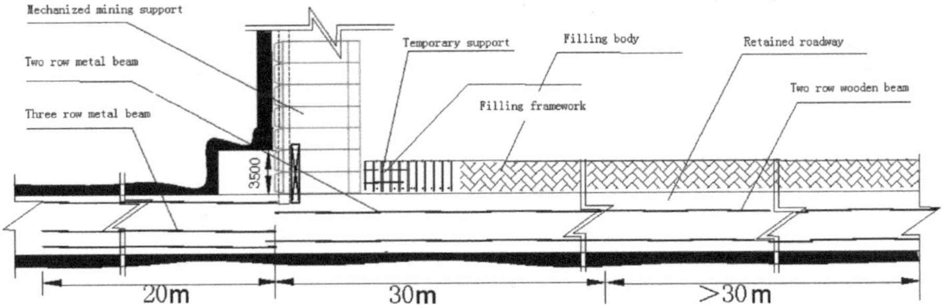


Figure 5. Reinforcing support design for haulage entry of No.512 (5) working face

(6) With the mining of working face, open timely gap in the face end, the length of gap is 5 m, the width of gap is 3.5 m, combined support with anchor, cable, metal strip and wire netting is used to support the roof of gap.

① Anchors, which are attached to metal strip and No.10 wire netting, are adopted to support the roof of gap, support parameters of anchor are as follows: the diameter of 22 mm, the length of 2500 mm, steel type of 20 MnSi, lengthened anchoring, line-row space of 800×800 mm;

② Cables between anchors are installed according to the forms of "2-1-2", "1" indicates that one cable is installed to the middle of gap, "2" indicates that row spacing between two cables is 1600 mm; the diameter of cable is 22 mm, its length is 6500 mm. The area of cable pallet is 300×300 mm. Cables are installed on the heels of diving.

③ If deformation and pressure of the roof is both very large, individual hydraulic prop of DZ series and metal articulated roof beam of HDJA-1000 series should be used to reinforce with metal beam. A beam with a prop supports the roof along the strike, row spacing between beams is 800 mm.

④ Wire netting or thick wooden panel is used to support the rib to protect the previous excavated rib from spalling rib on the side of working face near to the upside of gap, as is shown in the Figure 6.

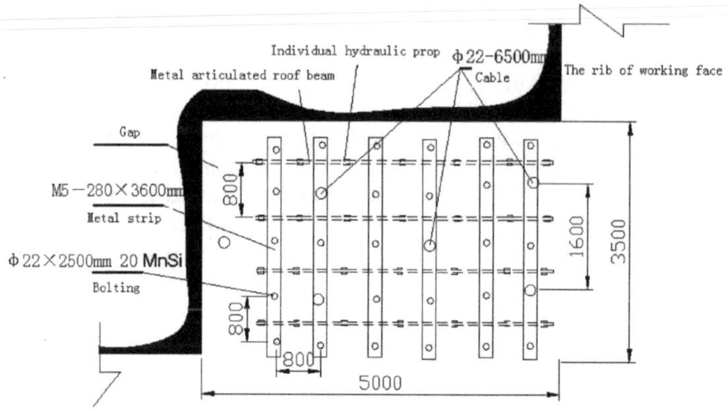


Figure 6. Support parameters of the roof of gap

## 4.4. *Roadway-side support of the first gob-side entry retaining*

Through a large number of laboratorial test, concrete glue material is suitable for filling body, main ingredients are as follows: silicate, sand, fly ash, water and additive. The result of strength experimental is shown that compressive strength of concrete at the age of one day and twenty eight days reaches 1 MPa and 14 MPa respectively, its strength can meet the needs of bearing.

Filling pump is of BSM1002E-type, made in Germany, filling pipeline matches with filling pump. Filling pipeline is laid along the underside of haulage entry, and fixed firmly to the floor.

According to the nature of filling material and reference, and through calculating, the width of filling body is taken as 2.5 m ultimately, as shown in the Figure 7.

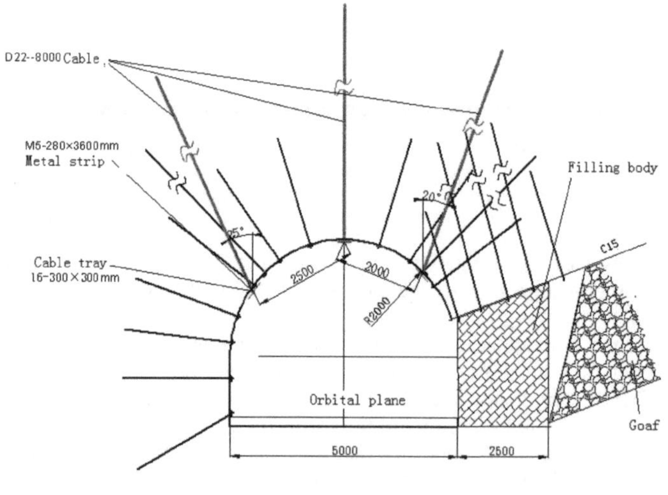


Figure 7. Cross-section figure of the first gob-side entry retaining

### 4.5. *Experimental result of the first gob-side entry retaining*

Currently, the length of the first gob-side entry retaining is over 1000 m in N0.512(5) working face, some pressure observation has been done during the period of the first gob-side entry retaining. The experimental result shows that roof-to-floor convergence is small, surrounding rock retains basically the integrity, the width of retained roadway is about 4100 mm, the height of retained roadway is about 2800 mm, surrounding rock deformation is within the extent permitted, the demands of return air and safety mining are satisfied. It is shown that roadway-in support design, roadway-side support design and reinforcing support design is reasonable, combined support with anchor, cable, metal strip and wire netting is of preferable support integrity, bearing capability and non-deformability of filling body is very strong, which can suit surrounding rock deformation.

## 5. CONCLUSIONS

(1) It is analyzed that the purpose of the second gob-side entry retaining and deformation characteristic of surrounding rock at different times, and the technique of controlling surrounding rock deformation of the second gob-side entry retaining is proposed.

(2) The technique of the gob-side entry retaining is applied successfully to No.512(5) working face in Xieyi mining of Huainan, which basically solves the problem of safe mining in No.C13 loose coal seam of low permeability and outburst, achieves successful experiences to control gas and mine relief layer, and produces obvious social and economic benefits.

(3) It is still necessary to apply the technique of the second gob-side entry retaining to haulage entry of No.512(5) working face in the future.

## ACKNOWLEDGEMENT

The project is sponsored by fund of Outstanding Academic Innovation Team of Anhui University of Science and Technology. Part of the research is supported by National Natural Science Foundation of China (No.50774001).

## REFERENCES

Hua Xin-zhu. 2002. Discussion on Technology of Backfill along Goaf Side of Gateway in Fully Mechanized Caving Face [J]. Mine Construction Technology 19(2): 31-34. (In Chinese)

Wu Shi-yue, Guo Yong-yi. 2001. Gas control of mechanical mining coal face of high yield using Y model ventilation manner [J]. Xi'an University of Science & Technology Journal 21(3): 205-208. (In Chinese)

International Mining Forum 2010, Liu et al. (eds) © 2010 Taylor & Francis Group, London, UK. ISBN 978-0-415-59896-5

# Numerical modeling of outbursts of coal and gas with a coupled simulator

Sheng Xue
*Key Laboratory of Coal Production and Methane Extraction, Anhui University of Science and Technology, Huainan, Anhui Province, China*

*CSIRO Earth Science and Resource Engineering, Pullenvale, QLD, Australia*

Y.C. Wang, X. Xie
*CSIRO Earth Science and Resource Engineering, Pullenvale, QLD, Australia*

ABSTRACT: An outburst of coal and gas is a major hazard in underground coal mining. It occurs when certain conditions of coal and rock stress, coal strength and gassiness in coal are met. A three-dimensional mechanical and fluid coupled numerical simulator, SimBurst, has been developed to simulate the process of the outburst. This paper describes the fundamentals of the simulator and the results of investigations on the effects of key coal seam and mining parameters on the initiation of the outburst, in particular the effects on gas content threshold values – a key index used to predict the occurrence of the outburst in Australia and China. The simulated results show that some parameters such as mining depth, coal strength, permeability and gas desorption/adsorption characteristics have significant effects on the threshold values, indicating that it is inappropriate to use the same gas content threshold value for outburst prediction in different underground coal mines, rather the value should be set according to the specific coal seam and mining parameters of each mine or mining panel.

KEYWORDS: Outburst, numerical modeling, SimBurst, gas content

## 1. INTRODUCTION

An outburst of coal and gas is the rapid release of a large quantity of gas in conjunction with the ejection of coal and possibly associated rock, into the working face or mine workings. The sudden and violent nature of the outburst is hazardous through the mechanical effects of particle ejection and by asphyxiation, poisoning and possible explosion from the gas produced. With an increase in depth of mining and production, outburst intensity and frequency tend to increase, although this trend is somewhat masked with advances in understanding the phenomena and application of new prediction technologies and control measures.

A key fundamental challenge to effectively predict and control the outburst is to understand its physical mechanism and influence factors of its occurrence. Many attempts have been made to derive a mechanism to explain the process of the outburst. A number of hypothesis and theoretical models have been proposed, including mainly cavity theory (Briggs 1920), pocket theory (Shepherd et al. 1981), dynamic theory (Farmer, Pooley 1967) and spherical shell destabilization (Jiang 1995). More detailed reviews can be found in literatures (Singh 1984; Hyman 1987; Lama, Bodziony 1998; Beamish, Crosdale 1998). However there is no single theory or hypothesis which can explain the whole process of the outburst. In terms of the factors contributing to the occurrence and development of the outburst, a lot of field observations have been made in the past several decades. The major recognized factors include gassiness of coal seams, geological structures, mechanical and physical properties of coal and stress conditions (Singh 1984; Lama, Bodziony 1998; Cao 2001, 2003). It is widely accepted that a coal seam is liable to an outburst under the following conditions: high gas content or pressure; the existence of complex geological structures such as folds, fracturing, faults, dykes, shear zones, changes in seam thickness, magmatic intrusions and mylonite zones; high stress regime, deep mining depth; high gas desorption rate; steeply inclined seam; thick seam; low coal strength and the less permeable seam.

To understand the physical mechanism and key influencing factors of the outburst occurrence, some attempts have been made to numerically model the outburst process. These include mainly a phase transformation model (Litwiniszyn 1985), a gas desorption and flow model (Paterson 1986), a boundary element model (Barron, Kullmann 1990), an airway gas flow model (Otuonye, Sheng 1994), a fracture mechanics model (Odintsev 1997), a simple finite element model (Xu 2006) and a plasticity model (Wold 2008). Despite these great efforts there is still no single numerical model to simulate the whole process of the outburst. It has proven to be difficult because the outburst is in fact a two-step process (initiation and development) and each step has its own characteristics and requires different approaches. As a first step, a numerical simulator, SimBurst, is developed to model the initiation process of the outburst.

## 2. SIMBURST DESCRIPTION

SimBurst is a three-dimensional numerical simulator developed by the authors. With this simulator, the process of mining-induced stress redistribution, changes of coal and rock permeability and fluid pore pressure, and coal and rock failure can be simulated to gain a detailed understanding of the outburst mechanism. It can also be used to undertake parametric studies in understanding the relative importance of key contributing factors in the outburst initiation process and help to determine the gas content threshold values - the minimum gas content to cause an outburst.

There are two important processes in outburst initiation: coal deformation and failure, and gas desorption and flow. The deformation and failure of coal are modeled with a code through geotechnical analysis of coal and rock and the gas desorption and flow are simulated with a code through analyzing coal-gas characteristics and fluid (water and gas) flow equations. These two codes are coupled to model the outburst initiation.

In modeling the deformation and failure of coal, coal is treated as a continuous medium and its mechanical behavior is numerically modeled as it reaches equilibrium or steady plastic flow. Application of the continuum form of the momentum principle yields the following equations of motion:

$$\sigma_{ij,j} + \rho b_i = \rho \frac{dv_i}{dt} \tag{1}$$

where $\rho$ is mass per unit volume of the medium; $b_i$ is body force per unit mass; and $dv_i / dt$ is material derivative of the velocity. Equation (1) is solved with a stress-strain law. The incremental stress and strain during a time step is governed by various elastic or elasto-plastic constitutive laws, which can be written in a general form as follows:

$$\Delta\sigma' = H(\sigma', \dot{\varepsilon}\Delta t) \tag{2}$$

where: $H$ is a given material functions; $\sigma'$ is the effective stress; $\dot{\varepsilon}$ is infinitesimal strain-rate tensor and $\Delta t$ is a time increment. The effective stress in Equation (2) is related to the total stress in Equation (1) by

$$\sigma' = \sigma + I\alpha P \tag{3}$$

where: $\alpha$ is Biot's effective stress parameter; $I$ is the unit tensor; and $P$ is pore pressure.

In simulating gas desorption and flow in coal, coal is considered as a dual-porosity/single-permeability system. Gas diffuses from the discontinuous coal matrix blocks into the continuous cleat system in coal. The basic equations governing fluid flow in the coal cleats (fractures) are mass conservation for gas and water:

$$\nabla \cdot \left[b_g M_g \left(\nabla p_g + \gamma_g \nabla Z\right) + R_{sw} b_w M_w \left(\nabla p_w + \gamma_w \nabla Z\right)\right] + q_m + q_g$$
$$= \left(d/dt\right)\left(\phi b_g S_g + R_{sw} \phi b_w M_w\right) \tag{4}$$

$$\nabla \cdot \left[b_w M_w \left(\nabla p_w + \gamma_w \nabla Z\right)\right] + q_w = \left(d/dt\right)\left(\phi b_w S_w\right) \tag{5}$$

where: $\nabla$ is gradient operator; $\nabla\cdot$ is divergence operator; subscript $g$ and $w$ stand for gas phase and water phase respectively; $M_n = kk_{rn}/\mu_n$ is phase mobility, where $k$, $k_{rn}$, $\mu_n$ are absolute permeability, relative permeability and viscosity respectively, $\mu_n$ is viscosity; $\gamma_n = \rho_n g$ is gas or water gravity gradient, where $\rho_n$ is phase mass density and $g$ is gravitational acceleration; $S_n$ is degree of saturation; $b_n = 1/B_n$ is gas or water shrinkage factor, where $B_n$ is formation volume factor; $t$ is time; $Z$ is elevation; $\phi$ is effective fracture porosity; $q_g$ and $q_w$ are the normal well source terms; $R_{sw}$ is gas solubility in water. Gas phase pressure $p_g$ and water phase pressure $p_w$ are related by capillary pressure $P_c$:

$$P_c = p_g - p_w \tag{6}$$

Water and gas saturation satisfy:

$$S_w + S_g = 1 \tag{7}$$

Equations (4)–(7) make up four equations and contain four unknown variables $p_g$, $p_w$, $S_g$ and $S_w$ hence it is a solvable system. The volume of adsorbed gas in the coal matrix is described by the Langmuir adsorption isotherms:

$$\frac{V}{V_L} = \frac{p_g}{P_L + p_g} \tag{8}$$

where $V$ is volume of gas adsorbed at pressure $P_g$ ; $V_L$ is Langmuir volume; $P_L$ is Langmuir pressure. The gas flow (rate) through the matrix is described mathematically by Fick's first law of diffusion expressed in the form:

$$q_m = \left(\frac{V_m}{\tau}\right)\left[C - C(p)\right] \tag{9}$$

where $C$ is average matrix gas concentration; $V_m$ is bulk volume of a matrix element; $p$ is gas pressure; and $\tau$ is gas sorption time.

Because the code for modeling the deformation and failure of coal and the code for simulating gas desorption and flow in coal are developed separately, when coupled together, the resultant equations cannot be solved simultaneously. In SimBurst, they are solved sequentially with coupling parameters passing to each equation at specific intervals. In the SimBurst, these two sets of codes are executed sequentially on compatible numerical grids and coupled through user-defined modules which serve to pass relevant information between the equations that are solved in the respective codes. The fundamentals of the coupling process are shown in Figure 1.

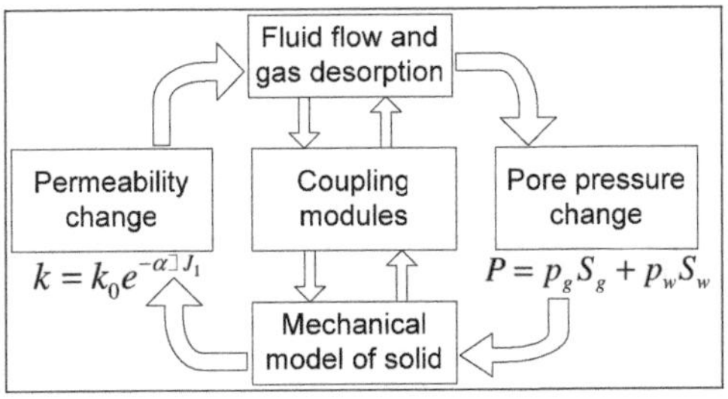


Figure 1. Basic process of coupling mechanical model with fluid model

## 3. PARAMETRIC STUDY

Gas content is an important factor in outburst proneness. Unless the gas content reaches a critical level, outbursts of gas and coal will not manifest themselves at all. Such a critical level is referred as the gas content threshold value and has been used as an index for outburst prediction in Australia and China (Xue 2007, 2008). The objective of this parametric study is to quantitatively determine the influence of key coal seam and mining parameters on outburst initiation, more specifically it aims to understand the effects of mining depth, coal strength, permeability and gas desorption/adsorption characteristics on the gas content threshold value. The key parameters and their selected values used in the study are: mining depth (200 to 1200 m), uniaxial compressing strength (UCS) of coal (2 to 12 MPa), coal seam permeability ($10^{-1}$ to $10^{-3}$ mD), Langmuir pressure (800 to 1200 KPa), and Langmuir volume (11 to 21 $m^3$/t). In simulation of each set of parameters, the gas content, which is linked to gas pressure through Langmuir adsorption isotherm, is gradually increased until an outburst occurs. The gas content corresponding to outburst occurrence is the threshold value for this set of parameters. The simulation results are shown and discussed below.

Figure 2 shows the relationship between mining depth and gas content threshold value for various UCS values of coal. Results from this Figure clearly indicate that the threshold value decreases with the increase of mining depth and the effects of mining depth are more pronounced for small UCS values, indicating that a deep mining seam is more prone to the outburst than a shallow mining seam for the same gas content in coal.

Figure 3 shows the effect of coal UCS on the gas content threshold value at various mining depths. It can be seen from this Figure that the threshold value increases with UCS for a given mining depth and its increase is more significant when UCS is small, indicating that a seam of low strength is more liable to the outburst than a hard seam for the same gas content in coal.

Figure 4 shows the effect of coal permeability on gas content threshold value for various mining depths. The results indicate that the threshold value increases with permeability. This agrees with the observation that coal with permeability less than $10^{-3}$ mD is highly outburst prone and coal with permeability greater than $10^{-1}$ mD is less prone to outbursts (Lama, Bodziony 1998).

Gas adsorption and desorption in coal can be characterized with Langmuir pressure ($P_L$) and Langmuir volume ($V_L$). Numerical simulations have been carried out to study the effect of $P_L$ and $V_L$ on the gas content threshold. The results indicate that the threshold value decreases with increase in $P_L$ (Figure 5) and decrease in $V_L$. The results are understandable from a typical gas adsorption isotherm curve that, for a given gas content, either large $P_L$ or smaller $V_L$ will increase the value of gas pressure and hence increase the risk of outburst occurrence.

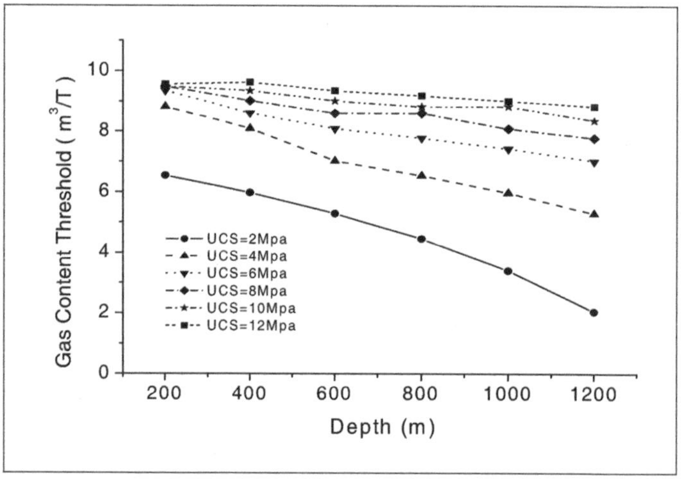


Figure 2. Effect of mining depth on gas content threshold value

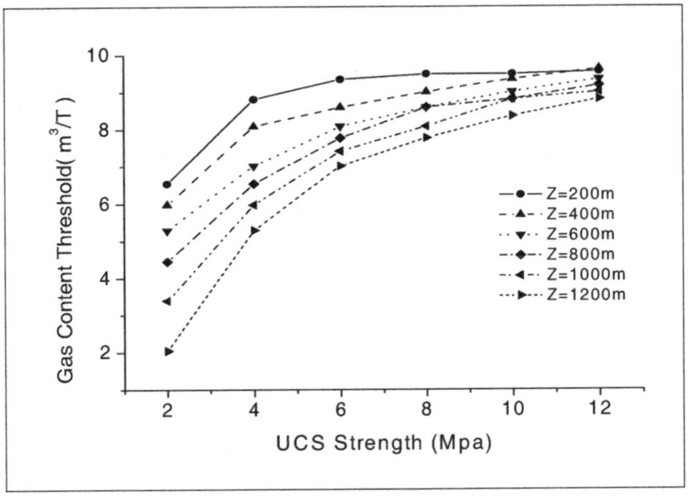


Figure 3. Effect of coal strength on gas content threshold value

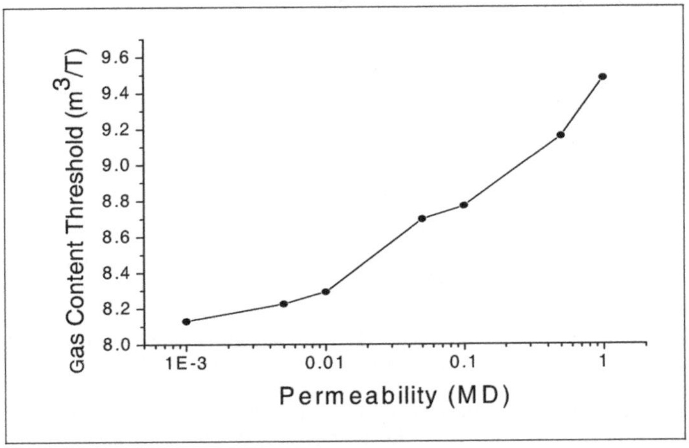


Figure 4. Effect of coal permeability on gas content threshold value

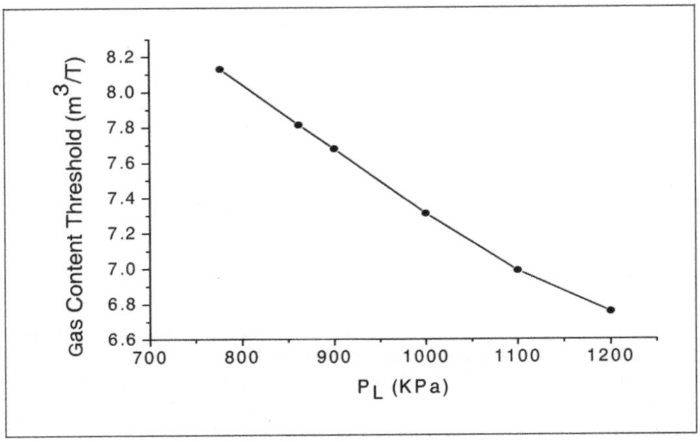


Figure 5. Effect of Langmuir pressure on gas content threshold value

## 4. CONCLUSIONS

A three-dimensional numerical simulator, SimBurst, has been developed by coupling a geotechnical analysis code with a fluid flow and gas desorption model to simulate the initiation process of an outburst. Within this simulator, two important interactive processes in outburst initiation: coal deformation and failure, and gas desorption and flow, are coupled. The simulator is used to gain a better understanding of outburst mechanism. It has also been used to undertake parametric studies to investigate the influences of key contributing factors in the outburst initiation process and help to determine the gas content threshold value. The results of the parametric studies show that the gas content threshold value is significantly influenced by mining depth, coal strength, seam permeability and gas adsorption/desorption characteristics, indicating that it is inappropriate to use the same gas content threshold value for outburst prediction in different underground coal mines, instead the threshold value should be set according to the specific seam and mining parameters of each mine or mining panel. The threshold value will decrease as mining depth increases, coal strength decreases, seam permeability becomes low, Langmuir pressure in-creases and Langmuir volume decreases. These simulated results are in general agreement with field observations.

It should be noted that the current version of SimBurst is limited to simulate the initiation process of the outburst, mainly because it is based on continuum media and no explicit mechanisms of fracture and fragmentation of coal is included. Despite these limitations, the simulated results suggest that the coupled simulator presented in this paper is capable of reproducing the most basic features of outbursts.

## REFERENCES


Barron K., Kullmann D. 1990. Modeling of outburst at #26 Colliery, Glace Bay, Nova Scotia, Part 2, proposed outburst mechanism and model. Mining Science and Technology 2: 261-268

Beamish B.B., Crosdale P.J. 1998. Instantaneous outbursts in underground coal mines: an overview and association with coal type. International Journal of Coal Geology 35: 27-55

Briggs H. 1920. Characteristics of outbursts of gas in mines. Trans. Instn. Min. Engrs. 61: 119–146

Cao Y.X., He D.D., Glick D.C. 2001. Coal and gas outbursts in footwalls of reverse faults. International Journal of Coal Geology 48: 47-63

Cao Y.X., Davis A., Liu R.X., Liu X.W., Zhang Y.G. 2003. The influence of tectonic deformation on some geochemical properties of coals – a possible indicator of outburst potential, International Journal of Coal Geology 53: 69-79

Farmer I.W., Pooley F.D. 1967. A hypothesis to explain the occurrence of outbursts in coal, based on a study of west Wales outburst coal. International Journal of Rock Mechanics & Mining Sciences 4: 189-193

Hyman D.M. 1987. A review of the mechanisms of gas outburst in coal, Bureau of Mines Information Circular 9155. US Department of the Interior

Jiang C.L., Yu Q.X. 1995. The hypothesis of spherical shell destabilization as the mechanism of coal and gas outbursts. Coal Safety 16: 17-25. (In Chinese)

Litwiniszyn J. 1985. A model for initiation of gas outburst. International Journal of Rock Mechanics, Mining Sciences & Geomechanics Abstract 22: 39-46

Lama R.D., Bodziony J. 1998. Management of outburst in underground coal mines. International Journal of Coal Geology 35: 83-115

Otuonye F., Sheng J. 1994. A numerical simulation of gas flow during coal/gas outbursts. Geotechnical and Geological Engineering 12: 15-34

Odintsev V.N. 1997. Sudden outburst of coal and gas - failure of natural coal as a solution of methane in a solid substance. Journal of Mining Science 33: 508-516

Paterson L. 1986. A model for outbursts in coal. International Journal of Rock Mechanics & Mining Sciences & Geomechanics Abstract 23: 327-332

Shepherd J., Rixon L.K., Griffiths L. 1981. Outbursts and geological structures in coal mines: a review. Intern. J. Rock Mechanics & Mining Sciences & Geomechanics Abstract 18: 267–283

Singh J.G. 1984. A mechanism of outburst of coal and gas. Mining science and technology 1: 269-273

Wold M.B., Connell L.D., Choi S.K. 2008. The role of spatial variability in coal seam parameters on gas outburst behavior during coal mining. International Journal of Coal Geology 75: 1–14

Xu T., Tang, C.A. Yang, T.H. Zhu, W.C., Liu J. 2006. Numerical investigation of coal and gas outbursts in underground collieries. Intern. J. Rock Mechanics & Mining Sciences 43: 905-919
Xue S. 2007. Gas content based outburst control technology in Australia. In: Proceedings of International Symposium on Coal Gas Control Technology, Huainan, China, Oct. 25-26, pp 405-412
Xue S. 2008. Development of gas content based outburst control technology in Huainan. In: Proceedings of International Conference on Coal Mine Gas Control and Utilization, Huainan, China, Oct. 23-24, pp 252-260

International Mining Forum 2010, Liu et al. (eds) © 2010 Taylor & Francis Group, London, UK. ISBN 978-0-415-59896-5

# The influence of sealing abandoned areas on climatic conditions in the room and pillar extraction method in copper mining

Stanisław Nawrat
*AGH University of Science and Technology, Krakow, Poland*

K. Soroko, S. Gola
*KGHM Polska Miedź S.A., Lubin, Poland*

ABSTRACT: In Polish copper mines ore extraction is carried out at depths of between 600 and 1200 m. Both depth and approved mining methods cause gradual deterioration of climatic conditions in the underground excavations. The assurance of further mining requires using appropriate prophylactic methods, such as: increasing of air stream volume in mining areas, suitable regulation of air flow and use of air cooling. This article demonstrates the changes of climatic conditions in the mining excavations and dependence on resistance of ventilation dams which are used in sealing abandoned area in the room and pillar method of copper ore deposit extraction.

## 1. INTRODUCTION

One of the factors which has influence on mining profitability of KGHM Polska Miedź S.A. is presently connected with opening out deeper parts of the deposit. At present extraction of the copper ore deposit is carried out at depths of between 600 and 1200 m and excavating deeper and deeper parts of the deposit depends on availability of technical means and efficiency of undertaken prophylactic activities to reduce the influence of natural hazards and doing so ensure miners work safety.

One of the natural hazards which at large depths of excavation influences work safety and in considerable way limits the possibility of opening out deeper parts of the deposit are high original rocks temperature and high air temperature. Extraction of the copper ore deposit in most mines of KGHM is carried out in difficult climatic conditions. Main factors creating thermal environment in mining headings are:

- geological conditions of deposit deposition , that is depth and connected with it geothermic gradient,
- implemented system of mining,
- large concentration of exploitation works,
- use of high power machines and devices.

The thermal conditions are also influenced by the structure of the mine ventilation system, which is characterized by: sufficient lengths of headings bringing fresh air into the exploitation faces, large number of air flow regulators and excessively extended shaft bottoms, which finally causes considerable air losses and air temperature increase. That is why keeping required by law (Rozporzadzenie 2002) climatic conditions of work in mining headings is provided by complex activities such as:

- increasing of air stream flow rate,
- shortening working time,
- use of self-propelled mining machines with air conditioned operator's compartments,
- cooling down the air brought into the headings.

The technical and organizational means used at present do not allow to control occurring the hazard of difficult thermal conditions in the mining headings at large depths. Such situation forces necessity to search for new methods and ways of reducing temperature hazard and this way enabling exploitation in deeper parts of the deposit.

In the mines of KGHM copper ore deposit is excavated using different modifications of room and pillar system, in which liquidation of the mined out area is mostly carried out by roof deflection and waste rock backfilling or hydraulic backfilling. Nevertheless, in the liquidated area there are voids through which flow the air and the air transports heat into the mining headings, making the climatic conditions worse. This paper shows the results of the research made in one of the KGHM's mines and the aim of the research was to define the possibilities of reducing heat emission from the worked out area into the exploitation headings.

## 2. CHARACTERISTICS OF THE EXTRACTION SYSTEM

In the mine where the research was done extraction of copper ore deposit is carried out using room and pillar system of mining with movable closing pillar and liquidation of the worked out area by roof deflection marked in a catalog as J-UGR-PS (KGHM 2007).

This system is dedicated to extract a bedded kind of deposit with thickness up to 6 m and decline not exceeding 8°. What is characteristic for such system is that it requires a small range of development works from which the exploitation front is activated (Fig. 1).

The deposit is extracted with rooms located perpendicularly to the exploitation front line and with stripes located parallel, leaving technological pillars supporting the roof over the worked out area. The length of the exploitation front, which is defined by the number of rooms, is determined based on the planned output of the mining section and usually varies between tens and several hundreds meters. The width of the exploitation front is determined by the number or pillars rows. Pillars measurements and location of their axes in relation to the exploitation front line are selected depending on: the strength parameters of the roof rocks, the deposit thickness and the rock mass pressure in the extracted area in such a way so they could cooperate with the roof in the final phase. The width of the exploitation fronts can vary between tens and several hundreds of pillars rows.

As the exploitation front advances the technological pillars located in the liquidation line are ripped by loading and haulage machinery and the remnant pillars are left to minimize the curvature of the deflecting roof layers. Liquidation of the technological pillars does not include this part of the area in which those pillars are left on purpose. They create a gradually getting longer operational closing pillar.

The width of the operational pillar is defined individually for each mining area depending on the strata deposition conditions in such a way that the stability of at least three headings located within this operational pillar is ensured during the mining area exploitation and during liquidation of the operational pillar.

The length of the operational pillar is a consequence of rock mass behavior and most of all the convergence. In one of the operational pillar's headings a belt conveyor is built, which is gradually lengthened as the exploitation front advances. Liquidation of the operational pillar takes place after the area is mined out and is done by ripping the technological pillars in the opposite direction to the so far direction of the front advancement.

The length of the exploitation fronts at most mining sections of the mine in which the research was done is forty rooms and is a result of: optimization of the self-propelled machines haulage road length, used technological cycle and strength parameters of the roof layers rocks.

Although as for the rock burst prophylactics the J-UGR-PS room and pillar mining system provides full safety of the underground staff , it makes serious difficulties to ensure proper thermal work conditions when the length of the exploitation front is tens rooms and when the opening width is minimum.

Such situation is mostly a consequence of:

- outflow of heat stream from the worked out area as a result of insufficient stopping or backfilling;

- necessity of several exploitation blocks in series venting depending on the geometry of the system as well as the arrangement of exploitation areas, main ventilation routes and localization of mechanical depression sources;
- difficulties in providing required intersections of mining headings;
- bringing the air into the exploitation front via the operational pillar, often along a heading with a belt conveyor installed there;
- outcropping considerable planes of the rock mass, due to the anti- rock burst prophylactics which requires maintaining at the width of the exploitation front at least three rows of pillars and three patent stripes near the solid;
- concentration of technological devices such as self-propelled Diesel machinery at one block (Soroko 2007).

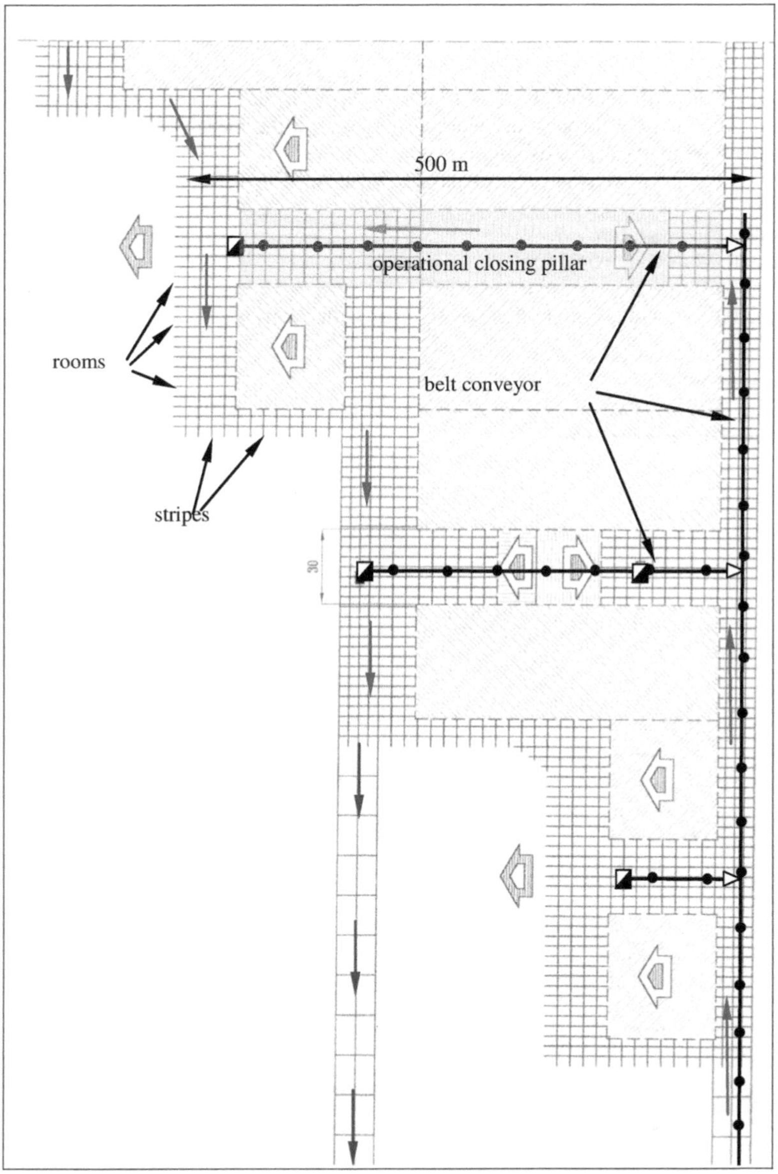


Figure 1. Scheme of the J-UGR-PS extraction system of model mining section

## 3. CHARACTERISTIC OF THE RESEARCHED AREA

The mining section is located at the depth of 850 m below sea level, where the original temperature of the rock mass rocks is 36.5°C. To the mining section is brought the air stream in the volume of 7000 $m^3$/min with temperature of 20.8°C – wet bulb and 16.4°C – dry bulb (Fig. 2). The air is transported into the exploitation front by headings which are isolated from the work-ed out area. When isolating the worked out area using gotten dams (green line on fig. 2) the temperatures at the exploitation front reach 34°C – dry bulb. Such situation occurs due to the fact that the air migrates through the leaks in the dams into the worked out area, becomes heated and comes back into the exploitation front "F" (Fig. 2). In Polish copper mining the maximum permitted temperature measured with dry bulb is 35°C. In case of exceeding permissible temperature exploitation is forbidden by law.

### 3.1. *Research method*

In order to research the influence of the leak tightness (type) of dams to isolate the worked out area and to limit the air flow through the worked out area on the thermal conditions at the mining section the following research was done:

a) experimental research to determine aerodynamic resistance for different dams' constructions:

T.1 – a dam made of heap gotten,
T.2 – a brick dam 0.24 m wide with single-sided plaster,
T.3 – a brick dam 0.60 m wide,
T.4 – a brick dam 0.60 m wide, from both sides covered with tekkflex.

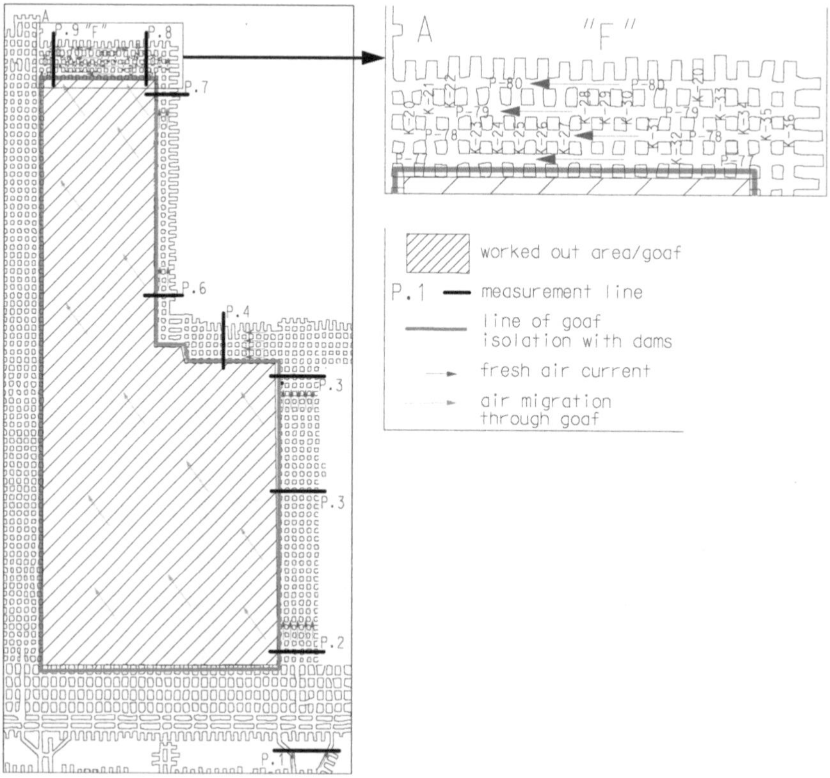


Figure 2. A mining section – object of the research

The worked our area isolation was carried out along the green line marked on figure 2 and the resistance of the used ventilation dams were enumerated using the following formula:

$$R_f = \frac{\Delta P}{\dot{V}} \tag{1}$$

where: $R_f$ = dam resistance, kg/m$^7$; $\Delta P$ = pressure difference (decrease of potential) at the dam, Pa; $\dot{V}$ = air volume stream, m$^3$/s.

b) model research was done using a package of the AutoWent numerical program (a mining section in the AutoWent program consisted of 2792 nodes and 5456 splits). The numerical model scheme of the mining headings system is shown in figure 3.

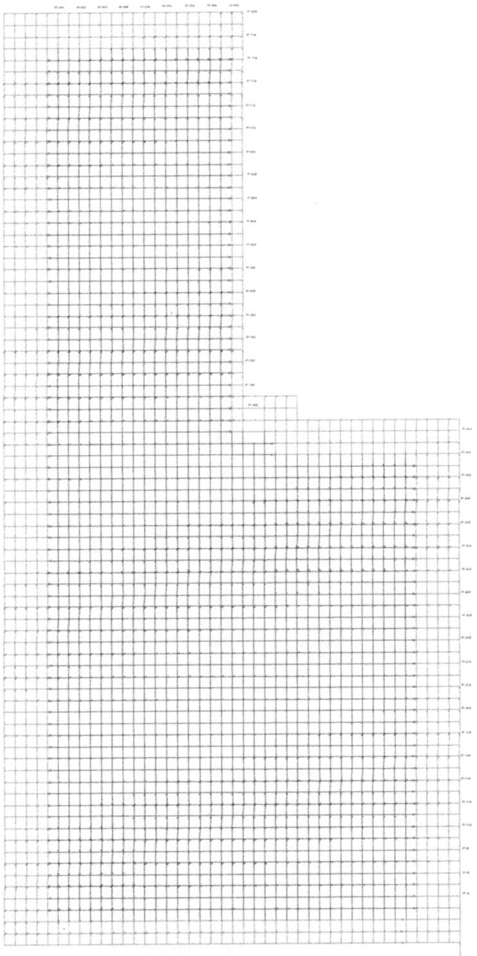

Figure 3. Numerical model scheme of the mining headings system

3.2. *Influence of dams' leak tightness on their aerodynamic resistance*

Experimental research made it possible to determine dams' aerodynamic resistance depending on the dam's building construction and the dam's type (Soroko unpublicized). The obtained resistance values of the analyzed types of ventilation dams are given in Table 1.

Table 1. Resistance values of ventilation dams

| Type of dam | Characteristics of dam | Determined resistance of dam $R_f$ $kg/m^7$ |
|---|---|---|
| T.1 | a dam made of heap gotten | 1.965 |
| T.2 | a brick dam 0.24 m wide with single-sided plaster | 60.196 |
| T.3 | a brick dam 0.60 m wide | 100.624 |
| T.4 | a brick dam 0.60 m wide, from both sides covered with tekkflex | 208.163 |

### 3.3. *Model research of dams' resistance influence on temperature distribution at mining section*

Model research of dams' resistance influence on temperature distribution at mining section was done using the AutoWent program and the temperature distribution is shown on Figures 3 and 4.

The calculations which were made have shown that with the increase of the isolated worked out area leak tightness, decreases the value of thermal energy which can be transported by 1 $m^3$ of the air flowing through the worked out area. It has substantial influence on the improvement of climatic conditions in the headings of the exploitation front and with respect to work efficiency, location of coolers of the central air conditioning or air-conditioned mining faces machinery.

Values changes of the thermal energy transported by 1 $m^3$ of air in relation to the type of ventilation dams used is shown on Figure 5.

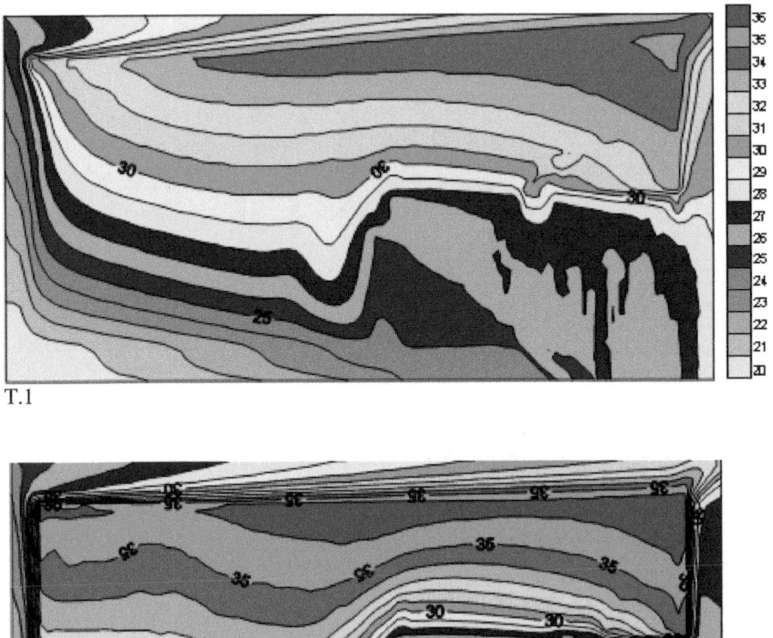


T.1

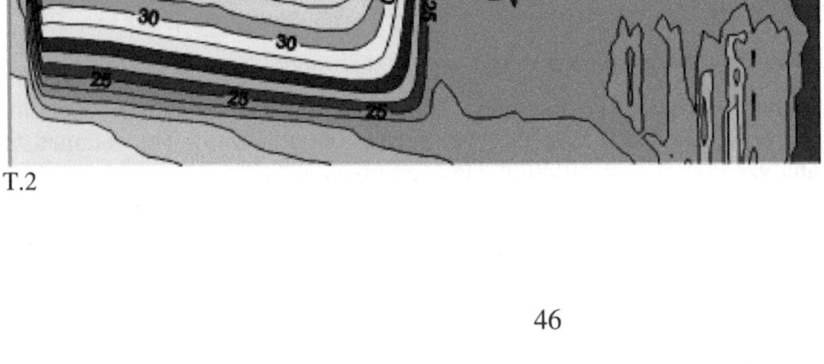


T.2

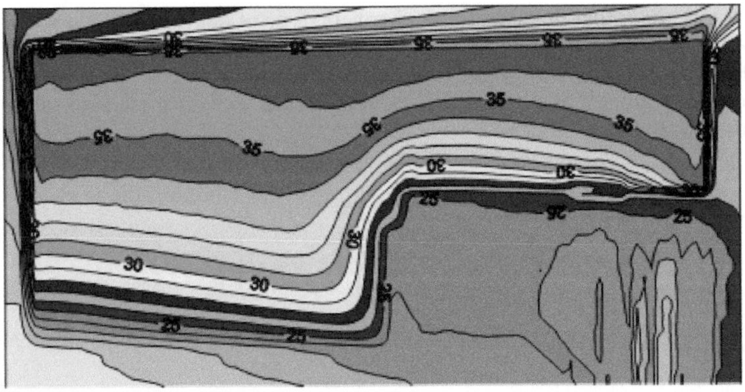


T.3

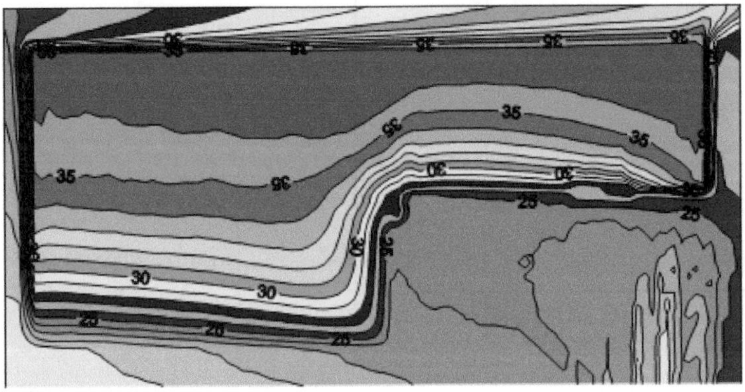


T.4

Figure 4. The temperature distribution in abandoned / worked out area sealed by given types of ventilation dams: T-1, T-2, T-3, T-4

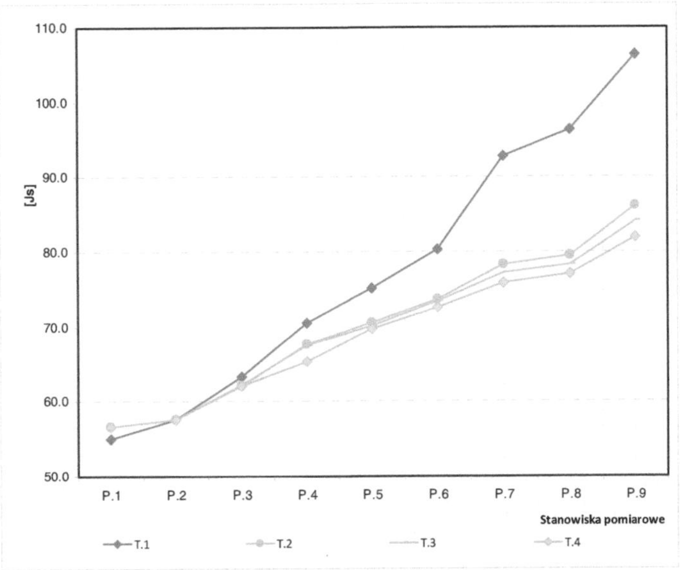


Figure 5. Value of thermal energy in 1 m$^3$ of mining headings for particular dams types

From the analysis' results arises that the use of more hermetic type of dams to seal the worked out area has gat considerable influence on improving thermal conditions in the mining headings of the researched mining section. On figures 6–9 there are shown changes of the air dry bulb temperature, which flows along the stripes P-77, P-78, P-79 and P-80 at the length of the front that is from the room K-36 to the room K-20 (Fig. 2).

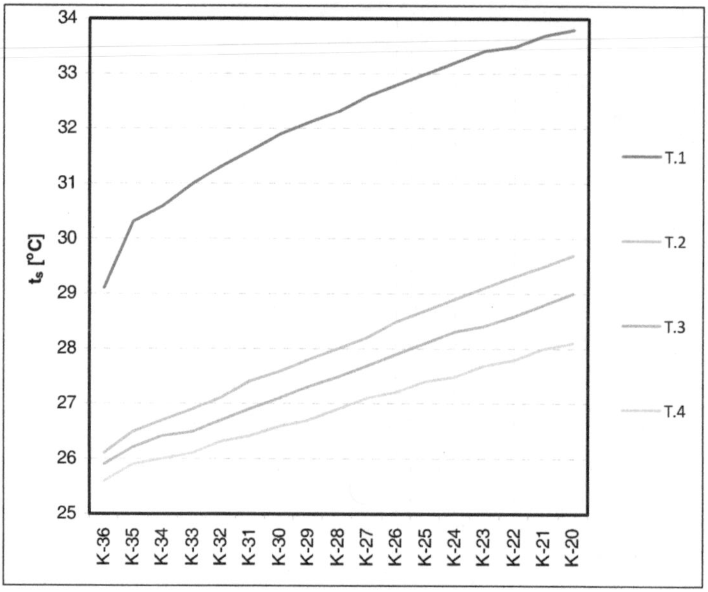


Figure 6. Changes of the mining air dry bulb temperature at the stripe P-77 – worked out area

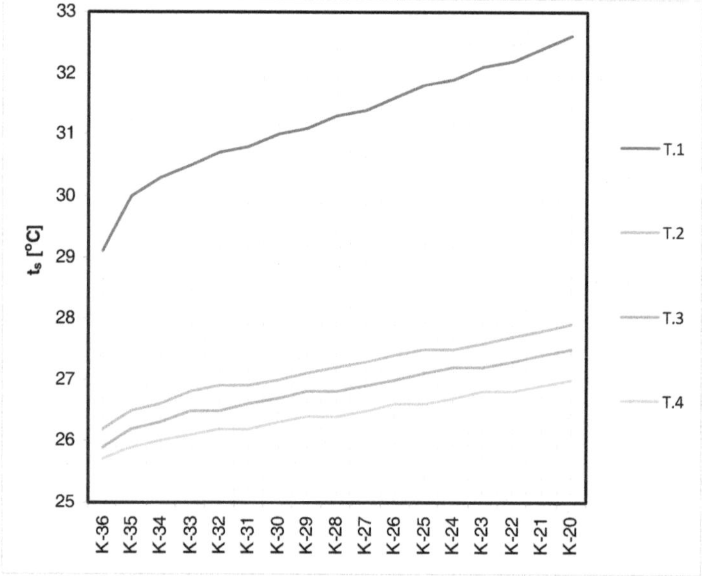


Figure 7. Changes of the mining air dry bulb temperature at the stripe P-78

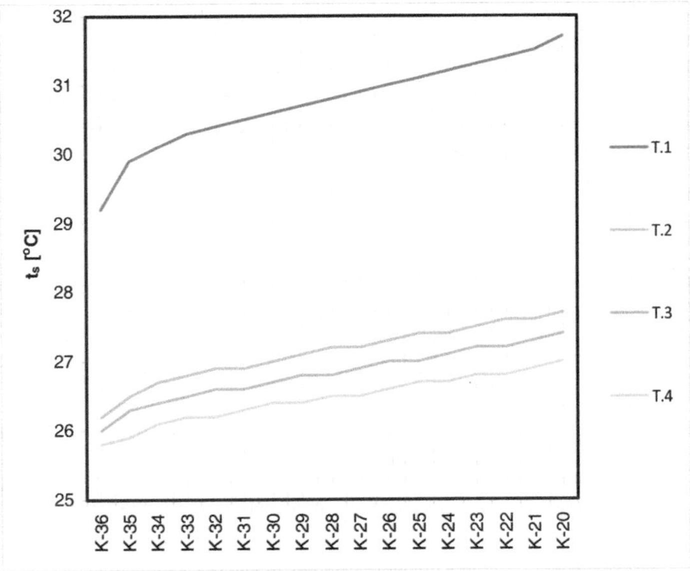


Figure 8. Changes of the mining air dry bulb temperature at the stripe P-79

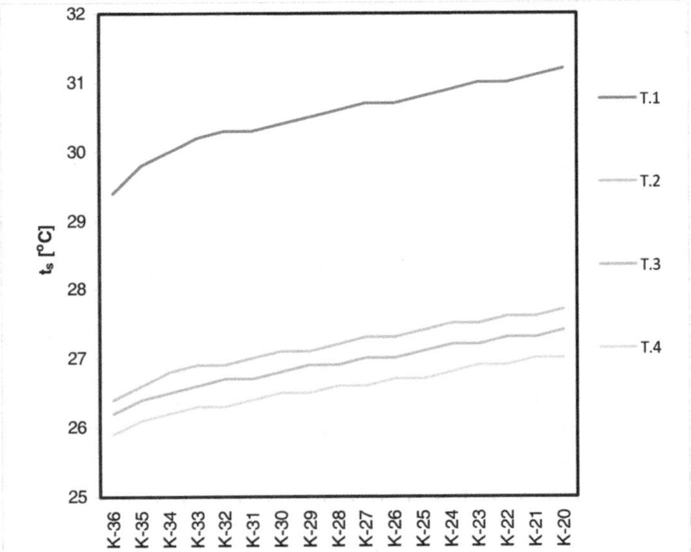


Figure 9. Changes of the mining air dry bulb temperature at the stripe P-80 – at solid

## 4. SUMMARY

In this paper there are shown calculation results of the worked out area sealing influence on the climatic conditions at the mining section with the copper ore deposit extraction by room and pillar system of mining. The authors' aim was to analyze the possibility of improving the climatic conditions at the front of the mining section by hermetic sealing, limiting at the same time transport of thermal energy from the worked out area.

The research which was done has shown that the increase of aerodynamic resistance of the dams isolating the worked out area significantly influences the decrease of air temperatures in the operative (active) mining headings, which contributes to improving safety as well as technological and economic efficiency at the mining section.

## REFERENCES


Soroko K. Wpływ izolowania zrobów na emisję ciepła i warunki cieplne w rejonie wydobywczym kopalni rud miedzi. Opracowanie niepublikowane. (In Polish)

Soroko K., Wysocka L., Gola S. 2007. Wpływ tamowania zrobów na długość frontu eksploatacyjnego w aspekcie występowania temperatury dopuszczalnej przy eksploatacji systemem J-UGR-PS. XIV Międzynarodowa Konferencja Naukowo-Techniczna – Górnicze Zagrożenia Naturalne 2007. Główny Instytut Górnictwa. Katowice. (In Polish)

KGHM 2007. Katalog systemów eksploatacji złóż rud miedzi dla kopalń KGHM Polska Miedź S.A. KGHM Polska Miedź S.A. w Lubinie. Grudzień 2001. (In Polish)

Rozporządzenie Ministra Gospodarki z dnia 28 czerwca 2002. W sprawie bezpieczeństwa i higieny pracy, prowadzenia ruchu oraz specjalistycznego zabezpieczenia przeciwpożarowego w podziemnych zakładach górniczych. (In Polish)

International Mining Forum 2010, Liu et al. (eds) © 2010 Taylor & Francis Group, London, UK. ISBN 978-0-415-59896-5

# Proecological technology utilization of methane from mines

Piotr Czaja, Stanisław Nawrat, Sebastian Napieraj
*AGH University of Science and Technology, Krakow, Poland*

ABSTRACT: Coal production is accompanied by the release of methane, which for security reasons is removed from the mine. Annually in Poland into the atmosphere with ventilation air is emitted around 580 million $m^3$ of methane. Methane is a greenhouse gas which has a potential 21 times greater than carbon dioxide. The maximum content of methane in ventilation air according to the Polish mining law amounts to 0.75%. The use of methane from ventilation air is very important from economic and environmental reasons. Efficient way to use low concentration air-methane mixtures is catalytic oxidation.

## 1. INTRODUCTION

Methane from mines (CBM) accompanying the primary mineral exploitation - coal and not received by the drainage, in the greater part is excreted into ventilation air, creating an air methane mixtures with different methane content (VAM).

Methane utilization from coal beds is very important because of following reasons:
- economic – Coal Beds Methane according to polish mining law is primary mineral,
- ecological – methane emission to the atmosphere make greenhouse effect.

In polish mines since many years underground drainage is development. Use of methane in heat and energetic plants increasing from 124 mln $m^3$ $CH_4$/year in 2000 year to 159,5 mln $m^3$ $CH_4$/year in 2009 year.

Big challenge for mining is utilization of Ventilation Air Methane (VAM). In coal mines, methane occurring in coal seams is emitted to ventilation air making methane-air mixtures with methane content from 0.0% to 0.75% (0.75% is max methane content in ventilation air according to polish mining law).

The world has conducted intensive research - development that led to the development of many technologies and devices, allowing to carry out the process of combustion of methane with a low concentration.

## 2. VAM RESOURCES IN POLAND

In Polish coal mines total methane emission growing since 2001, despite a decrease in the quantity of mines and coal production. Ventilation Air Methane and Drainage volume presents Table 1 and in Figures 1–3.

In 2009 total methane emission in polish coal mines was 1646.01 $CH_4$/min (855.71 mln $m^3$/year), with ventilation air 1133 $m^3$ $CH_4$/min (595.91 mln $m^3$/year), with drainage 494 $m^3$ $CH_4$/min (259.8 mln $m^3$/year).

Table 1. Ventilation Air Methane and Drainage volume since 1999 to 2009

| | Year | | | | | | | | | | |
|---|---|---|---|---|---|---|---|---|---|---|---|
| | 1999 | 2000 | 2001 | 2002 | 2003 | 2004 | 2005 | 2006 | 2007 | 2008 | 2009 |
| Total methane emmision (mln $m^3$/year) | 744,5 | 746,9 | 743,7 | 752,6 | 798,1 | 825,9 | 851,1 | 870,3 | 878,9 | 880,9 | 855,7 |
| Drainage (mln $m^3$/year) | 216,1 | 216,1 | 214,3 | 207,3 | 227,1 | 217,2 | 255,3 | 289,5 | 268,8 | 274,2 | 259,8 |
| Used methane (mln $m^3$/year) | 136,9 | 124,0 | 131,5 | 122,4 | 127,8 | 144,2 | 144,8 | 158,3 | 165,7 | 156,5 | 159,5 |
| Number of coal mines | 47 | 42 | 42 | 42 | 41 | 39 | 33 | 33 | 31 | 31 | 31 |
| Coal output (mln Mg) | 110,4 | 102,5 | 102,6 | 102,1 | 100,4 | 99,5 | 97,1 | 94,3 | 87,4 | 83,6 | 77,3 |

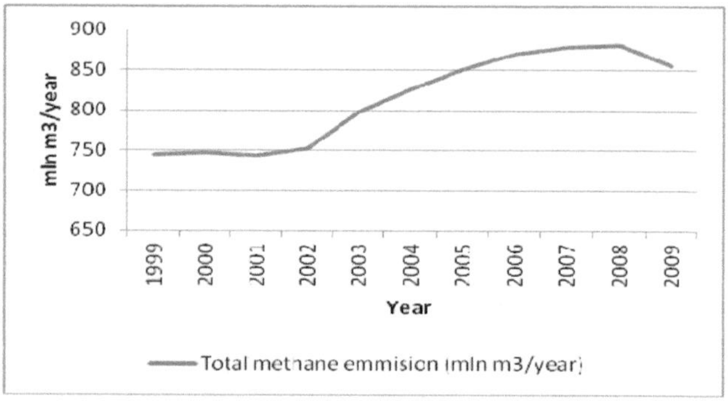


Figure 1. Total methane emission (mln $m^3$/year)

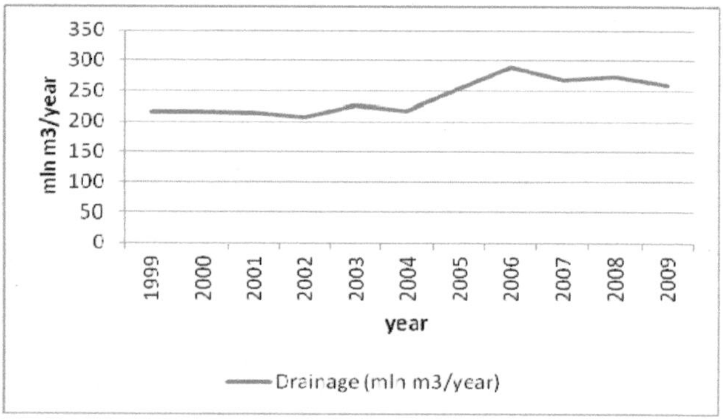


Figure 2. Drainage (mln $m^3$/year)

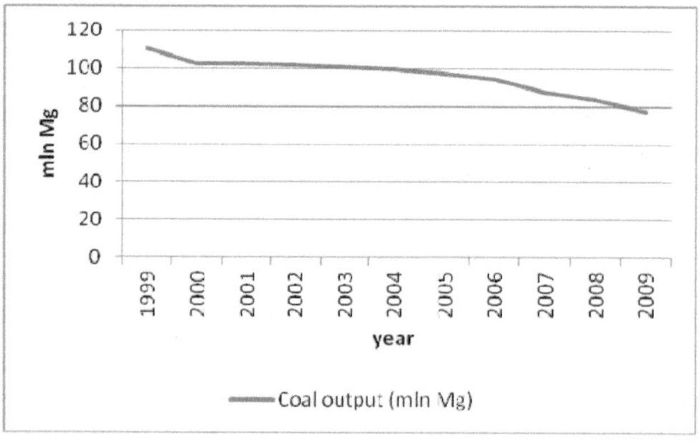


Figure 3. Coal output (mln Mg)

The annual resources of methane in ventilation air (VAM) in Polish coal mines are approximately 590 million m$^3$ and they are not used as fuel for heat and power plants.

In Poland, performed work on the economic utilization of methane from mine ventilation air to produce electricity and heat, but there is a few technical, technological and economical barriers hamper the development of use such technology

For many years, are conducted in Poland, analysis and research:

- in AGH University of Science and Technology in range of methane drainage and utilization of methane from ventilation air,
- in Wrocław University of Technology and Maria Curie-Skłodowska University in range of methane oxidation catalyst.

The current status of work in these Universities allowed to undertake research that will build equipment for the utilization of methane from coal seams, including the mine ventilation air.

## 3. PROJECT INSTALLATION FOR METHANE UTILIZATION FROM VAM

Measurable effect of the project should be to build and test following installations:

(1) IUMK – 1/0 Semi-laboratory installation allowing oxidize methane from 0.4 to 1.0% in air and produce gases containing mainly carbon dioxide – without the use of thermal energy generated (replacing methane emissions of carbon dioxide emissions is very beneficial because methane contributes to the greenhouse effect level 21 times greater than carbon dioxide)

(2) IUMK – 1/1 Semi-laboratory installation allowing oxidize methane from 0.4 to 1.0% in air and produce gases containing mainly carbon dioxide – with heat energy production 1 kW

(3) IUMK – 2/0 semi-technical, allowing oxidize methane from 0.4 to 1.0% in air from ventilation air and mine drainage (possibility of adding methane drainage) and to obtain gases containing mainly carbon dioxide – without the use of thermal energy (replacing methane emissions of carbon dioxide emissions is very beneficial because methane contributes to the greenhouse effect level 21 times greater than carbon dioxide).

(4) IUMK – 2/100 semi-technical, allowing oxidize methane from 0.4 to 1.0% in air from ventilation air and mine drainage (possibility of adding methane drainage) and to obtain gases containing mainly carbon dioxide - with heat energy production 100 kW

## 4. DEVICES FOR CATALYTIC UTILIZATION OF VAM

The basic layout of the use of VAM technology is presented in Figure 4.

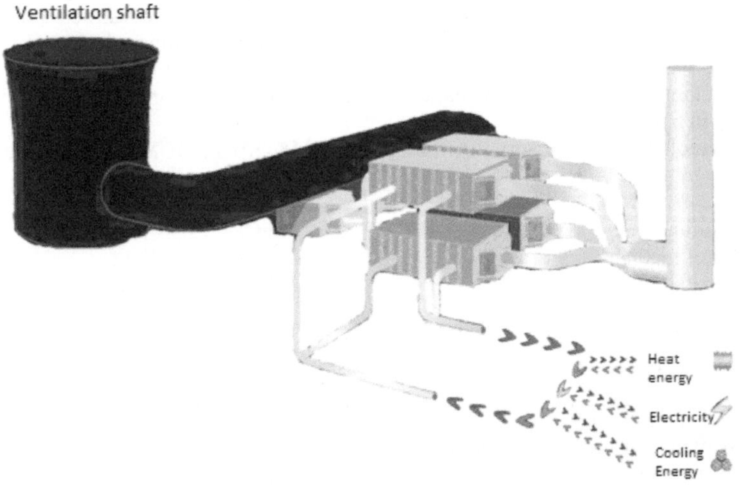


Figure 4. Layout of the use of VAM technology

The basic installation of devices to enable utilization of methane from underground coal mines are:

- devices for taking gas (air and methane) from coal mine ventilation shaft (VAM),
- drainage installations,
- pumps for transport VAM to reactor burning methane,
- mixers gas drainage with VAM,
- reactors oxidizing methane which produce gases containing mainly carbon dioxide and heat,
- heat exchangers for energy purposes such as heating or electricity production,
- chimney flues into the atmosphere,
- measuring equipment.

## 5. RESEARCH OF KRUM-1/0 AND KRUM-1/1 IN SEMI LABORATORY SCALE

Figure 5 presents the scheme of obtaining heat energy from the proposed catalytic methane oxidation reactor (KRUM -1). Oxidation of methane in the KRUM-1 occurs at a temperature of 300-600° C and the concentration of methane from 0.4–1.0%.

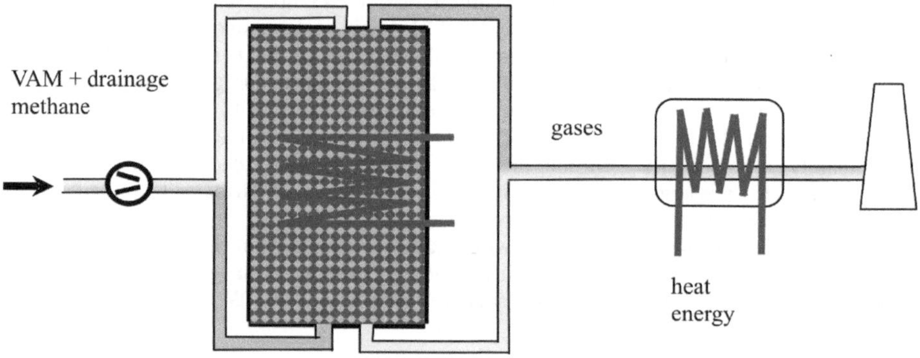


Figure 5. System to obtain heat from the catalytic reactor (KRUM -1)

The reactor may receipt heat from the reactor core by using heat exchanger or from gases.

Therefore, in order to receive heat from the reactor will be designed and constructed a heat exchanger WC-1, which will be tested to determine the thermodynamic optimum design.

Another challenge will be conducting research on the rationalization of work KRUM-1 in particular in terms of thermal insulation and heat use of the reactor to maintain the required temperature distribution in a reactor.

Technical and technological parameters for installation IUMK-1/0 and IUMK-1/1 with reactor KRUM-1:

- air volume — $Vp = 20\ m^3/h$;
- methane concentration — $zCH_4 = 0{,}4–1\%$;
- the calorific value of methane — $Wd = 35\ MJ/m^3$;
- the efficiency of oxidation reactor — $h = 50\%$;
- methane volume — $VCH4 = 0{,}2\ m^3/h$;
- theoretical thermal efficiency of the reactor (without loss) — $Qt = 7\ MJ/h$;
- theoretical power (without loss) — $Pt = 1{,}95\ kW$;
- efficiency — $Q = 3{,}5\ MJ/h$;
- power — $P = 0{,}97\ kW$.

The results of oxidation of methane in the reactor KRUM-1 and heat exchanger WC-1 with allow preparation of technological documentation “Installation for utilization of ventilation air methane from coal mines (IUMK-2/0 and IUMK-2/100)”. IUMK-1 installation is presented in Figure 6 and Figure 7.

Figure 6. Catalytic oxidation installation of VAM under construction

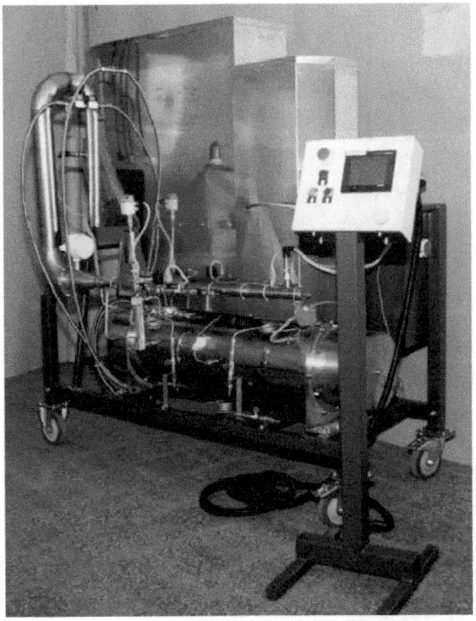

Figure 7. Catalytic oxidation installation of VAM

## 6. RESEARCH OF INSTALLATION IUMK-2/0 AND IUMK 2/100 IN SEMI TECHNICAL SCALE

Semi-technical researches will be conducted in selected coal mine. Based on the documentation "Installation for utilization of ventilation air methane from coal mines (IUMK-1)" will be built in a semi-technical scale device producing heat (IUMK -2) including:

- construction of pump installation for pumping VAM from ventilation shaft to reactor KRUM-100;

- project of gas mixer (MG-2) for mixing VAM and drainage methane;
- construction of gas mixer (MG-2) for mixing VAM and drainage methane;
- project of catalytic reactor (KRUM-100);
- construction of catalytic reactor (KRUM-100);
- project of heat exchanger (WC-2) with power 100 kW;
- construction of heat exchanger (WC-2) with power 100 kW;
- montage of installations elements for installation IUMK-100;
- tests of pumping installation (IPG-100) from ventilation shaft to reactor;
- tests of VAM and drainage mixer MG-100;
- research of catalytic oxidation in catalytic reactor KRUM-100 without heat exchanger;
- research of catalytic oxidation in catalytic reactor KRUM-100 with heat exchanger WC-2;
- elaboration of technical and technological documentation for KRUM-100 for industries apply;
- elaboration of technical and technological documentation for KRUM-100 catalyst for industries apply;
- elaboration of technical and technological documentation for IUMK-2/0 i IUMK-2/100 for industries apply.

Scheme of IUMK-100/0 installation is present in Figure 8. Scheme of IUMK-100/1 installation is presented in Figure 9.

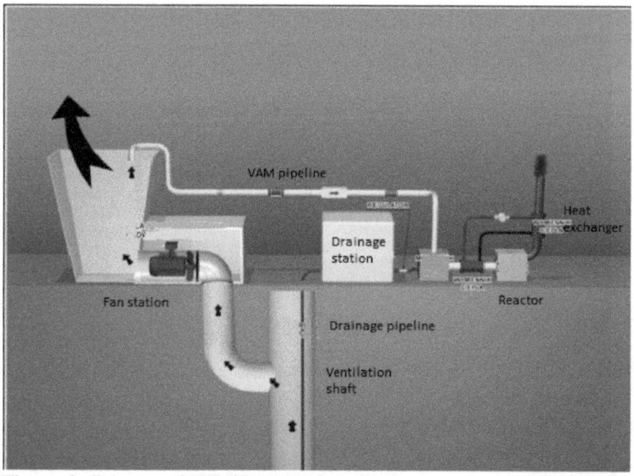

Figure 8. Scheme of IUMK-100/0 installation

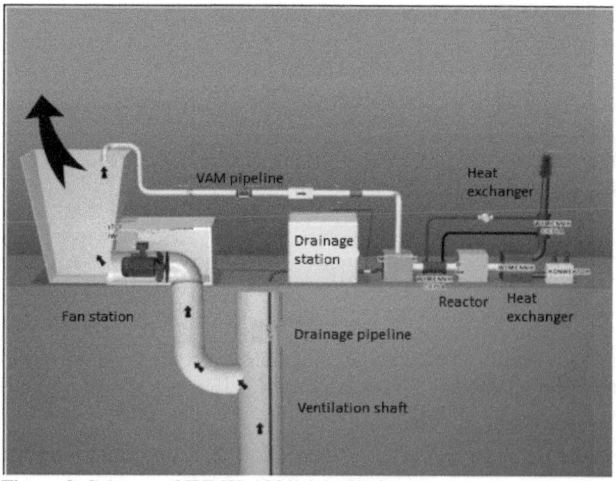

Figure 9. Scheme of IUMK-100/1 installation

Technical and energetic parameters of IUMK-2/0 and IUMK-2/100 in semi technical scale with reactor KRUM-100:

| | |
|---|---|
| – VAM volume | VVAM = 1000 – 6000 $m^3/h$; |
| – methane content in VAM | $zCH_4$ = 0.4 - 1%; |
| – the calorific value of methane | Wd = 35 $MJ/m^3$; |
| – the efficiency of oxidation reactor | h = 90%; |
| – methane volume | $VCH_4$ = 12 $m^3/h$; |
| – theoretical thermal efficiency of the reactor (without loss) | Qt = 420 MJ/h; |
| – theoretical power (without loss) | Pt = 116.7 kW; |
| – efficiency | Q = 380 MJ/h; |
| – power | P = 105 kW. |

## 7. CONCLUSIONS

After completion of the project in 2012 year will be possible to use: (1) installation IUMK – 2/0 to reduce methane emissions into the atmosphere (after burning will only be emitted carbon dioxide that contributes to the greenhouse effect as far as 21 times less than methane), (2) installation IUMK-2/100 for heat production, (3) technical and technological documentation for installation IUMK-2/0 and IUMK-2/100 allows to decide on the construction full scale industrial investment for producing heat and electricity power and limiting emissions of methane into the atmosphere and the greenhouse effect.


## ACKNOWLEDGEMENT

The article presents the assumptions implemented by AGH University of Science and Technology, Faculty of Mining and Geoengineering Project "Proecological technology utilization of methane from the mines".

The project is funded by the European Union, contract no. POIG.01.03.01-24-072/08-01.

INNOWACYJNA GOSPODARKA
NARODOWA STRATEGIA SPÓJNOŚCI

AGH

PROJEKT WSPÓŁFINANSOWANY ZE ŚRODKÓW UNII EUROPEJSKIEJ

*International Mining Forum 2010, Liu et al. (eds) © 2010 Taylor & Francis Group, London, UK. ISBN 978-0-415-59896-5*

# Numerical simulation of airflow in blind headings ventilated with freestanding fans

Marian Branny, Bernard Nowak
*AGH University of Science and Technology, Krakow, Poland*

ABSTRACT: Room and pillar headings are frequently ventilated with the use of freestanding fans. The range of penetration of an air stream generated by jet fan is determined by several parameters among them the major are initial parameters of stream and the place of fan's installation. In the paper are presented airflow patterns inside a heading which are involved by jet fans. The fans are placed in different positions at the entrance section of heading. A series of numerical tests was performed to determine the effect of different fan's positions on airflow and efficiency of headings ventilation. Three viscosity models of turbulence were used in calculations, mainly two equations model k-ε and k-ω SST and one equation of Spalart-Allmaras. Results were compared with some mine measurements.

KEYWORDS: Auxiliary systems, jet fans, CFD simulation

## 1. INTRODUCTION

Board and pillar headings are typically ventilated with the use of freestanding fans. Polish copper mines are using this method of ventilation with great success. Fans are typically installed at the inlet to the heading or a drift. In order to restrict the air re-circulation, the fans are installed on the side of air inflow in the heading. The ventilation efficiency depends on the range of air stream produced by the fan and its assessment procedure can be based on the existing distribution of parameters, such as airflow velocity, air temperature or concentration of hazardous gases. Other determinants of the air stream range include the initial parameters of the air flow, such as initial flow velocity, the scale of turbulence and diameter of the diffuser. The shape of the developed velocity field depends to a large extent on the fan location in the cross-profile of the heading.

## 2. TURBULENCE MODELS

Modelling of complex flow problems typically uses the CFD (Computational Fluid Dynamics) approach. Underlying the conventional models of turbulence is the Reynold's concept whereby motion of fluid is as superimposition of the average and fluctuating motion. Accordingly, the Navier-Stokes equations can be transformed to the form containing an extra term (a turbulent shear stress tensor) which cased that the model is unclosed.

The largest group of closure hypotheses includes turbulence models assuming the eddy viscosity (Boussinesq hypothesis). Of particular importance are one- and two-equation models in cluding the Spalart-Allamaras model involving one equation and the two-equation k-ε, k-ω models together with their modified versions. Turbulence models based on Reynold's averaging hypothesis (RANS) include the Reynold stress transport model put forward by Hanjalic. Another approach is adopted in the LES method (Large Eddy Simulation) which involves the filtering of smaller eddies and numerical solving of large eddy fields displaying certain anisotropy whilst small scale eddies are modeled analytically.

The new class of solutions is offered by the DNS (Direct Numerical Solution) approach whereby the Navier Stokes equations are solved directly, without averaging. However, that requires very dense numerical grids whose cells must be smaller than the tiniest scales of turbulence. It appears that potential applications of the DNS approach to the analysis of flows in engineering problems are still the matter of the future.

The LES method provides a major reduction of calculation time in relation to the DNS approach. This approach has received a great deal of attention lately, yet the method proves too costly to be used in practical applications.

CFD codes, widely used in modelling of engineering problems, are based on RANS (Reynolds Averaged Navier Stokes) viscous models. Their main drawback, however, is that they are not universal and hence the validation procedure will be required. As regards their application to mining, the research team of Wala (University of Kentucky) checked the adequacy of several RANS models and their applicability to airflow imaging and studies of distribution of methane concentration in blind headings aired by means of line brattices. 3D measurements of the velocity field are taken on laboratory models using the laser anemometry and point-by-point measurements of the velocity vector and methane concentration (NOIH Pittsburgh Research Laboratory) tested on 1:1 scale models.

In Poland no research has been conducted as yet that should be aimed at full validation of models, though attempts have been made to give credence to CFD codes basing on measurements in situ, but those are related only to measured global parameters or to a small number of quantities measured point-by-point in the flow zone (Krawczyk 2007, Nawrat et al. 2006, Podleszczyk et al. 2009, Branny 2003).

Research and investigation data collected so far suggest that application of two-equation models k-ε, k-ω and the Spalart-Allamaras model involving one equation to numerical modeling of gas flows in mine headings leads to most promising results. Of major importance are conclusions drawn on the basis of the full validation procedure carried out at the University of Kentucky (USA) and in the NIOSH Pittsburgh Research Centre, which confirmed the adequacy of the Spalart-Allmaras and SST k-ω models and proved their applicability to simulation of ventilation performance in the blind headings – however tested only when the line brattices are used.

This study summarizes the airflow data simulated in a heading ventilated with the use of a free-standing fan. The simulation procedure was applied to find out how the fan position should affect the volume of air flowing in the direction of the face front and the amount of air recirculation. The procedure uses the viscous turbulence models: the k-ε, SST k-ω and Spalart-Allmaras models. The first model is chosen chiefly because it is commonly used in practical applications, including the mining engineering problems (Krawczyk 2007, Poleszczyk et al. 2009, Aminossadati 2009); the other two are widely recommended in literature (Wala et al. 2007, 2009).

## 3. FLOW REGION

The ventilation region comprised a heading of 30 m length, 6.4 m width and 3.8 m height and the gallery in which the air current flows through the cross-section identical as in the heading. The average airflow rate in the heading is 2 m/s. The fan is mounted at the inlet to the heading, on the side of the fresh air inflow in the airway with the streamlined current. The air flow rate delivered by the fan is 4.4 $m^3/s$. The sketch of the flow region is shown in Figure 1. Data admitted in the calculations correspond to the test conditions prevailing during the studies of efficiency of ventilation in blind drifts using jet fans that were carried out in South Africa (Meyer 1993). The approach pursed was to measure the airflow velocity in selected points inside the heading and try to determine the range of the air stream penetration depth.

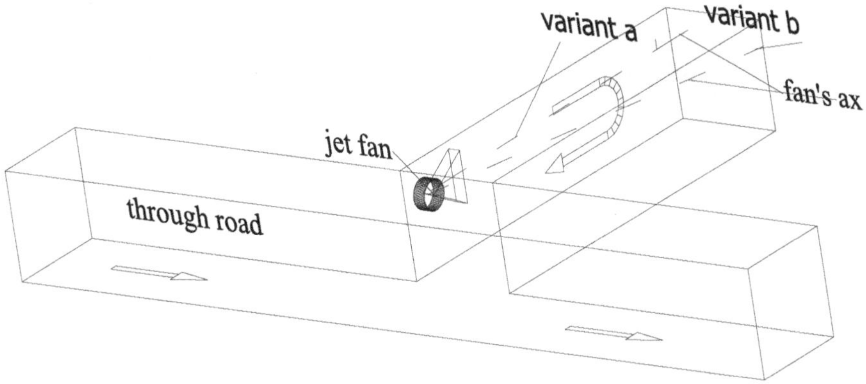


Figure 1. Flow region

4. THE RELATIONSHIP BETWEEN THE VOLUMETRIC FLOW RATE OF AIR STREAM SUPPLIED TO THE FACE ZONE ON THE FAN LOCATION

Several fan locations were considered in calculations. In all cases the fan was installed on the side of air inflow to the heading. Obviously, such positioning of the fan help limit the undesired effects of re-circulating flows. The fan is shaped like a cylinder of 0.5 m diameter and 1 m length. The fan operation is modeled by predetermining the fixed pressure step of 180 Pa in the cylinder cross-section, half way along its length. The following variants were considered:

- the fan axis located at the distance of 1.3 m, 1.0 m, 0.7 m from the wall sides and configured perpendicularly to the face,
- the fan axis positioned at the distance of 1.3 m from the wall sides and at the angle of 5.70 (the angle between fan axis and the perpendicular direction to the face).

Calculation results obtained for the variant 'a' are shown in Figures 2–4. Velocity field images obtained using the three turbulence models mentioned before are quite similar in qualitative terms though displaying certain quantitative differences. In qualitative terms the high level of agreement is achieved also for those variants when the fan was mounted closer to the walls in the heading: 0.7 m and 1.0 m (Fig. 2a). The stream of air flows towards the face along the walls, next to which a fan is installed while the return stream is swept along the opposite walls. As the fan is moved towards the axis of the heading, the stream of air produced by the fan moves towards the opposite wall side. When the fan is located at the distance of 1.3 m from the heading wall, a secondary vortex is formed in the face zone – the k-ω and Spallart-Allamaras models (Fig. 2b–4). With this fan position, the k-ε model yields a slightly difference image of velocity field. The air stream moves towards the opposite wall side but when flowing towards the face zone, it remains in the roof regions.

The location of a jet fan has a major influence on the amount of air flowing towards the face region. That is why it is recommendable to position the fan as close as possible to the walls sides in the heading, in the entrance section. In this fan configuration, the mixing of air stream with the air in the heading becomes less intense, the air penetration distance is enhanced and the axial speed tends to fall more slowly.

Figure 5 shows the air volume flowing towards the face region in the function of distance from the fan. Calculations were performed using the three selected turbulence models and for various fan configurations in the inlet section. There are major quantitative differences between the calculation data, particularly for the face region. The change of fan position, measured as the distance from the walls within the limits 0.7 and 1.3 m, leads to:

- increase of the volume of recalculated air in the 15-meter-long zone, starting from the fan,
- reduction of the air volume transported towards the face regions at distances in excess of 15 m.

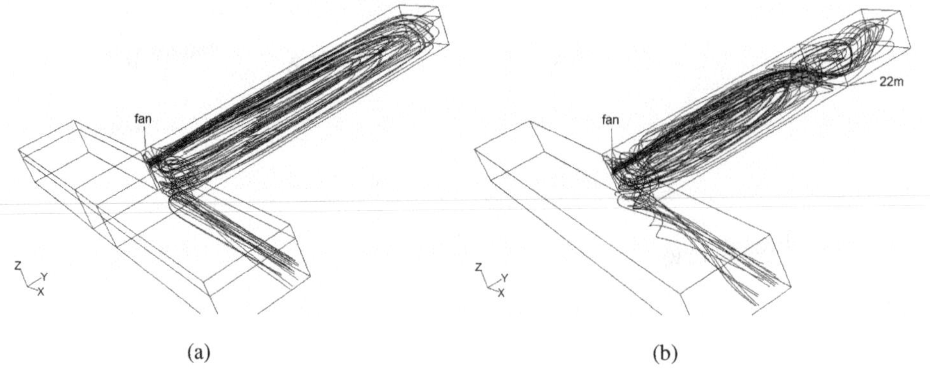


Figure 2. Path lines: distance of fan from the wall sides equals to a) 1.0 m (k-ε model), b) 1.3 m (SA model)

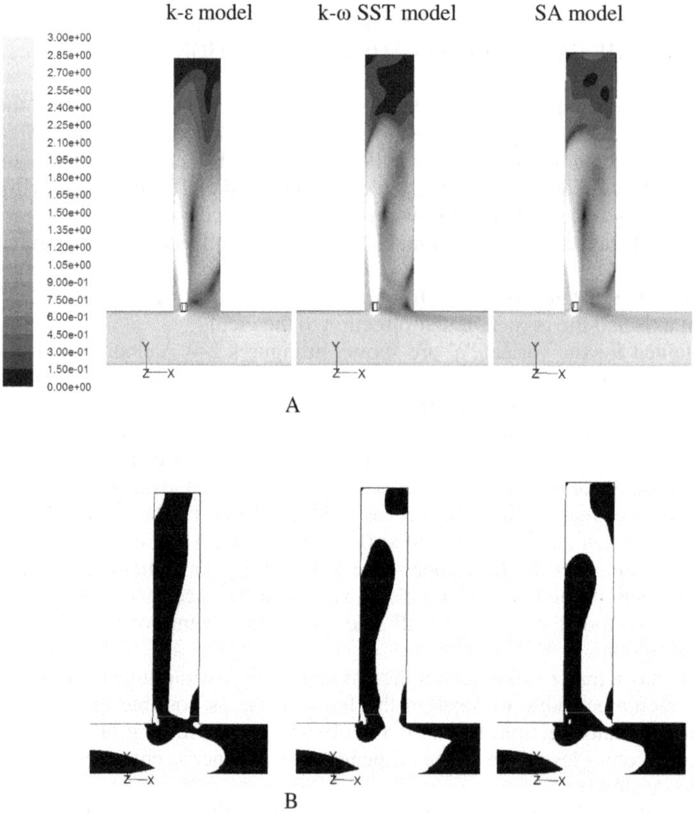


Figure 3. Images of velocity field in the horizontal plane in the fan axis (1.3 m from the roof)

A - modulus of velocity (grey scale in the range 0-3 m/s, white colour above)
B - black colour - areas of positive values of the velocity vector component in the direction of the y-axis; white colour- the area of negative values of this component

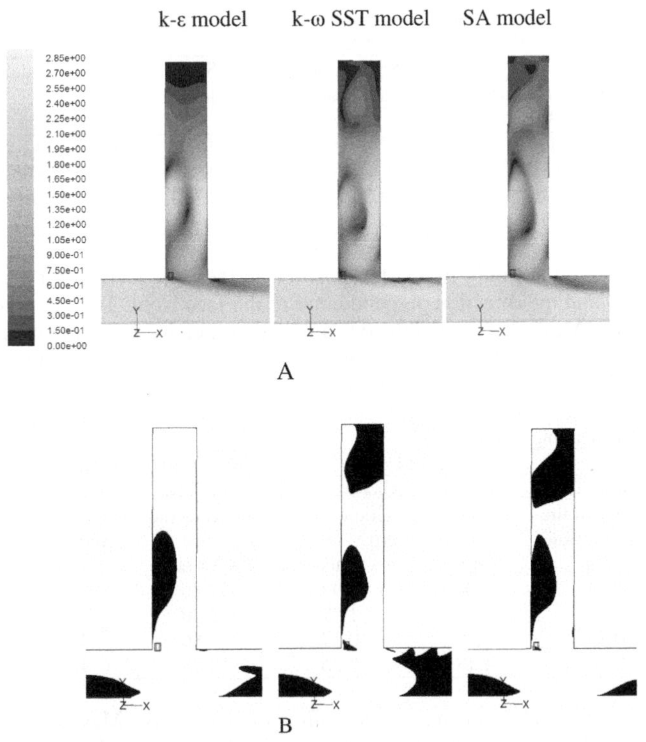


Figure 4. Images of velocity field in the horizontal plane in the fan axis (0.5 m from the floor)

A - modulus of velocity (grey scale in the range 0-3 m/s, white colour above)
B - black colour- areas of positive values of the velocity vector component in the direction of the y-axis; white colour- the area of negative values of this component

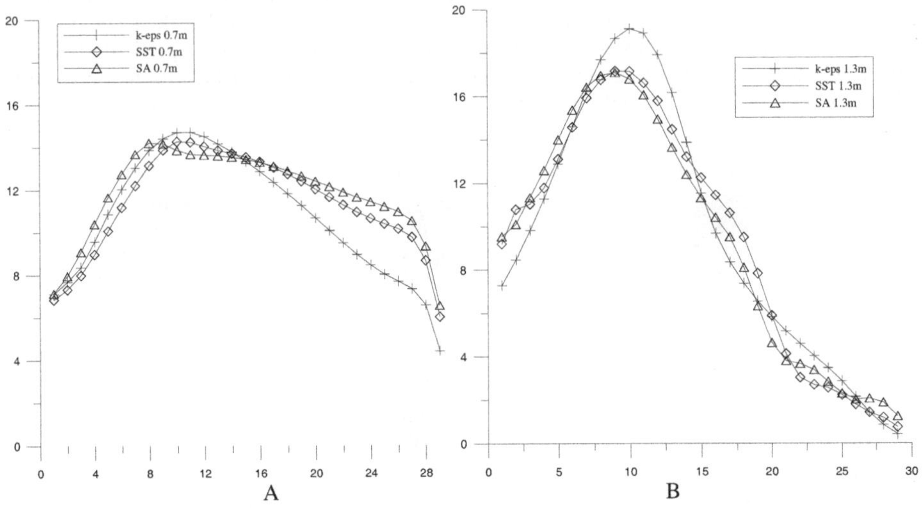


Figure 5. Air volume flowing towards the face region:
the distance between the fan and walls sides:
A - 0.7 m; B - 1.3 m

Table 1. Provides the predicted amounts of air in selected cross-sections of the investigated heading

| Distance from the fan /m | Volumetric flow rate /m$^3$/s | | | | | | | | |
|---|---|---|---|---|---|---|---|---|---|
| | k-ε model | | | k-ω SST model | | | Spalart-Allmaras model | | |
| | Distance between the fan and the heading wall sides /m | | | | | | | | |
| | 0.7 | 1.0 | 1.3 | 0.7 | 1.0 | 1.3 | 0.7 | 1.0 | 1.3 |
| 15 | 13.34 | 17.93 | 14.12 | 16.58 | 17.29 | 14.99 | 16.46 | 16.94 | 13.85 |
| 20 | 13.07 | 11.53 | 7.13 | 14.74 | 12.75 | 7.20 | 15.20 | 13.95 | 5.66 |
| 25 | 9.83 | 7.14 | 3.49 | 12.72 | 9.54 | 2.72 | 13.71 | 11.82 | 2.80 |
| 29 | 5.40 | 3.32 | 0.50 | 7.37 | 4.28 | 0.90 | 8.02 | 6.33 | 1.52 |

The velocity field image is affected not only by the fan position in the inlet section but also by the angle between the fan axis and the direction perpendicular to the face front. The fan used in tests performed in South Africa (Meyer 1993) had its axis directed perpendicular to the intersection edge of the face front and wall side (variant b, Figure 1). Selected numerical data of ventilation conditions for the variant "b" are compiled in Table 2.

Table 2. Parameters of airflow in the heading - variant "b"

| Distance from the fan /m | Model k-ε | | Model k-ω SST | | Model Spalarta-Allamarasa | |
|---|---|---|---|---|---|---|
| | Air flow towards the face /m$^3$/s | Maximal flow velocity in the cross-section /m/s | Air flow towards the face /m$^3$/s | Maximal flow velocity in the cross-section /m/s | Air flow towards the face /m$^3$/s | Maximal flow velocity in the cross-section /m/s |
| 15 | 5.90 | 1.74 | 7.86 | 1.97 | 7.36 | 2.04 |
| 20 | 4.64 | 0.97 | 4.63 | 1.09 | 3.30 | 1.01 |
| 25 | 1.27 | 0.34 | 2.49 | 0.57 | 1.86 | 0.44 |
| 29 | 0.54 | 0.16 | 1.18 | 0.26 | 0.73 | 0.21 |

Basing on results of measurements taken in mines, the author of that study (Meyer 1993) considers the zone of 25 m in length as the air penetration zone. Velocity contours obtained numerically (Figs 6 and 7) and their visualization using streamlines (Fig. 8) seems to confirm this assumption. The measured (maximal) flow velocity in the cross-section at the distance of 25 m from the fan is equal to 0.4 m/s. Numerically derived maximal flow velocities in this cross-section are: 0.34 m/s (model k-ε); 0.57 m/s (model k-ω SST) and 0.44 m/s (Spalart-Allamaras model).

In his research program Meyer (Meyer 1993) determined the rate of airflow recirculation (produced by the fan), it was estimated to be 5%. The amount of re-circulated air obtained numerically is typically twice as large. It is assumed now that the tracer gas enters the heading at a fixed rate (i.e. constant productivity of the tracer gas source). Tracer gas has the same physical properties as air. Basing on thus obtained field of tracer gas concentrations, particularly at the fan inlet, the flow rate of re-circulated air is determined. This parameter falls in the range from 9.3% (model k-ε) to 14.6% (model k-ω SST).

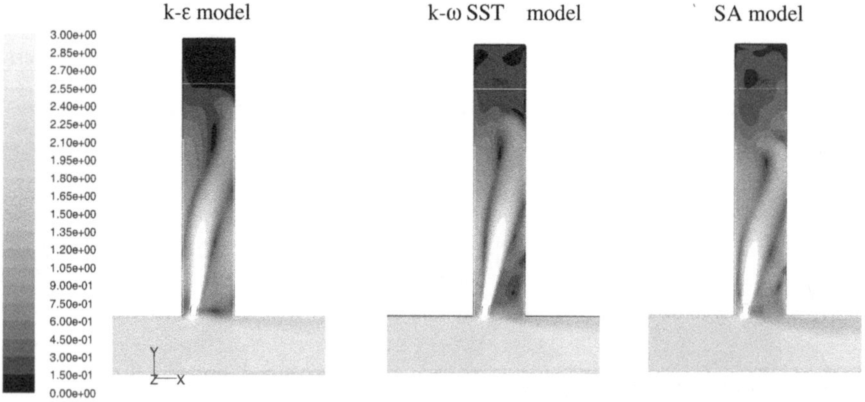


A

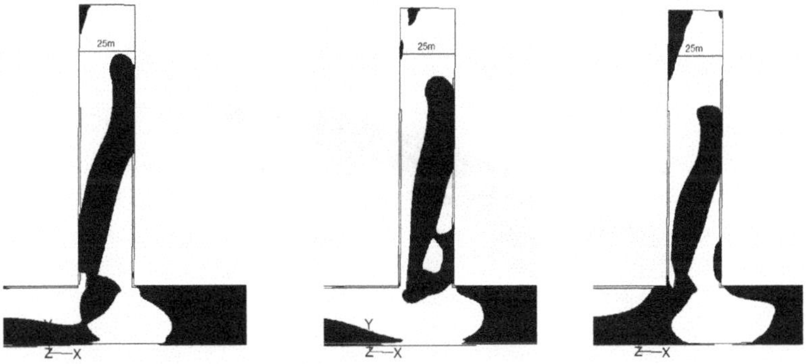

B

Figure 6. Images of velocity fields in the horizontal plane, in the fan axis
(variant 'b'- the fan mounted at the distance of 1.3 m from the wall sides, $5.7^0$ angle between the fan axis and the direction normal to the face)
A - modulus of velocity (grey scale in the range 0-3 m/s, white colour above)
B - black colour - areas of positive values of the velocity vector component in the direction of the y-axis; white colour - the area of negative values of this component

k-ε model k-ω SST model SA model

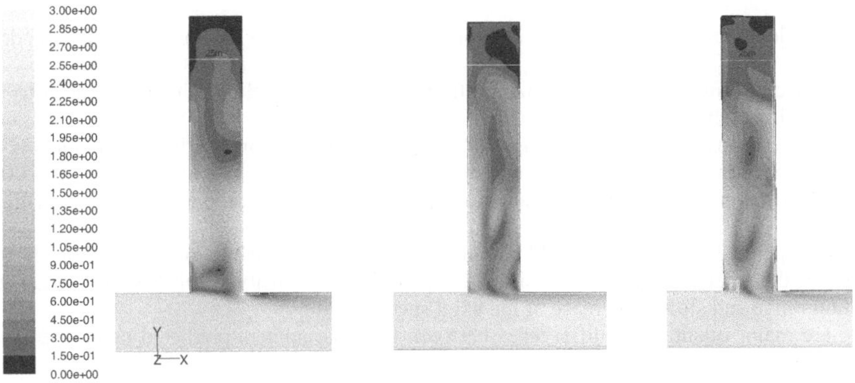


A

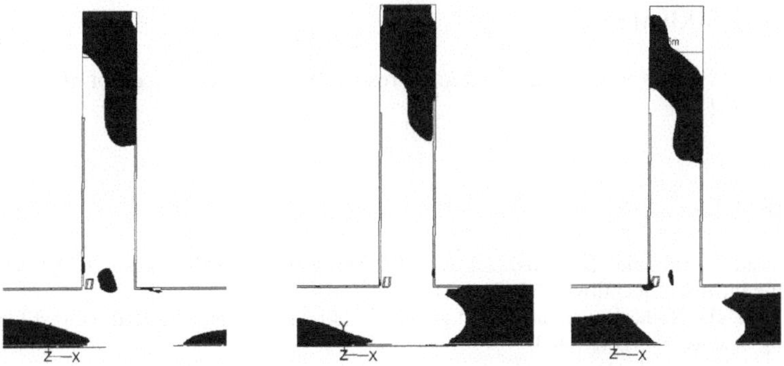

B

Figure 7. Images of velocity fields in the horizontal plane 0.5 m from the floor (variant 'b'- the fan mounted at the distance of 1.3 m from the wall, $5.7^0$ angle between the fan axis and the direction normal to the face)
A - modulus of velocity (grey scale in the range 0-3 m/s, white colour above)
B - black colour - areas of positive values of the velocity vector w component in the direction of the y-axis; white colour - the area of negative values of this component

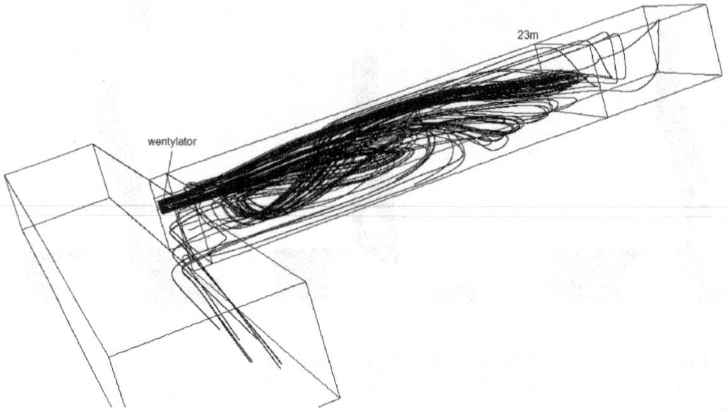


Figure 8. Streamlines inside the heading, variant "b" - turbulence model k-ε

## 5. CONCLUSIONS

Modeling of flows for practical engineering applications mostly uses viscous turbulence models RANS.

The major constraint of solutions obtained through RANS modeling is that they need to be verified experimentally.

In qualitative terms, velocity field images obtained by using the three turbulence models (k-ε, k-ω SST and Spalart-Allamaras model) agree well with literature data (Sułkowski et al. 2000, Myer 1993) however the quantitative evaluation will be possible after the experimental verification.

It is recommended that the fan should be mounted as close as possible to the wall sides in the headings, in the inlet section. This position ensures lower intensity of air mixing, larger air penetration range and slower decrease of the axial speed.

There are major quantitative differences between flow parameters predicted using the three tested models.

## ACKNOWLEDGEMENT

This study is sponsored by the statutory research funds, as a part of the research project 11.11.100.281,

## REFERENCES

Bogusławski A., Drobniak S., Tyliszczak A. 2008. Turbulencja – od losowości do determinizmu, Modelowanie Inżynierskie, Gliwice. (In Polish)

Branny M. 2003. Numerical simulation of airflow in blind headings ventilated with jet fans, Archives of Mining Science 48(4): 425-443

Fluent 2005. Fluent 6.1. Documentation, Flunet Inc.

Krawczyk J. 2007. Jedno i wielowymiarowe modele niestacjonarnych przepływów powietrza i gazów w wyrobiskach kopalnianych. Przykłady zastosowań. Archives of Mining Science (2). (In Polish)

Meyer C.F. 1993. Improving underground ventilation conditions in coal mines. SIMRAC, Project no. COL 029a

Podleszczyk E., Ligęza P., Skotniczny P. 2009. Metody termoanemometrycznego wyznaczania parametrów przepływu jako wspomaganie symulacji numerycznej procesu przewietrzania kopalni głębinowej. 5 Szkoła Aerologii Górniczej, Wrocław: 129-142. (In Polish)

Sułkowski J., Drenda J., Biernacki K., Domagała L. 1999. Opracowanie wytycznych stosowania wentylatorów wolnostrumieniowych do przewietrzania drążonych przodków o długości do 60 m. Politechnika Śląska, Gliwice. (In Polish)
Sułkowski J., Drenda J., Biernacki K, Gumiński A., Różański Z., Wierzbicki K., Musioł D. 2000. Określenie skuteczności przewietrzania drążonych wyrobisk górniczych, których długość nie przekracza 60 metrów przy zastosowaniu wentylatorów wolnostrumieniowych. Politechnika Śląska, Gliwice. (In Polish)
Tu J., Yeoh G.H., Liu Ch. 2008. Computational Fluid Dynamics. Elsevier
Wala A.M., Vytla S., Huang G., Taylor C.D. 2009. Study on the effect of scrubber operation on the face ventilation. 12th US/North American Mine Ventilation Symposium, Reno, USA: 281-289
Wala A.M., Vytla S., Taylor C.D., Huang G. 2007. Mine face ventilation: a comparison of CFD results against benchmark experiments for the CFD code validation. Mining Engineering (10): 49-55
Wierzbicki K. 2009. Modelowanie komputerowe rozkładu parametrów powietrza oraz koncentracji metanu w rejonie skrzyżowania z chodnikiem wentylacyjnym. 5 Szkoła Aerologii Górniczej, Wrocław: 111-119. (In Polish)

International Mining Forum 2010, Liu et al. (eds) © 2010 Taylor & Francis Group, London, UK. ISBN 978-0-415-59896-5

# Mining in Poland – history and future

Antoni Tajduś, Piotr Czaja, Marek Cała
*AGH University of Science and Technology, Krakow, Poland*

ABSTRACT: Poland is one of the countries rich in various minerals. There are significant deposits of hard coal and brown coal, copper ore, zinc and lead ore, rock salt, native sulphur, gypsum, kaolin, rock minerals, remedial and geothermal waters, as well as some deposits of natural gas and small volume of oil. The paper discusses in more detail the volumes of the resources, the problems related to their extraction and the perspectives of several of the most important Polish mine plants, that is: hard and brown coal, copper ore, zinc and lead ore, sulphur, natural gas and oil.

KEYWORDS: mining, exploitation, material resources, power safety

## 1. INTRODUCTION

Poland is one of the countries rich in various minerals. There are significant deposits of hard coal and brown coal, copper ore (production of copper, silver and smaller amounts of gold, lead, nickel and selenium), zinc and lead ore (production of zinc and lead), rock salt, native sulphur, gypsum, kaolin, rock minerals, remedial and geothermal waters, as well as some deposits of natural gas and small volume of oil.

The boom of the mining industry in Poland came for 1950s and continued until 1980s. After a period of intense exploitation, drop in or even abandoning of their extraction was recorded for many raw minerals (Ney, Galos 2008). This was due to, on one hand, the economic transformation of the country and implementation of the market economy principles. On the other side, it was related to low quality of some minerals as referenced against world standards, depletion of deposits, as well as environmental conditions and spatial management.

Currently, the mining industry in Poland is still a very significant area of economy. Poland is a leading world producer of hard coal (ranked 8th) and brown coal (ranked 7th) (Dubiński, 2008). These minerals guarantee power engineering safety of the country, as over 90% of electricity is generated in power plants based on them. In turn, exploitation of native deposits of copper and silver ore places Poland among the leading world producers of copper and silver. Other important branches of the mining industry are exploitation of rock salt as well as mining of rock minerals, especially with the intensely developing building industry in EU countries. Production of natural gas currently satisfies 40% of domestic needs, whereas that of oil – as little as 5% (Ney, Galos 2008). Moreover, exploitation of other raw minerals is conducted in Poland to a lesser extent, such as zinc and lead, ore of metals, chemical minerals and other. In case of some minerals, documenting new deposits is possible, as well as – in a longer perspective – development of new, unconventional methods of exploitation, which will allow their effective exploitation.

## 2. HARD COAL AND BROWN COAL

### 2.1. *Hard coal*

For dozens of years, solid fuels have been a source of original energy. Until early 20th century, coal was the basic source of original energy. Only in the last several dozen years, consumption of coal (especially of hard coal) in some countries of Europe was limited, with preferences for other sources of energy, i.e. oil, natural gas and nuclear energy and renewable energies of water and wind.

The basic market for coal in the world economy has been and will be in the sector of generating electricity, where the share of hard and brown coal in the span of 20-30 years will practically not change. Achieving this result will require increase in production of coal by over 30%, mostly in the coming 15 years. This immense increase in demand for electricity will be compensated with almost three-times increase in use of natural gas, energy from renewable sources and nuclear energy.

This forecast contradicts the opinions often expressed in the media, stating that using energy from traditional media (coal, oil and gas) is on the decline, in favour of nuclear energy, water and other energies.

Long-term forecasts for the year 2100 state that only after 2050 significant reduction in the share of gas and oil in production of energy will become the fact, with strong increase in the share of nuclear energy, biomass, and solar and wind energy. What is interesting is that by the year 2100 slow but continuous increase in consumption of coal for production of energy is planned. One may envisage new technologies to be developed to allow other usage of coal, such as e.g. coal gasification, production of liquid fuels from coal and production of hydrogen from coal, etc. There also is a possibility of developing new technologies to allow use of new not available deposits of coal without its extraction to the surface.

In Poland, coal plays major role in the fuel and power sector and is the guarantor of power safety. The definite prevalence of coal fuels in the structure of consumption of original energy results from relatively easy availability of rich deposits of brown and hard coal, with simultaneous lack of sufficient deposits of other fuels. For example, in 2007, the structure of consumption of fuels in the electrical power engineering industry ranks hard coal at 63.13%, with 32.63% share of brown coal and small, but ever increasing 3.02% share of gas.

The documented deposits of hard coal (as at 31 Dec 2006) amount to 41.996 bn Mg, almost 30% of which comes for deposits of coking coal. The developed deposits amount to 15.35 bn Mg, 5.057 bn Mg of which comes for industrial deposits. Operational extractable deposits, mostly due to the currently used exploitation systems, are 3.1 bn Mg. 18% of this comes for industrial and operational deposits in low-depth beds (up to 1.5 m) and 22% in protecting pillars. It is estimated that the undeveloped deposits still hold ca. 26 bn Mg of coal deposits (Dubiński 2008).

The analyses of the data on hard coal mining in Poland show that:

- the total extraction of hard coal (coking and power) in 2008 was 83.6 m Mg. This production dropped in the last 5 years from about 100 m Mg (coking coal included), that is by over 16 m Mg, (Fig. 1),
- after 2015, the number of active mine facilities will quickly drop due to depletion of operative deposits on active levels and due to the lack of investments in the coal industry. The decrease in the number of active mines results in further significant decrease in the output. It is expected that in 2028 it will bring up about 60 m Mg, including 48.373 m Mg for the power engineering industry,
- without significant investments, mostly in construction of new exploitation levels, extending or constructing new shafts, halting the decrease in the output from the Polish mines will not be possible. The investment decisions have to be made already now, as construction of a new mine can take several years, and a new shaft will take 3-5 years. If money is not found for the investments, despite quite significant deposits, the output from the Polish mines will drastically drop within a few years' time, which will result in major decrease in the level of power safety of Poland,
- the progressing intense ageing of mining personnel. In the next few years, over 1/3 supervision employees with secondary and higher education will leave the mines. To prevent posing hazard for safety of operation of mining plants, over 3,000 engineers should be employed in

the mines in the years 2009-2012. The actual situation is that the Polish educational facilities which provide for the needs of the entire mining industry (along with surface mining and borehole mining) produce not more than 300 graduates per year.

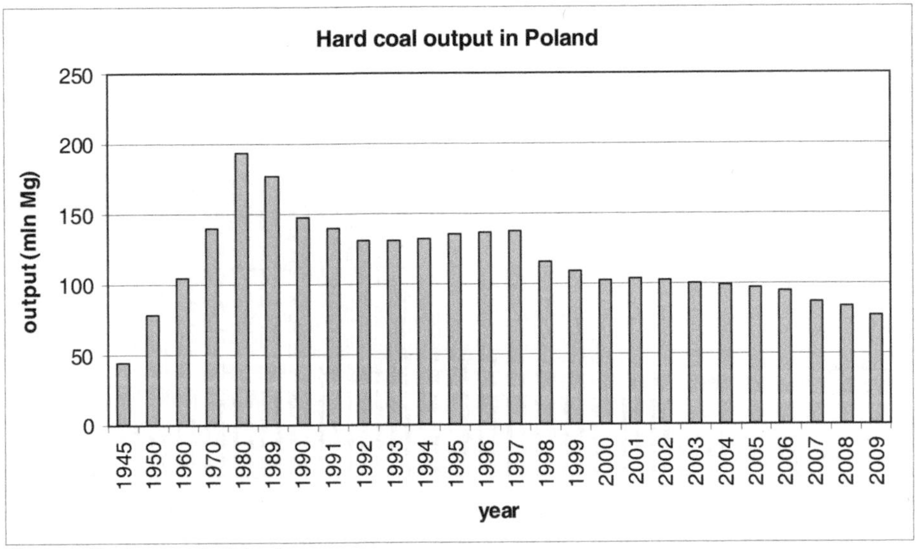


Figure 1. Coal output in Poland

It has to be stated that the proper economy with relatively large industrial deposits of hard coal (investments, protection of deposits, etc.), should ensure sufficient production of coal for the needs of Poland, as well as nearby countries in Europe.

Otherwise, major import of hard coal will be necessary, which will definitely reduce power safety of Poland and Europe as well. The first symptoms of this process are already noticeable. It is estimated that about 10 m Mg of hard coal was imported from Russia to Poland in 2008 (comparing 7 m Mg in 2007).

### 2.2. *Brown coal*

Countries with significant deposits of brown coal produce major amount of electricity from this fuel, even up to 70%. Currently in Poland, about 32-33% of the cheapest energy is produced from brown coal (with over 40% in professional heat generation power plants). Over 95% of the explained brown coal is used in professional power plants located close to surface mines, whose power output is from several dozen to 4,400 MW (Bełchatów). The plans may be that, in the perspective of the coming 20-30 years, brown coal will keep its strong position as safe and relatively cheap fuel for electrical power engineering in the world scale, including Poland.

Industrial production and use of brown coal in Poland was developed in the second half of the 20$^{th}$ century. In this time, production of coal increased from 4.3 m Mg in 1947 up to 73.3 m Mg in 1988, and stabilised at the level of about 60 m Mg (Fig. 2).

Production of brown coal in Poland is run mostly in four large surface mines (Bełchatów, Konin, Turów, Adamów) owned by the State Treasury. Three of these mines (Bełchatów, Konin, Adamów) are multi-surface mines. The total number of brown coal open pits is 10. A new brown coal open pit is at the stage of construction in Szczerców (KWB Bełchatów). A fifth, private-owned brown coal mine is also active in Sieniawa, but it extracts only small volume of coal (about 80,000 Mg/year) only for local needs. The deposits exploited are used for power engineering purposes.

The calculated deposits of brown coal in Poland are 13.7 bn Mg, including ca. 0.8 m Mg of bituminous coal, 2.5 bn Mg of briquetted coal and 1.5 bn Mg of low-temperature carbonisation coal.

The active and developed documented deposits hold 1.88 bn Mg of calculated deposits, including 1.49 bn Mg into industrial deposits (The Balance of Deposits of Minerals and Underground Waters in Poland, 2006).

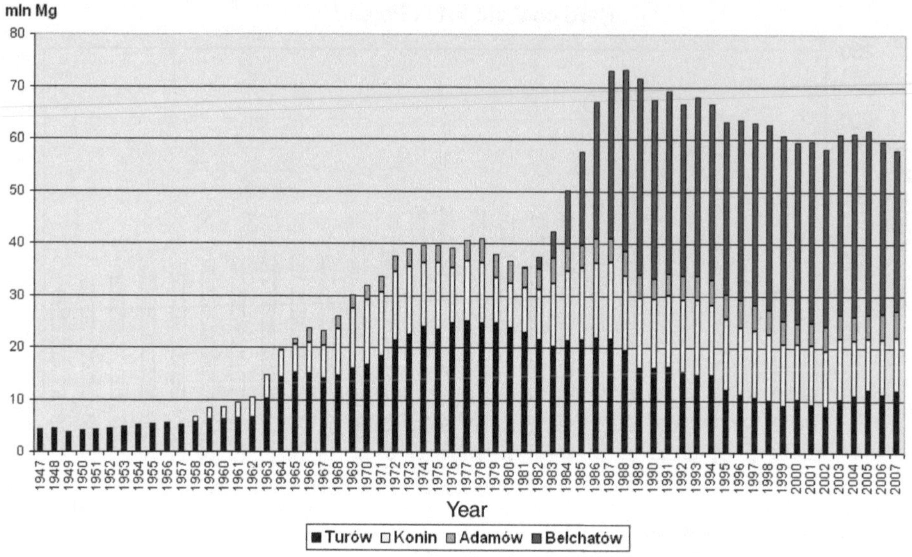


Figure 2. Production of brown coal from the origin of operations in Polish mines

Perspective deposits of brown coal in Poland are estimated at 60 to 140 bn Mg. Due to depletion of brown coal resources in active mines, in order to keep the assumed output level (60-66 m Mg) it is necessary to manage new deposits of brown coal in the coming 20 years. Within several years (in KWB Konin: after 2010) the extraction possibilities of the currently active brown coal mines will not meet the needs of professional power plants. Therefore, apart from the currently constructed Szczerców open pit with 38.0 m Mg output capacity and the small Drzewce open pit released for operational use in 2006, with 2.2 m Mg output capacity, it is necessary to develop new brown coal deposits in the Konin Coal Basin soon, and then, after 2020, to develop a new brown coal basin (Legnickie or Gubińskie Coal Basins). The planned production of brown coal from active and perspective mines is given on Figure 3.

The perspective coal basins and their deposits include (Gawlik et al. 2007):

- Legnica-Głogów, the largest perspective brown coal basin with ca. 40 bn Mg of brown coal deposits. This includes 5 bn Mg of documented calculated deposits in the Legnica and Ścinawa deposits, 10 bn Mg of (calculated) deposits preliminarily recognised in the Ścinawa-Głogów deposits and 25 bn Mg of non-calculated deposits.
- The Western Coal Basin with the main Gubin-Brody deposits. The documented calculated deposits of this coal basin amount to ca. 3 bn Mg, with 5 bn Mg of forecast deposits.

The Wielkopolskie Coal Basin with deposits in the so-called "Rów Poznański": Mosina, Czempin, Krzywin, Gostyń, Szamotuły, and other, with 4 bn Mg of documented calculated deposits and 6 bn Mg of forecast deposits. Part of these deposits (Czempin, Krzywin, Gostyń) are under agricultural land with high botanical class, though.

The total documented calculated brown coal deposits in these three coal basins are 12 bn Mg, and the forecast deposits with calculated parameters are ca. 21 bn Mg. Due to the long period of preparation and execution of mining developments (several years), the absolutely necessary development of the said deposits should start as fast as possible, so as to ensure keeping coal output after 2020 at the level of 60-70 m Mg/year, with the possibility of increase up to 80 m Mg/year. The values of documented calculated deposits of brown coal show that they will be sufficient for almost 200 years at the current output parameters.

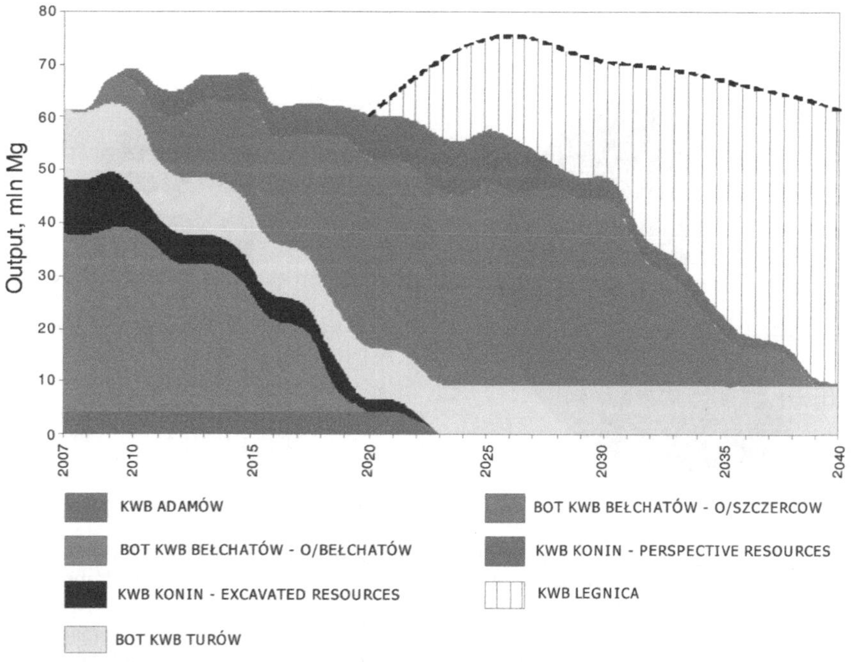


Figure 3. The planned production of brown coal from active and perspective mines in Poland

The following are arguments for keeping the current level of brown coal output and its increase in the further time horizon:

- significant deposits of documented and perspective brown coal deposits, favourably located in the regions with scarce other power raw minerals,
- modern, reliable, efficient, safe and effective techniques and technologies of open-pit deposit mining,
- the possibility of minimising unfavourable impact of mining activities on the environment, and, in case of their occurrence, effective and efficient repair methods (e.g. model reclamation),
- the possibility of implementation of ecologically "clean" technologies of electricity production with brown coal,
- well developed, of international renown, scientific, design and industrial facilities and highly qualified mining personnel.

## 3. NATURAL GAS AND OIL

Hydrocarbon energy media, i.e. natural gas and oil, are currently, apart from coal, the main power raw materials. In accordance with many forecasts, this image of the world power will continued in the 21$^{st}$ century, and the humanity will have to largely use these raw minerals. A statement could be even made of further domination of oil and, especially, of natural gas in the current century. The world population will reach the number of 9 bn in 2050 and supply of energy in this period for that number of people may be only provided with major share of hydrocarbon raw minerals, but also of coal.

Hydrocarbons are the second after coal important power engineering raw material in Poland and their share in the structure of consumption of original energy is 35.3% (Fig. 4, Polityka 2007). When compared with European Union countries, where it reaches 61.1%, it is small. Consumption of hydrocarbons in recent years was on the increase, yet with dropping progression. The increase rate for oil and natural gas were (2004) 5.5% and 6.6%, (2005) 3% and 1.6%, and (2006) as low as 0.5% and 0.7%, respectively.

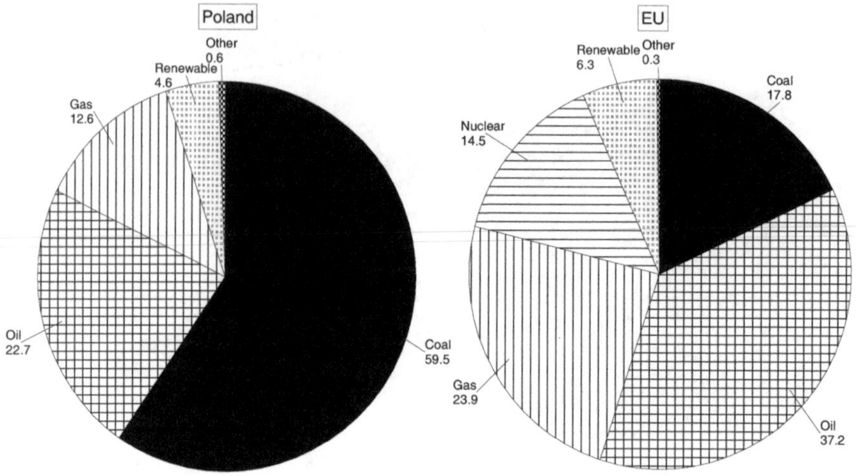


Figure 4. Structure of consumption of original energy (Polityka... 2007)

Domestic production of natural gas at 4.3 bn $m^3$ was sufficient (2005) to cover about 31% of demand for this fuel. The remaining portion of gas, i.e. 9.7 bn $m^3$, was imported, mostly from Russia (ca. 6.3 bn $m^3$). Moreover, contracts for supply of gas from Germany and Norway (for the total of ca. 0.8 bn $m^3$) were executed. Additionally, when major increase in demand for gas was reported, short-term supply of gas from Central Asia was started (about 2.5 bn $m^3$) (Report 2005).

Due to the increasing demand for natural gas in Poland, increase in the production of natural gas equals 5.1 bn $m^3$ in 2008 and then is planned up to the level of 5.5 bn $m^3$ per year (Ele, Sprunt 2006; Radecki 2005). In order to keep this production volume steady for several years at the same level, or even increase it, keeping the index for rebuilding natural gas deposits would be needed at the level of not less than 1.1 against production. The assumed production level is possible on the basis of the already available, discovered and documented deposits. Execution of this plan is not possible without increasing investment expenditures for managing exploration work.

The extractable deposits of natural gas at the end of 2006 were 238.6 bn $m^3$, including 143.1 bn $m^3$ in active mines and 95.2 bn $m^3$ in undeveloped deposits (Ney, Galos 2008). The industrial deposits in active mines reached the level of 75.1 bn $m^3$. In general, gas deposits are present in two regions, i.e. in Pogórze Karpackie (south-eastern Poland) - high methane content gas, and in Niż Polski (western Poland) - nitrated gas with admixture of sulphur. According to the specialists from the University of Science and Technology and the Polish Geological Institute, the unknown deposits of gas in Poland may even reach the level of from 700 to 1,200 bn $m^3$. On the basis of the documented deposits of natural gas, their sufficiency is at present calculated at about 25 years. It is forecast that the share of natural gas, according to the forecast for consumption of final energy in Poland, will increase from 11.9 bn $m^3$ in 2005 to 15.9 bn $m^3$ in 2030, whereas the forecast for demand for natural gas reports increase from 14.5 bn $m^3$ in 2005 to 23.5 bn $m^3$ in 2030. Domestic consumption in 2007 was almost 15 bn $m^3$.

When natural gas deposits are considered, methane in coal beds cannot be neglected. There are three Coal Basins in Poland: Górnośląskie Zagłębie Węglowe (GZW), Dolnośląskie Zagłębie Węglowe (DZW) and Lubelskie Zagłębie Węglowe (LZW). Due to its area, of over 5000 $km^2$, of which about 4500 $km^2$ is in the territory of Poland, Górnośląskie Zagłębie Węglowe takes a clear dominant position. It is related to the volume of geological deposits of coal, both calculated and industrial, the variety of lithological types of coal present there and production output of over 95%.

According to the Department of Geology and Geological Licences, methane in coal beds (MPW) is present mostly in the deposits of GZW. According to recent research, geological usable deposits of methane in GZW coal beds are assessed (as of the end of 2006) at about 95.3 bn $m^3$, including 29.8 bn $m^3$ in active mines, where industrial deposits are estimated at 4.8 bn $m^3$

(Ney, Galos 2008). In 2006, production of methane was 480 m $m^3$, and the volume of methane emitted to atmosphere along with mining air was 650 m $m^3$. For comparison purposes, it has to be noted that in 2004 production reached 250.88 m $m^3$ of methane with the output of 481.11 $m^3$/min, and 106.05 m $m^3$ was emitted to atmosphere (Nagy et al. 2006). Intense research is under way, aimed at development of management methods of this immense volume of methane.

In the years 1990-1996, several international companies, such as Amoco, Texaco, McCormic, Metanel – Poland, continued their projects in the scope of acquiring methane from coal beds. The interest in acquiring methane from coal beds in Poland was high, yet lack of spectacular successes clearly cooled down the interest of prospective future producers. The companies which started operations in this respect faced a number of obstacles, including:

- lack of a proper technology adjusted to the Polish conditions. Application of the methods which were successful under conditions of American deposits did not provide the results expected by the investors,
- lack of tax policy encouraging investments,
- the necessity of adjusting to economic conditions in the Polish coal and oil mining under close control of the government (especially in reference to price policy).

The calculated deposits of oil are at present as low as 21.6 m Mg (Nagy et al. 2006). Usable deposits as of the end of 2006 were estimated at the level of 23.9 m Mg, and industrial deposits in active mines at 15.1 m Mg (Ney, Galos 2008). Poland features small deposits located in the same regions as deposit of natural gas. Some of them are also located in the Baltic Shelf.

The current production of oil is at the level of 0.9 m Mg per year, including 0.6 m Mg on land and 0.3 m Mg from undersea deposits. The plans for increase of production on land refer to 1.1 m Mg, and when plans of Petrobaltic are included, that may increase in Poland up to the total of about 2 m Mg per year. Sufficiency of the deposits will depend on the production and on new discoveries, but it is currently estimated, with the current exploitation level maintained, at the level of 20-25 years. It is forecast that the share of oil products, according to the forecast for consumption of final energy in Poland, will increase from 20.52 Mtoe in 2005 up to 29.10 Mtoe in 2030 (1 Mg = 1 toe), whereas the forecast for demand for gas products reports increase from 22.1 Mtoe in 2005 up to 31.2 Mtoe in 2030 (Polityka 2007).

Oil, in about 95%, is imported from Russia, through the "Przyjaźń" pipeline running from the eastern borderline (Adamowo) to Płock, to the borderline with Germany, and farther on to the Schwedt refinery and to the Spergau refinery. The transport capacity of the pipeline is ca. 43 m Mg/year. The capacity of the western section of the pipeline is ca. 27 m Mg/year. The Polish PKN Orlen and Rafineria Gdańska (the Lotos Group) refinery plants receive about 18 m Mg/year whereas over 22 m Mg/year of oil is transported to Germany. The length of the Polish section of the pipeline is ca. 660 km (for the total length of over 2,500 km), and its branch – the "Pomorski " pipeline, with the length of ca. 200 km – connects Płock with Gdańsk (the refinery plant and the Naftoport reloading base in Port Północny). This is a reverse operation pipeline with effectiveness of ca. 30 m Mg/year (Gdańsk → Płock) and ca. 20 m Mg/year (Płock → Gdańsk). Placing the third line of the eastern section increases the capacity of the pipeline to 63 m Mg/year.

Small volume of oil is imported via Port Północny from the Northern Sea deposits and from Arab countries. Significance of the long-planned but not implemented (for political and economic reasons) construction of the Płock – Brody – (Odessa) pipeline is clear with this background.. At present, the Brody – Odessa pipeline is used for transport of oil from Russian deposits to Odessa (ca. 9 m Mg/year). Transport of oil from the countries of the Caspian Sea to Odessa and farther on through Brody to Płock would be a major diversification of import. The economic criterion should be decisive, with other premises of minor importance, like political ones.

Poland is not and will not be capable of ensuring major variation in the directions of gas supplies, unless it decides to incur excessively high costs. Therefore, active and rational joining politics and multi-dimensional power engineering activities of the European Union is absolutely necessary (Kolenda, Siemek 2008). Poland should aim at achieving the level of safety identical with that of the whole European Union.

In the Polish power engineering economy, due to the deposits, coal will continue to play significant role. This medium of energy ensures also high power engineering safety of the country. Combination of coal and natural gas may provide relatively high power engineering "freedom" of the country.

4. COPPER

The Polish deposits of copper ore are sediment deposits and occur in the rock from Cechsztyński (Permian) period, in the Sudety monocline, as well as in the Central Sudety Basin. The latter lost their economic significance.

The deposit area of copper ore in the Sudety monocline extends from Lubin in south east to Bytom Odrzański (the area is 60 km long and 20 km wide). It is in fact a single deposit operated by the mines of Lubin, Polkowice-Sieroszowice and Rudna, which are part of KGHM Polska Miedź S.A.

In 2006, the calculated deposits were assessed at 1961.27 m Mg of ore with content of 38.20 m Mg of copper and 104.660 Mg of silver. The calculated deposits of undeveloped copper ore are mostly present in the zone 1000-1250 m deep, reaching 1450 m in some places. Their development will be very difficult.

The perspectives of functioning of the domestic copper mining, until recently limited with the 2040 horizon, i.e. the date of the planned depletion of resources of the ore at the available depths in 2004, were radically changed. KGHM obtained the right to exploit the new Głogów-Głęboki-Przemysłowy deposit, which means extending the copper ore mining to 2030 or 2040. Reaching the full production capacity of the Głogów-Głęboki-Przemysłowy deposit is planned for 2020. Production of copper ore in 2008 reached 30.900 m Mg of ore with 526,806 Mg of metallic copper and 1193 Mg of silver and 0.902 Mg of gold (Figs. 5, 6 and 7).

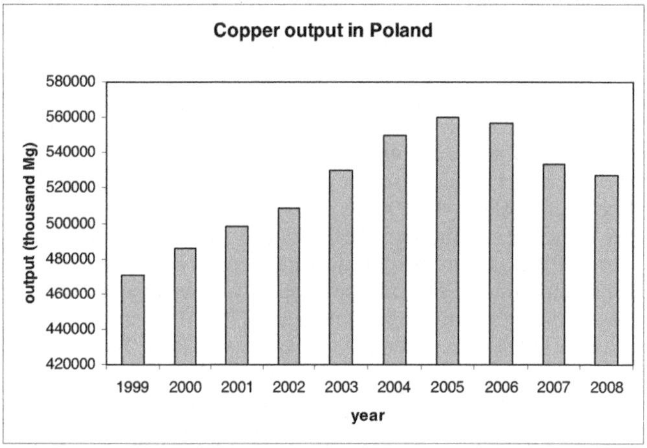

Figure 5. Copper output in Poland

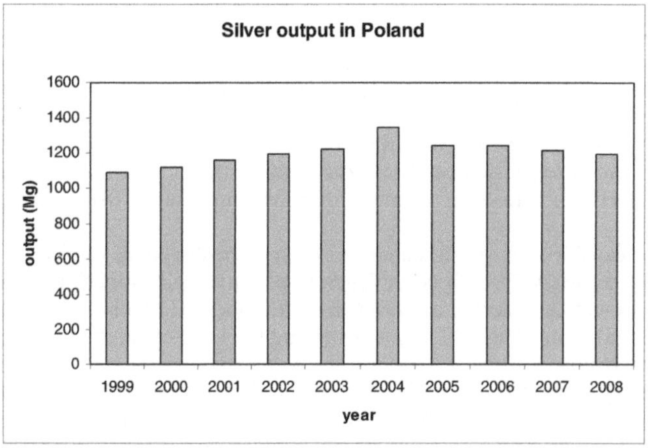

Figure 6. Silver output in Poland

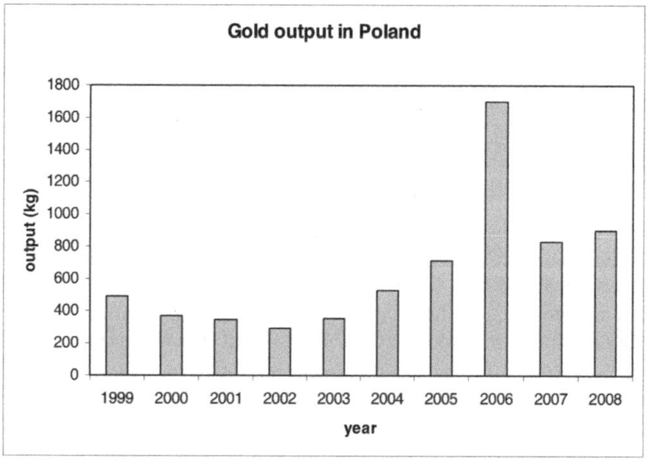


Figure 7. Gold output in Poland

## 5. ZINC AND LEAD

The area of deposits of zinc and lead ore in the northern and north-eastern periphery of Górnośląskie Zagłębie Węglowe is also called the Silesian-Krakow area. Industrial-significance Zn-Pb ore deposits are located there. The existing deposits are related to formation of carbonate rock built of Permian and Mesozoic rock monoclinally placed on Palaeozoic formations.

The calculated resources of zinc and lead ore as of the end of 2006 were 168.58 m Mg of ore with 6.54 m Mg of zinc and 3.01 m Mg of lead. The developed deposits hold 17.5% of the ore deposits. The industrial deposits in these include 22.38 m Mg of ore with 0.96 m Mg of zinc and 0.37 m Mg of lead.

The output of zinc and lead ore in Poland in 2007 was 4.176 m Mg of ore with 113,000 Mg of zinc and 52,000 Mg of lead (Fig. 8). The mining output of zinc and lead is not sufficient for the needs of the processing industry. The demand is also covered with imported concentrates. Major portion of zinc and lead production is exported.

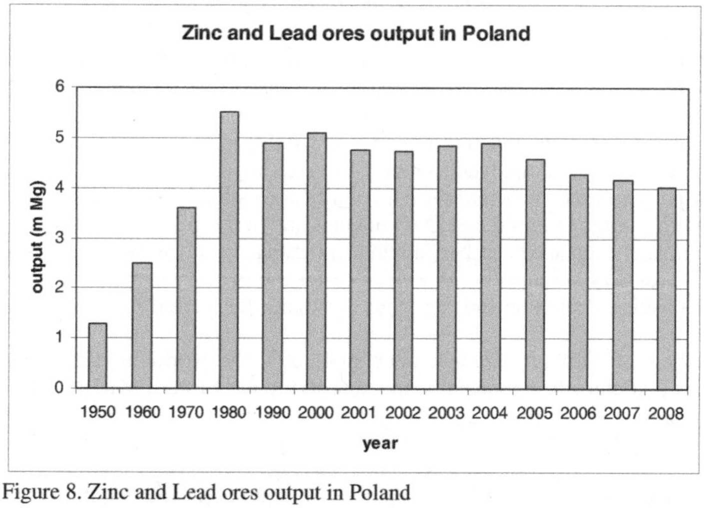


Figure 8. Zinc and Lead ores output in Poland

It is necessary to build a new zinc and lead ore mine Zakłady Górniczo-Hutnicze "Bolesław" S.A. for several years have been conducting activities aimed at starting a new mine close to the city of Zawiercie, where zinc and lead ore deposits were found and documented already in the 1970s. The "Zawiercie" deposit is currently the largest and richest of the undeveloped zinc and lead ore deposits in Poland, with the documented geological resources (as of the end of 2006) of 26.8 m Mg of ore with Zn content 5.5% and Pb content 2.2%.

## 6. SULPHUR

The documented the calculated Polish deposits of native sulphur are about 750 m Mg, which constitutes about 30% of all known world resources. The Polish deposits of native sulphur are located in the northern part of the Carpathian sink.

The largest output of sulphur at about 5.2 m Mg was reached in 1980. In the years 1961-2006, the total of 125.721 m Mg of sulphur was produced. About 80% of this was exported with high profitability (Fig. 9).

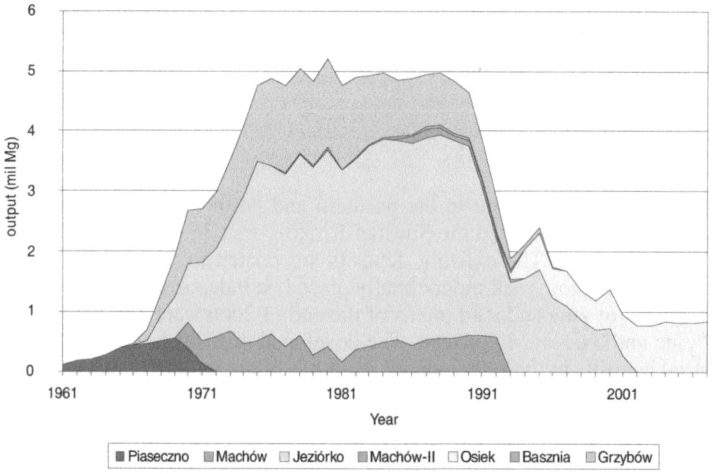


Figure 9. Output of sulphur in Polish mines

In the second half of 1991, the positive situation of sulphur in the world markets was crushed. The increased supply of cheap sulphur from recovery (sulphur obtained in the process of treating oil and natural gas) and the world economic recession resulted in abrupt decrease of price of sulphur. This inhibited further development of sulphur mining in Poland, and the final output was radically reduced. In the years 1992-2001, most of sulphur mines were liquidated (Fig. 9), and only the "Osiek" bore mine was left, with the output of ca. 800,000 Mg of sulphur per year. It is the sole operated sulphur mine in Poland and in the world. In this mine, a number of modern technical and process solutions are employed to reduce adverse impact on the environment. The closed circulation of process water was introduced, solidification of sulphur in blocks was eliminated and replaced with storage in liquid condition in tanks and in the granulated form.

Granulated sulphur completely replaced crushed sulphur in trading. Its advantage consists in reduced dusting in storage, reloading, transport, as well as easier use in further industrial processes. The mine records positive economic indexes, thus surviving the period of bad market conditions.

What are the perspectives of sulphur mining? Since mid-2007, the world markets show abrupt increase for prices of sulphur. It is especially apparent in the Far Eastern market, where prices of FOB Ruwais sulphur (UAE) offered by Abu Dhabi National Oil Co (Adnoc) in early March 2008 reached 640 USD/Mg, with as much as 1000 USD/Mg offered in mid-May by producers of sulphur.

There are opinions that the positive demand for sulphur will continue in the coming years, which results in evoking interest of investors in starting output facilities. Construction of new sulphur bore mines in Poland is realistic considering the short construction cycle which may be even shortened to 18 months when circumstances are positive.

## 7. SUMMARY

Most of the forecasts show that the global demand for original energy in the world in the first half of the 21st century will almost double. To meet these needs, development of production in all mined media of energy (hard and brown coal, oil, natural gas) and renewable energy are necessary.

In case of solid fuels, change in the structure of consumption of this fuel has been recorded for some time. The share of hard and brown coal in production of electricity is on the increase, while the share of hard coal used by minor industries and individual recipients is on the decrease. With high probability, one could state that coal will remain the basic fuel in the coming several dozen years, mostly for production of electricity, with simultaneous development of new technologies of incineration and reducing output costs and reducing emission of flue gas to atmosphere.

It seems that in Poland, after some years of neglecting advantages of coal, its position is again recognised (Germany understood it earlier, where almost 240 m Mg of hard and brown coal is consumed, while this value in Poland reaches only about 130 m Mg). The problem of the planned consumption of coal and gas in the structure of consumption of energy requires a rational solution, taking into account economic facts. Coal, both hard and brown, with rational management and observation of conditions for protection of the environment, should be the guarantor of power engineering safety and of relatively low prices of energy. Coal should be the ground for the basic power engineering industry, as well as large heat generating plants. Hard coal should be also used in industrial technological processes (coking plants, etc.). Due to ever increasing prices for oil and gas, using hard coal by minor recipients may be re-established.

Drastic reduction in the output capacity of the hard coal industry and stabilisation of brown coal output, with the increasing power engineering needs of the country, require increasing import of natural gas, oil and, which seems to be peculiar, hard coal. Even with significant increase in consumption of renewable energy, power engineering self-sufficiency will be definitely reduced in 2020 to the level of 60-65%, and down to slightly over 40% in 2030 (completion of operational use of most of active hard and brown coal mines). These trends should be taken into account and prevented right now. This forecast becoming true would become the hazard for the power engineering safety of Poland, therefore expansion of the existing and construction of new hard and brown coal mines should be considered even now. At the same time, one should be aware that costs of importing high volume of hydrocarbon fuels will definitely increase costs of electricity.

In the light of the forecasts, demand for electricity in Poland in the coming 20 years will significantly increase (especially after 2015). Hard and brown coal will continue to be one of the main fuels in our power engineering industry. The arguments for it come in the form of the deposits, definitely lower costs of production of electricity, technical preparation for continuing operational use of the developed deposits and development of perspective deposits. However, the actual market approach to production of electricity and finding the appropriate organisational form for management of new deposits will be the conditions for taking advantage of this opportunity, including also the solution to the problem of acquiring areas for the future exploitation. The decision on building nuclear energy plants shall also be made sooner.

For some time now, the international situation has been becoming more and more complex. At the same time, some EU countries attempt to increase the share of their own media in production of energy. The Polish power engineering industry based on coal should become one of the pillars of EU power safety.

## REFERENCES

Bilans Zasobów Kopalin i Wód Podziemnych w Polsce. 2006. PIG Warsaw. (In Polish)

Dubiński J., Turek. 2007. Prognoza wydobycia węgla kamiennego dla energetyki, w pracy zbiorowej „Uwarunkowania wdrożenia zero-emisyjnych technologii węglowych w energetyce" ed. M. Ściążko, Instytut Chemicznej Przeróbki Węgla, Zabrze. (In Polish)

Dubiński J. 2008. Węgiel kamienny jako paliwo XXI wieku dla energetyki oraz produkcji paliw płynnych i gazowych, www.komitetlegnica.agh.edu.pl. (In Polish)

Ele S., Sprunt E.S. 2006. Natural Gas – Image vs. Reality. Journal of Petroleum Technology, February

Gawlik L., Grudziński Zb., Lorenz U. 2007. Wybrane problemy produkcji i wykorzystania węgla brunatnego. V Międzynarodowy Kongres Górnictwa Węgla Brunatnego. Mat. Konf. AGH Uczelniane Wydawnictwa Naukowo-Techniczne, Krakow. (In Polish)

Kirejczyk J. 2008. Personal communication

Kolenda Z., Siemek J. 2008. Gazowa pułapka, Tygodnik Powszechny, 10 (3061)/2008. (In Polish)

Nagy S., Rychlicki S., Siemek J. 2006. Impact of inactive hard – coal mines processes in Silesian Coal Basin on greenhouse gases pollution. Acta Geologica Polonica 56 (2): 221-228

Ney R. 2006. Wybrane problemy polityki energetycznej Polski. Polityka Energetyczna Vol. 921/2006. Publ. IGSMiE PAN, Krakow. (In Polish)

Ney R., Galos K. 2008. Bilans polskich surowców mineralnych (energetycznych, metalicznych, chemicznych i skalnych), kierunki polityki przestrzennej w zakresie wykorzystania złóż, problemy ochrony złóż i terenów eksploatacyjnych; rekomendacja dla KPZK, www.min-pan.krakow.pl. (In Polish)

Polish power engineering policy by 2030. The government document of 10 Sept. 2007

Radecki S. 2005. Możliwości i warunki zwiększenia krajowej bazy zasobowej ropy naftowej i gazu ziemnego. II Krajowy Kongres Naftowców i Gazowników SITPNiG. Bóbrka 2005. (In Polish)

PGNiG S.A. 2005. Annual Report

J. Popczyk (ed.) 2004. Polityka energetyczna Polski do roku 2025 – założenia, Katowice, 2004. (In Polish)

www.pgi.gov.pl

www.epa.gov/coalbed

International Mining Forum 2010, Liu et al. (eds) © 2010 Taylor & Francis Group, London, UK. ISBN 978-0-415-59896-5

# The influence of natural hazards on occupational safety in Polish collieries

Zbigniew Burtan
*AGH University of Science and Technology, Krakow, Poland*

ABSTRACT: The coal mining sector in Poland is exceptional throughout the world in that underground mining operations are accompanied by nearly all natural hazards while the scale and intensity of their occurrence is constantly growing. That applies mostly to dangers associated with methane emissions, tendency of self-ignition of coal, seismic activity of the rock strata and climatic conditions.

This study briefly describes the coal deposits in Poland, highlighting the aspects that may lead to intensification of natural hazards and help predict their scale and tendency. The paper attempts an analysis of technological factors that might further enhance the methane-related hazards, fires and coal bumps, particularly in the context of monitoring and control methods. The analysis clearly identifies the need for safety precautions and preventive measures to be put in place in Polish collieries.

## 1. INTRODUCTION

Coal production in 31 Polish collieries reached the level of 74 million Mg in 2009. As many as thirty of those collieries are located in the Upper Silesia Coal Basin. Because of depletion of some coal reserves, limited production capacities of mines, increased production costs and lower demand for domestic coal, the coal production levels in Poland are now gradually decreasing and the number of operating mines is becoming lower, too (Table 1) (Report of the State Mining Authority 2010a).

Table 1. Characteristics of the coal mining sector in Poland in 2000-2009

| Year | 2000 | 2001 | 2002 | 2003 | 2004 | 2005 | 2006 | 2007 | 2008 | 2009 |
|---|---|---|---|---|---|---|---|---|---|---|
| Total output $\cdot 10^6$ [Mg] | 102.5 | 102.6 | 102.1 | 100.5 | 99.5 | 97.0 | 94.4 | 87.5 | 83.6 | 74.0 |
| Numbers of mines | 42 | 42 | 42 | 41 | 39 | 33 | 33 | 31 | 31 | 31 |

The reduction in the output level does not help to limit the scale of natural hazards, which are further intensified by adverse geological and mining condition and regional concentration of mining activities. Mining of available coal deposits in the Upper Silesia Coal Basin is now continued at increasing depth. Presently coal is mined up to the depth of 1150 m, in 7 collieries the mining operations are pursued below 1000 m. Nearly 90% of coal production comes from below 500 m and the average mining depth now approaches 700 m (State Mining Authority 2010a, 2010b).

Coal mining at increasing depths leads to stress increase in the rock strata, reduced strength of coal in the deposits and increased strength of neighbouring rocks, higher temperature of rocks, increased methane capacity of coal deposits. The strength of coal deposits decreases with depth and rock porosity becomes reduced, too (Konopko 2010), leading to reduction of their permeability to gas. Because of multi-seam structure of coal deposits, coal now has to be mined in geologically disturbed regions, in strata disturbed by previous mining activities and in residual parts of the coal deposits.

Coal is mostly mined by the longwall mining system with the roof control by caving-in. In 2009 only 2.9 million Mg were mined by the hydraulic stowing system, mostly because of surface protection requirements. That accounts for 3.8% of the total coal production (State Mining Authority 2010a). Coal mining as well as hauling, loading and coal handling operations are now fully mechanised and we observe high concentration of coal production. A large proportion of coal, accounting now for nearly 50% of the total output, is obtained from sub-level mining. Those conditions are responsible for the occurrence and intensification of natural hazards typically encountered in mines, such as methane emissions, coal dust explosions, fires, adverse climatic conditions, caving-in, coal bumps, rock and gas outbursts.

Radiation and water hazards threaten the Polish mines in a lesser degree. The distinctive feature of the mining sector on Poland is the tendency to simultaneous occurrence of several natural hazards which may occur jointly, further increasing the risk involved in mining operations (Burtan et al. 2008).

## 2. NATURAL HAZARD LEVELS IN POLISH COLLIERIES

### 2.1. *Methane emissions*

Methane emissions present a major threat to safe and effective coal mining in Poland. 23 out of 31 operating collieries are threatened by methane emissions, 15 collieries belong to the IV (the highest) category of methane emission risk. The proportion of coal production from gassy mines tends to increase steadily, now approaching 80%. In 2009 the methane emission hit the level of 855.7 million $m^3$, released in the consequence of mining operations. When related to the production levels, that yields the highest ratio reported recently - 11.1 $m^3/Mg$. In 2009 three major accidents were reported that were associated with the methane hazard (Table 2) (State Mining Authority 2010a).

Table 2. Methane hazard in 2000-2009

| Year | 2000 | 2001 | 2002 | 2003 | 2004 | 2005 | 2006 | 2007 | 2008 | 2009 |
|---|---|---|---|---|---|---|---|---|---|---|
| Methane bearing capacity·$10^6$ [$m^3$] | 746.9 | 743.7 | 752.6 | 798.1 | 825.9 | 851.1 | 870.3 | 878.9 | 880.9 | 855.7 |
| Methane emission [$m^3/Mg$] | 7.3 | 7.2 | 7.4 | 7.9 | 8.3 | 8.8 | 9.2 | 10.4 | 10.5 | 11.1 |
| Methane fires and explosions | 1 | 0 | 3 | 5 | 1 | 3 | 2(1*) | 4 | 3(1**) | 3 |

* explosion of methane and coal dust, ** fire, explosion of methane and coal dust

Accidents caused by methane fires and explosions involved other risks as well. Methane ignitions lead to fires whilst methane explosions often resulted in coal dust explosions. Methane fires and explosions are often triggered by fires and sudden methane emissions to the mine workings, sometimes they were caused by coal bumps.

Despite productivity cutbacks and reduction of the number of operating mines, the absolute level of methane bearing capacity of coal has been steadily going up since 2001. The enhanced level of methane hazard is associated with the mining depth as methane bearing capacity of coal deposits increases with depth. At the same time, lower rock porosity at larger depths limits the natural methane removal (Konopko 2010).

At greater depths a most undesired process is observed, namely dynamic release of free methane occurring under large pressure of overlying strata. Large mounts of free methane are present in geologically disturbed features, such as faults, caverns or voids, forming so called 'gas traps'. Mining operations continued in the vicinity of such gas traps can also results in methane and rock outbursts.

Methane hazard is further enhanced by sub-level mining and high concentration of mining operations and high daily rates of advance of the longwall systems (Burtan et al. 2008).

### 2.2. *Coal dust hazard*

The risk of coal dust explosion is present in all collieries as it is the consequence of coal mining, hauling and transport operations within the mine. Alongside methane emissions and endogenous fires, coal dust explosions are the major cause of catastrophic accidents. In recent years there have been but a few coal dust explosions (Table 3), owing to effective dust control strategies that were put in place. The last coal dust explosion not involving any other risks was reported in 2002, after 15 years from the previous explosion and the last fatal accidents initiated by a methane explosion were reported in 2006 and 2008 (State Mining Authority 2010a).

Table 3. Scale of coal dust hazard and dust control strategies in 2000-2009

| Year | 2000 | 2001 | 2002 | 2003 | 2004 | 2005 | 2006 | 2007 | 2008 | 2009 |
|---|---|---|---|---|---|---|---|---|---|---|
| Number of coal dust explosions | 0 | 0 | 1 | 0 | 0 | 0 | 1* | 0 | 1** | 0 |
| The amount of rock dust [Mg] per $10^6$[Mg] of produced coal | 516 | 494 | 746 | 699 | 657 | 655 | 632 | 756 | 814 | 966 |

* explosion of methane and coal dust, ** fire, explosion of methane and coal dust

Coal dust explosions are associated with the presence of methane. When methane concentration increases, the lower limit of coal dust tendency to explosion is reduced. Typically, during the mining, hauling and loading processes about 2% of the mined material turns into coal dust, so enhanced dust explosion risk is associated with mechanisation of most operations and increased concentration of coal production (Burtan et al. 2008).

### 2.3. *Fire hazard*

Endogenous fires are quite common in most collieries. Apart from the coal's natural tendency to self-ignition, there are many factors that further enhance the risk of a fire: increased mining depth affecting the primary state of stress (leading to the formation of fracture zones) and increased temperature of rocks. Mining of burst prone seams, rock bursts and the applied preventive measures might further enhance the fire hazard. Methane control measures increase the risk of a fire, because of the air supply along the headings. Besides, methane control methods are applied which adversely impact the methane behaviour (Zorychta, Burtan 2008).

Major factors that reduce the risk of an endogenous fire include: concentration of mining production enabling the faster coal extraction from the endangered zones and application of hydraulic stowage systems. In 2009 ten endogenous fires where reported, in the consequence the fire risk increased in relation to the previous year (Table 4) (State Mining Authority 2010a).

Table 4. Fire hazard in 2000-2009

| Year | 2000 | 2001 | 2002 | 2003 | 2004 | 2005 | 2006 | 2007 | 2008 | 2009 |
|---|---|---|---|---|---|---|---|---|---|---|
| Number of endogenous fires | 2 | 1 | 4 | 4 | 5 | 7 | 2 | 4 | 6 | 10 |

The reported fires occurred in coal seams from different categories of self-ignition behaviour (from I to IV), though a larger number of fires occurred in coal beds displaying low or medium tendency to self-ignition than in those categorised as class IV and V (fire-prone) (State Mining Authority 2010a).

The fire risk is associated with natural factors and mining conditions, which may further enhance or reduce the fire risk. It is worthwhile to mention that the risk of an endogenous fire has increased since 2000 and now we experience the joint occurrence of the methane risk, rock bursts and adverse climatic conditions.

### 2.4. *Adverse climatic conditions*

Another hazard, more intensive with the mining depth, is posed by adverse climatic conditions associated with high primary temperature of rocks and high air humidity.

Adverse climatic conditions, experienced now by more and more collieries, further deteriorate because of the concentration of production and the presence of powerful mining and hauling equipment. In 2009 the working shifts in 19 collieries (183 face operations) had to be shortened (Table 5). The proportion of coal production from regions of rock strata temperatures in excess of 28°C accounts for 30% and the average rock temperature at the lowest production level is equal to 45°C.

Table 5. Adverse climatic conditions in mines in 2000-2009. The number of mines and faces operated by shorter working shifts

| Year | 2000 | 2001 | 2002 | 2003 | 2004 | 2005 | 2006 | 2007 | 2008 | 2009 |
|---|---|---|---|---|---|---|---|---|---|---|
| Number of exposed mines | 20 | 20 | 21 | 21 | 27 | 22 | 21 | 18 | 20 | 19 |
| Number of mine workings | 92 | 98 | 97 | 143 | 152 | 173 | 182 | 180 | 255 | 183 |

In several collieries coal is planned to be mined below 1200 m, which means that mining operations will have to be pursued in rock strata whose primary temperature approaches 50°C and climatic conditions may become the major threat, affecting the miners' safety and practicability of the mining enterprise.

### 2.5. *Ventilation risk*

Ventilation risks are understood to be hazards associated with mine atmosphere when it is unsuited for breathing. This hazard is experienced in poorly ventilated headings, secured with dams to prevent fires or in the course of fire fighting actions. Accidents happen when miners enter the zone with too little oxygen in the atmosphere or when gas is released from behind stoppage dams to mix with unstable or slow airflows, further enhanced by joint occurrence of adverse climatic conditions. The accidents can also happen to rescue teams, taking part in rescue actions. The last reported accident caused by a ventilation hazard happened in 2005 (Table 6) (State Mining Authority 2010a). No life loss has been reported since 2006 associated with type of hazard.

Table 6. Scale of ventilation risk in 2000-2009

| Year | 2000 | 2001 | 2002 | 2003 | 2004 | 2005 | 2006 | 2007 | 2008 | 2009 |
|---|---|---|---|---|---|---|---|---|---|---|
| The number of accidents | 3 | 0 | 0 | 0 | 0 | 2 | 0 | 0 | 0 | 0 |

### 2.6. *Caving-in*

Caving-in, referred to as gravity-induced subsidence of rocks in the mine workings in the degree precluding the restoration of its original features within less than 8 hours, is a common occurrence in all collieries. In mine workings accidents are also caused by broken rocks, falling from the roof and wall sides, which do not disturb the mine operations. The caving-in hazard enhances when typical gravity-induced rock displacements turn into intermediate processes between the caving-in and dynamic rock displacements characteristic of rock bursts. In 2009 3 major caving-ins were reported, which is the largest number registered in recent years (Table 7) (State Mining Authority 2010a).

Table 7. Caving-in risk in 2000-2009

| Year | 2000 | 2001 | 2002 | 2003 | 2004 | 2005 | 2006 | 2007 | 2008 | 2009 |
|---|---|---|---|---|---|---|---|---|---|---|
| Number of caving-ins | 1 | 1 | 2 | 1 | 1 | 1 | 1 | 1 | 2 | 3 |

### 2.7. *Rock burst and bumps hazard*

Rock bursts and tremors are experienced now in a growing number of collieries. In 22 out of 31 currently operated collieries coal is mined from burst-prone seams, and 14 collieries are classified as those belonging to the highest category of risk. Though the proportion of coal production from burst-prone seams increased in recent years, the number of coal bumps has remained on the stable, relatively low level, in relation to that reported between 1988-1995 (State Mining Authority 2010b). Reduction of the number of rock bump occurrences to several incidents yearly (Table 8) is the result of production cutbacks and more effective coal bump control strategies.

Table 8. Scale of coal bump hazard in 2000-2009

| Year | 2000 | 2001 | 2002 | 2003 | 2004 | 2005 | 2006 | 2007 | 2008 | 2009 |
|---|---|---|---|---|---|---|---|---|---|---|
| Coal production of threatened seams $\cdot 10^6$ [Mg] | 37,2 | 37,4 | 41,8 | 42,3 | 39,2 | 41,6 | 42,1 | 40,5 | 41,9 | 34,3 |
| Proportion of coal production from seams threatened by coal bumps [%] | 36,3 | 36,5 | 40,9 | 42,1 | 39,4 | 42,9 | 44,6 | 46,3 | 50,1 | 44,4 |
| The number of coal bumps | 2 | 4 | 4 | 4 | 3 | 3 | 4 | 3 | 5 | 1 |
| Strata relief occurrence | * | * | 0 | 3** | 3** | 1 | 2 | 1 | 1 | 4 |

* no data available
**categorised as tremors causing accidents

In 2009 one occurrence of a coal bump was reported and in recent years the number of high-energy tremors (> $1\cdot10^5$[J]) leading to rock bursts and coal bumps went down (Table 9), still the yearly amount of released seismic energy tends to increase.

Table 9. Rock burst and coal bump risk in 2000-2009

| Year | 2000 | 2001 | 2002 | 2003 | 2004 | 2005 | 2006 | 2007 | 2008 | 2009 |
|---|---|---|---|---|---|---|---|---|---|---|
| Number of shocks and tremors $\geq 1\cdot10^5$ [J] | 1088 | 1137 | 1324 | 1524 | 974 | 1451 | 1170 | 885 | 883 | 741 |
| Number of shocks and tremors in relation to output levels - per $10^6$ [Mg] of produced coal | 10,6 | 11,1 | 13,0 | 15,2 | 9,8 | 15,0 | 12,4 | 10,1 | 10,6 | 10,0 |
| Average energy of a shock or tremor $\cdot 10^6$ [J] | 1,95 | 1,63 | 1,48 | 1,47 | 1,34 | 1,23 | 1,76 | 2,49 | 2,27 | 3,04 |
| Seismic energy [J/Mg] | 20,7 | 18,1 | 19,2 | 22,3 | 13,1 | 18,4 | 21,8 | 25,2 | 24,0 | 30,4 |

Besides, each year high-energy shocks and tremors lead to dangerous incidents categorised as strata relief occurrences whose impacts are less threatening than coal bumps. In 2009 four occurrences of rock strata relief were reported (State Mining Authority 2010a).

The seismic hazard is the consequence of the increased mining depth, geomechanical properties of coal seams and rocks and the thick-layer structure of the rock strata (particularly sandstone features and shales). Besides, coal is now mined in the proximity of tectonically and seismically disturbed zones, in the regions disturbed by previous mining activities and in residues. The need to maintain the high concentration of production, for economic reasons, plays some role, too (Burtan et al. 2008).

Shocks, tremors and coal bumps are often followed by methane emissions to the galleries and mine workings whilst coal bumps lead to the enhanced risk of endogenous fires, due to the stopping of the mining front. Furthermore, most rock burst and coal bump control strategies contribute to fire hazard, too.

## 2.8. *Rock and gas outburst hazard*

Rock and gas outbursts are regarded as the most dangerous natural hazards in underground mining. The collieries threatened in a large degree include the collieries situated in the Lower Silesia Coalfield, which were closed in the 1990s (State Mining Authority 2010b). Increased level of methane and coal outburst risk is encountered in collieries belonging to the Jastrzębie Coal Holding extracting coking coal, where major methane and coal outbursts were reported in 2002 and 2005 (Table 10) (State Mining Authority 2010a).

Table 10. Rock and gas outburst hazard in 2000-2009

| Year | 2000 | 2001 | 2002 | 2003 | 2004 | 2005 | 2006 | 2007 | 2008 | 2009 |
|---|---|---|---|---|---|---|---|---|---|---|
| Number of rock and gas outbursts | 0 | 0 | 1 | 0 | 0 | 1 | 0 | 0 | 0 | 0 |

To a large extent, rock and gas outburst hazard is a consequence of the increased mining depth and of continuing the mining activities in outburst-prone seams. The state of stress in the rock strata increases with depth, at the same time the strength of coal seams tends to deteriorate while their methane-bearing capacity increases and coal permeability to gas becomes lower, which further enhances the risk (Zorychta, Burtan 2008).

### 2.9. *Water hazard*

Water hazard involves uncontrolled water inflow to the mine workings, two cases only have been reported in the last ten years (Table 11) (State Mining Authority 2010a).

Table 11. Water hazard in 2000-2009

| *Year* | *2000* | *2001* | *2002* | *2003* | *2004* | *2005* | *2006* | *2007* | *2008* | *2009* |
|---|---|---|---|---|---|---|---|---|---|---|
| Number of water intrusions | 1 | 0 | 0 | 0 | 0 | 0 | 0 | 1 | 0 | 0 |

Unlike previously listed hazards, water hazard tends to decrease with depth as water-bearing capacity of rocks becomes lower and the volume of water reservoirs in workings becomes restricted. Effective control measures, involving the regular monitoring of water hazard and strict adherence to the mine safety procedures, has helped to limit this hazard (State Mining Authority 2010b).

## 3. THE INFLUENCE OF NATURAL HAZARDS ON THE NUMBER OF ACCIDENTS AT WORK

### 3.1. *Accidents caused by natural hazards*

Accidents caused by natural hazards are understood as conditions that pose the threat to human life or health, jeopardise the property of mining companies or disrupt the normal operations in the mines. Accidents reported in the period 2000-2009 are categorised with reference to the natural risks listed in Table 12 (State Mining Authority 2010a).

Table 12. The number of hazard occurrences in 2000-2009

| Year/ Natural hazard | 2000 | 2001 | 2002 | 2003 | 2004 | 2005 | 2006 | 2007 | 2008 | 2009 | 2000-2009 |
|---|---|---|---|---|---|---|---|---|---|---|---|
| Methane fire and explosions | 1 | 0 | 3 | 5 | 1 | 3 | 2(1*) | 4 | 3(1**) | 3 | 24 |
| Coal dust explosions | 0 | 0 | 1 | 0 | 0 | 0 | 0 | 0 | 0 | 0 | 1 |
| Endogenous fires | 2 | 1 | 4 | 4 | 5 | 7 | 2 | 4 | 6 | 10 | 45 |
| Ventilation and climatic risk | 3 | 0 | 0 | 0 | 0 | 2 | 0 | 0 | 0 | 0 | 5 |
| Caving-in | 1 | 1 | 2 | 1 | 1 | 1 | 1 | 1 | 2 | 3 | 14 |
| Coal bumps | 2 | 4 | 4 | 4 | 3 | 3 | 4 | 3 | 5 | 1 | 33 |
| Rock and gas outbursts | 0 | 0 | 1 | 0 | 0 | 1 | 0 | 0 | 0 | 0 | 2 |
| Water inflows | 1 | 0 | 0 | 0 | 0 | 0 | 0 | 1 | 0 | 0 | 2 |
| TOTAL | 10 | 6 | 15 | 14 | 10 | 17 | 8 | 13 | 15 | 17 | 126 |

*methane and coal dust explosions, **fire, methane and coal dust explosion

In 2009 17 dangerous accidents happened that were caused by natural hazards, most of these being endogenous fires. This figure is the largest in the whole period 2000-2009. Despite the production cutbacks in recent years, the total number of accidents caused by natural hazards tends to increase.

The analysis of contribution of particular natural hazards to dangerous situations in the mine (Fig. 1) reveals that in the last ten years the largest numbers of them were caused by endogenous fires (35.7%); coal bumps (26.2%) and methane explosions (19.0%). The least impacts are produced by coal dust explosions (0.8%), rock and gas outbursts (1.6%) and water inflows. It appears that endogenous are responsible for the largest number of accidents.

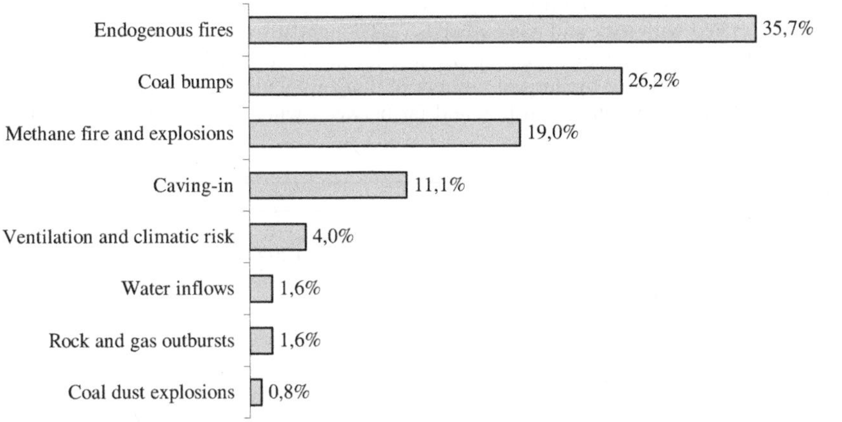


Figure 1. Proportion of hazard-induced dangerous occurrences reported in 2000-2009

Although the fire risk has increased recently, the frequency of fire occurrence has been vastly reduced since the post-war period. In the 1950s nearly 400 fires were reported each year, then this figure went down and in the 1970s it stabilised below 30 fires per year (State Mining Authority 2010a). One has to bear in mind, however, that the number of hazardous occurrences associated with coal bumps is underestimated because the impacts they produce are sometimes minor and do not cause any life loss, so they are frequently classified as the strata relaxation processes.

### 3.2. *Life loss caused by natural hazards*

The scale of fatalities caused by the occurrence of particular natural risks is shown in Table 13 (State Mining Authority 2010a). The caving-in statistics also includes the accidents caused by broken rocks falling from the roof or wall sides.

Table 13. Life loss due to natural hazards reported in 2000-2009

| Year/ Natural hazard | 2000 | 2001 | 2002 | 2003 | 2004 | 2005 | 2006 | 2007 | 2008 | 2009 | 2000-2009 |
|---|---|---|---|---|---|---|---|---|---|---|---|
| Methane fire and explosions | 0 | 0 | 4 | 4 | 0 | 0 | 23 (23*) | 0 | 8 (2**) | 20 | 59 |
| Coal dust explosions | 0 | 0 | 10 | 0 | 0 | 0 | 0 | 0 | 0 | 0 | 10 |
| Endogenous fires | 1 | 0 | 4 | 3 | 0 | 0 | 0 | 0 | 0 | 0 | 8 |
| Ventilation and climatic risk | 2 | 0 | 3 | 1 | 0 | 3 | 0 | 0 | 0 | 0 | 9 |
| Caving-in | 3+7 | 0+4 | 1+0 | 0+3 | 0+1 | 1+2 | 1+1 | 2+3 | 1+4 | 1+2 | 37 |
| Coal bumps | 0 | 2 | 3 | 2 | 0 | 1 | 4 | 0 | 0 | 0 | 12 |
| Rock and gas outbursts | 0 | 0 | 0 | 0 | 0 | 3 | 0 | 0 | 0 | 0 | 3 |
| Water inflows | 0 | 0 | 0 | 0 | 0 | 0 | 0 | 0 | 0 | 0 | 0 |
| TOTAL | 13 | 6 | 25 | 13 | 1 | 10 | 29 | 5 | 13 | 23 | 138 |

*methane and coal dust explosions, **fire, methane and coal dust explosion

23 life losses were reported in 2009 that were caused by the occurrence of natural risks; 20 people were killed by methane explosions. The largest number of fatalities (29) was reported in 2006, out of that number 23 were killed by a methane and coal dust explosion. A relatively high number of accidents was reported in 2002, including 25 fatalities, 10 of which were caused by a coal dust explosion. In the remaining years the number of reported fatalities was relatively low.

The analysis of the hazard-induced fatalities reported in the years 2000-2009 (Fig. 2) reveals that the largest number of dangerous incidents happened because of methane explosions, in two cases involving also coal dust explosions. Fallen rocks and coal bumps are responsible for several life losses; their relevant contributions being given as 26.8% and 9.4%. The lowest number of fatalities is associated with rock and gas outbursts (2.2%) whilst water inflows in recent years did not cause any life loss. As regards the total life loss levels in the Polish mining sector, methane emissions are responsible for the largest number of dangerous incidents. However, the accidents involving methane emissions seem few and far between when compared to other hazard types.

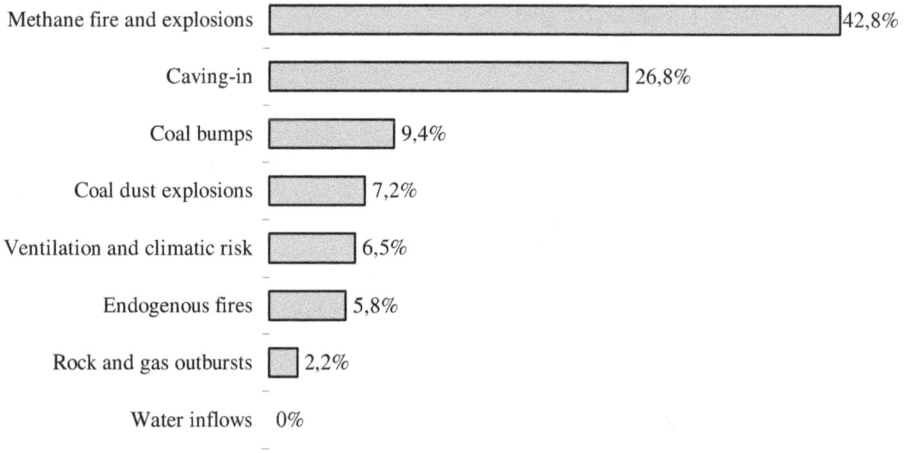


Figure 2. Number of fatalities and proportion of involved risk types in 2000-2009

### 3.3. *Natural hazards and the overall number of fatalities*

The number of fatalities being the consequence of natural hazards in relation to the total number of fatalities reported between 2000 and 2009 is shown in Table 14 (State Mining Authority 2010a).

Table 14. Fatalities reported in 2000-2009

| Year / number of fatalities | 2000 | 2001 | 2002 | 2003 | 2004 | 2005 | 2006 | 2007 | 2008 | 2009 | 2000-2009 |
|---|---|---|---|---|---|---|---|---|---|---|---|
| Total | 28 | 25 | 33 | 29 | 11 | 15 | 45 | 16 | 25 | 36 | 263 |
| Caused by natural hazards | 13 | 6 | 25 | 13 | 1 | 10 | 29 | 5 | 13 | 23 | 138 |
| Proportion of incidents caused by natural hazards [%] | 46.4 | 24.0 | 75.8 | 44.8 | 9.1 | 66.7 | 64.4 | 31.3 | 52.0 | 63.9 | 52.5 |

In 2009 the proportion of hazard-induced fatalities was 64%. The largest number of occurrences (76%) were reported in 2002, relatively high levels were reported in 2005 (67%), 2006 (64%) and 2008 (52%). The largest number of fatalities were caused by explosions of methane and coal dust, the smallest proportion of fatalities caused by natural factors was reported in 2004 (9%). In the analysed period 2000-2009 the number of fatalities associated with natural hazards displays a growing tendency.

The contribution of natural risk factors to the catastrophic life losses reported in the Polish coal mining sector in the last ten years accounts for 52.5% (Fig. 3). The remaining fatalities were caused by technical and human factors.

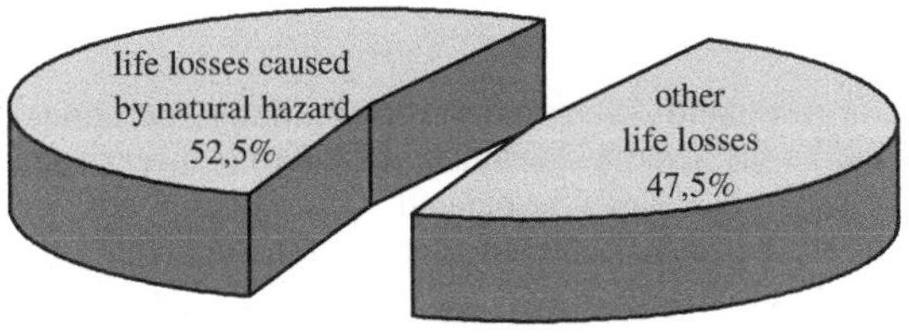


Figure 3. Proportion of fatalities caused by natural factors to the total number of life losses in 2000-2009

3.4. *Fatalities caused by natural risks in relation to coal production levels*

One of the major parameters used in evaluating the work safety is the accident rate per one million ton (Mg) of produced coal. The production levels in 2000-2009 and the accident rate indices in relation to all reported fatalities and dangerous occurrences caused by natural hazards are compiled in Table 15 (State Mining Authority 2010a).

Table 15. Number of fatalities in relation to 1 million ton of produced coal in 2000-2009

| Year | 2000 | 2001 | 2002 | 2003 | 2004 | 2005 | 2006 | 2007 | 2008 | 2009 | 2000-2009 |
|---|---|---|---|---|---|---|---|---|---|---|---|
| Coal production $\cdot 10^6$ [Mg] | 102.5 | 102.6 | 102.1 | 100.5 | 99.5 | 97.0 | 94.4 | 87.5 | 83.6 | 74.0 | 943.7 |
| All accidents per $10^6$ [Mg] | 0.27 | 0.24 | 0.32 | 0.29 | 0.11 | 0.15 | 0.48 | 0.18 | 0.30 | 0.46 | 0.28 |
| Accidents due to natural risks per $10^6$ [Mg] | 0.13 | 0.06 | 0.24 | 0.13 | 0.01 | 0.10 | 0.31 | 0.06 | 0.16 | 0.31 | 0.15 |

In 2009 the accident rate index was equal to 0.31 and, like in 2006, reached the highest level in the last ten years. The accidents rate in 2000-2009 was 0.13 and the level of fatalities due to natural risks tends to increase in the analysed period, in relation to production levels. However, when analysed over a longer time period, the frequency of fatal accidents tends to decrease as the highest accident rate was reported in the post-war years (9.09% in 1946). In 1975 it fell below 1.0 and stabilised there and in 1991 it fell below 0.5 (State Mining Authorities 2010b).

3.5. *The number of fatalities due to natural risks in relation to the number of employees*

Another major parameter illustrating the accident rate in the mining sector is the index expressed as the number of fatalities per 1000 employees. This index (taking into the staff in mining companies and sub-contractors) related to the overall number of fatalities and accidents due to natural hazards in 2000-2009 is shown in Table 16 (State Mining Authority 2010a).

Table 16. Number of fatalities in proportion to 1000 employees reported in 2000-2009

| Year | 2000 | 2001 | 2002 | 2003 | 2004 | 2005 | 2006 | 2007 | 2008 | 2009 | 2000-2009 |
|---|---|---|---|---|---|---|---|---|---|---|---|
| All accidents [per 1000 employees] | 0.18 | 0.14 | 0.24 | 0.20 | 0.08 | 0.11 | 0.32 | 0.12 | 0.19 | 0.26 | 0.18 |
| Accidents caused by natural hazards [per 1000 employees] | 0.08 | 0.03 | 0.18 | 0.07 | 0.01 | 0.07 | 0.21 | 0.04 | 0.10 | 0.17 | 0.10 |

In 2009 this index approached 0.17 and alongside the years 2002 and 2006 reached the highest level in the last ten years. Throughout the entire period 2000-2009 the index remained on the level 0.17 and its growing tendency can be clearly observed. Similar to the accident rate related to coal production levels, the accident rate related to the number of employees tends to decrease throughout the analysed period (State Mining Authority 2010b).

## 4. CONCLUSIONS

The distinctive feature of the mining sector in Poland is that underground mining operations are threatened by nearly all natural hazards whilst the scale and intensity of their occurrence seems to be growing each year. That applies mostly to hazards associated with methane emissions, self-ignition of coal seams, seismic activity of the rock strata and adverse climatic conditions.

The analysis of scale and tendency of natural hazards and associated accidents and fatalities leads us to the following conclusions:

- Coal in Poland is now mined in difficult mining and geological conditions. The mining depth is increasing and so is the methane bearing capacity of coal seams. The strength of the rock strata and rock permeability to gas tend to increase, as well. For that reasons mining activities are constrained, which affects the selection of the mining methods. Those conditions are responsible for the occurrence and intensification of natural hazards typically encountered in mines: methane emissions, fires, climatic conditions, caving-in, rock bursts and bumps and rock and gas outbursts.
- Economic factors in favour of the high concentration of coal production do not help towards limiting the methane emissions, coal dust risk, adverse climatic conditions and coal bumps while the sub-level mining operations on the large scale further enhance the methane risks, making the potential impacts of natural hazards still more gruesome.
- In recent years these hazards have intensified, particularly the methane emissions, fires, climatic impacts, coal bumps and rock and gas outbursts. Besides, they now often occur jointly, particularly methane emissions, coal dust explosions and rock bursts. Despite production cutbacks, there is a growing number of dangerous accidents caused by natural hazards, particularly endogenous fires.
- Nearly half the fatalities (52.2%) reported in the coal mining sector in 2000-2009 were caused by natural hazards, methane emissions were responsible for the largest number of accidents. The analysis of accident and fatalities rate suggests that the number of accidents caused by natural hazards tends to increase, both in relation to the production level and to the number of employees.

The high level of natural risks, growing tendency of their occurrence and the high number of fatalities they cause shall encourage the improvement of the existing preventive and control measures and prompt the development of new monitoring techniques and systems.

While planning the future of the coal mining sector in Poland, one has to bear in mind that (Zorychta, Burtan 2008) "*the modern mining sector should guarantee high efficiency, low production costs and, first of all, high safety features*".

## REFERENCES

Burtan Z., Zorychta A., Chlebowski D. 2008. Trends in the development of preventive measures against natural hazard in Polish collieries. 21st Word Mining Congress. Session "Underground Mine Environment". Kraków-Katowice-Sosnowiec

Konopko W. 2010. O zagrożeniach i bezpieczeństwie pracy w kopalniach węgla kamiennego w GZW. Szkoła Eksploatacji Podziemnej. Wydawnictwo IGSMiE PAN. Kraków. (In Polish)

Wyższy Urząd Górniczy. 2010a. Stan bezpieczeństwa i higieny pracy w górnictwie w 2009 roku. Katowice. (In Polish)

Wyższy Urząd Górniczy. 2010b. Zagrożenia naturalne i górnicze w podziemnych zakładach górniczych. Materiały Departamentu Górnictwa. Katowice (unpublished). (In Polish)

Zorychta A., Burtan Z. 2008. Uwarunkowania i kierunki rozwoju technologii podziemnej eksploatacji złóż w polskim górnictwie węgla kamiennego. Gospodarka Surowcami Mineralnymi. Tom 24, Zeszyt 1/2. Kraków. (In Polish)

*International Mining Forum 2010, Liu et al. (eds) © 2010 Taylor & Francis Group, London, UK. ISBN 978-0-415-59896-5*

# Three-dimensional numerical simulation of spontaneous combustion in goaf of fully mechanized top-coal caving longwall

Weimin Cheng, Gang Wang, Jiacai Zhang, Lulu Sun
*University of Science and Technology, Qingdao, Shandong Province, China*

Gang Wang, Jun Xie
*Commonwealth Scientific and Industrial Research Organization, Brisbane, Australia*

ABSTRACT: In the fully-mechanized mining top-coal caving face spontaneous combustion accident, high temperature ignition sources are mostly away from the floor at a certain height over the spatial position, rather than the position of the flat floor currently considered. The spontaneous combustion in the three-dimensional goaf was studied. By using CFD software – Fluent, it establishes the spontaneous combustion three-dimensional mathematical model of the loosen coal in the goaf, and simulates the spontaneous combustion process in the model. The simulation results show that along the height direction of goaf, from the floor up, the oxygen concentration decreases progressively. The distribution of low oxygen concentration is just like an "O" district, while the distribution of high is just like a "U" zone. In the rear of the fully-mechanized mining top-coal caving face goaf, the spontaneous combustion not only exists along the strike direction but also in the vertical direction. The intersecting area in both directions constitutes the actual spatial spontaneous combustion "three zones" of the goaf.

KEYWORDS: Goaf, three-dimensional space, spontaneous combustion, numerical simulation

## 1. INTRODUCTION

By using the fully-mechanized mining top-coal caving technique, mining intensity increased continuously and coal spontaneous combustion risk increased significantly (Xian et al. 2001). At present, the research on regularity of spontaneous combustion of coal at home and abroad mainly concentrated on flow field simulation of goaf rear and the demarcation of goaf 2-d "three zones". Research on regularity of spontaneous combustion of coal goaf 3d space is relatively weak (Cui 2001). In the process of conduct coal goaf spontaneous combustion accident, most of the high temperature burning point area or high temperature zone must be located in certain height space apart from floor. This phenomenon cannot be reasonably explained by "2-d". To make certain of the development and changes of coal goaf spontaneous combustion and the boundary of spontaneous combustion dangerous areas, it must be studied from the "3-d" space to provide theoretical basis for full-mechanized top-coal caving mining goaf spontaneous combustion control (Li et al. 2004).

## 2. THREE-DIMENSIONAL MATHEMATICAL MODEL OF LOOSEN COAL MASS SPONTANEOUS COMBUSTION IN GOAF

Regard coal goaf air leak as incompressible gas seepage in porous media, ignore the radiation heat transfer, thermal expansion, soret and defour effect, disregard the water evaporation and the latent heat effect of gas desorb, and ignore the mechanical dispersion caused by leaking airflow pulsation. The goaf spontaneous combustion mathematical model can be established as formula (1) (Xie 2008):

$$\begin{cases} \frac{\partial}{\partial x}(K_x \frac{\partial p}{\partial x}) + \frac{\partial}{\partial y}(K_y \frac{\partial p}{\partial y}) + \frac{\partial}{\partial z}(K_z \frac{\partial p}{\partial z}) = 0 \\ \frac{\partial c}{\partial t} + u_x \frac{\partial c}{\partial x} + u_y \frac{\partial c}{\partial y} + u_z \frac{\partial c}{\partial z} = D_x \frac{\partial^2 c}{\partial x^2} + D_y \frac{\partial^2 c}{\partial y^2} + D_z \frac{\partial^2 c}{\partial z^2} + S_\Omega \\ \rho \frac{\partial}{\partial t}(gradu_i) + \rho div[u_j(gradu_i)] = \mu div(gradu_i) - divp + \rho f_i \\ \rho_e c_e \frac{\partial T}{\partial \tau} = \lambda_e (\frac{\partial^2 T}{\partial x^2} + \frac{\partial^2 T}{\partial y^2} + \frac{\partial^2 T}{\partial z^2}) - \rho_g c_g (u_x \frac{\partial T}{\partial x} + u_y \frac{\partial T}{\partial y} + u_z \frac{\partial T}{\partial z}) + q(T) \end{cases} \tag{1}$$

where: $K_x$, $K_y$, $K_z$, are osmotic coefficient, $p$ is air leakage pressure, Pa; $c$ is $O_2$ volume concentration, %; $u_i = K_i \frac{\partial p}{\partial x}$, $i$ is denoted by $x$, $y$, $z$; $u_x$, $u_y$, $u_z$ are $x$, $y$, $z$ direction speed of leaking airflow, m/s; $grad() = \partial()/\partial x + \partial()/\partial y + \partial()/\partial z$ ; $D_x$, $D_y$, $D_z$ are diffusivity of $O_2$ in loosen coal along three dimensional, m$^2$/s, generally recognized $D_x$+$D_y$=$D_z$; $S_\Omega$ is consumption of oxygen; suffix $g$, $e$ is gas and coal respectively; $\lambda_e$, $c_e$, $\rho_e$ is equivalent thermal conductivity, equivalent specific heat and equivalent density of test coal related to porosity $n$ respectively; $T$ is absolute scale, $K$; $f_i$ is cohesive force component; $q(t)$ is source item of coal oxidation.

Mainly reason of loosen coal spontaneous combustion is that the oxidation of remained coal generate heat. In loosen coal, transfer process of the heat and quality is very complex. According to energy conservation principle, the mathematical model to describe the temperature field, oxygen concentration and leaking airfield is established. Though it simplifies many secondary factors, the solution of temperature and oxygen concentration distribution of loose coal by three-dimensional differential equation is still very difficult. In the actual conditions of fully mechanized caving face, loosen coal spontaneous combustion model can usually be solved by combining with the finite difference method. Taking Xing Longzhuang Colliery 1307 fully mechanized caving face for example to analysis.

## 3. RESEARCH ON MATHEMATICAL MODEL PARAMETERS BASED ON FLUENT

The model parameters is determined according to the experiment data and the distribution of underground pressure, porosity distribution, wind pressure distributions and residual coal characteristic of 1307 fully mechanized caving face of Xing Longzhuang Colliery while exploiting No.3 coal seam.

### 3.1. *Oxygen diffusion coefficient of loosen coal mass in goaf*

In known that the oxygen diffusion coefficient is $D_0$ when the temperature is $T_0$, and the pressure is $P_0$, then oxygen diffusion coefficient of loosen coal in the goaf of Xing Longzhuang Colliery in No. 3 coal seam can be calculated as formula (2) (Wen 2003):

$$D = (-0.171 + 0.934\phi) D_0 \left(\frac{T}{T_0}\right)^{1.5} \frac{P_0}{P} \tag{2}$$

where: $\phi$ is the heat flux of micro unit leaded from external through interface.

### 3.2. *Permeability coefficient of goaf*

The permeability coefficient of goaf is relevant to the lithology of coal roof, the compressive strength, distribution of underground pressure in goaf and such factors. The numerical fitting of permeability coefficient of goaf is as formula (3):

$$K=\begin{cases}2.57\times10^{-3} & y\le 21\\ 1.605\times10^{-2}(0.0001y^2-0.0042y+0.405)^2 & 21<y\le150\\ 3.61\times10^{-4} & y>150\end{cases} \tag{3}$$

where: $y$ is the positive direction of $y$ axis, means coal seam strike.

### 3.3. *Oxygen consumption velocity of remained coal in goaf*

According to the test data of oxygen consumption velocity of No.3 coal seam in Xing Longzhuang Colliery, which was measured by large-scale coal oxidation equipment, the oxygen consumption velocity is as formula (4):

$$S_\Omega=V(T,C)=1.46\times10^{-3}\left[0.44754 - 0.711\ 53\ln(\frac{\mathrm{d}_{50}}{\mathrm{d}_{\mathrm{ref}}}+0.5)\right]\cdot\frac{C}{C_0}\cdot e^{\frac{8.81\times(T+273)}{300}} \tag{4}$$

where: $T$ is time; $C$ is oxygen concentration; $C_0$ is constant; *dref* is reference diameter (take 1 cm).

### 3.4. *Heat release intensity of remained coal in goaf*

According to the oxidation experiment of coal samples and the actual situation of 1307 working face, the heat release intensity of remained coal in goaf (Xu 2001, Deng et al. 1999, Wen et al. 2001) is as formula (5):

$$\begin{aligned}q(T)&=\psi(d_{50})\cdot\frac{C}{C_0}\cdot q_0(T)\\ &=1.68\times10^{-5}\left[0.44754 - 0.711\ 53\ln(\frac{\mathrm{d}_{50}}{\mathrm{d}_{\mathrm{ref}}}+0.5)\right]\cdot\frac{C}{C_0}\cdot e^{\frac{11.4\times(T+273)}{300}}\end{aligned} \tag{5}$$

where: $\psi(d_{50})$ is influence function of particle size of Yanzhou mining coal layer 3; $C_0$ is the oxygen concentration 21% in fresh wind; $C$ is measured oxygen concentration; $q_0(T)$ is calorific value intensity in fresh wind.

### 3.5. *Boundary conditions*

Definite conditions: $\frac{dc}{dt}=0,\frac{du}{dt}=0$; initial conditions:
- oxygen concentration in working face:
  $C|_{t=0}=C_0=0.231$;
- airflow temperature:
  $T|_{t=0}=T_0$;
- air pressure in working face:
  $p = 105504.3 - 0.1459x$.

## 4. 3-D NUMERICAL SIMULATION OF SPONTANEOUS COMBUSTION IN GOAF OF FULLY MECHANIZED TOP-COAL CAVING LONGWALL

### 4.1. *The outline of 1307 fully mechanized top-coal caving face and goaf*

The air intake of 1307 fully mechanized top-coal caving working face of Xing Longzhuang Colliery of Yanzhou Mining Group is 875 $m^3$/min. The strike length (depth) of goaf is 400 m, and inclination length (face length) is 210 m. The thickness of coal seam is 9.27 m, cutting/ /drawing ratio is 1:2.09, and top coal recovery ratio is 70%. The height of caving zone is about 16.7 m according to the site observation. The coal mining face is integrated coal face, around which there was no stoping goaf. The angle of coal seam is 5–7°. In the coal goaf, the remained coal distribution is that thickness is small in the middle and large near the cross-heading.

### 4.2. *Geometric model of 1307 fully mechanized top-coal caving face goaf*

Geometric model of 1307 fully mechanized top-coal caving face goaf is three-dimensional solid. Its size is 400 m in length, 210 m in width, and 16.7 m in height. It is shown in Figure 1. On the geometric model, use hexahedral grid to mesh. Grid size is 1.5 m in x and y direction, 0.15 m in *z* direction of remained coal area and 0.5 m in *z* direction of rock formation. As shown in Figure 2.

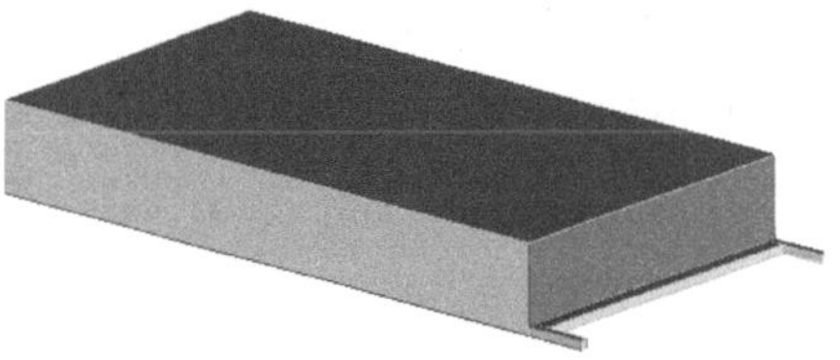

Figure 1. Numerical simulation geometric model of 1307 face gob

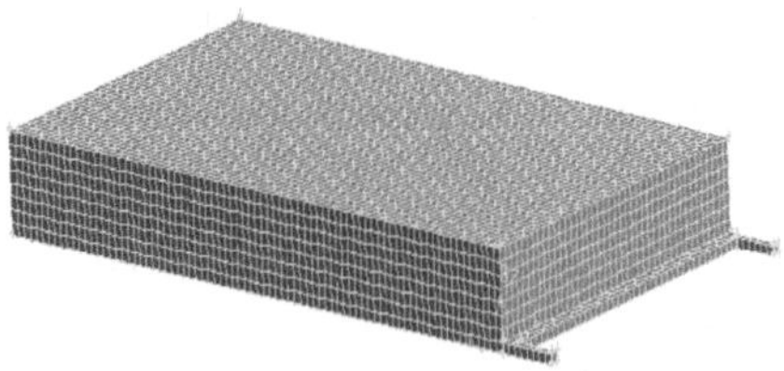

Figure 2. The mesh partition schema of geometric model of 1307 gob

### 4.3. *Simulation results and analysis*

The main purpose of numerical simulation for 1307 fully mechanized top-coal caving face is to determine the oxygen distribution within the scope of the goaf space by simulation calculation, and analyse the spontaneous combustion of goaf three-dimensional space according to the dynamic distribution of oxygen.

In the simulation, key factors which influence spontaneous combustion such as air quantity, the length, the height, state of rock falling, the thickness of the remained coal goaf and so on had been comprehensively considered. It reflects the environment conditions of the goaf coal spontaneous combustion. The simulation results were shown in Figures 3–10 (Fluent 2001).

Figures 3–6 show horizontal profile along height direction of 1307 fully mechanized top-coal caving face goaf. As shown in the Figure: along the *z* direction of the goaf, from the floor up, the oxygen concentration decreases progressively.

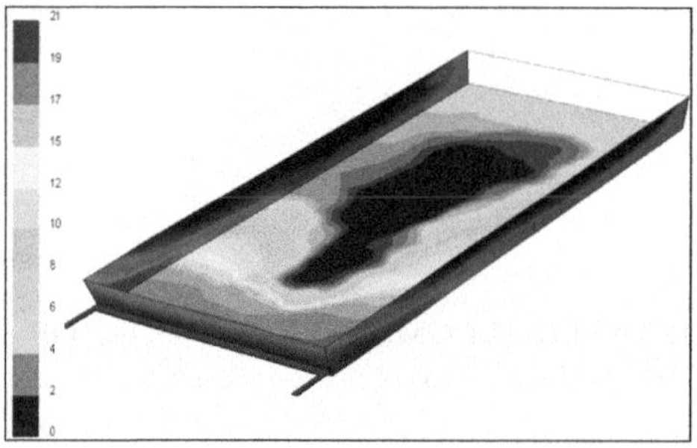


Figure 3. Numerical simulation results of oxygen distribution in 1307 gob (upper floor 0.5 m)

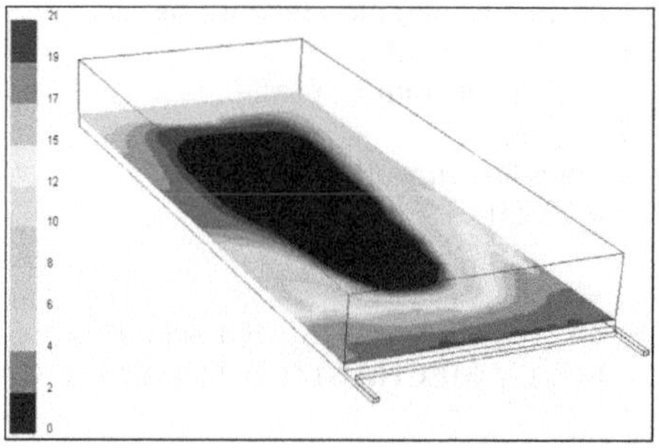


Figure 4. Numerical simulation results of oxygen distribution in 1307 gob (upper floor 2 m)

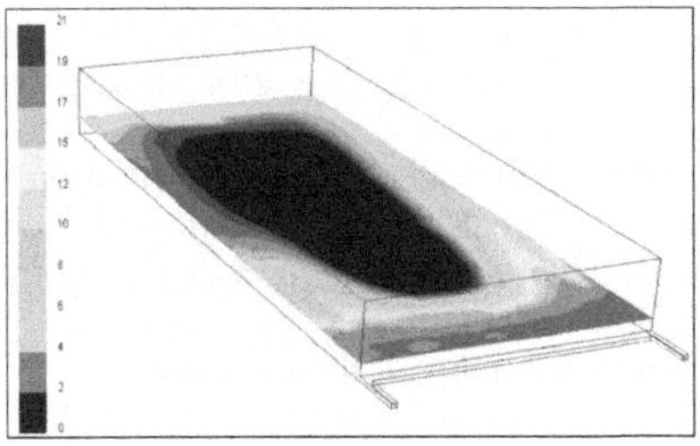

Figure 5. Numerical simulation results of oxygen distribution in 1307 gob (upper floor is 3 m)

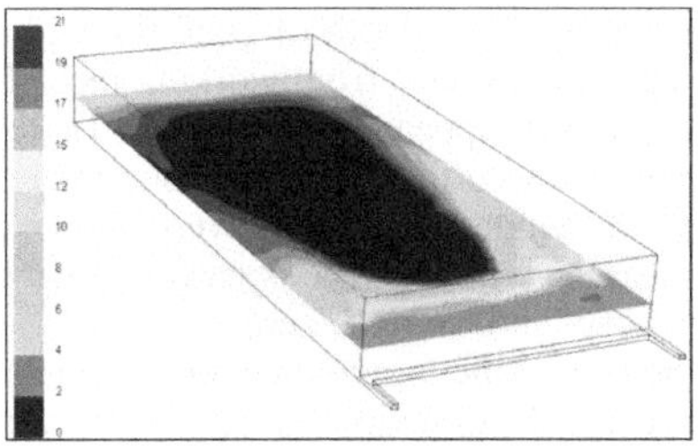

Figure 6. Numerical simulation results of oxygen distribution in 1307 gob (upper floor is 4 m)

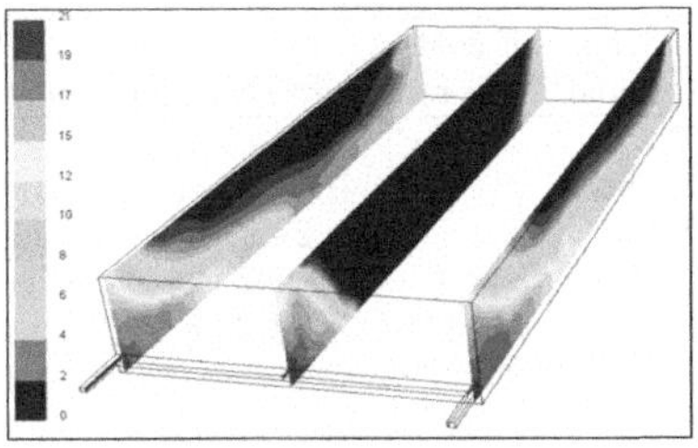

Figure 7. Vertical profile of oxygen distribution in 1307 gob along incline (towards intake 0,105, 210 m)

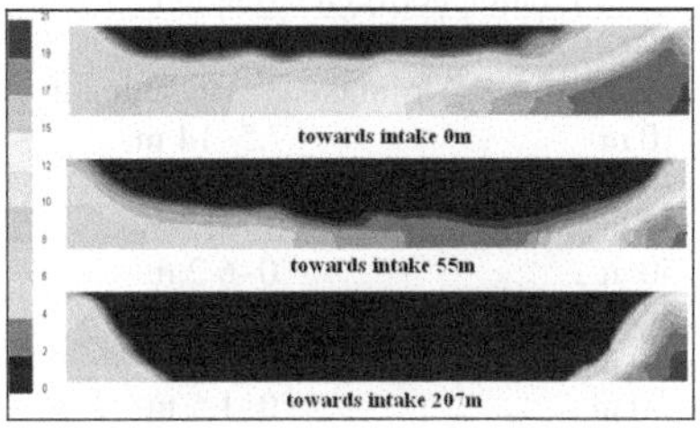


Figure 8. The vertical profile of oxygen distribution in 1307 gob along incline

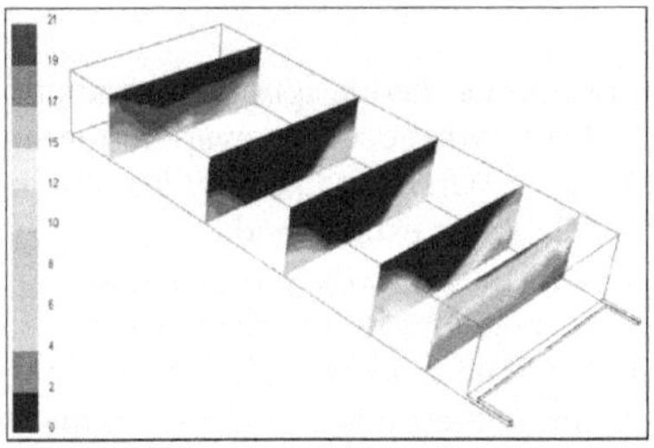

Figure 9. The vertical profile of oxygen distribution in 1307 gob along longwall

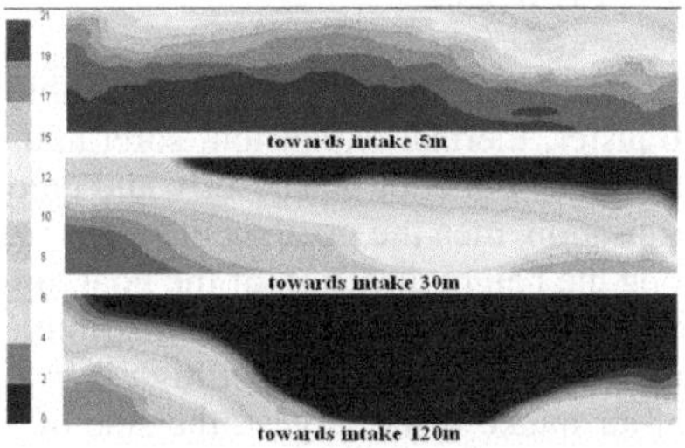


Figure 10. The vertical profile of oxygen distribution in 1307 gob along longwall

The distribution of low oxygen concentration is just like an "O" district, while the distribution of high is just like a "U" zone. The oxidation in the side of air intake is the largest, the side of air-reture takes the second place, and the width of middle of working face is the narrowest. The oxidation zone range on the different height is shown in Table 1.

Table 1. The relation between oxidation zone range and space of 1307 gob (in vertical direction)

| Height distance from floor | the oxidation zone range of goaf | | |
|---|---|---|---|
| | Air intake crossheading in goaf | Middle of goaf | Air return crossheading in goaf |
| 0.5 m | 33.5~126.3 m | 14.5~55.2 m | 26.5~83.3 m |
| 1 m | 36.5~165.6 m | 17.6~68.5 m | 28.6~95.7 m |
| 2 m | 30.3~113.6 m | 16.2~63.8 m | 24.1~75.2 m |
| 3 m | 26.2~72.7 m | 13.2~40.5 m | 19.5~50.6 m |
| 4 m | 16.7~50.6 m | 11.8~33.8 m | 15.3~29.7 m |

As were shown in Figures 7–8, along the working face sloping direction, from air intake side to air-reture side, the distribution of the oxidation zone change from wide to narrow then to wide.

As were shown in Figures 9–10, the nearer to the working face, the higher the oxygen concentration is. The distribution of oxygen concentration at different height in coal goaf is that the closer with working face and floor, the higher the oxygen concentration is in goaf space. With the increment of the depth of goaf, the width of the oxidation zone becomes narrow gradually. Until a certain distance from working face, there is an asphyxia zone. The oxidation zone at vertical height is disappeared as shown in Table 2.

Table 2 showed the oxidation zone range is divided by oxygen concentration distribution in the vertical height of goaf. As were shown in Table, in the vertical height, the oxidation zone range near the working face is narrow, then gradually turns wide, and to a certain distance turns narrow again until turns to zero.

In the rear of the top-coal caving face goaf, the spontaneous combustion "three zones" is not only in the strike direction but also in the vertical direction. The intersection areas in both directions constitute the actual spontaneous combustion three-dimensional of the goaf.

Table 2. The relation between oxidation zone range and space of 1307 gob (in strike direction)

| Distance from working face /m | The oxidation zone range of goaf | | |
|---|---|---|---|
| | Air intake crossheading in goaf | Middle of goaf | Air return crossheading in goaf |
| 10 m | 7.5~14 m | 6.3~12.8 m | 4.2~8.3 m |
| 20 m | 2.9~10.3 m | 2.1~8.6 m | 1.2~5.7 m |
| 30 m | 1.9~7.5 m | 1.9~5.3 m | 1.2~5.7 m |
| 50 m | 0~6.2 m | 0~2.1 m | 0~3.2 m |
| 100 m | 0~2.5 m | - | 0~1.2 m |
| 120 m | 0~2 m | - | - |
| 150 m | 0~1.5 m | - | - |
| 160 m | 0~1.1 m | - | - |

## 5. CONCLUSIONS

Regard coal goaf air leak as incompressible gas seepage in porous media, ignore the radiation heat transfer, thermal expansion, soret and defour effect, disregard the water evaporation and the latent heat effect of gas desorb, and ignore the mechanical dispersion caused by leaking airflow pulsation, then the goaf spontaneous mathematical model can be established.

Along the height direction of the goaf, up to the floor in working face, the oxygen concentration decreases progressively. The distribution of low oxygen concentration is just like an "O" district, while the distribution of high is just like a "U" district. The oxidation zone range in the side of air intake is the largest, the side of air return takes the second place, and the middle of working face is the narrowest.

In the height of fully mechanized caving face, there exists the spontaneous combustion three-dimensional. The nearer working face, the wider the oxidation zone is. With the increase of depth of goaf, the width of the oxidation zone gradually turns narrow. Until a certain distance from working face, there is an asphyxia zone, and then the oxidation zone at vertical height is disappeared.

## REFERENCES:


Cui Hong-yi, Wang Zheng-ping, Wang Hong-quan. 2001. Technology and application of early stage prediction for natural ignition in coal seam. Coal Mine Safety (12). (In Chinese)

Deng Jun, Xu Jing-cai, Li Li, Zhang Xing-hai, Wen Hu. 1999. Investigation on the Relation of the Rate of Oxygen Consumption and the Size of Coal Sample. Journal of Xi'an Jiaotong University 33 (12): 106-108. (In Chinese)

Fluent Inc. 2003. FLUENT User Defined Function Manual. Fluent Inc.

Fluent Inc. 2003. FLUENT User's Guide. Fluent Inc.

Li Shu-gang. 1996. Analysis of Characteristics on Rock coefficient of bulk increase of LTCC Goaf. Shangxi Coal Technology (4): 19-21. (In Chinese)

Li Zong-xiang, Wang Ji-ren, Zhou Xi-hua. 2004. Numerieal Simulation of Gas Drainage During Open Region Movement in Goaf. Journal of China University of Mining & Technology 33(1): 74-78. (In Chinese)

Wen, Hu, Xu, Jing-cai, Xue, Han-ling. 2001. Effect Factor Analysis of Heat Liberation Effect of Coal Self-Ignite Oxidation. Coal Conversion 24(4): 60-62. (In Chinese)

Wen, Hu. 2003. Study on Experimental and Numerical Simulation of Coal Self-ignition Process. Xi'an University of Science & Technology. (In Chinese)
Xian, Xue-fu, Wang, Hong-tu, Jiang, De-yi, Liu, Bao-xian. 2001. The Summarization of the Investigation on Coal Mine Fire Prevention & Fire Extinguishing Techniques in China. Engineering Science 3(12): 28-29. (In Chinese)
Xie, Jun. 2008. Study on Rules of Spontaneous Combustion in Three-Dimensional Space of LTCC Goafs. Shandong Qingdao: Shandong University of Science and Technology. (In Chinese)
Xu, Jing-cai. 2001. Theory of Decision on spontaneous combustion Risk Regional in coal seam. Coal Industry Press: 176-178. (In Chinese)

*International Mining Forum 2010, Liu et al. (eds) © 2010 Taylor & Francis Group, London, UK. ISBN 978-0-415-59896-5*

# The stability of soft rock roadway supporting and the analysis of fractal feature of surrounding rock crack

Xiangdong Zhang, Mingxing Zhang, Jianlai Xu
*Liaoning Technical University, Fuxin, Liaoning Province, China*

ABSTRACT: The rock burst is one of the major problems in deep mining. How to predict the rock burst effectively and to reduce the disaster caused by the rock burst, has great significance. The rock burst is affected by many complex factors, so the forecast of rock burst intensity is a nonlinear, high dimensional, multiclass pattern recognition problem with small samples. A support vector machine is suitable for solving this pattern recognition problem. In this paper, the principle of support vector machine was introduced, the main influence factors of rock burst were given, and a new forecast method of rock burst intensity based on one-against-one SVM classification was presented, through learning the small training samples collected from a mine, the complicated nonlinear mapping relationship between degree of rock burst and its affected factors was established by the proposed method. The case study shows that the method is feasible, easy to be implemented. So the proposed method is very attractive for a wide application in forecasting rock burst.

KEYWORDS: Soft rock roadway, supporting, fractal, loose circle, stability

## 1. ENGINEERING SITUATION

Soft rock is widely distributed in china, many mines have met the problem of soft rock in the development of coal resources (Lin 1999), it not only brings great economic losses but also effects the construction speed, threat the softy of constructers, it urgent to solve the problem of soft rock supporting.

The mining area of GaoJiaLiang is located in the southeast that is 8 km far from the Ordos city, it is a new developed mining area, which is typical geological mine field, the surrounding rock has low strength, joint fissure is very well developed and swelling deformation is very large, supporting is quite difficult. Coal mine formation from the old to the new development are: Yanchang formation of upper Triassic (T3y), yan' an formation of early-middle Jurassic (J1-2y), middle Yurassic (J2), zhidan group of early Cretaceous strata (K1zh), upper, Tertiary (N2) and Quaternary(Q). The whole geological structure is monoclinic structure of near level of occurrence that tilt to the southwest, the formation dip is less than 5°.

## 2. SOFT ROCK ROADWAY SUPPORT STRUCTURE DESIGN

(1) Determination of the minimum width of main haulage roadway.

Comprehensively considered with the factors of the cutting of top coal and bottom coal of the upper and lower trigonum, make sure that the wide of haulage drifts is not less than 5200 mm, arched radius is 2600 mm, the height of vertical is 1600 mm, the total height is 4200 mm.

(2) Determination of the parameters of bolt support.

Bolt used resin bolt, rod body used thread steel with $\varphi$22 mm. The anchoring force of this bolt is great, which is more than 80 KN on average. This bolt has high strength, small aperture and convenient construction.

- The length of bolt

According to suspension theory and avoid pressure arch theory (Bai, Du 2001), determine the length of bolt is 2.6 m.

- The distance between the bolts

According to the stability calculation of rock mass in anchor spacing (Hou, Gou 2001), the distance between the bolts take 800 mm, bolt uses quincunx arrangement mode.

(3) The selection of anchor agent.

Using the resin medicine-one, diameter is $\varphi$23 mm (bore bit with $\varphi$25 mm) the length is 35 cm, with two types of sub-fast (CK type) and fast (k type).Place the sub-fast and fast anchor agent for volume one in each drilling.Make sure that the anchoring force is more than 80 KN.

(4) The ratio of shotcrete and the determination of the thickness of shotcrete layer.

The determination of the thickness of shotcrete layer: the thickness of shotcrete layer is estimated by the reinforcing function of shotcrete rockbolt mesh, consider the possibility of construction and reduce the resilience amount of injection process,the thickness of one injection is not less than twice the maximum diameter of aggregate materials, which is 20~40 mm. So use the thickness of shotcrete layer d=150 mm (the thickness of primary jetting is 50 mm, the thickness of re-jetting is 100 mm).

(5) Metal net, steel strip and pallet.

Using welding of round steel with $\varphi$8 mm, whose mesh size is 100 mm×100 mm, each mesh size is 3100 mm×1600 mm. Using the anchor plate with cast iron tablets of $\varphi$150 mm. Using round steel strip, which is welded together by round steel with $\varphi$12 mm, width is 100 mm.

## 3. MONITORING DATA ANALYSIS OF SOFT ROCK ROADWAY SUPPORT

That the deformation monitoring of soft rock roadway wall rock is one of important link for success, field monitoring can provide basic data for the optimization of support parameter and engineering quality management(Wang, Hou 2000), make sure the time secondary support reasonably. So monitoring work is an essential component of soft rock roadway support.

### 3.1. *Test point arrangement*

According to Preliminary filed investigation results, as the five roadway in which the thickness of the sandstone rock is different, particle size is becoming thicken from top to bottom, there is fine sandstone in the top and gritstone in the bottom. The main components of particles are soil particles, quartz, mica and a small amount of debris. Huge bedding-thickness joint fractures are all not development. Cementation is very loose, what leads to weather when it encounter air, soften rapidly when encounter water, then slime. This layer is pervious aquifer, which has little water, but it has water leaking on rock face. As the effects of water and cementation loose and other comprehensive reasons, this rock soften, slime rapidly, gangue is become the appearance of "tofu pulp", it is typical non-swelling soft rock. To monitor the sections whose condition of surrounding is poor and whose seepage is large in the five roadways, squat of the roof and convergence value.

### 3.2. *Analysis of results of typical convergent section*

According to the monitoring results,select monitoring results of the typical 1# and 2# sections among the monitoring section to analysis.

(1) The results of monitoring section 1#.

Section 1# is the crowned section which has deformation monitored when digged. Roof is mudstone and sandstone, section is crowned section, it appears tatter top and rib spalling in part.

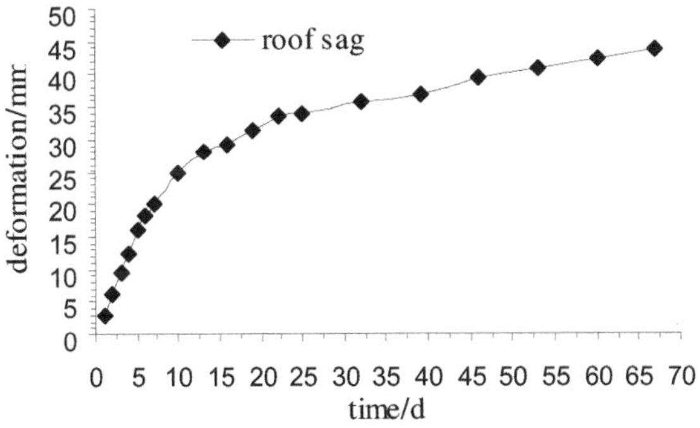


Figure 1. 1# sections of roof subsidence curve with time

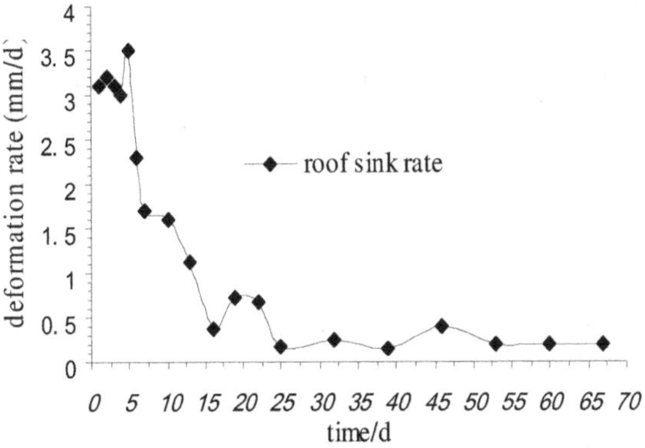


Figure 2. 1# sections of roof falling rate of curve with time

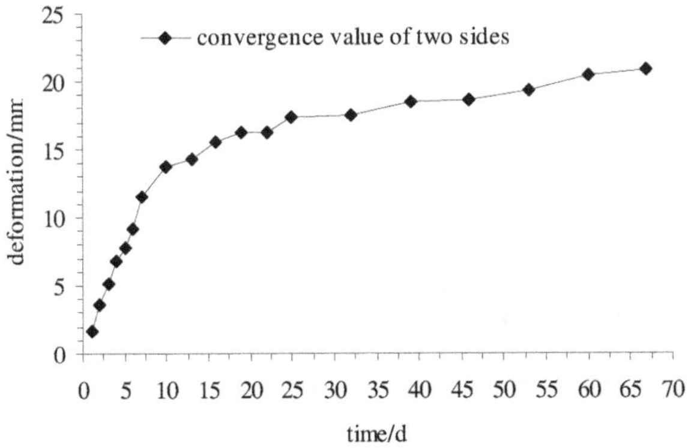


Figure 3. 1# sections of side convergence curve with time

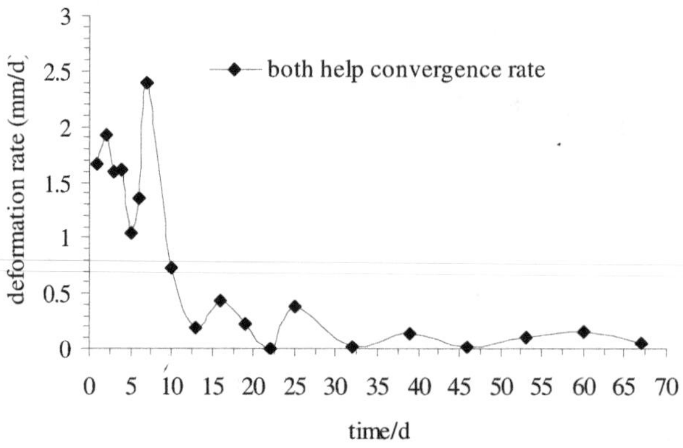


Figure 4. 1 # sections of side convergence curve with time

From above pictures the largest squat of monitoring section1# is 43.7 mm, the largest rate of settling is 3.50 mm/d, the average rate of settling is 1.37 mm/d; the largest relative convergence value is 20.85 mm, the convergence rate is no more than 2.40 mm/d at last, respectively as shown in Figure 1, Figure 2, Figure 3 and Figure 4.

(2) The results of monitoring section 2#.

Section 2# is rectangular section which has deformation monitored after excavation and support 45d.

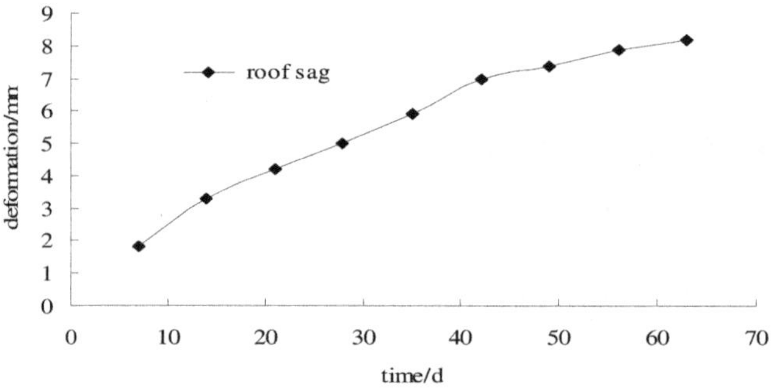


Figure 5. 2# sections of roof subsidence curve with time

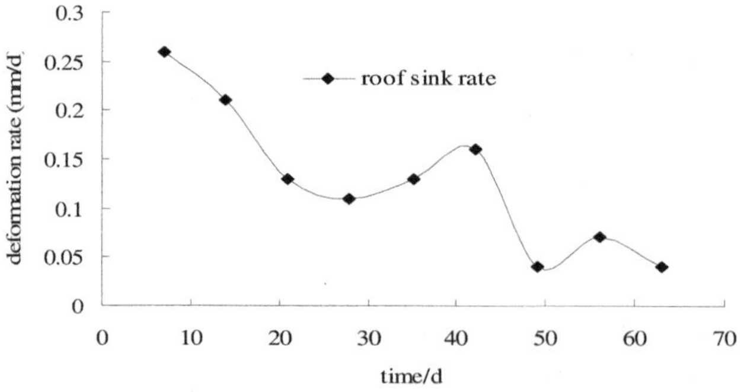


Figure 6 .2 # sections of roof falling rate of curve

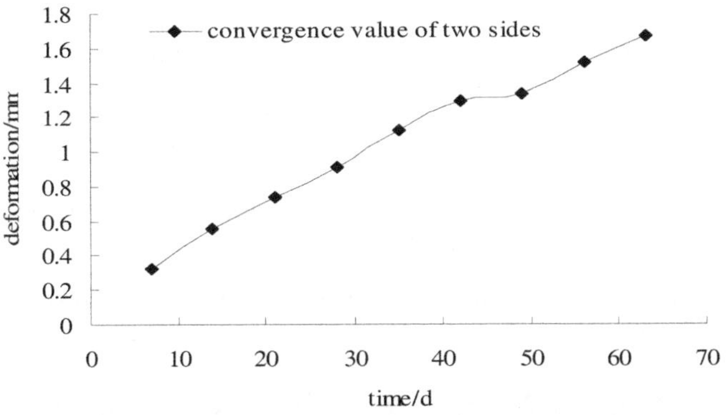


Figure 7. 2# sections of roof subsidence curve with time

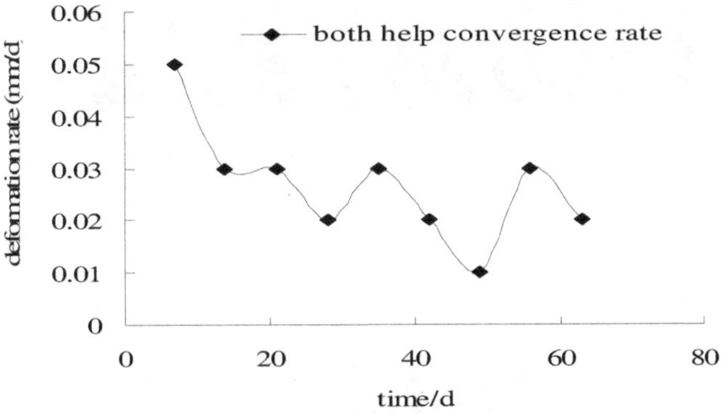


Figure 8. 2 # sections of side convergence rate of curve

From above pictures the largest squat of monitoring section 2# is 8.2 mm, the largest rate of settling is 0.26 mm/d, the average rate of settling is 0.18 mm/d, the largest relative convergence value is 1.67 mm, the convergence rate is no more than 0.05 mm/d at last, respectively as shown in Figure 5, Figure 6, Figure 7 and Figure 8.

The actual measurement of surrounding rock deformation shows that surrounding rock deformation has the characteristics with large deformation, large initial deformation rate, deformation of long duration, roadways stability is poor (He et al. 2002).

Global stability of roadway section has been strengthen obviously after the roadway section was changed from rectangular to circular arch, the squat of the section of which surrounding rock has serious crushing and serious seepage is larger than other monitoring. Roadway excavation is about 10d compared with stationary phase; that in the initial period of tunnel excavation, roadway deformation is rapid, deformation is large (Liu et al. 2004). Squat of roof at around 45-50d can reach stability, walls of tunnel at around 35-40d can reach stability, convergence of walls of tunnel is lower than roof, the stability of walls of tunnel is earlier than roof (Wang, Hou 2003).

The largest squat of roof is 43.7 mm, the largest level convergence value is 20.85 mm, the level convergence value is lower than the squat of roof. Roadway deformation rate has trend to decline after support 45d, but the formation is sustainable, and has long duration.

## 4. THE FRACTAL FEATURE AND THE ANALYSIS OF SOFT ROCK ROADWAY SURROUNDING ROCK SYSTEM

The research results of Xie Heping show that the more crushing the rock is, the larger the fractal value is (He et al. 2005). Through the studies on the microscopic pore of soft rock roadway surrounding rock, the fractal feature of particle size distribution and the relationship of rock failure of GaoJiaLiang in ordos city, using fractal geometry theory to calculat the fractal dimension of surrounding rock materials, quantitative characterize the pore characteristics of surrounding rock materials, response to the complexity of the composition of rock material, the severity of structural damage and the trend of fracture system, determine the damage zones of surrounding rock , to provide a reliable theoretical basis for mining area (Bai, Hou, 2006).

Drilling mining trough a 7 m deep hole by bits of $\varphi$32 mm in mining area 20307, then make the sample of image acquisition by drilling peep instrument, take one sample point per 0.5 m to analyze, after the filter treatment for the pattern through Photoshop "Poster Edges", extract the meso-crack.

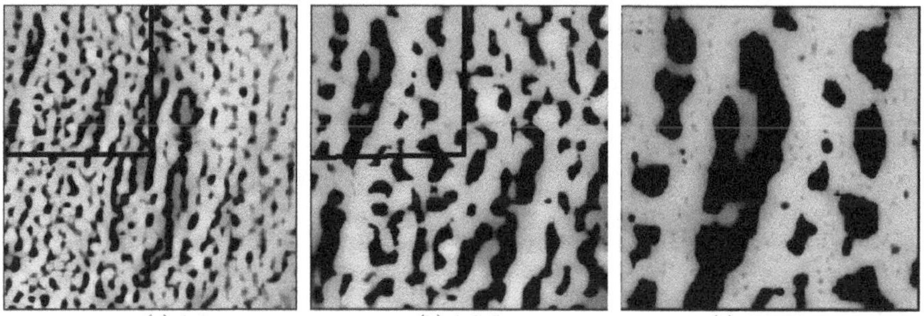

(a) 1:1 (b) 1:0.5 (c) 1:0.25

Figure 9. 0.5 m deep hole depth distribution of rock fracture characteristics of self-similarity

Figure 9 is the macro-comparison chart of the meso-crack after it was fractionated gain at the hole depth of 0.5 m, it is easy to see that the crack distribution of surrounding rock has self-similar characteristics, and the fractal feature is good. Figure 10 is the grid number (N) of the cocer crack when the hole depths(r) are different.

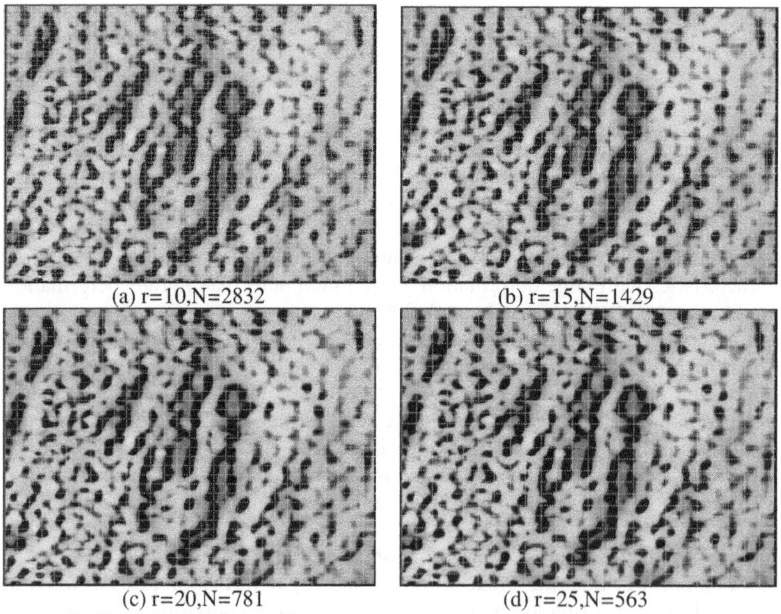

(a) r=10,N=2832 (b) r=15,N=1429

(c) r=20,N=781 (d) r=25,N=563

Figure 10. Hole depth 0.5 m trellis

Take the hole deep of 0.5 m for example, the relationship between r and N can be obtained, such as Figure 11 and Figure 12 show.

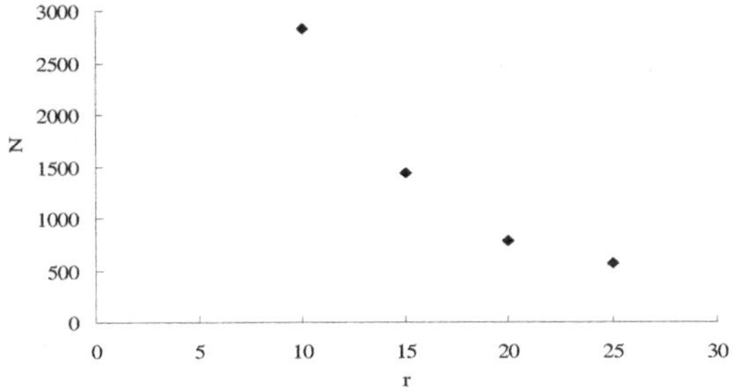


Figure 11. Plot of r and N

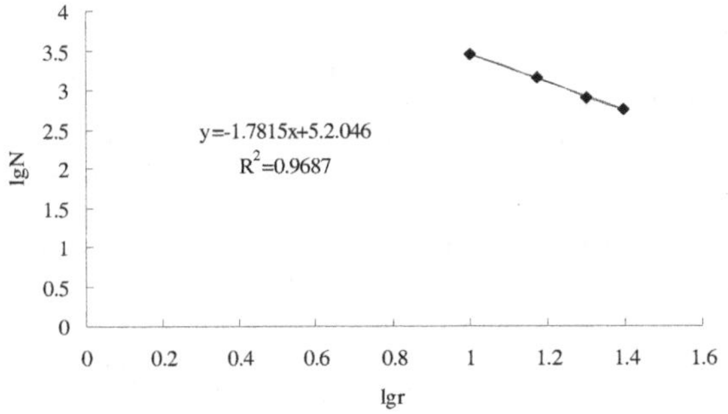


Figure 12. Logarithmic curve of r and N

Figure 12 shows that lg N and lg r showed a significant linear relationship, it shows rock meso-crack has self-similarity, its self-similarity is 1.7815, and similarity coefficient is 0.9687. Take the field analysis of the drillings for soft rock roadway surrounding rock by drilling peep instruments, to say that there are good fractal properties in distribution characteristics of surrounding rock, the fractal value can be the important parameters what evaluate the fragmentation degrees of roadway surrounding rock, stability and guiding construction (Hurt 1994, Yang et al. 1996).

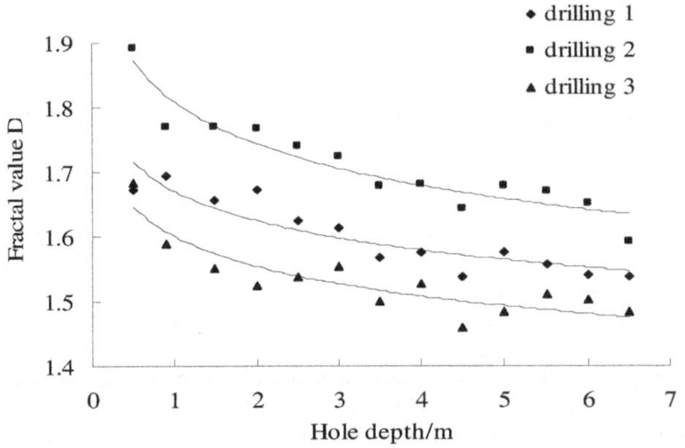


Figure 13. Borehole at different depths of the fractal dimension fitting curve

Figure 13 is the fitting curve of the surrounding rock fracture distribution, it is easy to see that fractal value is changing while the hole depth is changed, we can find the regularity from it, determine the size of the broken zone, that is the thickness range of loose circle.from the Figure, it can be found that the thickness range of loose circle is between 1.4~2.0 m, this thickness can provide support parameter for roadway support, it has great technical and economic benefits.

Figure 13 shows that the fractal values of roadway surrounding rock fracture in this mining area are distributed between 1.5~1.9, the coefficient of self-similar is upon 0.95, that is to say crack distribution of roadway surrounding rock has great self-similarity nature,what is typical fractal feature. Fractal theory has provided reliable theoretical basis for quantitative description and analysis of roadway surrounding rock.

## 5. CONCLUSIONS

1) Mining surrounding rock deformation has characteristics such as large deformation, large ini tial deformation rate, deformation of long duration, roadway stability is poor.
2) The whole stability is strengthened after the sections of roadway are changed from rectangular to.
3) The sectional subsidence whose rocks crush and seepage seriously.
4) The relative stable period of roadway is around 10d,that is the initial period of tunnel excavation, roadway deformation is rapid, deformation is large, roof sag can achieve stability when it is around 45~50d, two sides can achieve stability when it is around 35~40d, convergence value of two sides is lower than roof convergence, also the stability of two sides is earlier than roof.
5) The largest roof convergence is 43.7 mm, the largest convergence value is 20.85 mm, the level convergence value is lower than roof convergence.
6) It can avoid a lot of experimental analysis with the human operator error when take the field analysis of the drillings for soft rock roadway surrounding rock, response to the flawy development of roadway surrounding rock, analytical results show that its distribution features have great fractal properties.
7) The thickness range of loose circle is between 1.4~2.0 m, this thickness can provide sup-port parameters for roadway support, it has great technical and economic benefits (Croney 1997), also provide new attempt for detection of surrounding rock loose circle.

## ACKNOWLEDGEMENT

This work was supported by Liaoning Provincial Science and Technology Research Laboratory Projects under Grant No. 2008S114, the National Natural Science Foundation of China under Grant No. 50978131, and Higher School Talent in Liaoning under Grant No. 2008RC23.

## REFERENCES

Bai Jian-biao, Du Mu-min. 2001. On bolting support of roadway in extremely soft seam of coal mine with complex roof. Chinese Journal of Rock Mechanics and Engineering 20(1): 53-55. (In Chinese)

Bai Jian-biao, Hou Chao-jiong. 2006. Control Principle of Surrounding Rocks in Deep Roadway and Its Application. Journal of China University of Mining & Technology 35(2): 145-146. (In Chinese)

Croney P. 1997. Location of block release mechanisms in tunnels from geological data and the design of associated support. Computer Methods in Tunnel Design: 97-119. (In Chinese)

He Man-chao, Lu Xiao-jian, Jing Hai-he. 2002. Characters of surrounding respectfully rockmass in deep engineering and its non-Linear dynamic-mechanical design concept. Chinese Journal of Rock Mechanics and Engineering 24(8): 1215-1224. (In Chinese)

He Man-chao, Xie Heng-ping, Peng Suping. 2005. Study on rock mechanics in deep mining engineering. Chinese Journal of Rock Mechanics and Engineering 24(16): 2803-2814. (In Chinese)

Hou Chao-jiong, Gou Pan-feng. 2001. Mechanism study on strength enhancement for the rocks surrounding roadway supported by bolt. Chinese Journal of Rock Mechanics and Engineering 19(3): 342-345. (In Chinese)

Hurt K. 1994. New developments in rock bolting. Colliery Guardian 7: 133-143
Lin Yu-liang. 1999. The research on several theoretical problems about engineering mechanic soft rock. Chinese Journal of Rock Mechanics and Engineering 18(6): 690-693. (In Chinese)
Liu Quan-sheng, Zhang Hua, Lin Tao. 2004. Study on stability of deep rock roadways in coal mines and their support measures. Chinese Journal of Rock Mechanics and Engineering 23(21): 78-97. (In Chinese)
Wang Wei-jun, Hou Chao-jiong. 2000. Optimum of supporting parameter of soft rock roadway and engineering practice. Chinese Journal of Rock Mechanics and Engineering 19(5): 647-648. (In Chinese)
Wang Wei-jun, Hou Chao-jiong. 2003. Stability analysis of coal pillar and immediate bottom of extraction opening. Rock and Soil Mechanics 24(1): 75-76. (In Chinese)
Yang R., Bawden W. F., Katsabanis P. D. 1996. A New Constitutive Model for Blast Damage. International Journal of Rock Mechanics and Mining Science & Geomechanics Abstracts 33(3):245-254. (In Chinese)

International Mining Forum 2010, Liu et al. (eds) © 2010 Taylor & Francis Group, London, UK. ISBN 978-0-415-59896-5

# Study on pre-splitting blasting for gas drainage rate improvement in heading excavation

Hongtu Wang
*State and Local Joint Engineering Laboratory of Methane Drainage in Complex Coal Gas Seam, Chongqing University, Chongqing, China*

*Key Laboratory for Exploitation of China Southwestern Resources & Environmental Disaster Control Engineering, Ministry of Education, Chongqing University, Chongqing, China*

Xiaogang Fan, Zhigang Yuan, Houxue Xu
*Key Laboratory for Exploitation of China Southwestern Resources & Environmental Disaster Control Engineering, Ministry of Education, Chongqing University, Chongqing, China*

ABSTRACT: To solve the problem of low excavation speed and reduce the high risk of coal and gas outburst due to low gas draining rate in heading excavation, taking a mine excavation heading as an example, the test of deep-hole pre-splitting blasting was processed. The amount of gas draining after pre-splitting was compared to the direct draining, and the effective radius of draining was surveyed. The test result showed that the draining rate of K1 coal seam reached 37.13%, which reduced the draining time by 60% 10 days after pre-splitting, and the gas desorption index exceeded the standards until the fifth blasting. It took 12 days for the draining rate to reach 34.28%, which reduced the draining time by 60%, and the gas desorption index exceeded the standards when the third blasting processed. Obviously, the effect of gas draining was improved by pre-splitting blasting. The effective draining radius of K1 and K4 were 4.5 m and 5.2 m respectively after pre-splitting blasting.

KEYWORDS: coal mine, heading excavation; pre-splitting blasting; draining radius

## 1. GENERAL INSTRUCTIONS

Coal and gas outburst was an extremely violent dynamic effect of mines, which seriously affected the normal production and the safety of staff. It was one of the worst disasters in coal mine. However, must of the domestic coal and gas outburst mine coupled with the characteristic of low penetrability. When driving the coal roadway and draining the gas directly, the draining rate could not meet requirement in a shot time. So the risk of coal and gas outburst in heading face could not sufficiently reduce which affected the driving speed and succeed of ex-ploitation and driving. As a new technology of looseness and antireflection to the coal seam, pre-splitting blasting was widespread used in high gas mine. It could increase the length and breadth of the stratum several times in range around the blasting drill, and improve the penetrability for the coal seam in the influence range of pre-splitting blasting (Gong et al. 2008). So the gas draining level was increased, and the draining time was reduced (Liu et al. 2007). At the same time, the stress concentration ribbon was moved forward, and the risk of coal and gas outburst was weakened. Consequently, the driving speed and security of coal roadway was improved (Cai et al. 2007).

At the pre-splitting blasting, the disposal and explosive mass of blasting drills influenced the crack scope and the advance of gas drainage directly. Aim at the problem, the author carried out the test of pre-splitting blasting at the time of driving the coal roadway. The amount of gas drainage after pre-splitting was compared to the amount of direct draining, and the draining radius was surveyed for different blasting drills. The research had guidance significance for the field pre-splitting blasting, and provided reference and basis for other mine executing the pre-splitting blasting.

## 2. COAL SEAM AND GAS CONSERVATION CONDITION IN TEST REGION

There were two workable coal seam in test region of the coal mine executing pre-splitting blasting, which was $K_1$ and $K_4$. The thickness of $K_1$ was 1.03 ~ 5.86 m, and averaged 2.92 m which was all workable. The thickness of $K_4$ was 0.09 ~ 1.22 m, and averaged 0.61 m which was partially workable. The layer distance between $K_1$ and $K_4$ was 30.8 m. The inclination of coal measure stratum was 14 ~ 16°. The absolute gas emission rate of coal mine was 54.8 $m^3$/min, and the relative gas emission rate was 44.6 $m^3$/t. The primitive gas pressure of coal seam in test region was 2.75 MPa by field measuring, and the permeability coefficient of coal seam was $1.01\times10^{-4}$ ~ $4.96\times10^{-3}$ $m^2/MPa^2$.d, belonging to coal seam of harder gas drainage.

## 3. ARRANGEMENTS OF DRAINAGE AND BLASTING DRILLS

The gas drainage and blasting drills were arranged along the direction of coal roadway driving. Out of the contour line of the coal roadway, the controlling scope was 5 m in up-incline and 3 m in down-incline, and amount to 450 $m^2$. There were 14 drills to construction, which combined deep and shallow drills. The depth of drills was about 4.6~45.2 m, and the aperture was 75 mm. In order to increase the permeability of coal seam, one or two drills out of the 14 drills were chosen to be blasting drills in each round of driving. The disposal of drills was shown in Figure 1.

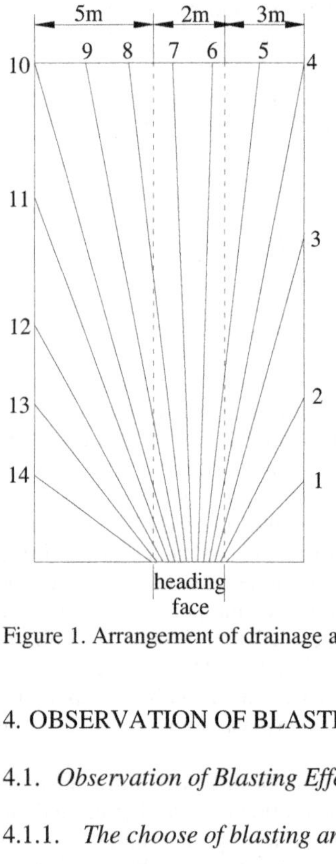


Figure 1. Arrangement of drainage and blasting boreholes

## 4. OBSERVATION OF BLASTING EFFECT

### 4.1. *Observation of Blasting Effect in K1 Coal Seam*

#### 4.1.1. *The choose of blasting and observation method*

In order to review the effect of pre-splitting blasting on gas drainage, as shown in Figure 1, one of the 14 drills was selected to be blasting hole. The other 13 were chosen as draining drills. The amount of gas drainage, the time of drainage and the cuttings desorption index after pre-splitting were compared to analysis. The blasting hole selected was the 7 # drill in contour of the front of coal roadway driving. The depth of the drill was 45 m. The length of charge was 30.4 m.

4.1.2. *Comparative Analysis of blasting effect before and after blasting*

After pre-splitting blasting in transport lane of $K_1$ coal seam, the amount of gas draining after blasting was compared with direct drainage, as shown in Figure 2.

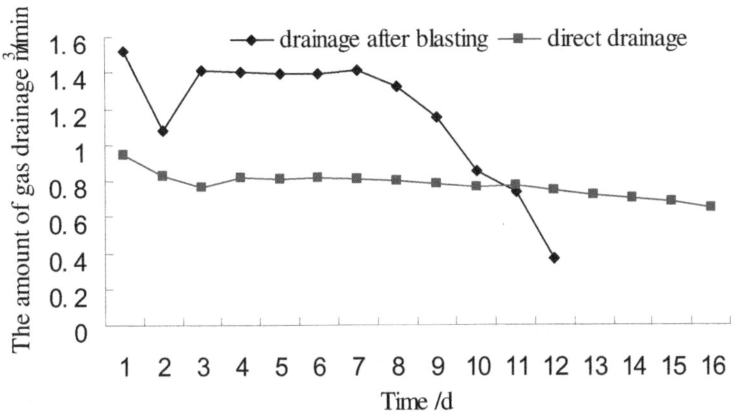


Figure 2. The amount of gas draining comparation of blasting and direction in $K_1$ coal seam

When driving without pre-splitting blasting but just pre-draining, it took 25 days for the draining rate to reach 30%. But after taking pre-splitting blasting, through 10 days draining, the gas rate decayed from beginning draining 42% to demolishing 5%. The instantaneous gas flow decayed from beginning draining 1.53 $m^3$/min to demolishing 0.33 $m^3$/min. The total amount of gas draining was 1.8312 Million $m^3$. The pre-draining rate got to 37.13%. The time of draining was cut down 60%. The draining rate was improved 7.13%.

4.1.3. *The cuttings desorption index contrastive analysis*

Before the driving of coal roadway, studied on cuttings desorption index, the results showed that without pre-splitting blasting, the gas desorption index exceed the standards when constructed to the third blasting. The maximum value could get to 1.223 ml/g·$min^{1/2}$.

After practicing deep hole pre-splitting blasting draining, this could drive 5 small blasting. The efficacy of indicators is 0.643 ml/g·$min^{1/2}$ appeared that the gas desorption index exceed the standards at the fifth small blasting (total driving is 32 m).

4.2. *Observation of Blasting Effect in K4 Coal Seam*

4.2.1. *The selection of blasting drills*

Taking into account the penetrability of $K_4$ coal seam was very low, and the difficulty of gas drainage was large relatively. There were 2 drills selected as the pre-splitting blasting holes. They were 5, 7# in Figure 1 respectively. The depth of the holes was 45 m and 45.2 m respectively. The length of charge was all 30 m. The other 12 were as draining holes.

4.2.2. *Comparative Analysis of blasting effect after pre-splitting*

When did pre-splitting blasting in transport lane of $K_4$ coal seam, the gas flow of blasting cycles and draining directly was shown in Figure 3.

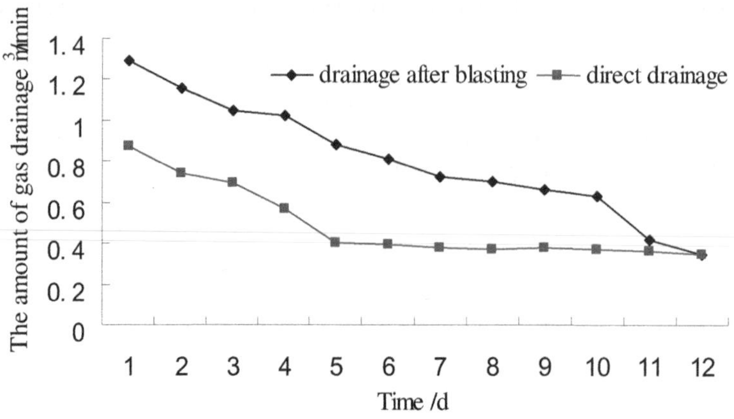


Figure 3. The amount of drainging comparation of blasting and direction in $K_4$ coal seam

After pre-splitting blasting, the total draining time was 12 days. The gas rate decayed from beginning draining 31% to demolishing 4%. The instantaneous gas flow decayed from beginning draining 1.21 $m^3$/min to demolishing 0.24 $m^3$/min. The pre-draining rate got to 34.28%. When driving without pre-splitting blasting but just pre-draining, in order to get to the same pre-draining rate, it took more than 30 days. It reduced 18 days by carrying out pre-splitting blasting.

### 4.2.3. *The cuttings desorption index contrastive analysis*

If taking the traditional emission outburst technology, the gas in driving process sometimes was at the critical value. This made it difficult to drive and construct.

After practicing deep hole pre-splitting blasting draining, this could drive 3 small blasting cycles. The efficacy of indicators is 0.768 ml/g·min$^{1/2}$ appeared that the gas desorption index exceed the standards at the third small blasting cycle.

Thus, reliable protection was provided by pre-splitting blasting for the safety of driving in the coal. Prominent sign did not occur in driving process. The gas rate was generally between 0.1 and 0.6.

## 5. OBSERVATION OF PRE-SPLITTING BLASTING DRAINING RADIUS

The gas drainage radius was an important parameter which influenced the effect of drilling gas drainage. Generally speaking, the longer the drainage time, the larger the drainage radius (Chen, Guo 2008; Clarson, Karacan 2001). But in order not to affect the efficiency of the driving face and ease mining to succeed, we must draw out more gas in a limited time.

The above analysis showed that after pre-splitting blasting, drainage results had improved markedly in driving face. Drainage radius was related with the permeability of coal seam (Wang et al., 2008), gas content and pressure (Liu et al., 2009), and many other factors (Liu Y. et al., 2007), and therefore if needed to design reasonable drainage radius of the drainage to ensure the best results, it must be put investigation after the pre-splitting blasting.

### 5.1. *Observation of Drainage Radius in $K_1$ Coal Seam*

At the time of the driving of coal roadway in $K_1$ coal seam, that conducted on the impact of drainage area study which pre-splitting blasting effected on the drainage. There were 3 cycles for deep hole pre-splitting blasting were conducted. Each cycle was layout a blast hole and a inspection hole, hole label was showed Figure 1. The Settings of the blast hole and inspection hole and the gas flow before and after blasting were showed in Table 1.

Table 1. The amount of gas draining before and after pre-splitting blasting of $K_1$ coal seam

| Cycle | Blasting borehole | | Inspection borehole | | Charge investigation department and borehole spacing (m) | Gas flow in studied borehole ($m^3$/min) | | Incremental increase of blasting |
|---|---|---|---|---|---|---|---|---|
| | Borehole No. | Borehole length (m) | Borehole No. | Borehole length (m) | | Before blasting | After blasting | |
| 1 | 5 | 30.4 | 7 | 45 | 3.2 | 0.016 | 0.036 | 125% |
| 2 | 7 | 38 | 3 | 33 | 4.5 | 0.018 | 0.041 | 128% |
| 3 | 8 | 38 | 3 | 33 | 5.1 | 0.018 | 0.025 | 39% |

As could be seen from Table 1, after the first two pre-splitting blasting cycles, gas drainage at inspection holes increased significantly by 125% and 128% respectively and got to more than double. After the third pre-splitting blasting, the gas flow at inspection holes while had increased, but the increase in volume compared to the previous two circles was little. Department under the charge blasting holes and the inspection holes could determine the effective radius was 4.5 m at the pre-splitting blasting of $K_1$ coal seam.

### 5.2. *Observation of Drainage Radius in K4 Coal Seam*

The Observation of pre-splitting blasting drainage radius in $K_4$ coal seam was conducted in one driving working face. There were 3 deep holes pre-splitting blasting were conducted. Each test were layout a blast hole and a inspection hole, hole label was showed Figure 1.

The Settings of the blast hole and inspection hole and the gas flow before and after blasting were showed in Table 2.

Table 2. The amount of gas draining before and after pre-splitting blasting of $K_4$ coal seam

| Cycle | Blasting borehole | | Inspection borehole | | Charge investigation department and borehole spacing(m) | Gas flow in studied borehole ($m^3$/min) | | Incremental Increase of blasting |
|---|---|---|---|---|---|---|---|---|
| | Borehole No. | Borehole length (m) | Borehole No. | Borehole length (m) | | Before blasting | After blasting | |
| 1 | 8 | 38.2 | 10 | 45.1 | 2 | 0.026 | 0.058 | 123% |
| 2 | 7 | 38 | 4 | 45 | 5.2 | 0.029 | 0.062 | 114% |
| 3 | 9 | 38.3 | 4 | 45 | 5.6 | 0.036 | 0.054 | 50% |

As could be seen from Table 2, the antireflection effect at the first and second blasting was more obvious. Gas drainage at inspection holes increased by 123% and 114% respectively, while the third does not increase significantly compared to the first two, a 50% increase. Department under the charge blasting holes and the inspection holes could determine the effective radius was 4.5 m at the pre-splitting blasting of $K_1$ coal seam.

## 6. CONCLUSIONS

1) Pre-splitting blasting could make a range of coal around the blasting hole produce a large number of fissures, and increased gas permeability, thereby enhanced the gas drainage volume. And it was conducive to reduce the conspicuous risk at the front of driving face.
2) The draining rate of $K_1$ coal seam after pre-splitting blasting reached 37.13% in just 10 days, which reduced the draining time by 60%, and the gas desorption index exceeded the standards until the fifth blasting cycle. It took 12 days for the draining rate of $K_4$ to reach 34.28%, which reduced the draining time by 60%. And the gas desorption index exceeded the standards when the third blasting cycle processed. Obviously, the effect of gas draining was improved by pre-splitting blasting.

3) Field test study showed that pre-splitting blasting was only within a certain range to improve gas drainage volume. The effective draining radius of $K_1$ and $K_4$ coal seams were 4.5 m and 5.2 m respectively.

## ACKNOWLEDGEMENT

Parts of the research were performed under grant of the Natural Science Foundation of China (Grant no.50774106), the National Key Fundamental Research and Development Program of China (Grant no.2005CB221502), the Natural Science Innovation Group Foundation of China (Grant no.50621403), the technical plan project of land housing administration (2009-01) and Scientific and technical project of Sichuan coal mine group (2009-08).

## REFERENCES

Cai Feng, Liu Zegong, Zhang Chaoju. 2007. Numerical simulation of improving permeability by deep-hole pre-splitting explosion in loose-soft and low permeability coal seam. Journal of China Coal Society 32(5): 499-503. (In Chinese)

Chen Hao. 2008. Application od deep borehole blasting technology in draining and excavating heading face. Coal Technology 27(7): 92-94. (In Chinese)

Clarkson, C. R. 2001. Binary gas adsorption/desorption isotherms: Effect of moisture and coal composition upon carbon dioxide selectivity over methane Bustin R M. International Journal of Coal Geology 42(4): 241-271

Gong Min, Huang Yihua, Wang Desheng. 2008. Numerical simulation on mechanical characteristics of deep-hole pre-splitting blasting in soft coal bed. Chinese Journal of Rock mechanics and Engineering 8(27): 1674-1681. (In Chinese)

Guo Deyong, Pei Ruhai, Song Jiancheng. 2008. study on spliting mechanism of coal bed deep-hole cumulative blasting to improve permeability. Journal of China Coal Society 33(12): 1381-1385. (In Chinese)

Karacan, C O. 2001. Adsorption and gas transport in coal microstructure: Investigation and evaluation by quantitative X-ray CT imaging . Okandan E. Fuel 80(4): 509-520

Liu Haibo, Cheng Yuanping, Song Jian-cheng. 2009. Pressure relief, gas drainage and deformation effects on an overlying coal seam induced by drilling an extra-thin protective coal seam. Mining Science and Technology 19(6): 724-729. (In Chinese)

Liu Jian, Liu Zegong, Shi Biming. 2007. Study on roadway excavation rapidly in the low permeability outburst coal seam. Journal of China Coal Society 32(8): 827-831. (In Chinese)

Liu Yuzhou, Li Xiaohong. 2007. Safety analysis of stability of surface gas drainage boreholes above goal areas. Journal of Coal Science & Engineering 13(2): 149-153. (In Chinese)

Wang Liang, Cheng Yuanping, Li Feng-rong. 2008. Fracture evolution and pressure relief gas drainage from distant protected coal seams under an extremely thick key stratum. Journal of China University of Mining & Technology 18(2): 182-186. (In Chinese)

International Mining Forum 2010, Liu et al. (eds) © 2010 Taylor & Francis Group, London, UK. ISBN 978-0-415-59896-5

# Analysis of the economic strength index model based on wet fiber reinforced shotcrete

Aixiang Wu, Chunlai Wang, Li Li, Bin Han, Shaoyong Wang
*Civil and environmental engineering college, University of science and technology, Beijing, China*

ABSTRACT: The wet fiber reinforced shotcrete is high-cost. In order to ascertain the relationship between the quantity of materials, strength and cost of wet fiber reinforced shotcrete, to obtain the reasonable and optimal mix ratio of wet fiber reinforced shotcrete, and further to reveal the rule of strength changed with different materials. The experimental research was carried out using MTS servohydraulic rock mechanics testing system. Based on the ASTM C 1 550 test standard, the strength change in the process of uniaxial compression and flexural resistance was investigated at the different ages of wet fiber reinforced shotcrete. The influence of main components quantity on the increased strength in wet fiber reinforced shotcrete was analyzed, especially the quantity of cement, plastic fiber, water reducer, accelerator, etc. A new concept of economic strength index (ESI) was put forward, and the equations of ESI about uniaxial compressive and flexural were got. The relationship was illustrated between the quantity of materials, strength and cost of wet fiber reinforced shotcrete. Results show that, when the cost is equal, high quantity of cement and water reducer could reduce ratio of water to cement and significantly enhance the uniaxial compressive and flexural strength of shotcrete at the same time according to the ESI equation. However, the increasing quantity of fiber improves the cost, enhances the residual strength of shotcrete layer and the supporting capacity of cracked shotcrete layer, but does not increase the compressive strength. In conclusion, to satisfy the supporting strength, the ESI equation effectively decreases the cost in design the mix ratio of wet fiber reinforced shotcrete.

## 1. INTRODUCTION

Mineral resources are essential part of nature resources, which guarantee the national safety and the development of economy to a lager extent. Recently, there are about one third of open-pit mines to be mined with the model of open-cast and underground combination mining, which has been proved to be an important strategy for resources development in China (Chen et al. 2009, Han et al. 2009, Kang et al. 2009, Ziad et al. 1993, Pang et al. 2007). However, in complex geological conditions with broken rock mass, the support research is not enough in the combination mining model. The wet fiber reinforced shotcrete is an effective method to solve this problem (Malmgren et al. 2005, Sangpil et al. 2006, Jeon et al. 2006, Guan et al. 2007, Yao et al. 2009). The support need provide not ony the high compressive strength and the flexural strength, but also the high ability of absorbing the elastic deformation energy (Wang et al. 2004, Wang 2004, Tang et al. 2004, Wang et al. 2006). Under the influence of financial crisis, the cost of support must be saved now. Therefore, it has become exceedingly important to study the relationship between the mixing amount of materials, strength and cost, to obtain the reasonable and optimal mix ratio of wet fiber reinforced shotcrete, and to know the strength change with cost of different materials. Based on the above mentioned factors, the experiment of energy changes were studied in the whole process of unaxial compressive and Flexural resistance with MTS servohydraulic rock mechanics testing system. The model of economic strength index (ESI) was established. Using this model, the relationship was analyzed between the quantity of materials, strength and cost of wet fiber reinforced shotcrete.

## 2. THE GEOLOGICAL CONDITIONS OF MINE

Jinfeng Mine is the largest, modern and single gold mine in China. In this mine, the engineering geological condition is complex. The mineralized surrounding rocks near the roof and floor are mostly medium thick sandstone and thin extremely crushed clay stone layer. The orebody is dominated by the fault zone with the nonuniform distributed grade. The roof and floor rock mass are unstable and easily collapsed because the rock mass is broken (RQD=15~60%, f=1~6), weathered, and argillizated. The attitude of orebody is controlled by F3 fault. The tendency of F3 fault is northeast, and the dip (45° ~ 70°) is vary with the fault gradient. The attitude of F3 fault bottom wall changes much. The dislodged rock mass easily collapses when the mining drift through the slight inclines or horizontal rock layer.

The rock support is confronted with a new challenge because the structure of surrounding rock mass is layered, crushed, and the local abundant groundwater. The groundwater makes the rock mass dilatation and corrosion, resulting in the secondary stress of the rock bolts and wet fiber reinforced shotcrete. The rock bolts lessens the function by the corrosive water.

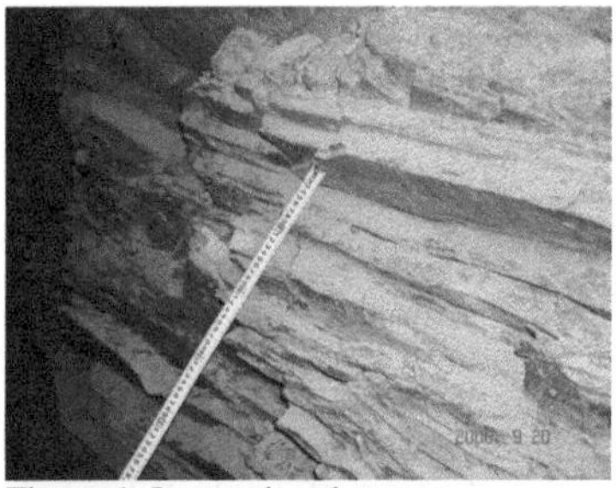

Figure 1. Layered rock mass

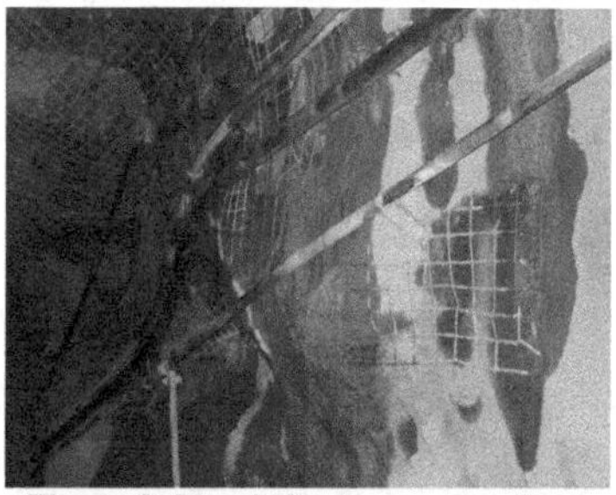

Figure 2. Local abundant groundwater

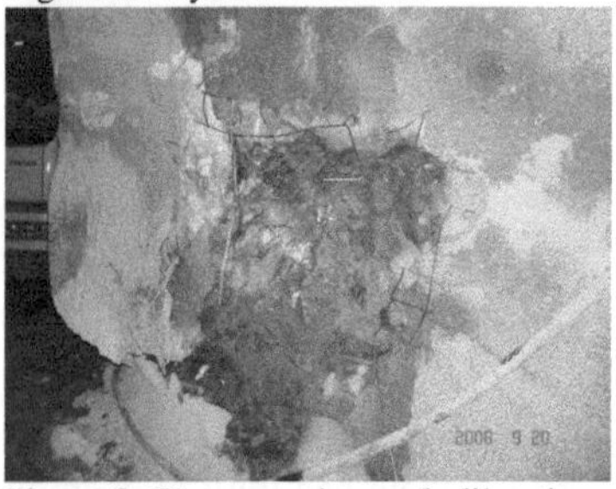

Figure 3. Water causing rock dilatation

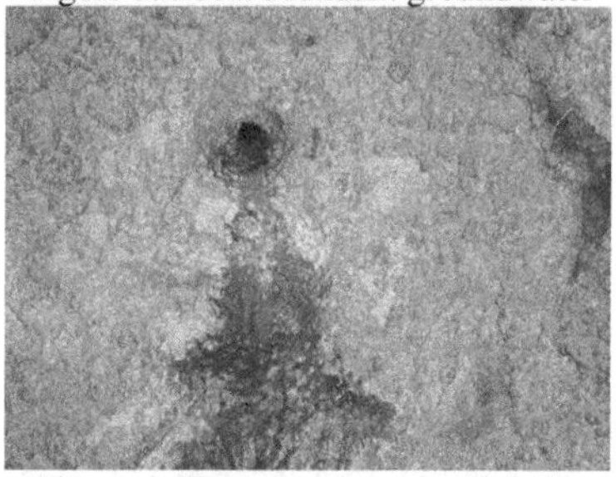

Figure 4. The corrosion of rock bolts

## 3. TESTING STUDY

### 3.1. *Samples and test system*

#### 3.1.1. *Samples*

According to the relevant rules(CAECS 1996, ASTM 2005), the samples used for uniaxial compressive strength are cube with 100 mm length, and the samples used for uniaxial Flexural strength are disk with 800 mm diameter and 80 mm thickness. The designed strength grade is C30. The cement is ordinary portland cement which grade is 425 produced by industry in Zhenfeng country, Qianxinan district, Guizhou province. The aggregate is rubble that the maximum diameter is 10 mm. Removing the mould after 24h, hydrostatic curing age about 28d, and then the samples are processed based on rule of reference(CAECS 1996).

### 3.2. *Loading equipment*

All of testing carried out using MTS815 servohydraulic digital and stiff rock mechanics testing machine. The testing system is based on visual operating software of Windows platform, which could record the figure of current time, load, stress, displacement, strain, the graph of load-displacement and stress-strain, and so on. The loading method applies equal displacement and the speed of loading is $2\times10^{-3}$ mm/s. The computer controls the whole process of testing and save all of the graph and figure.

Table 1. The compressive strength data

| NO. | Cement (kg/m$^3$) | Plastic fiber (kg/m$^3$) | Water cement ratio | Sand ratio | The compressive strength (MPa) | | |
|---|---|---|---|---|---|---|---|
| T1 | 400 | 1 | 0.48 | 0.69 | 33.3 | 33.2 | 34.5 |
| T2 | 400 | 4 | 0.51 | 0.79 | 31.8 | 32.5 | 30.4 |
| T3 | 400 | 7 | 0.54 | 0.89 | 17.6 | 16.2 | 18 |
| T4 | 425 | 1 | 0.51 | 0.89 | 19 | 19.6 | 20.5 |
| T5 | 425 | 4 | 0.54 | 0.69 | 34.3 | 32 | 35.5 |
| T6 | 425 | 7 | 0.48 | 0.79 | 32.7 | 33.5 | 33 |
| T7 | 450 | 1 | 0.54 | 0.79 | 34.5 | 33.9 | 33.9 |
| T8 | 450 | 4 | 0.48 | 0.89 | 22.3 | 23.1 | 22.9 |
| T9 | 450 | 7 | 0.51 | 0.69 | 37 | 39.6 | 38.8 |

Table 2. The flexural strength testing for wet fiber reinforced shotcrete in-site

| NO. | Cement (kg/m$^3$) | Plastic fiber (kg/m$^3$) | Water cement ratio | Sand ratio | Flexural strength (MPa) | | |
|---|---|---|---|---|---|---|---|
| T1 | 400 | 1 | 0.48 | 0.69 | 5.31 | 5.26 | 5.08 |
| T2 | 400 | 4 | 0.51 | 0.79 | 3.80 | 3.71 | 3.34 |
| T3 | 400 | 7 | 0.54 | 0.89 | 3.43 | 3.20 | 3.57 |
| T4 | 425 | 1 | 0.51 | 0.89 | 2.79 | 2.93 | 4.03 |
| T5 | 425 | 4 | 0.54 | 0.69 | 5.72 | 5.72 | 5.95 |
| T6 | 425 | 7 | 0.48 | 0.79 | 4.12 | 4.35 | 4.44 |
| T7 | 450 | 1 | 0.54 | 0.79 | 3.29 | 3.75 | 4.35 |
| T8 | 450 | 4 | 0.48 | 0.89 | 3.89 | 3.89 | 4.35 |
| T9 | 450 | 7 | 0.51 | 0.69 | 5.95 | 5.72 | 6.31 |

Table 3. The main materials unit price of wet fiber reinforced shotcrete

| Components | Cement (RMB/Kg) | Fiber (RMB/Kg) | Water reducer (RMB/Kg) | Accelerating agent (RMB/Kg) | Remarks |
|---|---|---|---|---|---|
| Unit price | 0.53 | 29 | 11.4 | 4.55 | |

### 3.3. *Data analysis*

The effect factors for the pumpability and strength of wet fiber reinforced shotcrete include the quantity of cement, aggregate, fiber, additives, the particle-size ingredient of aggregate, and the types of additives. According to the reasonable testing and the data analysis, the optimal project was obtained. Considering of the above effect factors, the orthogonal experimental design method is adopted. The final design scheme is finished by choosing orthogonal design of $L_9(3^4)$ in Table 1. There are total 9 schemes, and each scheme has 3 compressive samples and 3 flexural samples. The optimal mix ratio is gained by ascertaining factors, levels, and multi-scheme optimization. The detail results have showed in the following Tables 1 and 2.

## 4. STUDIES TO THE MODEL OF ECONOMIC STRENGTH INDEX

### 4.1. *A new concept of the Economic Strength Index model*

According to the effect factors of wet fiber reinforced shotcrete strength, such as cement, plastic fiber, water-reducing agent and accelerating agent, a new concept economic strength index (ESI) is introduced to compare the effects of different materials with the strength of wet fiber reinforced shotcrete. The so-called ESI refers to the increased amout of shotcrete strength when unit cost increase of each material. At the same time, the main mechanical indexes of wet fiber reinforced shotcrete include uniaxial compressive strength indicators and flexural strength indicators. Therefore, the generalized ESI can be divided into uniaxial compressive economic strength index (CESI) ($f_{ci}$) and flexural economic strength index (FESI) ($f_{fi}$):

$$f_{ci} = \frac{S_{ci}}{p_i w_i} \tag{1}$$

where: $f_{ci}$ – the uniaxial Compressive Economic Strength Index, MPa/RMB; $S_{ci}$ – the increased uniaxial compressive strength with the cost of materials for $c_i$ RMB; $p_i$ – the unit price of materials, RMB/kg; $w_i$ – the increased amount of materials, kg.

$$f_{fi} = \frac{S_{fi}}{p_i w_i} \times 100\% \tag{2}$$

where: $f_{fi}$ – the FESI of the materials, MPa/RMB; $S_{fi}$ – the increased uniaxial flexural strength of concrete with the cost of materials; $p_i$ – the unit price of materials, RMB/kg; $w_i$ – the increased amount of materials, kg.

The new concept of ESI analyzes the growth rate of the concrete strength which is caused by the increase of different concrete materials under the condition of the same cost increment. It has attributed to maximize the optimization and mechanics of the best mix under the same material costs, that is to say, the cheapest mix is filtered when the mechanical indicator reaches.

### 4.2. *The calculation of the economic strength index model*

According to the definition of ESI, the CESI in 7 days is calculated:

$$f_{cc} = \frac{S_{cc}}{p_c w_c} \tag{3}$$

where: $f_{cc}$ – the CESI, MPa/RMB; $S_{cc}$ – the increased amount of uniaxial compressive strength with the added concrete for $w_c$ kg, that is $S_{cc}$ = 39.27–37.60 MPa; $p_i$ – the unit price of cement, RMB/kg, 0.53 RMB/kg; $w_c$ = the increased amount of concrete, $w_c$ = 480–430 = 50 kg.

Table 4. Economic strength index Unit (MPa/RMB)

| Age | Cement | | Plastic fiber | | Water reducing agent | | Accelerator agent | |
|---|---|---|---|---|---|---|---|---|
| | Uniaxial compression | Flexural | Uniaxial compression | Flexural | Uniaxial compression | Flexural | Uniaxial compression | Flexural |
| 7 days | 0.062 | 0.067 | -0.008 | 0.004 | 0.442 | 0.094 | -0.047 | 0.004 |
| 28 days | 0.283 | 0.005 | -0.005 | 0.002 | 0.148 | 0.022 | -0.039 | 0.0002 |

### 4.3. *Analysis of ESI model*

The testing results are used to analyze the ESI model. From the CESI in 7 days age, the following conclusions were obtained:

(1) The water reducer index of CESI is the highest. When the cost is the same, the increment of water reducer contributes to the enhancement of the compressive strength largely and the reduction of water cement ratio.

(2) The plastic fiber index of CESI is negative, that is to say, the increment of plastic fiber don't enhance the uniaxial compressive strength.

(3) The accelerating agent index of CESI is negative, which is, the increment of accelerating agent could be controlled to assure the early strength of wet fiber reinforced shotcrete.

From the analysis of CESI in 28 days age, in Table 4, the following conclusions were reached:

(1) The CESI cement reaches to 0.283 MPa/RMB. It can help to improve the concrete strength with the increment of cement, for the cement reaches to 0.283 MPa/RMB.

(2) Inputting the same cost, it is more conducive for improving permanent strength of concrete when the cement increases much more than water reduced agent. On the contrary, the increase of water reduced agent can help to enhance the early strength of concrete.

(3) The increase of plastic fibers content makes the negative growth of uniaxial compressive strength of concrete.

(4) In order to reduce the amount of accelerating agent content, the water-cement ratio should be strictly controlled in the process of concrete configuration. With the increase of accelerating agent content, the resilience rate is reduced. When the concrete concentration is too low, it is necessary to increase the accelerator agent for avoiding high rebound.

From the analysis of FESI in 7 days, the following conclusions were acquired:

(1) The maximum of the FESI is water-reducing agent, it is followed by cement, plastic fibers, and accelerating agent. For increasing FESI in 7 days age, the best way is to improve the water-reducing agent or to increase the amount of cement.

(2) To improve the flexural strength in 7 days age, it is limited to increase plastic fibers. The plastic fibers index of FESI is 0.004 MPa/RMB.

From the analysis of FESI in 28 days, the following conclusions were acquired:

(1) The maximum FESI is concrete. From an economic point of view, the increase of cement improves the permanent wet shotcrete flexural strength, but the increase of water-reducing agent improves the early strength.

(2) The FESI of plastic fibers is 0.005 MPa/RMB. It is limited for plastic fibers to improve the flexural strength.

The optimized mix ratio was obtained after analysis ESI and using effect in site, as follow:

Table 5. The optimized mix ratio

| Cement ($kg/m^3$) | Aggregate (kg) | Plastic fibers ($kg/m^3$) | Water reducing agent (%) | Accelerator agent (%) | Water-cement ratio |
|---|---|---|---|---|---|
| 430 | 1862 | 6 | 0.9 | 6 | 0.42 |

## 5. CONCULSIONS

1) Under the condition of the equal cost, high addition of cement and water reducer can enhance the uniaxial compressive and flexural strength of wet fiber reinforced shotcrete.
2) There is little function to enhance the uniaxial compressive and flexural strength by the addition of plastic fibers. However, the flexural strength is improved when the quantity of plastic fiber is from 3 $kg/m^3$ to 6 $kg/m^3$, the proper quantity is 6 $kg/m^3$.
3) The uniaxial compressive strength reduces because of plastic fibers addition. However, when the structures of shotcrete are destroyed, the plastic fibers can improve the capacity of absorbing the plastic deformation energy. At the same time, it is important to use high fibers of the support for broken rock mass in Jinfeng mine.


## ACKNOWLEDGEMENT

The research is sponsored by the Key Technology P&D Program supporting project of the National 'Eleventh Five-Year-Plan' of China (No.2006BAB02A01).

## REFERENCES


ASTM. 2005. Concrete standards. Standard test method for flexural toughness of fiber reinforced concrete (Using Centrally Loaded Round Panel). ASTM C 1550-2005

Chen Xiao-xiang, Gou Panfeng, Yan Anzhi. 2009. The research on deformation mechanism and harmonization supporting technique of bolts and anchorage cables in high stress boadway of deep mine. The 7th International Symposium on Rockburst and Seismicity in Mines. New York: Rinton Press

China association for engineering construction standardization(CAECS). 1996. Standard test method for steel fiber concrete. Beijing: China Planning Press. (In Chinese)

Guan Z C, Jiang Y J, Tanabasi Y, et al. 2007. Reinforcement mechanics of passive bolts in conventional tunneling. International Journal of Rock Mechanics and Mining Sciences 44(4): 625-636

Han Bin, Wu Shuanjun, Li Hongye. 2009. High-Efficient wet shotcrete technology and its application in underground mines. Metal Mine 395(5): 23-26. (In Chinese)

Jeon, S., You, K., Park, B., H-G Park. 2006. Evaluation of support characteristics of wet-mixed shotcrete with powder type cement mineral accelerator. Tunnelling and Underground Space Technology 21 (3): 425-426

Kang Hongpu, Wang Jinhua, Gao Fuqiang. 2009. Stress distribution characteristics in rock surrounding heading face and its relationship with supporting. Journal of China Coal Society 34(12): 1585-1593. (In Chinese)

Malmgren, Lars, Nordlund E., Rolund S. 2005. Adhesion Strength and Shrinkage of Shotcrete. Tunnelling and Underground Space Technology v20 (n1): 33-48

Pang Jiangyong, Daofu. 2007. Support technology of polypropylene fiber concrete shotcrete layer and ite application to Guqiao coal mine. Chinese Journal of Rock Mechanics and Engineering 26(5): 1073-1078. (In Chinese)

Sangpil, Lee, Donghyun, Kim, Jonghyun, Ryu, et al. 2006. An experimental study on the durability of high performance shotcrete for permanent tunnel support. Tunnelling and Underground Space Technology 21 (3-4): 1-6

Tang Haiyan, Wang Chunlai, Li Shulin. 2004. Experimental study on the uniaxial compressive toughness of SFRS. Proceedings of Sixth National Conference on MTS materials Testing, Beijing: Press of University of Science and Technology Beijing. (In Chinese)

Wang Chunlai, Xu Bigen, Li Shulin, et al. 2004. Press test of SFRC under different confined pressures. Mining Research and Development 24(4): 4-7. (In Chinese)

Wang Chunlai, Xu Bigen, Li Shulin, et al. 2006. Study on a constitutive model of damage of SFRC under uniaxial compression. Rock and Soil Mechanics 27(1): 151-155. (In Chinese)

Wang Chunlai. 2004. Study on the mechanical properties and damage of support materials and supporting techniques for drifts under condition of high stress. Changsha Institute of Mining Research. (In Chinese)

Yao Gaohui, Wu Ai-xiang, Wang Hongjiang. 2009. Research on the application of large efficient mechanized shotcrete-bolt support technology in underground mine. Metal Mine 400(10): 126-130. (In Chinese)

Ziad B., Jack Z. 1993. Properties of polypropylene fiber reinforcedconcrete. ACT Material Journal 90(6): 615-618

*International Mining Forum 2010, Liu et al. (eds) © 2010 Taylor & Francis Group, London, UK. ISBN 978-0-415-59896-5*

# Energy mechanism of sulphide dust explosion in sublevel mining drawing outlet for high sulphur metal ore mine

Yunzhang Rao
*Jiangxi University of Science and Technology, Ganzhou, Jiangxi Province, China*

Sujing Huang
*Jiangxi Copper Group Co., Ltd., Dongxiang, Jiangxi Province, China*

Jianping Zhang, Guangzhe Xiao
*Jiangxi University of Science and Technology, Ganzhou, Jiangxi Province, China*

ABSTRACT: This paper studies the ignition of sulphide dust explosion in the sublevel mining drawing outlet of high-sulphur metal mine on the basis of chemical reaction heat and energy diffusion mechanism. The study shows that the major ignition of the sulphide dust explosion is the combustion of oxidized high sulphure ore, lintel being the major explosion position.

KEYWORDS: Metal ore mine, sulphide dust explosion, spontaneous ignition

## 1. INTORDUCTION

There have been reports of sulphide dust explosion in metal mines in Europe, South Africa, Australia and the Former Soviet Union (Ye et al. 1995). And several explosion accidents also occurred in some high sulphide and flammable mines in China resulting in great casualties (Ye et al. 1995, Rao et al. 2005). For example, 10 explosion accidents led to 5 deaths and 10 injuries in Songshushan Copper Mine (Anhui Province) during 1970 to 1974. In 1978, the explosoin in Xilin Lead and Zinc Mine (Heilongjiang Province) and Tongkeng Tin Mine (Guangxi Autonomous Region) caused 4 deaths. In 1979, the explosion in Tongguanshan Copper Mine (Anhui Province) caused 4 deaths. In 2002, the explosion in Tongshan Copper Mine (Anhui Province) caused 2 deaths. The latest 2 explosion accidents in Dongxiang Copper Mine (Jiangxi Province) killed 7 workers.

The researches on the explosion mechanism of sulphide dust explosion in the high sulphure metal mine have been far from enough for the scarcity of such kind of explosion. However, about 20~30% sulphur iron ores and more than 10% non-ferrous ores or multi-metallic sulphur ores can be categorized into high sulphur ores (Wu 2002), which are likely to bring about ore spontaneous combustion and dust explosion. So the study on the explosion mechanism is of great significance.

## 2. CONDITIONS OF SULPHIDE DUST EXPLOSION AND ITS ANALYSIS

The sulphide dust explosion should meet the following conditions simultaneously (Zhang et al. 1996, Liu 1995, Li 2003, Bartechnact 1997):

(1) adequate oxygen (air);

(2) dust of adequate concentration and dispersiveness floated in the air;

(3) adequate ignition energy.

As can be seen clearly, at least one condition from the above three should be prevented in order to avoid the sulphide dust explosion.

Condition (1) should naturally be met, for miners can't work in the oxygen-scarcity surroundings, which also violate the relative security regulations. Condition (2) can be easily met, for dust is naturally produced in mining process. Technology measures can be taken to reduce the concentration degree. The major ignition energy in condition (3) includes man-made fire, electronic sparks and sulphure ore spontaneous combustion. Man-made fire and electronic sparks can be excluded. However, the high-sulphur ore, especially the pyrogelite, oxidates and burns easily, which is the most important and most effective ignition way of the sulphide dust explosion.

## 3. KEY REACTION FORMULA OF SULPHIDE ORE AND HEAT EFFECT

The sulphide minerals which lead to the spontaneous combustion of high-sulphur mainly includes: $FeS_2$, FeS and $Fe_7S_8$. These minerals enjoy distinctive oxidation response model and heat effect under different temperature and humidity conditions (Wan 1998, Jiang et al. 2003, Li et al. 2004, Huang 2000, Zhang et al. 2004).

(1) The major reaction formulas and heat effects of $FeS_2$, FeS and $Fe_7S_8$ under the conditions of normal temperature (40°C) and humid surroundings.

$$2FeS_2 + 7O_2 + 2H_2O = 2FeSO_4 + 2H_2SO_4 + 2588.4KJ \tag{1}$$

$$4FeS_2 + 15O_2 + 14H_2O = 4FeO_3 + 8H_2SO_4 + 5092.8KJ \tag{2}$$

$$4FeS_2 + 15O_2 + 8H_2O = 2Fe_2O_3 + 8H_2SO_4 + 5740.5KJ \tag{3}$$

$$FeS + 2O_2 + H_2O = FeSO_4 \bullet H_2O + 914.4KJ \tag{4}$$

$$2Fe_7S_8 + 31O_2 + 2H_2O = 14FeSO_4 + 2H_2SO_4 + 12590.0KJ \tag{5}$$

(2) Under the conditions of normal temperature (40°C) and dry surroundings, almost no $SO_2$ is released in the oxidation reaction of $FeS_2$ and FeS, whose major reaction formula and heat effect are shown as follows:

$$FeS_2 + 2O_2 = FeSO_4 + SO + 750.7KJ \tag{6}$$

$$12FeS + 11O_2 = 5FeSO_4 + Fe_2O_3 + 11SO + 3241.8KJ \tag{7}$$

(3) When the temperature is higher than 60°C, $SO_2$ produced in the oxidation process of $FeS_2$, FeS and $Fe_7S_8$ increases quickly, whose major reaction formula and heat effect are shown as follows:

$$4FeS_2 + 11O_2 = 2Fe_2O_3 + 8SO_2 + 3312.4KJ \tag{8}$$

$$FeS_2 + 3O_2 = FeSO_4 + SO_2 + 1047.0KJ \tag{9}$$

$$4FeS + 7O_2 = 2Fe_2O_3 + 4SO_2 + 3219.9KJ \tag{10}$$

$$4Fe_7S_8 + 53O_2 = 14Fe_2O_3 + 32SO_2 + 18077.4KJ \tag{11}$$

The above reactions are all heat exothermic. The produced heat accumulates continuously when it can't diffuse into the air. The oxidation process of sulphide minerals initiates from the low temperature condition and accelerates with the increased temperature of sulphure ores and surrounding medium. Spontaneous combustion of sulphure ores occurs in 2 or 3 days, which produces spontaneous combustion fire.

## 4. EXPLODING POCESS OF SULPHIDE DUST

Sulphide dust explosion acutely occurs under the condition of sulphure ore spontaneous combustion, whose process can be shown in the following steps:

(1) Heat energy is added to the surface of the dust particles, which increases the temperature of dust particles gradually.

(2) The molecule on the surface of dust particles decomposes or carbonizes under the heat, which forms gas around the particles.

(3) The gas produced in $\alpha$ mixes with air, forming explosive gas, which burns simultaneously.

(4) Energy produced by burning encourages the discomposition of dust particles, which produces flammable gas continuously.

(5) Some parts acutely conflagate and explode.

As can be seen from the above process, the major source for the sulphide dust explosion is the energy produced by sulphide dust oxidation is far greater than the dispersed heat, which makes the mining dust floating in the air oxidates drastically, burn or even explode.

## 5. ENERGY MECHANISM OF SULPHIDE DUST EXPLOSION IN THE SUBLEVEL MINING DRAWING OUTLET

As can be seen from Figure 1, the drawing outlet (the position of lintel) is exactly the meeting point of the rising high concentration dust cloud and the falling high-energy spontaneous combustion ore. This position is most likely to explode for it can meet the three requirements of sulphide dust explosion: adequate oxygen, sulphide dust of appropriate concentration and diffusion, adequate ignition energy.

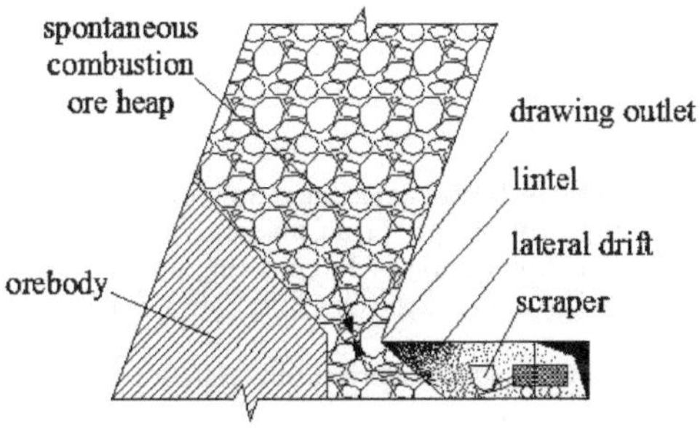


Figure 1. Sketch of sulphide dust explosion position in the sublevel mining drawing outlet

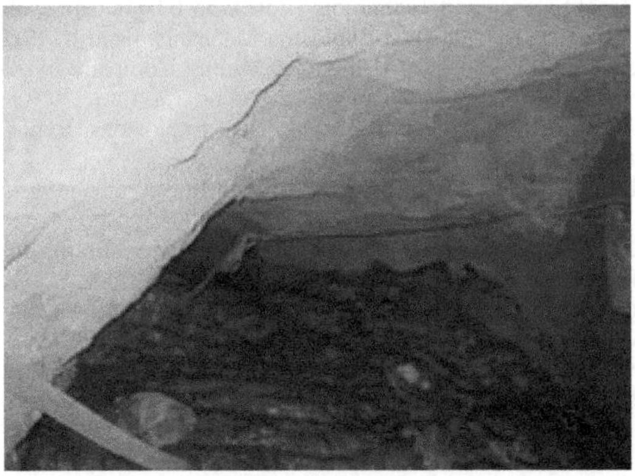

Figure 2. Locale of sulphide dust explosion in Dongxiang Copper Mine

If there is inadequate ignition energy in the drawing outlet ( the ore is not spontaneous combustion ), the dust particles won't conflagate, let alone explode; if the dust concentration in the drawing outlet is low, the energy in the dust can be diffused quickly, thus the dust won't explode either; if adequate ignition energy and dust of appropriate concentration exist in the drawing outlet at the same time, the dust particles are likely to conflagate and explode; if the ignition energy maintains at a fixed value and there is also relatively high concentration dust in the mine outlet, the dust won't burn for the time being. However, while the mining dust changes from high concentration state to low concentration state for the sedimentation and diffusion of mining dust, there will be one moment when the ignition energy meets the conditions of dust conflagration, and there exists a relatively high concentration of dust in the drawing outlet, which makes the mining dust conflagate and explode.

## 6. CONCLUSIONS

Sulphide dust explosion of high-sulphide metal mines poses great threat to the production security and life security. Through the analysis of explosion conditions and explosion process of sulphide dust, this paper studies the explosion mechanism of sulphide dust explosion in the sublevel mining and shows that the most important ignition of the sulphide dust explosion is the spontaneous combustion of the oxidized sulphide ore by applying heat and energy mechanism of chemical reaction. The sulphide dust explodes in the appropriate concentration.

## ACKNOWLEDGEMENT

Thanks to: Jiangxi Provincial Natural Science Foundation of and Jiangxi Copper Group Co., Ltd. for their generous financial support; to Mr. Pan Jianping and Mr. Cai Wei, Mr. Huang Benwen for their preliminary work on the paper.

## REFERENCES

Huang Yuejun. 2000. Study on the Spontaneous Combustion of High Temperature Sulphride Ores and Its Control Technology. Non-ferrous Mining and Matellurgy 16(1): 13-15. (In Chinese)

Jiang Zhongan, He Zhuojun, Du Cuifeng. 2003. Study on Application of Fireproofing and Fire- extinguishing Technology Using Fire-retardant Lime-Gel in high sulfur deposit. China Safety Science Journal 11(10): 31-33. (In Chinese)

Li Shixiang. 2003. Sulphur Dust Explosion Accident Analysis and Countermeasure. Safety Environment Health 3(8): 2-3. (In Chinese)

Li Zijun, Gu Desheng, Wu Chao. 2004. Dangerousness Assessment of Ore Spontaneous Combustion in High Temperature High Sulfur Deposits. Metal Mine 52(5): 57-59. (In Chinese)

Liu Zhenkun. 1995. Dustproof in Mine. Hunan Coal Science and Technology, 23(2): 58-62. (In Chinese)

Rao Yunzhang, Wu Hong, Huang Sujin. 2005. Study on Sulfide Dust Explosion Mechanism and Its Prevention Scheme. Jiangxi University of Science and Technology, Dongtong Mining Company of Jiangxi Copper Co., Ltd., 10. (In Chinese)

Bartechnact W. 1997. Preventive and Design Measures for Protection Against Dust Explosions. Explosion-proof Electric Machine 17(3): 23-44. (In Chinese)

Wan Bing. 1998. The Possibility Distinguished Spontaneous Combustion of Sulfide Mine. World mining bulletin 14(11): 31-34. (In Chinese)

Wu Changfu. 2002. Mining High Sulphur and Metal Mineral-deposit Reason of Fire and Measure for Put out a Fire. Mining Safety & Environmental Protection 29(2): 21-22. (In Chinese)

Ye Hongwei, Wang Zhiguo. 1995. Mining High Sulphur Mineral-deposit Especial Disaster and Mechanism Take Place. Non-Ferrous Mining and Metallurgy 26(4): 38-42. (In Chinese)

Zhang Hong, Zhang Chunsheng. 2004. Principle of Pyrite Spontaneous Combustion and Its Prevention. Copper Engineering 19(3): 53-54. (In Chinese)

Zhang Ziqiang, Shao Fu. 1996. The Condition and Precaution of Dust Explosion. Nonferrous Metals Design 21(1): 10-12. (In Chinese)

International Mining Forum 2010, Liu et al. (eds) © 2010 Taylor & Francis Group, London, UK. ISBN 978-0-415-59896-5

# Influence of pore structure on low temperature oxidation of coal

Guanglong Dai
*Anhui University of Science and Technology, Huainan, Anhui Province, China*

*Key Laboratory of Coal Mine Safety and Efficient Exploitation of Ministry of Education, Huainan, Anhui province, China*

ABSTRACT: Based on fractal characters of coal pore structure, mercury intrusion method was used to study pore structure in coal. Experiment of low temperature oxidation of coal was done, in order to study relationship between low temperature oxidation of coal and pore structure in it. The results showed that total pore volume, specific surface area of coal and fractal dimension value decrease with coal metamorphism degree increase. Temperature increase rate and heat release rate in coal oxidation process decrease with decrease of pore volume in coal, specific surface area and fractal dimension value. Specific surface area plays a key role in coal oxidation when pore volume in coal samples is the same.

KEYWORDS: Coal spontaneous combustion, low-temperature oxidation, pore fractal structure

## 1. INTRODUCTION

Y.S.Nugroho (Nugroho et al. 2000) studied influence of particle size and physical structure (including pore volume and surface area) of Indonesian coal on critical temperature, activation energy, heat release and pre-exponential factor by using cross point temperature method. The result showed that the smaller of particle size, the bigger surface area and the bigger porosity is, the lower critical temperature is. The relationship between pore structure and susceptibility to spontaneous combustion of different kinds of coal was studied by Oreshko (Bushuev et al. 2003). He found, that the rate of coal oxidation in low temperature process increases with pore volume increase. Schmidt thought that coal oxidation rate is proportional to cubic of pore surface area (Schmidt et al. 1984). Bouwma pointed out that pore structure is the key factor in process of coal low temperature oxidation, coal with big pore volume is easier to be oxidized. Thus susceptibility to spontaneous combustion is big (Bouwmam et al. 1980). Thus it can be seen, pore structure has great influence on low temperature oxidation of coal. Based on the analysis of coal pore structure, difference of susceptibility to spontaneous combustion of different kinds of coal was qualitatively studied, in order to deeply understand intrinsic factors influencing on spontaneous combustion.

## 2. PORE STRUCTURE IN COAL AND ITS FRACTAL CHARACTERISTICS

In the study Wu Jun's method (Wu Jun 1993) was adopted, in which pore structure in coal is classified in four grades: big pore, middle pore, transitional pore and micro-pore.

Research results showed that coal is a fractal body with plenty of cranny and pores. The fracture of surface has the character of fractal, so the distortion porosity and permeability. Volume and superficial area of the coal is fuzzy, and fractal can figure out the true character of coal pore. Irregularity degree can be described by dimension division.

Table 1. Pore type of coal and its character

| Classification | Pore diameter (nm) | Pore character | Type of gas diffusion pore |
|---|---|---|---|
| Ⅰ (big pore) | >1000 | Most is tubular and plane | gas cubage diffusion type |
| Ⅱ (middle pore) | 100 ~ 1000 | Most is tubular and plane, some is unparallel plane | |
| Ⅲ (transitional pore) | 10 ~ 100 | Most is unparallel, and some is ink bottle like | gas molecule diffusion type |
| Ⅳ(micro-pore) | <10 | Most is unparallel and ink bottle like | |

If we want to know the object which volume is *V*. we can use small ball whose radius is *r* to fill it, and the number of the small ball is

$$N(r) = V/(1.33\pi r^3) \propto 1/r^3 \tag{1}$$

If we use small ball to measure the pore volume $V_m$ of coal, that is $V_m = N \times 1.33\pi r_0^3$. Under the definition of fractal, the relationship of radius of the small ball $r_0$ and the number of the ball needing filled the pore of coal is $N = cr_0^{-D}$. Thus:

$$V_m = 1.33\pi c r_0^{3-D} \tag{2}$$

Where, $r_0$ is the minimum scale of coal pore. *c* is constant. *d* is fractal dimension number.

The relationship of volume of pore $V_m$ and the radius *r* of pore is $Vm \propto r_0^{3-D}$, and differential function is $dVm/dr \propto r_0^{2-D}$, thus:

$$\log(dV_m / dr_0) \propto (2 - D)\log(r_0) \tag{3}$$

Including formula (1), another formula can be drawn in follow (Muhua et al. 1985):

$$\log(dV_m / dp) \propto (d - 4)\log p \tag{4}$$

That means dimension of pore can be deduced by the double-logarithm relationship of $dV_m/dp$ and p. if the relationship of $\text{Log}(dV_m/dp)$ and $\text{Log}(p)$ is line, that means distribution of pore has the character of fractal. And the dimension value D of pore, which is the slope of the line, can be gained, that is D = 4+K, where K is the slope of the line.

Normally, BET gas absorption method and high pressure mercury method are used to measure the character of coal pore. BET gas absorption method which $N_2$ or $CO_2$ is adsorbed at low temperature is a classical method to study the structure of coal surface. While the distribution of coal pore can be measured by the method of pressure mercury (Tan Muhua et al. 1985), so it is used in the study at test center of China University of Mining and Technology. Coal sample is 0.5~0.75 mm, before put into Pore Sizer 9310 micro-pore analysis equipment, which is made in USA, it must be vacuum evaporated for 5 h at 110$^0$C. So the curve of pore volume and mercury pressure is drawn, the same of pore radius and the surface. And the dimension of fractal is calculated.

## 2.1. *Result of Experiment*

The relationship of mercury pressure and accumulated pore volume is shown in Figure 1. There are much more pore in lignite and then gas coal, gas-fat coal. And anthracite has least pore.

Based on the result of test, pore volume and surface of different radius of coal is calculated in Table 2 and Table 3. The relationship of carbon and pore volume and surface of different metamorphic coal is shown in Figure 2. Frequency column figure of different range of pore radium and pore volume is shown in Figure 3. Frequency column figure of different range of pore radium and surface of coal is shown in Figure 4. Regressive curve of $\log(dV_m/dP)$ versus log(P) for lignite and gas coal is shown in Figure 5 and regressive curve of $\log(dV_m/dP)$ versus $\log(P)$ for gas-fat coal and anthracite is shown in Figure 6. Fractal dimension number of coal samples was calculated on the basis of Figure 5, Figure 6 and Formula (4) (Table 3).

Table 2. Characteristic parameter of pores in coal

| Coal Sample | Pore volume /mm$^3$/g | | | | | Pore surface /m$^2$/g | | | | |
|---|---|---|---|---|---|---|---|---|---|---|
| | Big pore | Middle pore | Transitional pore | Micro-pore | Total volume | Big pore | Middle pore | Transitional Pore | Micro-pore | Surface |
| | V1 | V2 | V3 | V4 | V | S1 | S2 | S3 | S4 | S |
| Lignite | 136 | 62 | 448 | 346 | 992 | 0.0066 | 0.1183 | 9.8535 | 15.9736 | 25.9520 |
| Gas coal | 90 | 24 | 142 | 80 | 336 | 0.0039 | 0.0376 | 2.7423 | 3.7651 | 6.5489 |
| Gas-fat coal | 89 | 43 | 140 | 64 | 336 | 0.0047 | 0.0539 | 2.8645 | 2.9117 | 5.8348 |
| Anthracite | 79 | 24 | 127 | 59 | 289 | 0.0005 | 0.0222 | 2.5668 | 2.7580 | 5.3475 |
| | V1/Vt | V2/Vt | V3/Vt | V4/Vt | | S1/St | S2/St | S3/St | S4/St | |
| Lignite | 13.7 | 6.3 | 45.2 | 34.9 | 100 | 0.03 | 0.46 | 37.97 | 61.55 | 100 |
| Gas coal | 26.8 | 7.1 | 42.3 | 23.8 | 100 | 0.06 | 0.57 | 41.87 | 57.49 | 100 |
| Gas-fat coal | 26.5 | 12.8 | 41.7 | 19.0 | 100 | 0.08 | 0.92 | 49.09 | 49.90 | 100 |
| Anthracite | 27.3 | 8.3 | 43.9 | 20.4 | 100 | 0.01 | 0.42 | 48.00 | 51.58 | 100 |

Table 3. Basic parameters of pore diameter, porosity and fractal dimension

| Coal sample | total volume of mercury intrusion | Total specific surface area | Median diameter of pore volume | Surface median diameter | Average diameter of pore (4V/A) | Bulk density | True density | Dimension | Porosity |
|---|---|---|---|---|---|---|---|---|---|
| | (mm$^3$/g) | (m$^2$/g) | (μ m) | (μ m) | (μ m) | (Kg/m$^3$) | (Kg/m$^3$) | (D) | (%) |
| Lignite | 992 | 25.9521 | 0.0142 | 0.0098 | 0.0153 | 1.1548 | 1.3041 | 3.24 | 11.449 |
| Gas coal | 336 | 6.5489 | 0.0277 | 0.0100 | 0.0205 | 1.3124 | 1.373 | 3.20 | 4.414 |
| Gas-fat coal | 336 | 5.8348 | 0.0326 | 0.0108 | 0.0230 | 1.3306 | 1.3928 | 3.16 | 4.466 |
| Anthracite | 289 | 5.3475 | 0.0254 | 0.0106 | 0.0216 | 1.3838 | 1.4415 | 3.13 | 4.003 |

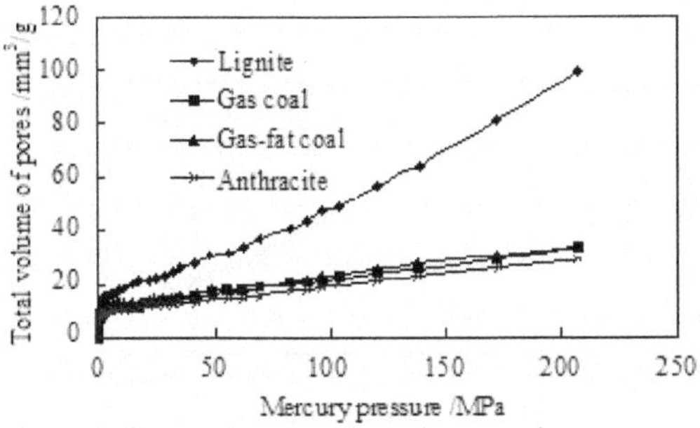

Figure 1. Curve of mercury intrusion porosimetry tests of coal

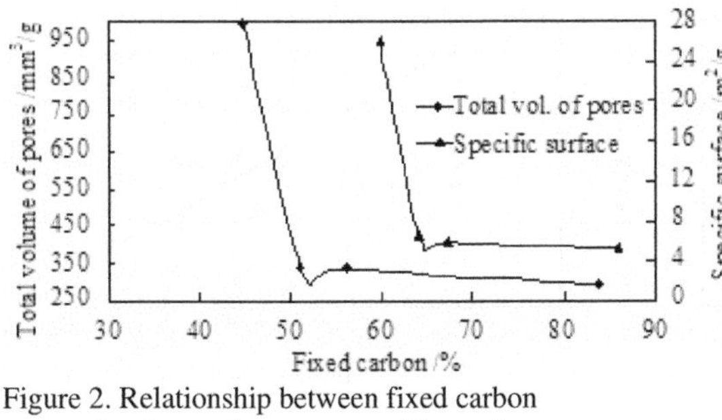

Figure 2. Relationship between fixed carbon and total volume of pores or specific surface

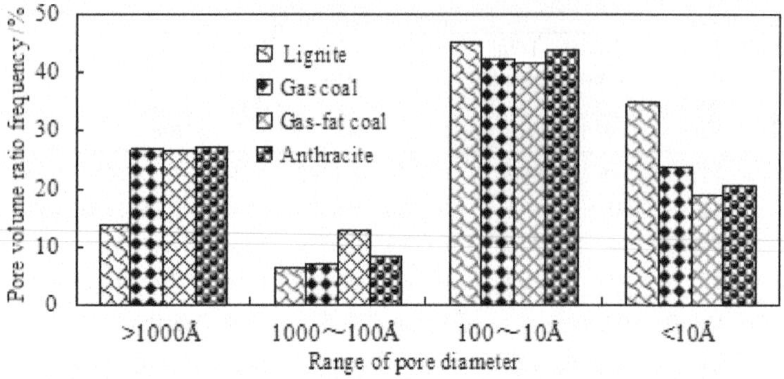


Figure 3. Histogram of pore volume ratio frequency in different range of pore diameter

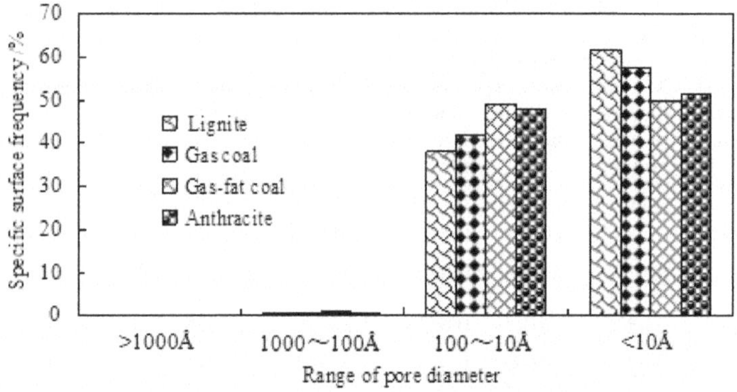


Figure 4. Histogram of specific surface frequency in different range of pore diameter

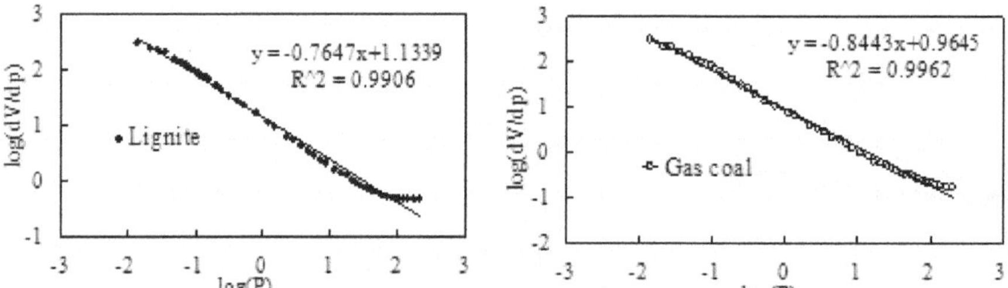


Figure 5. Regressive curve of log(*dVm/dP*) versus log(*P*) for lignite and gas coal

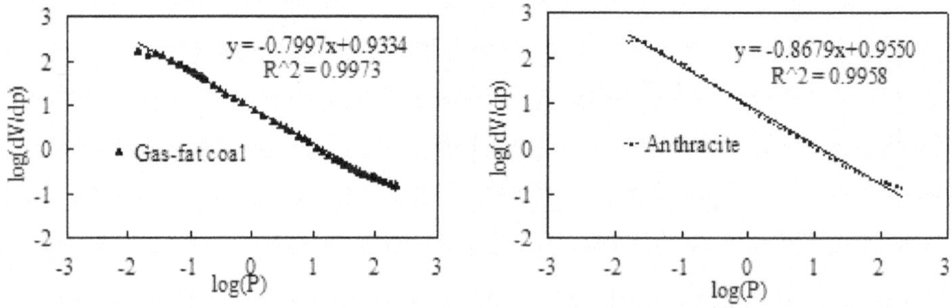


Figure 6. Regressive curve of log(*dVm/dP*) versus log(*P*) for gas-fat coal and anthracite

### 2.2. *Discussions*

1) Distribution of pore volume and specific surface area

From the result of experiment it can be seen, that pore volume decreases with coal metamorphic degree decrease. Lignite pore volume is biggest, and then gas coal, and gas-fat coal. Anthracite pore volume is smallest. Pores in lignite mainly belong to transitional pore and micro-pore. Pores in anthracite mainly belong to transitional pore.

Total specific surface area decreases with coal metamorphic degree. In anthracite and gas coal, main part of the pore is micro-pore, and then transitional pore. But there is no obvious different of specific surface area about micro-pore and transitional pore in gas-fat coal and anthracite. That is means specific surface area is main constituted from micro-pore and transitional pore. And indicated that oxygen and coal was combined on the surface of micro-pore and transitional pore.

2) Character of coal pore diameter distribution

By using mercury displacement method, average diameter of coal pore, median diameter of pore volume, median diameter of specific surface area, and the biggest diameter of connected pore can be calculated. The order of average diameter of coal pore is lignite<anthracite<gas coal<gas-fat coal. The order of median diameter of pore volume is lignite<blind<gas coal<gas-fat coal. And the order of median diameter of specific surface area is lignite<gas coal<blind< <gas-fat coal.

3) Fractal dimension of coal

Regressive curve correlation coefficient of $\log(dV_m/dp) \sim \log(P)$ is over 0.99. That means that surface structure of coal samples has fractal character. And surface structure of coal can be characterized by fractal dimension. The value of dimension decreases with coal metamorphic degree increase. That means that diversity of coal oxidization ability is caused by different surface roughness and nonuniformity, resulting in difference of coal oxidization ability.

## 3. THE RELATIONSHIP BETWEEN PORE STRUCTURE IN COAL AND ITS LOW TEMPERATURE OXIDATION

The relationship of total pore volume, specific surface area, the value of fractal dimension, temperature raise rate of low temperature oxidation, rate of heat release, and the spontaneous combustion period is shown in Table 4.

Table 4. Relationship between coal pore structure and low temperature oxidation

| Coal sample | Incubation period (d) | Total pore volume ($mm^3/g$) | Total specific surface ($m^2/g$) | Porosity (%) | Rate of temperature rise (°C/h) | Rate of heat release (W/(min·g)) | Dimension (D) |
|---|---|---|---|---|---|---|---|
| Lignite | 18 | 992 | 25.95 | 11.449 | 12.79 | 25.59 | 3.24 |
| Gas coal | 45 | 336 | 6.55 | 4.414 | 0.77 | 0.0232 | 3.20 |
| Gas-fat coal | 90 | 336 | 5.83 | 4.466 | 0.53 | 0.0133 | 3.16 |
| Anthracite | no fire | 289 | 5.35 | 4.003 | 0.18 | 0.0053 | 3.13 |

Note: Rate of temperature raise and rate of heat release is the average value gained by experiment of low temperature oxidation at the adiabatic condition of 40 ~ 70°C

Coal spontaneous combustion period and ability of self-heat at low temperature oxidation of coal is related to total pore volume and specific surface area and the value of fractal dimension. Increasing with coal metamorphic degree, coal's total pore volume, specific surface area and fractal dimension is decrease. The sort of coal total pore volume is lignite > gas coal = gas-fat coal > anthracite. The sort of coal specific surface area is lignite > gas coal > gas-fat coal > anthracite. And the sort of coal fractal dimension value is lignite > gas coal > gas-fat coal > blind. On the contrary, the rate of self-heat spontaneous oxidation and heat release is decrease, and real spontaneous combustion period is increase. The bigger the value fractal dimension is the more pore the coal is. And the more surface of the coal is the more oxidizing ability is. So it is easier spontaneous combustion.

Small difference of specific surface area can lead to great difference of coal low temperature oxidization ability. Total specific surface area of gas coal is 6.55 $m^2/g$. And gas-fat coal is 5.83 $m^2/g$. So the difference of the two is 0.72 $m^2/g$. But there is great difference between low temperature oxidization ability and actual incubation period. The rate of temperature raise and heat release of gas coal is higher than gas-fat coal. And the sort of real spontaneous combustion period of different kind coal is gas coal < gas-fat coal < anthracite. Thus specific surface area is the key factor of coal spontaneous combustion, if the coal has the same pore volume. Because the larger the surface is the more probability of combining coal and oxygen is, and the more serious fire hazard in coal mines.

## 4. CONCLUSIONS

Based on the research, relationship between pore structure in coal and its low temperature oxidation was obtained. Coal pore volume, total specific surface area, and fractal dimension value decreases with coal metamorphism degree increase. Coal oxidization rate and heat release rate decrease with decrease of total pore volume, specific surface area, and fractal dimension value. Specific surface area is the key factor influencing on coal spontaneous combustion, when pore volume in coal is the same. The larger coal surface area is, the more probability coal combining with oxygen has. Coal spontaneous combustion potential is decided by coal pore structure.

## REFERENCES

Bushuev V.A., Oreshko A.P. 2003. X-ray specular reflection under the conditions of extremely asymmetric noncoplanar diffraction from a bicrystal [J]. Kristallografiya 48(2): 212-219

Bouwmam R., Ivo L.C. 1980, Freriks [J], Fuel 59(5): 315-322

Fu Xuehai, Qin Yong, Xue Xiuqian, et al. 2001. Research on Fractals of Pore and Fracture-Structure of Coal Reservoirs[J], Journal of China University of Mining & Technology (Natural Science) 30(3): 225-228. (In Chinese)

Hu Song, Sun Xuexin, et al. 2003. Inner-pore structure change of Huainan coal char particle during combustion [J], Journal of Chemical Industry and Engineering (China) 54(1): 107-111. (In Chinese)

Nugroho, Y.S. McIntosh, A. C. Gibbs, B. M. 2000. Low-temperature oxidation of single and blended coals, Fuel (79): 1951-1961. (In Chinese)

Su Bo, Wang Kuijun, Zhang Xinghua 1999. Study on characteristics of pore fractal structure in coal [J], Safety in Coal Mines (1): 38- 40. (In Chinese)

Schmidt, P. Kalliat, W. Chul Y. 1984. Small angle X-ray scattering of the submicroscopic porosity of some low-rank coals [A], ACS Symposium Series: 79-94

Su Xianbo, Chen Jiangfeng, Sun Junmin. et al. (eds) 2001. Geology of Coalbed Methane and Its Exploration and Development: 15-40. Beijing: Science Press. (In Chinese)

Tan Muhua, Huang Yunyuan (eds) 1985. Surface Physical Chemistry, Beijing: China Architecture & Building Press: 56-58. (In Chinese)

Wu Jun. 1993. Study on relationship between characteristics of coal micro-pores and migration or storage of oil and gas [J], Science China (Part B) 23(1): 77-78. (In Chinese)

International Mining Forum 2010, Liu et al. (eds) © 2010 Taylor & Francis Group, London, UK. ISBN 978-0-415-59896-5

# Research on overburden strata deformation and protective effect with far-distance lower protective layer

Biming Shi
*Anhui University of Science and Technology, Huainan, Anhui, China*

ABSTRACT: Based on finite element calculation method, the dynamic process of far-distance protective layer mining is simulated by using the RFPA application system, through which the regulation of overburden strata rupture and movement, and the deformation characteristics of protected seam are gain. By investigating the gas pressure of protected seam, the permeability changes and the deformation laws of coal seam, the effects of outburst prevention by protected seam exploitation is analyzed. The experiment results at site were consistent with those results shown above, and the function mechanism were analyzed of the upper coal for optimizing collocation of unloading pressure holes and eliminating coal and gas outburst danger.

KEYWORDS: Far-distance lower protective layer, coal seam deformation, gas pressure, coal and gas outburst

## 1. INTRODUCTION

Coal and gas outburst is a dynamic phenomenon that broken coal and gas violently jet from the internal of coal to mining free-space suddenly under the effect of geophysical field which is mainly on the interaction between ground stress and coal gas. Technology of preventing seam mining is an effective method to prevent and control outburst risk which have been proven by lots of practice and established by the form of regulations, and it have be widely used at home and abroad. The purpose of preventing seam mining is relieving protected seam pressure, releasing the elastic potential energy of coal and increasing the permeability of coal seam, which is beneficial for gas emission from protected seam and decreasing coal internal energy. It is a conducive to achieve mine safety production by investigating the characteristic of deformation and displacement of protected seam and analyzing the protective effect with long-distance protective layer mining.

## 2. GENERAL SITUATION OF WORKING FACE

The elevation for wind roadway and machine roadway in one mine protective layer ($B_{11}$) is between -600 m and -650 m, inclined length is 200 m, strike length is 1840 m, the average thickness of coal seam is 1.9 m and the average angle about 9°. The primitive content of gas on coal seam is about 5.5 $m^3/t$ and the gas relative emission rate is 5.23~7.32 $m^3/t$, which belong to low gas coal seam. The average thickness of protected seam ($C_{13}$) is 6.0m,the primitive gas content ,pressure and the original permeability coefficient of coal seam is respectively 12~22 $m^3/t$, 4.4 MPa and $3.92\times10^{-2}$ $m^2/(MPa^2.d)$. Coal seam have outburst hazard and methane overrun in the process of mining which occurs many times in adjacent area. Currently it mainly adopt the way which discharge upper of mined area by using roof drills along strike, to reduce gas emission of upper corner in working face. Simultaneously discharging unpressure relief in coal

sion of upper corner in working face. Simultaneously discharging unpressure relief in coal seam to reduce gas content, gas pressure and gas elastic potential energy. But the coal seam belongs to low permeability and high outburst hazard ,so gas drainage effect is not obvious, then it is necessary to take effective measures to increase coal seam permeability, combined with occurrence condition of the mine coal seam to exploit coal seam $B_{11}$ and protect coal seam $C_{13}$. The average normal distance between the coal seam $B_{11}$ and the coal seam $C_{13}$ is 66.7 m, belong to long distance protection layer. As Figure1 is strata of floor coal system in coal bed roof.

| Stratum | True Thickness /m | Total thickness /m | Lithology |
|---|---|---|---|
| P | 15.0 | 15.0 | sandstone |
| | 5.0 | 20.0 | mudstone |
| | 3.0 | 23.0 | sandy mudstone |
| | 6.0 | 29.0 | Coal seam $C_{13}$ |
| | 3.0 | 32.0 | sandy mudstone |
| | 2.5 | 34.5 | fine sandstone |
| | 3.5 | 38.0 | mudstone |
| | 2.9 | 40.9 | siltstone |
| | 6.5 | 47.4 | mudstone * |
| | 2.5 | 49.9 | medium grain sandstone |
| | 2.0 | 51.9 | siltstone |
| | 8.0 | 59.9 | medium grain sandstone |
| | 13.8 | 73.7 | sandy mudstone |
| | 1.7 | 75.4 | siltstone |
| | 3.2 | 78.6 | mudstone |
| | 3.8 | 82.4 | siltstone |
| | 4.4 | 86.8 | fine sandstone |
| | 2.6 | 89.4 | medium grain sandstone |
| | 1.1 | 90.5 | siltstone |
| | 2.3 | 92.8 | sandy mudstone |
| | 1.3 | 94.1 | clay rock |
| | 2.1 | 96.2 | sandy mudstone |
| | 1.9 | 98.1 | coal seam $B_{11b}$ |
| | 1.9 | 100.0 | sandy mudstone |

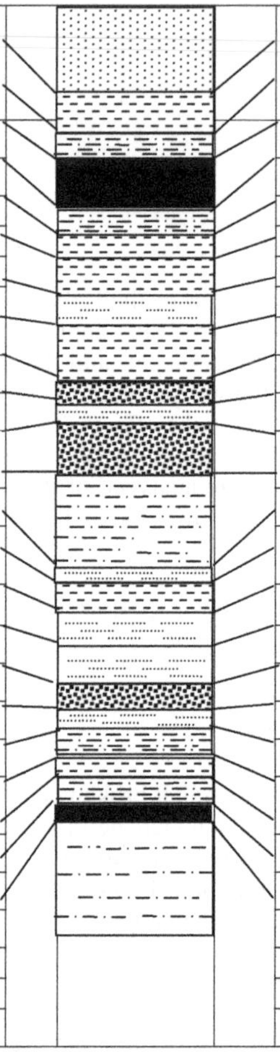

Figure 1. Histogram of coal system strata in seam roof and floor

## 3. ANALYSIS OF NUMERICAL SIMULATION

### 3.1. *Model*

Numerical model uses two-dimensional plane strained model. According to specific reality of research, it main simulate strata movement between protective seam layer ($B_{11}$) and Protected seam ($C_{13}$), deformation of protected seam and permeability of Coal Seam, there are 16 layers in cylindrical rock model.

Using uniformly distributed load instead of weight in the upper strata of model, the left and right bottom boundary are fixed, the horizontal length and the vertical length of the model is respectively 260 m, and 120 m, divided into 260 × 120 units, specifically shown in Figure 2, the mining method excavation by steps was been used in protective layer, mining all high and each step excavate 5 units, that is 10 m, collapse method of roof management .

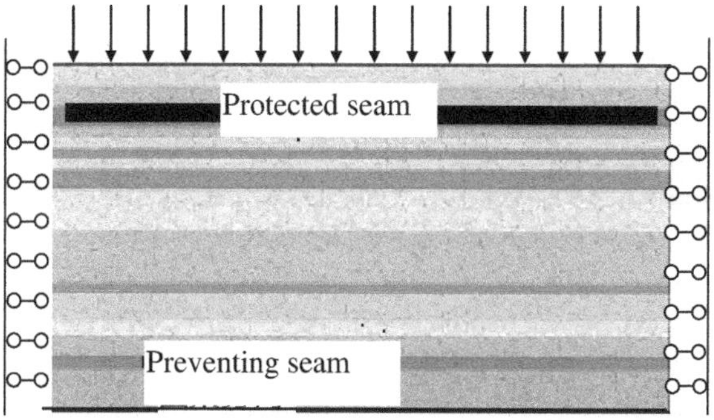


Figure 2. Numerical Simulation Model

### 3.2. *The result of numerical simulation and analysis*

#### 3.2.1. *Overburden rocks collapse and fracture evolution regularity*

Figure 3 is the numerical simulation result of dynamic development process of overburden rocks collapse with protective layer. For the limited space, just 6 representative steps of distribution map for damage elastic modulus are given. With the analysis of Figure 3, it can be obtained that overburden strata suspend after the formation of open-off cut, immediate roof emerge discharging swelling deformation by excavation, and overburden strata bend because of gravity action as work face promoting. When work face promote around 40 m, the "false plastic rock beam" form in middle of overlying strata. Because top plate of $B_{11b}$ coal seam is compound roof, which mixed by mudstone and sand mudstone, mainly with sand mudstone, and not easy to collapse, so the overburden strata appeared large range movement as work face continue promote to 60 m. And the basic strata on the top of the front and back coal well of mined-out area appeared shear failure and tensile damage fracture.

After the first breakdown of basic roof, as work face promoting, the range of abscission layer of enlarge constantly. When promote to 80 m, the top plate fracture in the end and middle, the top roof appear "secondary crack" and fall-down, that is the first periodic weighting phenomenon. At the same time, it appear abscission layer fracture and vertical fracture on top of caving zone. With work face promoting, immediate roof fall on their own influenced by mining, the overburden strata constantly emerge destruction movement under principal stress and shear stress, and abscission layer and vertical fracture develop upward which influence secondary overlying layer. As work face continue promoting, overburden strata appears periodic motion characteristic, while the range of caving movement and facture development tended to be stable on the vertical.

### 3.3. *Regulation of protected seam deformation*

Figure 4 shows the relationship between deformation of the protected layer and face promotion of the protective layer. In the figure the horizontal axis is for the face relative position, vertical axis be for coal deformation of protective layer. Some conclusion can be drawn through the analysis of Figure 4:

(1) Mining range have great effort on protected layer. At the begin of mining of protective seam, protective layer exploitation has little influence on coal mass deformation of protected seam. When actual mining arrived 60 m, the maximum swelling deformation value is 48 mm. With the advance of working face protected layer swelling deformation value increases gradually, the maximum swelling value reached 190 mm, which is about 31.7‰ of protected coal seam thickness.

With the advance of protecting seam face, the maximum principal stress of the protected seam has been restored to some extent, and the expansive deformation of coal seam gradually decreased. They tended to be stable until they reached a certain value. In a certain range of goaf that is behind the coal face of the protection layer, the deformation of coal seam whose stress do not recover remained higher value.

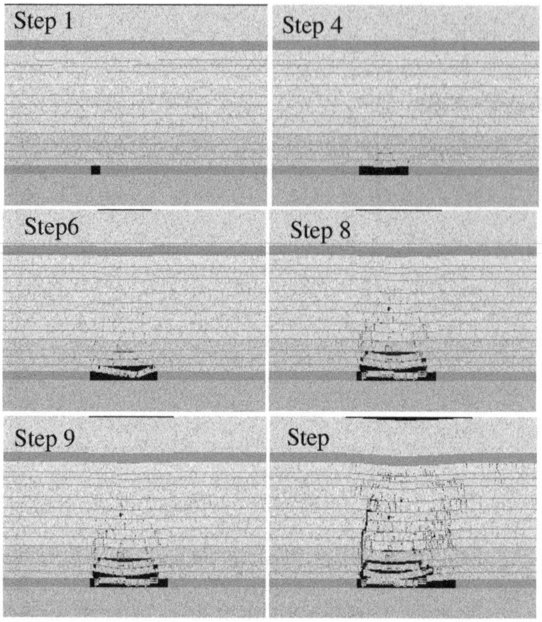


Figure 3. Dynamic development process of overburden strata withprotective layer mining

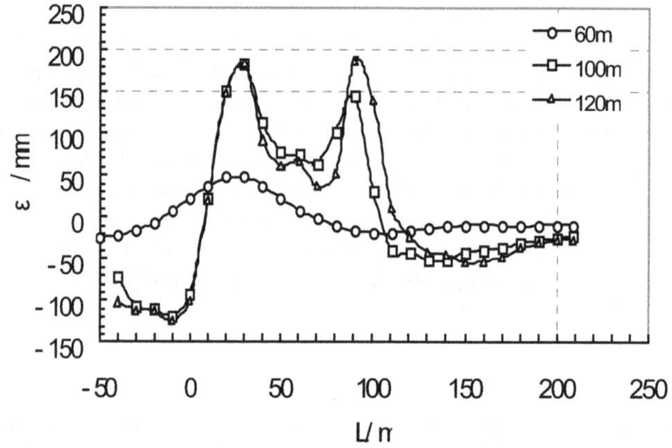


Figure 4. Distribution characteristic of the thickness changes along with the change of mining excavation range

(2) Overburden pressure has been transferred because of the protecting seam exploitation. In the coal pillar of the protected seam, which is over the cut-hole of the protective layer, the maximum compression deformation of the coal seam is up to 124 mm, is about 20.7‰ of $C_{13}$ coal seam thickness. The maximum compression of the protected seam which is in front of the working face is 51.5 mm, is about 8.58‰ of $C_{13}$ coal seam.

(3) With the advance of protecting seam face, the protected seam deformation presented the discipline of compression, expansion, swelling decreased to a stable variation. As can be seen from Figure 4, the maximum swelling deformation of coal is at a distance of 40 m in front of protective seam's open-off cut and 20 m behind the coal wall of working face, while the coal deformation expansion in the center of goaf was smaller. It indicates that there has a stable in-

crease of coal deformation zone in the protected seam in bending subsidence area after long-distance protection seam mining. The coal fractures are developed, so it's good for gas transport and gas drainage in coal seam.

## 4. THE MINING EFFECT OF LONG DISTANCE PROTECTIVE SEAM

### 4.1. *Gas suction for pressure relief of the protected seam*

According to testing layout, comparison of gas drainage of surface drilling in distressed zone, roof drilling along the strike, and drillings floor roadway of coal seam, we decide to use the drillings floor roadway of coal seam pressure relief. that is, located in sandstone (its thickness is 10~20 m) of the floor of the protective seam of coal $C_{13}$ it is 50 m away from the main return airway. The drainage roadway's direction is consistent with the strike direction of rock strata. The roadway length is 950 m. The section area of the roadway is 6.0 $m^2$. A drilling field is arranged every 10min the pressure region and every 40 m in the pressure relief region, so there are 3 drill fields in the pressure region of 2322(3) face and 9 drill sites in the pressure relief region of 2121(3) face. There are 39 drill sites arranged in the relief region of 2121(3) / 2322(3) faces in total. The area of drilling field is 6.16 $m^2$ with anchor supporting. A group of drilling field (4) fan-shaped borehole are drilled in the pressure relief area, its diameter is 91mm, borehole spacing is 40m, borehole 1 # and 4 # are 20 m away from the upper and lower gateways, drilling into the $C_{13}$ roof 0.5 m. The layout of drill field is shown in Figure 5, drilling parameters are shown in Table 1.

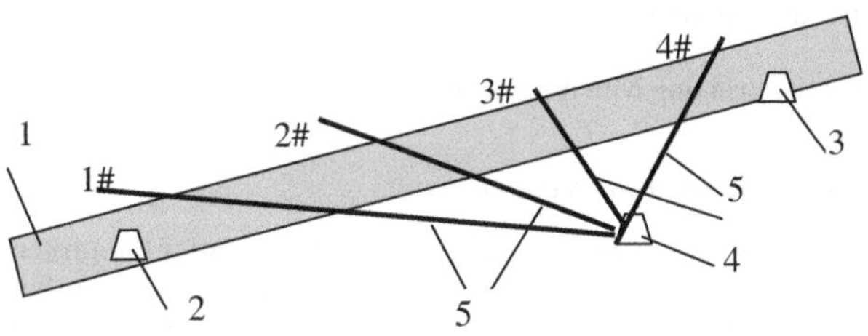


Figure 5. Sketch map of the boreholes layout
1 - coal seam $C_{13}$; 2 - lane machine; 3 – airway;
4 - floor drainage tunnel; 5 - boreholes of gas drainage

Table 1. Parameters of boreholes

| Borehole No. | Angle between borehole and coal seam /°C | Borehole length /m |
|---|---|---|
| 1 | 7 | 129.5 |
| 2 | 12 | 73.3 |
| 3 | 47 | 21.4 |
| 4 | 20 | 45.6 |

### 4.2. *The variation of gas pressure in protective seam*

11 boreholes are arranged to determine gas pressure of coal seam $C_{13}$, in which 4 pressure measurement boreholes are arranged along the strike direction at drainage roadways in floor of coal seam $C_{13}$, 7 pressure measurement boreholes along the inclination direction of the coal seam. Taking the results of borehole 4# in the drilling field 40# for example, it is shown in Figure 6 that the gas pressure measured in the borehole of protective seam varying with the advance of face. With the advance of protective seam, when the face is advanced at distance of strike projection of pressure measuring borehole of about 100 m, the gas pressure started from 4.4 MPa to decline about 80 m away from the pressure measurement borehole started to drop dramatically and when it is 62 m away from the pressure measurement borehole, the gas pres-

sure dropped to zero. Mainly due to pressure measurement borehole is sealed with cement mortar, under the action of the advance stress concentration of the stope, the surrounding rock occur fracture and pressure measurement borehole is connected with the bottom of drainage roadway. When the face pass by pressure measurement holes at distance of about 400 m, due to stress recovery makes rock fracture close again, pressure measurement borehole of gas pressure increased from zero gradually, keep long-term stabilize at 0.4 MPa. The value represented the residual gas pressure after drainage of coal seam gas in the protected seam, gas pressure is significantly reduced below the specified requirements.

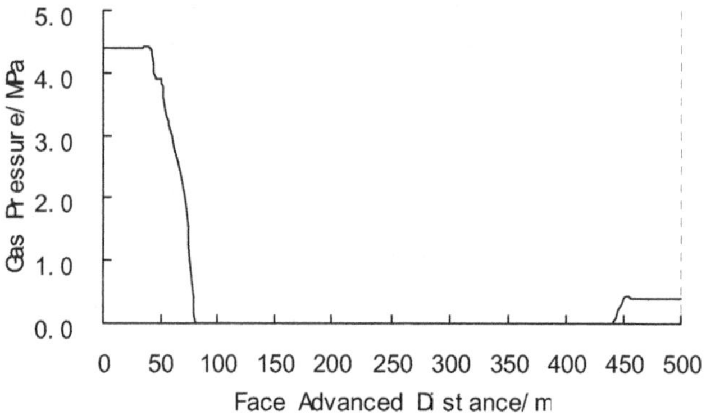


Figure 6. The gas pressure of protective seam changing with the face advance

### 4.3. *The variation of permeability coefficient of protected coal seam*

The original permeability coefficient of protected seam is $0.01135 m^2/(MPa^2 \cdot d)$, after the protective seam is mined, pressure relief caused by the protected seam, and after the pressure relief, the permeability coefficient is $32.687\ m^2/(MPa^2 \cdot d)$, 2880 times larger than the original permeability coefficient.

### 4.4. *The relative deformation laws of protected coal seam*

The deformation of coal seam $C_{13}$ is measured by the base point method, that is, installing measuring points respectively in the roof and floor rock of coal seam by deep boreholes to determine the deformation of coal seam by observing the relative displacement of the two measuring points. The measuring results show that during coal seam B11 mining, the maximum compression deformation of coal seam $C_{13}$ is up to 27 mm, the maximum swelling deformation is 210.44 mm. The maximum relative compression deformation of coal seam is 3.37‰. The maximum swelling deformation is 26.3‰. This shows protective seam mining leads to a greater decline of the protected seam stress, and the fracture in the coal seam increases. The measuring results are basically close to the numerical simulation.

## 5. ANALYSIS ON OUTBURST PREVENTION EFFECTS OF PROTECTED SEAM WITH PROTECTIVE SEAM MINING

In order to examine the outburst hazard of protected seam face, 3 checkpoint of gas outburst index were set along the top, middle and lower of the protected seam face regularly which can be used to measure the maximum drilling cuttings quantity and desorption velocity of drilling gas. Through many measurements, the desorption velocity of drilling gas in coal seam is less than 3.7 l/min, the maximum drilling cuttings quantity is less than 5.5 kg/m, they both are less than critical value that stipulated. There is no dynamic phenomenon during extraction. It is show that the coal and gas outburst risk of protected seam is effectively prevented by long distance protective seam mining combined with the drainage of pressure relieved methane.

## 6. CONCLUSIONS

After long distance protective coal seam is mined, the elastic potential energy of protected seam is reduced, the permeability of coal seam is increased, and it create the condition for coal methane simultaneous extraction. Combined with the drainage of pressure relief methane in coal seam, it will reduce the gas pressure greatly, decrease the gas content in coal seam, and increase coal mass strength. Thereby eliminate or reduce the outburst hazard of protected seam, and it realize the safety and high efficiency mining in the outburst coal seam.

## ACKNOWLEDGMENT

This work was supported by General Program of National Science Foundation(50874005).

## REFERENCES

Cheng Yuanping, Yu Qixiang, Yuan Liang. 2003. Gas extraction techniques and movement properties of long distance and pressure relief rock mass upon exploited coal seam. Journal of Liaoning Technical University 22(4): 483-486. (In Chinese)

Fu Yufang, Liang Zhengzhao, TANG Chun-an. 2000. Numerical simulation on influence of mesoscopic heterogeneity on macroscopic behavior of rock failure. Chinese Journal of Geotechnical Engineering 22(11): 705-710. (In Chinese)

Shi Biming, Yu Qixiang, Wang Kai. 2006. Experimental Study on regulation of dynamic evolution about overlying coal seam permeability with long distance protective layer. Chinese Journal of Rock Mechanics and Engineering 25(9): 1917-1921. (In Chinese)

Wang Yan, Wang Lujun. 2008. Study on Mining-induced Fracture Distribution of Coal and Rock Mass near Protective Layer Mined. Coal mine safety 39(1): 11-13. (In Chinese)

Yang Tianhong, Tang Chunan, Xu Tao, et al. 2004. Seepage Characteristics in Rock Failure - Theory, Model and Applications. Science Press. (In Chinese)

International Mining Forum 2010, Liu et al. (eds) © 2010 Taylor & Francis Group, London, UK. ISBN 978-0-415-59896-5

# Control technology of structure coupling and engineering application to inclined-shaft under abundant-water conditions

Yongping Wu, Youfu Zeng, Hongwei Wang
*Xi'an University of Science and Technology, Xi'an, Shanxi Province, China*

Ping Ye
*Qingshuiying Coal Mine, Shenhua Ningxia Coal Industry Group Co., Ltd., Yinchuan, Ningxia, China*

ABSTRACT: Deformation mechanism of roadway in abundant water-soft rock environment is more complicated than the general one, and the stability of surrounding rock for this roadway, both of which are not effectively controlled by the single and traditional supporting systems of mine workings, thus badly affecting the efficiency and safety of coal production. On the basis of controlling over the stability of surrounding rock of the No. 1 return slant at Qing shui ying coal mine, the inherent mechanism of the loss of the pre-stressed bolting and loosening between bolts and surrounding rock were analyzed, and the results showed that it resulted from the effects of groundwater, the weak rock mass, the non-coupling structure, and the irrational parameters of anchor cable. The practice in situ testifies that a key point of control soft rock roadway was controlled water, under the condition of blanking off water, and controlling technology of multi-media structure coupling based on high-strength and pretension system, which can effectively control the soft rock roadway. In addition, the stability of surrounding rock, which was coupling support is predicted. As the expected results are obtained in the industrial test, the stability of surrounding rock for coal roadway is effectively controlled and this technique can be used to control the stability of surrounding rock for similar underground constitution.

KEYWORDS: Abundant water-soft rock environment, inclined-shaft, anchorage capacity invalidation, structure coupling

## 1. INTRODUCTION

Along with the ever-increasing enlargement of social energy needs, coal mining enters to deep or more difficult mining coal seam (Gao 2007, Wu et al. 2006; Wu et al. 2009; Lai et al. 2010) gradually, which has the complex geological condition and the great ground stress. Following the mining intensity promotion and scale expansion, the mine workings section is also constantly larger, and the deterioration on surrounding-rock environment of roadway is caused by the large section, the complicated and Protean geological condition, the influence of groundwater and intense mining disturbance and so on. At present, the combined support is generally carried on the mine workings maintenance in view of the complex environment, such as bolt-mesh-anchor support, rigid support, anchor-net-spray support, bolt-grouting support and so on (Peng et al. 2009, Zeng 2006). However, the support system of the roadway destroy, surrounding rock instability of mine workings, and even the major accident of mine flooding and roof fall are caused by the reason of insufficient survey and analysis to the site geological condition of surrounding rock, support structure non-coupling, unreasonable parameters, nonstandard construction technology and so on. Therefore, the support of surrounding rock of large section roadway under complex environment has become the one of most difficult technical problems, which needs to be solved during the mine construction in Western China and safety and high efficiency mining.

This article aims at the complex soft rock environment of the No. 1return slant at Qing shuiying coal mine in Ningxia Province, which has lofty and crushing rock and rich fracture and ground water. From the angle of controlling the water flowing fracture growth and enhancing the integrity of support structure, the structure coupling support scheme-the bolt-mesh-anchor + +grouting reinforcement section support proposed, its success provided a beneficial model to the similar condition and was worth further promoting.

## 2. ENGINEERING SITUATION

### 2.1. *Geological conditions*

The cross section of return slant in Qingshuiying coal mine is the straight wall semicircular arch. Its net width is 4800 mm, net height is 4210 mm, and net area is 16.73 $m^2$. The design length of the slant is 1366 m, with its average angle is 20°. It was opened from the correspond ground location of sixth-fifth coal seam and entered into the second coal seam at the +1250 m level, then excavated along the coal bed to the +860 m level. The construction length along the 2nd coal seam is 918 m, the surrounding rock intensity is low, and the geological condition is complicated and diversified with joints and fractures. The overburden silt rock containing rich ground water, and the formation lithology primarily is composed of siltstone, shale cemented medium granular sandstones or mudstone, which are extremely easy to soften and slime after meeting water. The main physical mechanical parameters of surrounding rock at this sector slant were shown in Table 1.

Table 1. The physical-mechanical parameters of inclined shaft

| No. | Lithology | Density (kg/$m^3$) | Thick ness (m) | Elastic modulus (GPa) | Poisson ratio | Friction angle (°) | Compressive strength (MPa) | Tensile strength (MPa) |
|---|---|---|---|---|---|---|---|---|
| 1 | siltstone | 2900 | 6.68 | 9 | 0.14 | 38 | 3.95 | 2.80 |
| 2 | Medium sandstone | 2170 | 1.3 | 9 | 0.10 | 35 | 1.69 | 0.66 |
| 3 | mudstone | 1800 | 3.11 | 5 | 0.30 | 28 | 1.54 | 0.40 |
| 4 | $2^{nd}$ coal seam | 1460 | 7.88 | 2 | 0.32 | 25 | 2.00 | 0.20 |
| 5 | siltstone | 2900 | 4.92 | 9 | 0.14 | 38 | 3.95 | 2.80 |

### 2.2. *Original support design scheme*

The original support form of this slant is bolt-mesh-spurting supporting, the main support parameters are: Roof bolt: The MSGLW-335/20 common steel bolt is used, the specification is Φ20×2400 mm, the row & line space is 800 mm×800 mm, the four MSK2335 resin rolls are used to lengthen anchor, and the specification of steel support plate is 150×150×10 mm. Side bolt: The glass round steel bolt is used, the specification is Φ16×2000 mm, the row & line space is 800 mm×800 mm, the specification of wooden support plate is 400×200×50 mm and the two MSK2335 resin rolls are used to anchor. Roof cable: The specification of pre-stressed steel strand is Φ17.8×6000 mm, the row & line space is 2000×2400 mm, the seven MSK2335 resin rolls are used to anchor each root, the lowest breaking strength is 260 KN, the retightening force is 120-210 KN; and the anchor cable joist uses $14^{\#}$ channel steel with length 350 mm. Metal net: made by Φ6.5 round steel, the specification of metal net is 4050×900 mm, the specification of mesh is 150×150 mm. Anchor beam: Assembly welded by two Φ14 round steel, the specification is 3600×100 mm. Gunite concrete: the concrete type is C20, and the thickness is 100 mm.

### 2.3. *Deformation and failure of inclined shaft*

Since the slant was excavated along the floor of the $2^{nd}$ coal seam, the roof coal couldn't be remained to protect roof for the thinned coal seam, the vary degree spurt crack, shedding, subsidence, partial collapse and water spraying, all of which occurred along the arch crown. The roadway sides fell off, the serious floor heaven and even partial caving occurred (Fig. 1), which had serious influence to normal operation of person and vehicle in slant. The main performances are: 1) The destruction is in locality, mainly concentrated in sector with bad geological condition and the roadways' intersection. 2) The roof spurts occurred deformation under extrusion, partial collapse and subsidence, the roof caving happened when the partial collapse was serious. 3) With the slant stairs inclined to left side when the serious floor heaven occurred, the drainage cracked and deformed, the maximum floor heaven reached above 1400 mm. 4) The roadway sides spurt cracked and bulged outward seriously. 5) The anchor section occurred cohesion failure, the anchor cable was pulled out or occurred shearing failure. 6) The shedding and partial collapse often went with water spraying.

(a) Roof shedding, partial collapse and floor heaven

(b) Partial collapse after roof caving

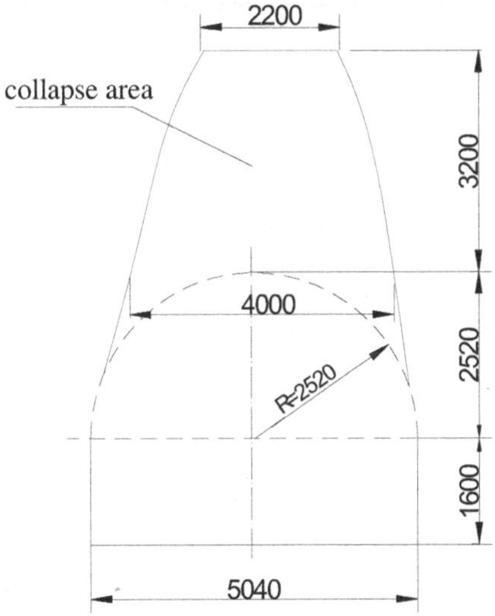


(c) Sketch of roof caving

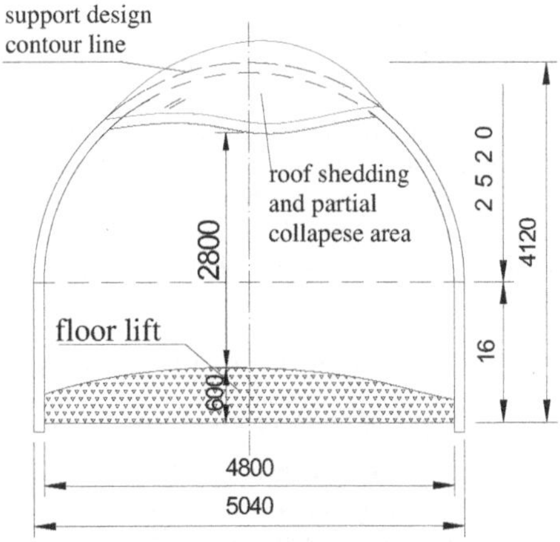


(d) Sketch of roof shedding, partial collapse and floor heaven

Figure 1. The scene and pencil sketch of surrounding rock failure in gyrus wind inclined shaft

## 3. ANALYSIS OF INCLINED SHAFT DEFORMATION AND FAILURE

The surrounding rock failure and deformation of roadway were the results of combined action of various factors, the main factors including lithology, section size, ground water, in-site stress dynamic pressure, support structure and so on, which all affected the surrounding rock stability of this section inclined shaft.

1) Surrounding rock mechanical properties. Around the inclined shaft is the 2nd coal seam, mudstone, sandstone and siltstone, its mechanical properties such as compressive strength, shearing strength and the project characteristic is bad. The roof rock is loose and broken and easy to cave; the joint crack develops and the intensity is low in sidewall coal seam; the easy break and swelling soft rock of floor has low residual strength and long-term strength, the surrounding rock deformation and failure show strong militancy and the strain softening under low confining pressure.

2) The big shaft span. The original design span of shaft was 5.3 m, because the actual construction of the inclined shaft local region was improper, the span reached above 5.7 m and the intersection may be even bigger. It let safety factor decline and caused deformation and failure of the surrounding rock of inclined shaft.

3) Water effect. The cable passed through roof aquifuge and then entered into overburden siltstone aquifer, the anchor end couldn't completely seal up the hole section in aquifer, so the aquifer water discharge into the anchor area, and cause that the intensity of rock mass mainly composed of charcoal mudstone and sandy mudstone, the friction coefficient and the elasticity coefficient of the rock and structural plane reduced, and rock mass volume expansion, and even the mudding disintegration. Finally, the anchor section occurred cohesion failure and the anchor cable failed.

4) The floor stability. The floor of shaft is made by mudstone with low strength, the strata joint crack develops and the water is very easy to enter into the weak strata, which can cause the strata to soften, disintegration and swelling, so the floor surrounding rock loses intensity and the rapid deformation happens. When the roof pressure is big, the surrounding rock of roadway floor appears stress concentration and plastic deformation and buckling failure and floor heaven, then the roadway roof stability is influenced immediately.

5) Support structure non-coupling. The choice of roadway support pattern and parameters, reasonable or not, has a direct relation to the working state of support system and the stable degree of surrounding rock. If the support system's performance were not adapt to the surrounding rock deformation characteristic, the support system would fail and the rock deformation rate would be increased and the stability would drop. The Q335 common steel bolt was used, but the anchor bolt pretightening force was lowered and even to zero in the shaft (generally 6–10 kN), it could not play the bolt initiative support role fully controlling early deformation the surrounding rock and the roof separation. Along with surrounding rock deformation of roadway, the cable force increased rapidly and the bolt force increased slowly, the elongation ratio of cable was lower than that of bolt, so the cable anchor end occurred shearing failure or the cable was sheared by deformed anchor beam under stress concentration, which caused the shotcrete shedding and the bearing and deformation of cable and bolt uncoordinated. Finally, the entire support system destroyed, and the slant surrounding rock failed.

## 4. STRUCTURE COUPLING SUPPORT SCHEME

### 4.1. *The principle of multi-medium structure coupling control technology*

The Multi-medium structure coupling control technology (Zeng et al, 2009) (Fig. 2) is one complex coupling system support method, which takes the bolt support as a foundation and takes other medium reinforcement as auxiliary support, such as cable, grouting, rigid support and so on. It can display each kind of support, the respective support and bearing function of each kind of supports fully, and also can make each kind of supports compliment each other to control the deformation stability of roadway. According to the surrounding rock deformation characteristics of roadway under abundant-water conditions, the structure coupling support countermeasures were taken patiently. During the period of driving to secondary supporting, the own bearing capacity of surrounding rock and the flexible characteristic of primary support structure such as the shotcrete layer, the net, bolt and so on. Under the condition of guarantying effective support resistance, the common coupling deformation of support structure and coal rock mass released partial energy of swelling and deformation, and realized adaptive yield of the outer structure; According to the deformation rule and in-site dynamic observation of this kind roadway, grouting, cable and rigid support are promptly taken to strengthen the support before surpassing the biggest permission deformation of surrounding rock. It can avoid the surrounding rock mechanics soften by invading water, protect and strengthen bearing capacity maximally. Finally, the positive initiative compound support system was formed through the material and structure performance coupling of bearing structures such as support body, surrounding rock and so on, "the complex coupling system" was formed by coupling of the system multi-choice performance and the complex stress, deformation, displacement characteristic of surrounding rock, which reformed the project characteristic of complex surrounding rock environment and realized the goal of controlling stability of inclined-shaft under abundant-water conditions.

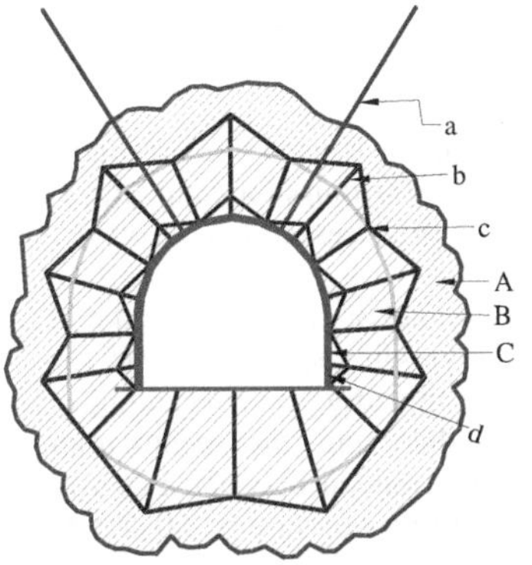


a - cable; b - grouting bolt; c - non-grouting bolt; d - rigid support;
A - grout diffusion and reinforce scope ;
B - anchor rock arch of bolt; C - combined arch formed by anchor net, steel belt and rigid support

Figure 2. Control principle of multi-media structure coupling

### 4.2. *Supporting measures and groundwater control*

According to the deformation failure reason of return slant and the multi-medium structure coupling control technology principle, the groundwater preventing measures and supporting strategies are taken to control surrounding rock stability under the water-rich soft rock environment in the sector passed by the slant.

(1) The groundwater preventing measure

While doing well in taking technical measures such as lead, interception, discharge and so on, the key is to seal off groundwater in the overburden siltstone. The concrete preventing measures are as follows: ① In the process of installing anchor cables, the length of anchor end is prolonged to achieve sealing hole section in aquifer by increasing the anchor agent pitch number, the water plugging apparatus, which is easy to swell after meeting water, is installed on the anchor cable body at the intersection point of roof aquifuge and it's lower rock (its aperture appropriate matches to anchor cable aperture, length is approximately 150–200 mm), and can prevent the water flowing off along anchor cable hole. ② The surrounding rock seepage condition and mechanics characteristic are improved by grouting reinforcement, the conductor formed by fracture is sealed off, which prevents the mechanics of roof rock weakening by water invasion.

(2) The supporting strategy

In order to achieve the mutual coupling of each kind of supports and form a complex coupling system, the following adjustments has made to the original support form: ① High strength and pre-stress and torque anchor is taken to enhance the anchor pretightening force and reduce elongation ratio of the cable body. The stilt is added in the anchor cable body, which can increase the anchor cable system's elongation ratio and realize synchronous bearing and support of bolt and cable and prevent shearing failure of the anchor cable. ② The addition grouting bolt was taken to reinforce the roof and sides of roadway, which improved mechanics of the structural plane and the structure and enhanced the intensity and self-bearing ability of surrounding rock; Simultaneously bolt-shotcrete support is added to form the multi-layered effective combined archs, which can form an integral bearing structure system by full grouted bolt, so the effective bearing scope of support structure was expanded and integrity and bearing capacity was enhanced. ③ The reinforcement belt with big rigidity and good integrity and strong bearing capacity was formed by adding the U section supports, which shared the partial loose deformation loads and caused the tunnel-surrounding deformation distributed uniform. The support roof strata and surrounding rock arching mechanism were improved by controlling the roof stress concentration and serious submersion phenomenon effectively.

### 4.3. *Support design adjustment*

Based on the original support design scheme and the water-rich soft rock environment in the sector passed by the slant, the adjustment is made as follows:

1) Roof bolt: The *Q*500 high strength and pre-stress and torque anchor is used and arranged between the U section support stagger, the row & line space is 800 mm×800 mm, the yield strength is 18 t, the tensile strength is bigger than 23 t, and the pre-stressed is not lower than 40 kN.
2) Roof anchor cable: The stilt and the water shutoff are added to anchor cable base and end respectively, the most yielding distance of the stilt is 30 mm, the yielding spot is 26–30 t.
3) The entire cross section arrangement of controllable pressure injection anchors along roadway radial direction is shown in Figure 2, the aperture is Φ45 mm, the specification is *Φ*27×2500 mm, the row & line space is 1600 mm×1600 mm, arranged between two row of bolts, the hole is sealed by quick swelling cement-roll, the grouting pressure is 2.0–2.5 MPa, the pre-stressed is not lower than 20 kN; The steel bolts and internal grouting bolts (specification are as same as above) are increased at base angle with interlaced arrangement, range distance is 1000 mm, the angle between bolt and floor is 30–45°.
4) U section support: Composed of three section 29U steel, arrange distance is 800 mm.
5) Other support parameters maintain invariable.

## 5. NUMERICAL SIMULATION ANALYSIS

Under the function of adjusted multi-medium structure coupling support, the displacement, stress and plasticity area distribution of inclined shaft were shown in Figure 4. Obviously, roof displacement decreased, the floor and roadway's side's displacement were under control basically. The stress shift to interior through roof deformation, the surrounding rock stress of roadway increased and the stress release zone reduced, the surrounding rock bearing capacity was enhanced, all of which played a positive role in maintaining roadway's stability. The roof plasticity area of roadway was small, plastic yield happened at the floor in roadway for stress concentration, but the plasticity area was within controllable scope. By this taken, the multi-medium structure coupling support design scheme was reasonable and reliable, which was based on high-intensive pre-stressed bolt and waterproof-yield-voltage sharing anchor cable.

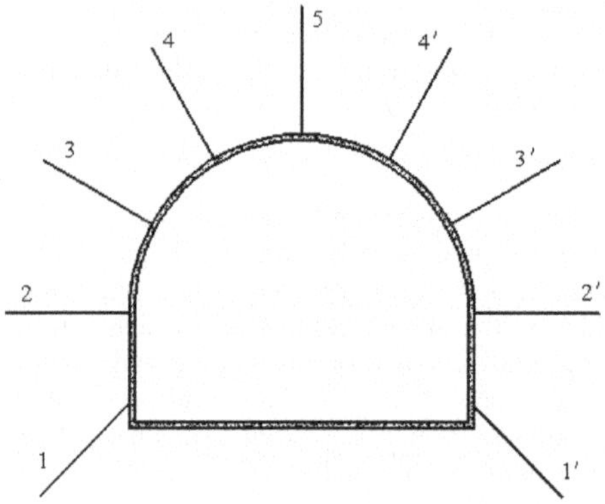


Figure 3. Adjusted design of supporting roadway

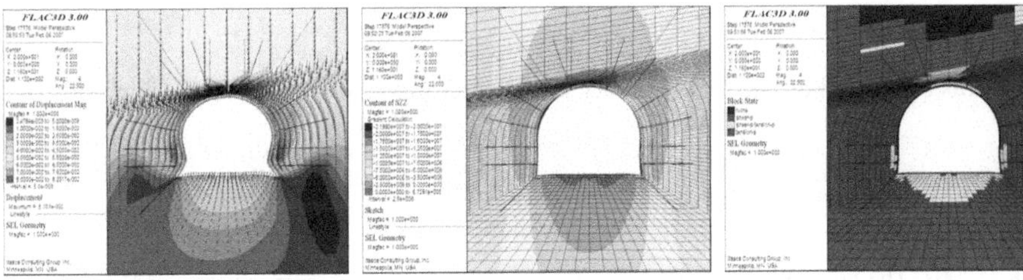

(a) Displacement distribution (b) Stress distribution (c) Failure area distribution

Figure 4. Displacement, stress and plasticity area distribution of inclined shaft

## 6. ANALYSIS OF SUPPORTING EFFECT

To test multi-medium structure coupling support effect of the bolt-mesh-anchor + grouting reinforcement + U section support, roof bedding separation, surface displacement, anchorage force and pretightening force of the anchor cable are observed and detected in the testing section roadway during inclined shaft driving.

The results of roof bedding separation and surface displacement shows that surrounding rock convergence of roadway is generally fiercest in 10 d after driving and tends to be stable one month later. The roof accumulative convergence, the roadway's sides relative deformation and the floor heaves are relatively 38 mm, 46 mm, 62 mm in 100 days, the average deformation rates are respectively 0.38 mm/d, 0.36 mm/d and 0.62 mm/d, and the relative deformation rates decrease continuously. This indicates that the roof accumulative convergence and the roadway's sides relative deformation are not obvious, the roadway's sides and floor deformation tend to be stable basically.

The anchor anchorage force meets the specification completely by pulling out test detecting in the testing section. The anchor cable pretightening force is monitored by anchor dynameter, anchorage force tends to be stable after anchor installed about 1 month, and pretightening forces of bolt and cable anchor respectively achieve 220 KN and 100 KN, which meet the specification. This indicates that the multi-medium structure coupling support form can full use own bearing capacity of surrounding rock and anchor active support function, the support structure, composed of bolt and cable, shotcrete layer, U section support and so on, can achieve compatible deformation and bearing together, which has obtained satisfactory supporting effect.

## 7. CONCLUSIONS

1) The poor mechanical properties of surrounding rock, large shaft span, support structure non-coupling and water effect are the primary causes of surrounding rock delamination, subsidence, partial collapse and roof fall of return air shaft in Qingshuiying coal mine.
2) The multi-medium structure coupling support, which is the bolt-mesh-anchor + grouting reinforcement + U section support, is adopted to enhance the intensity and own bearing capacity of surrounding rock and the support structure integrity enormously for that it both can display the load bearing function of each support fully and form an active and initiative compound support system.
3) The numerical analysis and industrial test indicated that the multi-medium structure coupling support technology of bolt-mesh-anchor +grouting reinforcement + U section support is tested successfully and solved the support problem of inclined-shaft under abundant-water conditions in Qingshuiying coal mine, and it also provides a beneficial model to the similar condition and is worth further promoting.

## ACKNOWLEDGEMENT

This work was supported by the New Century Excellent Talents in University of China (NCET-04-972).

## REFERENCES

Gao Jin-hai. 2007. The bolting theory and application study in the mining entries with complicated roof. Beijing: China Coal Industry Publishing House: 1-3. (In Chinese)

Lai Xing-ping, Wu Yong-ping, Cao Jian-tao, Fan Yong-ning, Zhang Yan-li, Cui Feng. 2010. Experiment on rock-mass deformation of large scale 3D-simulation in complex environment. Journal of China Coal Society 35(1): 31-36. (In Chinese)

Wu Yong-ping, Huang Chao-hui, Zeng You-fu, Wang Chao. 2006. Analysis of collapse of surrounding and stability control in soft rock roadway in a deep mine. Science Technology and Engineering 6(20): 3280-3283. (In Chinese)

Wu Yong-ping, Wu Xue-ming, Zhang Jian-hua, Cai Xiao-mang, Wu Shao-Xue. 2009. Stability analysis of inclined-shaft in abundant-water conditions. Mineral Engineering Research 24(1): 35-38. (In Chinese)

Peng Lin-jun, Zhao Xiao-dong, Song Zhen-qi, Li Li-gang. 2009. Predicting and controlling of water-inrush from roof in coal mines. Journal of Xi'an University of Science and Technology 29(2): 140-145. (In Chinese)

Zeng You-fu. 2006. Research on compound support of soft rock roadway under high stress and at the complex rock condition. Xi'an: Xi'an University of Science and Technology. (In Chinese)

Zeng You-fu, Wu Yong-ping, Hai Xing-ping, Wei Cheng. 2009. Analysis of roof caving instability mechanism of large-section roadway under complex conditions. Journal of Mining & Safety Engineering 4(26): 423-427. (In Chinese)

International Mining Forum 2010, Liu et al. (eds) © 2010 Taylor & Francis Group, London, UK. ISBN 978-0-415-59896-5

# Construction technology of the inclined shaft through thick wind-blown sand seam

Ying Xu, Yong Sun
*School of Civil Engineering and Architecture, Anhui University of Science and Technology, Huainan, Anhui Province, China*

ABSTRACT: The stratum movement failed to be kept in control, the steel brackets were broken out of torsion, and even the workers got covered up by the fallen roof when the advance support of pipe shed made of the regular steel bracket was utilized twice to tunnel through thick wind-blown sand stratum. Concrete line was unable to be poured, and construction was interrupted due to large roof collapse. To get over it, annular soil of upper bench is firstly excavated in shaft arch. The steel brackets are erected and the outer metal nets are suspended. Then the preliminary 100 mm thick concrete is shot, and the roof is supported by over-length advance condulet canopy. The inner metal nets are suspended; another 200 mm-thick shotcrete is done. The workface is sealed by grouting; the sand stratum is reinforced by condulet grouting. The achievements of smooth tunneling provide valuable experiences for the similar project.

KEYWORDS: Advance ductule, pipe shed grout, wind-blown sand stratum, inclined shaft drivage, short section drivage and lining

## 1. GENERAL ENGINEERING SITUATION

The technical characteristics of the inclined shaft of Hongliulin Coal Mine are as following: slope angle is 5 degrees, overall length is 1315 m, and of which, 1030 m long is in topsoil; Shaft driving basal area is 28.32 $m^2$; rough width is 6.6 m and the clear one is 5.6 m; The rough height is 5.5 m and the clear one is 4.2 m. Drilling data in geological report shows that there is slender wind blown sand stratum from the shaft surface until 588 m below, with some irregular lime concre-tions, soil and little water inside the sand stratum.

After 217 m long open channel excavated undercut tunnel was built by means of short session driving and lining, and the $18^{\#}$ double T-iron was taken as preliminary support. The first large collapse took place after 13.5 m undercutting, so open cutting method was used once more till the other 30 m construction completed, then the second collapse happened after another 13.5 m regular undercutting finished, next, 30 m long tunneling openly was excavated.

Considering steelbracket pipe shed roof support is used for the former undercutting it fails to control the loosening of sand stratum which makes the steel bracket broken out of torsion, and even gets the builders covered up by fallen roof.

Permanent concrete line fails to be done too, and open cutting becomes the last resort. Till then, tunneling has developed for 315 m in length and 35 m in vertical depth. The wind blown sand stratum, the supplementary exploration suggests, remains 240 m left in length ahead, so tunneling was forced to stop to find safe and reliable construction technology.

## 2. SLECTION OF CONSTRUCTION SCHEME

Open cutting method usually is the preferred option when incline shaft through in wind-blown sand stratum with small thickness. However, adoption of open cutting technology in the stratum with high thickness and length may lead to big project quantities, increased investment and delayed construction period. At the same time, such special undercutting techniques as freezing method, shield method, jet grouting pile method, and big pipe shed method, etc. feature not only professional construction team needed, complicated technology and high cost, but uncertain application results.

The Thirtieth Engineering Department Affiliated to China Coal Mining Group, Anhui University of Science and Technology, the owner and the construction supervising unit co-convened the project demonstration session to seek the safe and reliable tunneling technology suitable to smooth shaft tunneling in the wind blown sand stratum of 558 m thickness in Hongliulin Coal Mine. In the end, small pipe shed grouting is utilized for advance support and metal netting sprayed with concrete is for temporary support. In this technology, small pipe shed functions as advance grille. Grouted sand stratum possesses enhanced degree of wetness and caking ability, which changes the physical property of the loose stratum. The sand roof collapse is successfully overcome by the combined methods when short section driving and building proceeds. Circular cut begins from the arch of incline shaft left the core soil where it is, so advance collapse distance caused by big slumping angle is reduced in loose sand stratum.

The tunneling step adopted benefits workface support and offers a platform to operate as well. The combined steel bracket support with metal netting sprayed concrete effectively resists the compression from the upper loose sand and offers reliable support to tunneling. As a result, the third undercutting attempt witnesses success to shaft excavation through the thick wind-blown sand stratum. Good result is achieved of society and economically.

## 3. CONSTRUCTION METHODS

### 3.1. *Entrance slope protection and construction of advance pipe shed*

Before undercutting the open channel of the secondary inclined shaft is driven at 45 degrees of slope angle, and 75 degrees within the outline of the excavated section. Next, metal netting sprayed concrete is built to protect the face and lateral surface of slope in stability. The thickness of sprayed concrete is 100 mm and 200 mm within 5 m long limit from the outline of the ex-cavated section, and the shotcrete is used as closing layer for pipe grouting.

Metal net with specification of Φ14 mm×1500 mm is fixed into the 45 degree slope by means of twisted steel anchor, and the 75 degree one by grouting bolt. Pipe shed construction space is left when open cutting is 4 m away from entrance. The 1.5 m concrete arch with added Φ14 mm double steel bar in it from the end is thickened for convenience of pipe striking.

Pipes in roofing support are seamless steel tubes with specification of Φ76 mm×6 mm. The 7.5 m long pipes are arranged in two lines to ensure the formation of grouting certain. A 7.5 m long pipe is made of three 2.5 m tubes with their heads being oblique open tip so that conduct pipes are driven in stratum with less resistance. There are grouting holes, with their diameters of 6~8 mm in the pipes 1 m away from the end, and the holes are arranged in 150 mm pitch row, with shape of quincunx. Three 2.5 m tubes are welded with Φ60mm perforating tubes to form 7.5 m long pipes when driving.

The inner row of pipes, with space length of 250 mm, is arranged on the circle 200 mm away from tunneling section, and the outer with space length of 268 mm, is arranged on the circle 450 mm away from tunneling section. Either the inner or the outer pipes on the wall is with space length of 400 mm. The outward separation angle between pipes and shaft axis is 2 degrees. The range of effective design slurry spreading is 300 mm, and the relevant parameters are shown on figure1 below.

Four $18^{\#}$ I beam brackets are firstly set to guide pipe construction. Holes for conduct pipes are drilled by the pneumatic coal borer, and one pipe is stricken in just after its hole finished. The striking power comes from single hydraulic prop combined with handheld pneumatic rock drill, and the alternative power is large tonnage jack in case of big resistance met. Sands in pipes must be timely blown out to reduce driving resistance after one section pipe finished, and pipe weld operation with perforating tube takes place in time of near completion of the former one stricken, then striking is resumed till design depth.

Workface is closed with C20 shotcrete after all pipe shed finished, and the thickness of spray-up is not less than 300 mm. It is vital to ensure grouting effect that pipe ends, metal brackets and slope surface are sprayed into one integral unit. All sands in pipes are blown out when sprayed concrete reaches due strength to get well prepared for grouting.

Superfine cement slurry and modified sodium silicate slurry are adopted instead of normal cement slurry to overcome low permeability of fine sands. The water cement ratio of superfine cement slurry equals to one to a half or one to one. And grouting pressure is controlled within 1.5 MPa, for excessive pressure could damage the sprayed concrete sealed layer. The grouting machine is YSB-2 squeeze mortar pump. One inch bamboo steel tube is used as the orifice tube, and it is winded with flax silk and stricken in the pipe. There are steel thread drag hook welded with orifice tube and guide tube respectively, and the drag hooks are tied tightly with iron wire to guarantee the safety of grouting.

The sequence of pipe grouting is: The inner row is done first, and then the outer one does. Intermittent grouting is firstly performed in the single number pipes, next the double does. The short pipe shed should be constructed when the long ones in tunnel opening have done for 4 to 5 m long (200 mm thick netting sprayed concrete being the outer support), that is, the long pipe shed and the short one overlaps for 2 to 3 m.

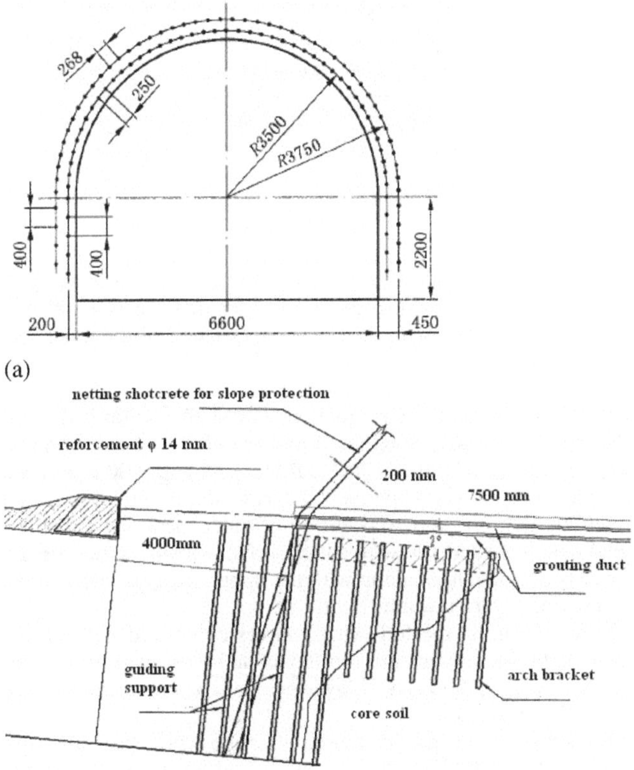

Figure 1. The schematic diagram about entrance section layout of pipes, lining and reinforcement, construction of guiding support and pipe shed

### 3.2. *Advance pipe shed construction within cave*

In cave the small guiding pipe of the pipe shed is Φ42 mm×3 mm seamless steel pipe, with its length of 2.4 to 3.0 m, and there is one row of pipes on every holder. The small guiding pipes are arranged on the tunnel section outline of the shaft, the spacing interval bet each pipes is

200 mm, and the outward separation angle between pipes and shaft axis is 15 degrees. The effective spreading range is 200 mm in design.

The locating hole is placed in the center of the web plate of double T iron when in-cave pipes are fixed. Either electric coal driller or pneumatic coal driller is adopted for drilling operation in the located holes. The pipes are stricken in with hand hold pneumatic rock borer or even with club hammer. Additionally, 8 pipes (each side 4 pipes) are fixed in balance at the bilateral bottom of arch brackets, and each pipe is welded with the arch bracket. All pipe grouting is operated at the same time to strengthen the sand stratum stability, and the arch bracket subsidence can be effectively avoided when the follow up wall excavation and landing leg are done. The arrangement and building of the in-cave pipe shed is shown on figure 2 below.

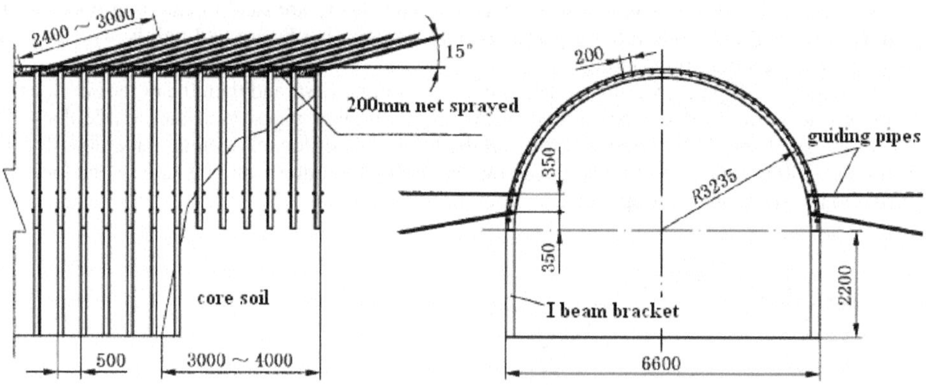


Figure 2. The in-cave pipe shed arrangement

3.3. *Shaft tunneling and lining*

The sequence of shaft tunneling and lining is as following: Circular soil excavation of the arch at the upper step→ erection of arch steel brackets→ suspension of outer metal nets→ 100 mm preliminary arch sprayed concrete→ small guiding pipes stricken→ suspension of the inner metal nets→ another 100 mm arch sprayed concrete and workface closed by grouting→ pipe grouting.

Short bench tunneling with core soil left is used in construction. The shaft section is divided into two steps, the arch belongs to upper bench with length of 3 to 4 m in favor of core soil reservation and operating. The outer support is made of I beam steel brackets with 500 mm space length and 200 mm thick netting sprayed layer. The upper bench excavation progresses for 500 mm in length, and it is time for setting arch brackets, suspension of outer metal nets, preliminary spraying, and then inner metal nets, repeated spraying and workface closing spray, finally pipe grouting is operated. The lower bench advances along with the upper one excavation, then the bracket legs are fixed and net spraying layer on wall is built.

The key factors of reliable construction lie in the following courses completed without any stop–erection of the arch brackets in open slot around core soil (concrete bearer put on), suspension of outer metal nets and preliminary concrete layer sprayed. Under the circumstances the exposure to air of sands is reduced to the minimum.

The inner plain concrete with thickness of 300 mm is poured after the workface with 20 m delay. The concrete mixture is conveyed by HBT-30 concrete pump, and a pouring segment is 4.5 m long. One grouting pipe is buried with 9 m space interval in arch at a time of its concrete pouring. The pipes are used for later grouting to guarantee tight combination between the inner plain concrete (permanent support) and the outer (nets sprayed layer). Such construction courses as the upper bench driving and lining, the lower bench bracket legs fixed and concrete pouring in the rear, are parallel operated to step up the construction speed. The shaft construction effect of anchor, netting and spraying support with small guiding pipe grouting in the windblown sand stratum is shown on figure 3.

(a) Small guiding pipe advance grouting increases with the cohesive strength of the windblown sand stratum

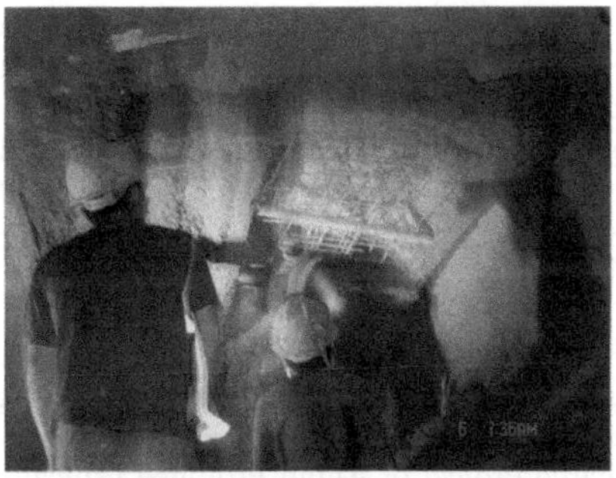

(b) Supplementary sands excavation in cave mini-type digging machine

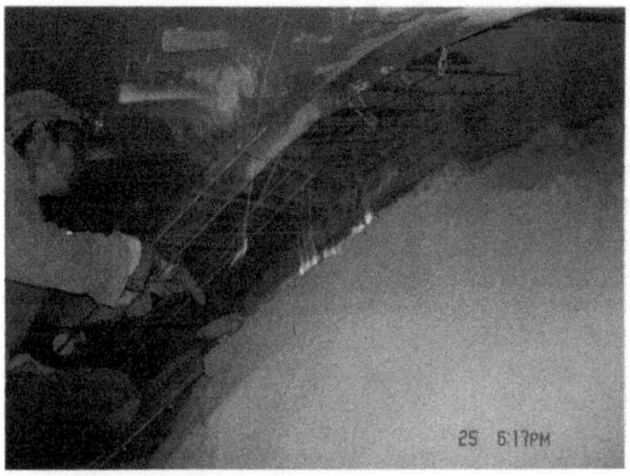


(c) Left core soil offers effective support for stratum excavation workface

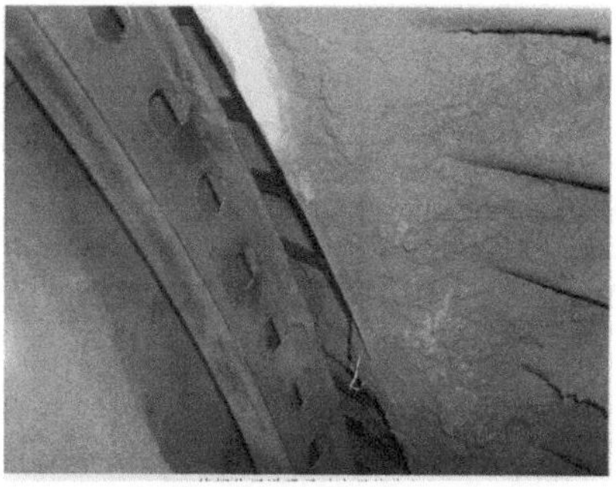

(d) Pipe shed strengthens the stability of sand

(e) Suspension of the outer metal nets

(f) The combining site between the mold of secondary concrete support and anchor, netting and spraying support

Figure 3. The construction technology of the anchor, netting and spraying support combined with small guiding pipe grouting.

## 4. CONCLUSIONS

It is the first time in China that the deep incline shaft is built in the over-length windblown sand stratum. As a matter of fact, the loose sands falling-in can not be gotten over if the pipe shed is fixed without grouting. The cohesive strength of sands is greatly increased if either superfine cement grouting or modified sodium silicate grouting is made after the completion of pipe shed. Since the loose sand stratum falling is kept in control such problems are effectively solved as the unable entering into cave caused by sand stratum falling, workers buried due to collapse, steel bracket damage in distortion and secondary concrete lining unable to set. Tunneling footage is up to 80 m per month, and the highest is 3.0 m in a day. The construction technology is successively created to benefit a lot the similar project, especially, the inclined shaft construction experience is accumulated in the desert district of northwest China.

## ACKNOWLEDGEMENT

The project is co-supported by National Natural Science Foundation of China (50774002), PhD. Programs Foundation of Institutions of Higher Education (200803610002), Natural Science Foundation Innovation Teams of Universities in Anhui Province (TD200705), Natural Science Research Project of Anhui Province Education Department (KJ2009B142) and Natural Science Foundation for Major Project of Anhui Provincial Colleges and Universities (KJ2010 ZD03).

## REFERENCES

Gao You-liang, Li JU, Zhao Qiang. 2009. Slope Shaft Construction Method by Grouting with Advanced Pipe through Aeolian Sand Seam. Coal Mining Technology 14 (2): 46-47. (In Chinese)

Jia Bao-liang, Zhang Qing-zhong. 2007. Rapid Construction of the Shaishang Main Inclined Shaft with Large Section in Zhongxing Coalmine. New Progress on Mine Engineering - Proceedings of Academic Conference on China Mining Construction. (In Chinese)

International Mining Forum 2010, Liu et al. (eds) © 2010 Taylor & Francis Group, London, UK. ISBN 978-0-415-59896-5

# Forecast of rock burst intensity based on one-against-one SVM classification methods

Yongkui Shi, Jiansheng Shao
*Shandong University of Science and Technology, Qingdao, Shandong Province, China*

ABSTRACT: The rock burst is one of the major problems in deep mining. How to predict the rock burst effectively and to reduce the disaster caused by the rock burst, has great significance. The rock burst is affected by many complex factors, so the forecast of rock burst intensity is a non-linear, high dimensional, multiclass pattern recognition problem with small samples. A support vector machine is suitable for solving this pattern recognition problem. In this paper, the principle of support vector machine was introduced, the main influence factors of rock burst were given, and a new forecast method of rock burst intensity based on one-against-one SVM classification was presented, through learning the small training samples collected from a mine, the complicated nonlinear mapping relationship between degree of rock burst and its affected factors was established by the proposed method. The case study shows that the method is feasible, easy to be implemented. So the proposed method is very attractive for a wide application in forecasting rock burst.

KEYWORDS: Rock burst, SVM, MATLAB, forecast

## 1. INTRODUCTION

At present, many coal mines are in deep mining, and a series of technological problems are encountered in deep coal mining, the rock burst is one of the major problems in deep mining, how to predict the rock burst effectively and to reduce the disaster caused by the rock burst, has great significance.

The rock burst prediction methods have fuzzy clustering, neural networks, SVM method, etc. fuzzy clustering method has subjectivity in determine the weight value of categories indicators; traditional neural networks algorithms are based on a large number of samples, there is no reliable theoretical guidance in determining the structure of the networks, the generalization of the neural networks is relatively poor; support vector machine forecast method has the following advantages:

(1) Support Vector Machine algorithm is very stable.

(2) Support vector machine algorithm is based on structural risk minimization principle, it has a strong generalization ability.

(3) Support vector machine can avoid the curse of dimensionality problem. Therefore, support vector machines method is more suitable for rock burst prediction. In this paper, a one-against-one SVM classification method was presented to establish the rock burst forecast model (Wang 2007).

## 2. THE PRINCIPLE OF SUPPORT VECTOR MACHINE

Supposing a given learning sample set is $(x_i, y_i)$ $(i=1,2,\ldots,n)$. Where $x_i \in R_d$ are learning samples, $n$ is the number of samples, $y_i \in (+1, -1)$ is the category label of sample. Binary classification is to find a function $g(x_i)$, through the discriminant function $f(x)=sgn(g(x_i))$, can infer the $y_i$ value which corresponding to $x_i$.

The general expression form of the discriminant function is $f(x)=sgn(w \cdot x+b)$, the corresponding separating hyperplane is: $w \cdot x+b=0$, where $w \in R_n$ is the hyperplane's normal vector, $b \in R$ is bias.

In the case of nonlinear separated, to solve the parameter $w$ of discriminant function is to solve the following optimization problem:

$$\min\left(\frac{1}{2}|w|^2 + c\sum_{i=1}^{n}\xi_i\right)$$

Constraint conditions is:

$$y_i\left[\left(w \cdot \varphi_{x_i}\right)+b\right] \geq 1-\xi_i, \xi_i \geq 0 \quad (i=1,\ 2,\ \ldots\ n)$$

where: $\xi_i$ is the slack variable, through the function $\varphi$, the learning samples $x_i$ are mapped to a higher dimensional space, SVM can find a separating hyperplane which separate the two types with the largest interval in this high-dimensional space, $c$ is the penalty coefficient.

Using the Lagrangian optimization method, the above problem can be transformed into its dual problem in order to solve this optimization problem, the finally discriminant function is as follows:

where: $\xi_i$ is the slack variable, through the function $\varphi$, the learning samples $x_i$ are mapped to a higher dimensional space, SVM can find a separating hyperplane which separate the two types with the largest interval in this high-dimensional space, $c$ is the penalty coefficient.

Using the Lagrangian optimization method, the above problem can be transformed into its dual problem in order to solve this optimization problem, the finally discriminant function is as follows:

$$f(x) = sign\left(\sum_{i=1}^{n} a_i y_i (\varphi_{(x)} \cdot \varphi_{(x_i)})\right) + b \tag{1}$$

where: $a_i(i=1,2 \quad sv,\ sv \leq l)$ are the non-zero Lagrange multipliers, $sv$ is the number of support vector; $l$ is the total number of samples $i$ and $j$ (Sun, Li 2010).

According to the theory of functional, you can always find a kernel function $K(x, x_i)=\varphi_x \cdot \varphi_{(x)}$ which meets the mercer condition, then the SVM discriminant function is:

$$f(x) = sign\left(\sum_{i=1}^{n} a_i y_i K(x \cdot x_i)\right) + b \tag{2}$$

At present, the more frequently used kernel functions have linear kernel, polynomial kernel function and RBF kernel function.

## 3. ESTABLISHMENT AND TESTING OF ROCK BURST INTENSITY FORECAS MODEL

In the forecast of rock burst intensity, select the values of mining depth, roof rock's nature, structural complexity, coal seam's dip, coal seam's thickness etc. eight major factors as sample input vectors.

According to requirements in the "Provisional regulations on rock burst coal seams safe mining", the rock burst intensity is divided into three degree: I level - serious rock burst risk zone (strong); II level - medium rock burst risk zone (middle); III level - no rock burst zone (no) use the risk degree: strong, middle, no, as the target output. Rock burst prediction is a three-class classification (Jiang 2005), three-class classification problem can be decomposed into three binary classification problem, the steps of establish and test the forecast model of rock burst intensity is as follows:

Step 1: Select the sample set, the samples will be divided into two parts: the training samples and the testing samples.

Step 2: Training to get three one-against-one binary classification model.

Step 3: Use the testing samples to test the model, put the testing samples into the classifier to classify, the final predicted result is determined by the vote, if $sign((w^{ij})^T \phi(x) + (b^{ij}))$ says $x$ is in the $i$ th class, then the vote for $i$ th class is added by one. Otherwise, the $j$ th class vote is added by one. Finally we predicate that $x$ is in the class with the largest vote (He, Li 2009). The classification results were compared with the actual situation, to test the accuracy of the model prediction.

## 4. CASE STUDY

### 4.1. *Sample set selection*

Select ground pressure samples of Xinwen coal property for research, collected 25 groups of rock burst samples, these datas are divided into two parts: 17 groups of training samples, 8 groups of testing samples. The training and testing datas are given in Table 1.

Table 1. Training samples and testing samples

| Sample number | Mining depth (m) | Roof rock's nature | Structural complexity | Coal seam's dip (°) | Coal seam's thickness (m) | Mining method | with pillar or not | Blasting or full-mechanized mining | Risk degree |
|---|---|---|---|---|---|---|---|---|---|
| 1 | -540 | 1 | 1 | 10 | 3.1 | 1 | -1 | 1 | No |
| 2 | -560 | 1 | 1 | 13 | 2.6 | 1 | -1 | 1 | No |
| 3 | -549 | 1 | 2 | 11 | 2.7 | -1 | -1 | 1 | No |
| 4 | -556 | 1 | 1 | 11 | 2.3 | 1 | -1 | 1 | No |
| 5 | -563 | -1 | 1 | 9 | 2.2 | -1 | -1 | 1 | No |
| 6 | -602 | 1 | 2 | 12 | 2.6 | 1 | 1 | 1 | No |
| 7 | -573 | -1 | 1 | 11 | 2.2 | -1 | -1 | 1 | Middle |
| 8 | -522 | 1 | 2 | 9 | 2.1 | 1 | -1 | 1 | Middle |
| 9 | -660 | -1 | 2 | 16 | 3.2 | 1 | -1 | 1 | Middle |
| 10 | -609 | -1 | 3 | 16 | 2.9 | 1 | 1 | -1 | Middle |
| 11 | -629 | -1 | 2 | 18 | 3.1 | -1 | 1 | -1 | Middle |
| 12 | -642 | -1 | 2 | 17 | 3.0 | 1 | 1 | 1 | Middle |
| 13 | -702 | 1 | 2 | 19 | 3.1 | -1 | -1 | 1 | Strong |
| 14 | -674 | 1 | 3 | 18 | 3.2 | -1 | 1 | -1 | Strong |
| 15 | -739 | -1 | 3 | 17 | 2.8 | -1 | -1 | -1 | Strong |
| 16 | -690 | 1 | 3 | 19 | 3.5 | -1 | 1 | -1 | Strong |
| 17 | -725 | 1 | 3 | 16 | 3.0 | -1 | 1 | -1 | Strong |
| 18 | -529 | 1 | 2 | 9 | 2.3 | 1 | -1 | 1 | - |
| 19 | -622 | -1 | 2 | 15 | 2.8 | -1 | 1 | -1 | - |
| 20 | -732 | 1 | 3 | 16 | 3.1 | -1 | 1 | -1 | - |
| 21 | -654 | -1 | 3 | 18 | 3.0 | 1 | -1 | 1 | - |
| 22 | -579 | 1 | 1 | 10 | 2.0 | -1 | -1 | 1 | - |
| 23 | -680 | 1 | 2 | 18 | 3.2 | 1 | 1 | -1 | - |
| 24 | -590 | 1 | 2 | 12 | 2.5 | 1 | 1 | 1 | - |
| 25 | -698 | 1 | 3 | 20 | 3.2 | -1 | -1 | -1 | - |

In the actual training and testing, the qualitative description in Table 1 need to be transformed to the quantitative variable, for example, in the "Roof rock's nature", "1" represents sandstone, "-1" represents mudstone, in the "Structural complexity", "1" represents simple, "2" represents moderate, "3" represents complex, in the "Mining method","1"represents longwall, "-1" represents shortwall, in the "With pillar or not","1"represents yes, "-1" represents no, in the "Blasting or full-mechanized mining", "1" represents full-mechanized, "-1" represents blasting, in the "Risk degree", "-" represents "waiting to be predicted" (Shu 2009).

### 4.2. *Establishment of the model with training samples*

Train the former 17 groups of training samples in Table 1 according to "one-against-one" algorithm, write the corresponding matlab program. After several attempts, determine the model of support vector machines, the kernel function is RBF kernel function $k(x,y)=exp\left[-\frac{|x-y|^2}{\sigma^2}\right]$, parameter $\sigma$=280, penalty coefficient C→ +∞ (Zhu, 2006). The final Matlab program is(use the first binary classifier as example, the program is based on SVM toolbox):

```
>>X = [-540 1 1 10 3.1 1 -1 1;-560 1 1 13 2.6 1 -1 1; -549 1 2 11 2.7 -1 -1 1; -556 1 1 11 2.3 1 -1 1; -563 -1 1 9 2.2 -1 -1 1; -602 1 2 12 2.6 1 1 1; -573 -1 1 11 2.2 -1 -1 1; -522 1 2 9 2.1 1 -1 1; -660 -1 2 16 3.2 1 -1 1; -609 -1 3 16 2.9 1 1 -1;-629 -1 2 18 3.1 -1 1 -1; -642 -1 2 17 3.0 1 1 1];
>>Y = [-1; -1; -1 ;-1 ;-1;-1;1;1;1;1;1;1];
>>C=Inf;
>>ker='rbf';
>>global p1
>>p1=280;
>>[nsv alpha bias] = svc(X,Y,ker,C)
```

Where, in the "Y",-1 represents "No",1 represents "Middle". Similarly, in the "Y" of second binary classifier, -1 represents "No",1 represents "Strong"; in the "Y" of third binary classifier, -1 represents "Middle",1 represents "Strong".

### 4.3. *Test of the model with testing samples*

Put the 8 groups of testing datas of Table 1 into the three binary classifier to forecast the testing samples' final class, the Matlab program is(use the first binary classifier as example):

```
>>P = [-529 1 2 9 2.3 1 -1 1; -622 -1 2 15 2.8 -1 1 -1; -732 1 3 16 3.1 -1 1 -1; -654 -1 3 18 3.0 1 -1 1;-579 1 1 10 2.0 -1 -1 1 ; -680 1 2 18 3.2 1 1 -1 ; -590 1 2 12 2.5 1 1 1;-698 1 3 20 3.2 -1 -1 -1];
>>predictedY=svcoutput(X,Y,P,ker,alpha,bias)
```

The classification results of testing samples is given in Table 2. In Table 2, we can see that in the classification results of the 8 groups of testing samples, only one group is wrong, the others are correct. it shows that the SVM prediction results are approximate consistent with the actual situation, the generalization of this model is good.

Table 2. Classification values of predicted samples.

| Sample number | The output of 3 sub-classifiers 1(No or Middle) | 2(No or Strong) | 3(Middle or Strong) | Risk degree Prediction result | Actual situation |
|---|---|---|---|---|---|
| 18 | -1(No) | -1(No) | -1(Middle) | No | No |
| 19 | 1(Middle) | -1(No) | -1(Middle) | Middle | Middle |
| 20 | 1(Middle) | 1(Strong) | 1(Strong) | Strong | Strong |
| 21 | 1(Middle) | 1(Strong) | -1(Middle) | Middle | Middle |
| 22 | -1(No) | -1(No) | -1(Middle) | No | No |
| 23 | 1(Middle) | 1(Strong) | 1(Strong) | Strong | Strong |
| 24 | -1(No) | -1(No) | -1(Middle) | No | Middle |
| 25 | 1(Middle) | 1(Strong) | 1(Strong) | Strong | Strong |

## 5. CONCLUSIONS

The rock burst is one of the major problems in deep mining. How to predict the rock burst effectively and to reduce the disaster caused by the rock burst, has great significance.

The forecast of rock burst intensity is a nonlinear, high dimensional, multiclass pattern recognition problem with small samples, support vector machines is suitable for solving this pattern recognition problem. In this paper, a new forecast method of rock burst intensity based on one-against-one SVM classification was presented, and the prediction model is constructed. The complicated nonlinear relationship between the degree of rock burst and its affected factors was established by the proposed model. The application to the practical engineering shows that the method is feasible, easy to be implemented. So the proposed method is very attractive for a wide application in forecasting rock burst.

## REFERENCES


He Xi, Li Xue-ming. 2009. M-SVM algorithm in the oil detection, Signal Processing. (In Chinese)

Jiang An-nan. 2005. Forecast of rock burst based on pattern recognition by least square support vector machine. Chinese Journal of Rock Mechanics and Engineering. (In Chinese)

Shu Guo-shao. 2009. Forecast of rock burst intensity based on Gaussian process machine learning. Journal of Liaoning Technical University (Natural Science). (In Chinese)

Sun Yu-feng, Li Zhongcai. 2010. The application of support vector machine method in the coal and gas outburst analysis. China Safety Science Journal. (In Chinese)

Wang Guo-sheng. 2007. Research on theory and algorithm for support vector machine classifier. PhD thesis of Beijing University of Posts and Telecommunication. (In Chinese)

Zhu Yi-ding, Ma Wen-tao. 2006. Tunnel classification method of support vector machine. Journal of Mining and Safety Engineering. (In Chinese)

International Mining Forum 2010, Liu et al. (eds) © 2010 Taylor & Francis Group, London, UK. ISBN 978-0-415-59896-5

# Force analysis of the material handled by belt conveyor at loading point and in accelerated area

Linjing Xiao, Hui Sun, Kewen Sun
*Shandong University of Science and Technology, Qingdao, Shandong province, China*

ABSTRACT: During the transportation, the forces exerted on the material handled by the belt conveyor at the loading point and in the accelerated area are very complicated. The forces are analyzed and some mistake is found in the calculations in ISO5048.1989 (E). Firstly, the frictional resistance $F_{Af}$ brought by material's momentum at the loading point is not taken into consideration. Secondly, the formulae of the resistance $F_f$ and $F_{gL}$, which are between material and baffle plates in the accelerated area and non-accelerated area respectively, are false. Then the right formulae of $F_{Af}$, $F_f$ and $F_{gL}$ are given.

KEYWORDS: Belt conveyor, additional resistance, inertial friction, frictional resistance

## 1. INTRODUCTION

Belt conveyor is an important transport in the production of the coal mine. At the loading point and in the accelerated area, the five forces exerted on the material handled are as follows: 1) artificial friction resistance $F_{Af}$ resulting from the momentum of the material handled; 2) slope resistance $F_{st}$; 3) inertial friction resistance $F_{bA}$ at the loading point and in the accelerated area; 4) frictional resistance $F_f$ between the material handled and the baffle plates in the accelerated area; 5) driving force $F_v$ exerted on the material by the belt.

Their formulae in ISO5048:1989(E) are as following:

$$F_f = \frac{\mu_2 I_V^2 \rho g l_b}{(\frac{v+v_0}{2})^2 b_1^2} \tag{1}$$

$$F_{bA} = I_v \rho (v - v_0) \tag{2}$$

$$F_{gL} = \frac{\mu_2 I_v{}^2 \rho g l}{v^2 b_1^2} \tag{3}$$

$$l_{b,\min} = \frac{v^2 - v_0^2}{2g\mu_1} \tag{4}$$

where: $\mu_2$ – friction coefficient between the material and the baffle plates; $l_b$ – length of the accelerated area; $I_v$ – transport capacity, m$^3$/s; $\rho$ – loose bulk density of the material handled; $v$ – speed of the belt; $v_0$ – speed of material in the direction of belt movement at the loading point; $b_l$ - width between baffle plates; $\mu_l$ - friction coefficient between the material and the belt; $F_{gL}$ - frictional resistance between material handled and the baffle plates; $l_{b\,min}$ - the minimum value of $l_b$ in Equation (1).

Equation (1) to Equation (4) will be discussed below and proved to be incorrect.

## 2. ANALYSIS ON THE FRICTIONAL RESISTANCE $F_{Af}$ AT THE LOADING POINT

In Figure 1, $\beta$ is the angle of the velocity component; $\gamma$ is the angle of the discharge plough; $\delta$ is the slope angle of transportation route. For uphill conveyor, $\delta > 0$; for downhill conveyor, $\delta < 0$. Usually $\gamma$ is set greater than $\delta$ (i.e. $\gamma > \delta$) to handle material easily.

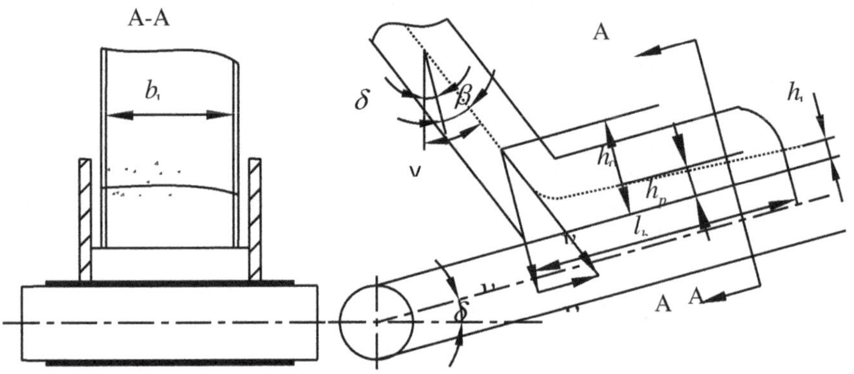


Figure 1. Schematic diagram of the loading area

The relations between $v_m$, $v_f$, and $v_0$ conform to the parallelogram rule. When the material bumps to the belt, the vertical speed $v_f$ reduces to zero during the time $\Delta t$ quickly. Therefore the momentum of material $p$ and the impulse of the material $I$ are obtained as follows.

$$p = I_V \rho \Delta t (v_f - 0) = I_V \rho \Delta t v_f \tag{5}$$

$$I = N_f \Delta t \tag{6}$$

where: $N_f$ – impact force; $p$ – momentum of material handled; $I$ – impulse of the material.

According to momentum conservation law, the Equation (5) equals to the Equation (6). So we get the following formulae.

$$N_f = I_V \rho \Delta t \cdot \frac{v_f}{\Delta t} = I_V \rho v_f \tag{7}$$

$$F_{Af} = N_f \mu_1 = I_V \rho v_f \mu_1 \tag{8}$$

It is improper to define $F_{bA}$ in ISO5048:1989 (E). In fact, the force $F_{bA}$ that is called the inertial friction resistance is the inertial force of the material handled in the accelerated area.

The force $F_{Af}$ is the artificial friction resistance resulting from the momentum of the material handled. It should be considered in ISO5048:1989 (E).

## 3. ANALYSIS ON THE FORCES EXERTED ON THE MATERIAL HANDLED IN THE ACCELERATED AREA

### 3.1. *Calculation of the stress* $\sigma_y$

After the material reaches the belt in the accelerated area, the speed of material $v_0$ rises to $v$ because of belt's frictional force. In this case,

$$\frac{q'_G}{q_{G0}} = \frac{v_0}{v'} \tag{9}$$

where: $q_{G0}$ – mass per meter of the material handled at the speed of $v_0$; $v'$ – speed of material handled; $q'_G$ – mass per meter of the material handled at the speed of $v'$.

To simplify the calculation we suppose that the free surface of the material handled is flat, the formulae are given as follows.

$$\frac{q'_G}{q_{G0}} = \frac{\rho b_1 h}{\rho b_1 h_0} = \frac{h}{h_0} \tag{10}$$

$$h = h_0 v_0 \cdot \frac{1}{v'} \tag{11}$$

where: $h$ – material's height at the speed of $v'$; $h_0$ – material's height at the speed of $v_0$.

We know that $q_{G0} = \rho b_1 h_0$. In this case:

$$I_v \rho = q_{G0} v_0 = \rho b_1 h_0 v_0 \tag{12}$$

$$h_0 = \frac{I_v}{b_1 v_0} \tag{13}$$

$$h_1 = h_0 v_0 \frac{1}{v} \tag{14}$$

where: $h_1$ – the final height in the accelerated area.

To simplify the calculation, the average material height $h_p$ is used in the total length $l_b$ in the accelerated area.

$$h_p = \frac{1}{v - v_0} \int_{v_0}^{v} h dv' = \frac{h_0 v_0}{v - v_0} \ln \frac{v}{v_0} \tag{15}$$

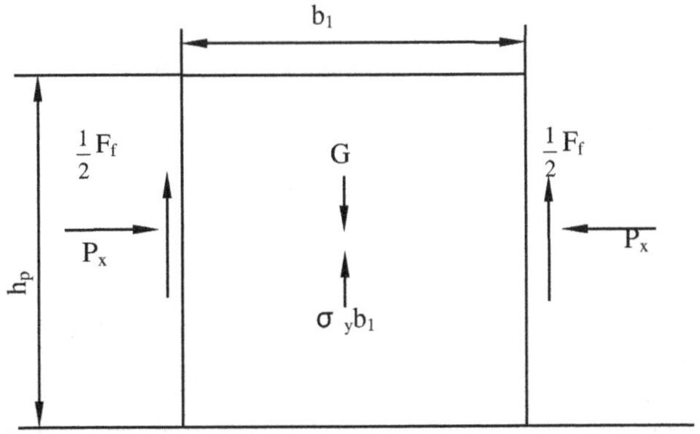


Figure 2. Diagram for calculating $\sigma_y$ ($\sigma_x$ = const)

Figure 1 and Figure 2 show that both sides have baffle plates. The equilibrium equation of the vertical forces is obtained by:

$$\sigma_y b_1 + F_f = G \tag{16}$$

where: $G$ – weight of the material handled, $G = \rho h_p b_1 g \cos\delta$; $\sigma_y$ – stress which the belt exerts on the material handled.

$$F_f = 2P_x\mu_2 = 2h_p\sigma_x\mu_2 = 2\tau h_p\sigma_y\mu_2 \tag{17}$$

where: $\tau$ – coefficient of side pressure; $\psi$ - inertial friction angle of the material; $\sigma_x$ – stress which the baffle plates exert on the material; $P_x$ – force which the baffle plates exert on the material.

$$\sigma_y = \frac{\rho h_p g \cos\delta}{1+2\tau\zeta\mu_2} \tag{18}$$

where: $\zeta$ – the ratio of the height to width of material handled.

Equation (18) is the simplified calculation of $\sigma_y$ because the calculating procedure mentioned above is based on the premise that $\sigma_x$ is a const while $\sigma_x$ changes with the height $y$ (Figure 3).

But there is a little difference between the value of $h_0$ and $h_1$, and the error of calculation is minor. Therefore, $\sigma_y$ is often calculated by Equation (18).

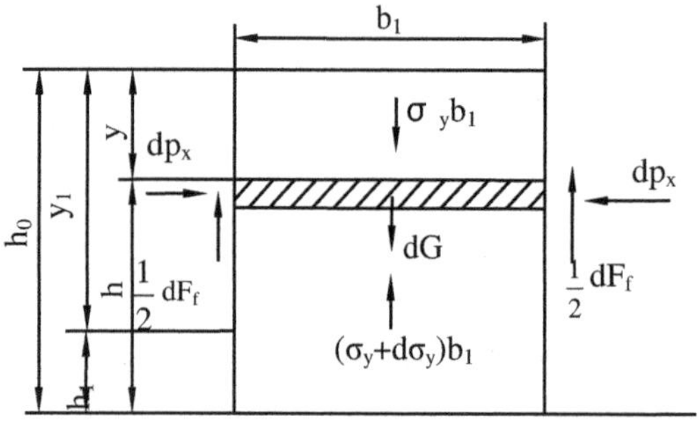


Figure 3. Diagram for calculating $\sigma_y$ ($\sigma_x$= variable)

### 3.2. *Calculation of the acceleration 'a' of material handled*

The driving force is the only force that makes $v_0$ reach $v$.The formulae are as follows.

$$F_{v0} = P_y\mu_1 = \mu_1 b_1\sigma_y \tag{19}$$

$$F_{f0} = 2P_x\mu_2 = 2\tau\mu_2 h_p\sigma_y \tag{20}$$

$$F_{st0} = b_1 h_p \rho g \sin\delta \tag{21}$$

$$\mathrm{d}E = \mathrm{d}\ (\frac{1}{2}b_1 h_p \rho v'^2) = b_1\rho h_p v' dv'$$

where: $F_{v0}$ – driving force per meter of the belt; $P_y$ – the force which the belt exerts on the material handled; $F_{f0}$ – resistance per meter of the baffle plates; $F_{st0}$ – slope resistance per meter; d$E$ – differential quantity of the material's kinetic energy per meter.

If all the artificial forces mentioned above make the material move the distance $dl$, the differential equation of material is obtained by

$$\mathrm{d}W = (F_{v0} - F_{f0} - F_{st0})\,\mathrm{d}l$$

With $\mathrm{d}W = \mathrm{d}E$, we can get the following equation.

$$F_{v0} - F_{f0} - F_{st0} = b_1 \rho h_p a = q_p a \tag{22}$$

where: $q_p$ – average mass per meter of material; $a$ – acceleration of material handled. Let introduce useful factors into Equation (22), the equation is obtained as following.

$$a = [\frac{\mu_1 - 2\tau\zeta\mu_2}{1 + 2\tau\zeta\mu_2}\cos\delta - \sin\delta]g \tag{23}$$

3.3. *Calculation of the forces in the accelerated area*

If both sides of Equation (22) are multiplied by $l_b$, Equation (24) is obtained as following.

$$F_v = F_{st} + F_f + F_{bA} \tag{24}$$

where:

$$F_v = \mu_1 q_p l_b g \frac{\cos\delta}{1 + 2\tau\zeta\mu_2} \tag{25}$$

$$F_f = 2\tau\zeta\mu_2 q_p l_b g \frac{\cos\delta}{1 + 2\tau\zeta\mu_2} \tag{26}$$

$$F_{st} = q_p l_b g \sin\delta \tag{27}$$

$$F_{bA} = q_p l_b a = I_v \rho t \frac{v - v_0}{t} = I_v \rho (v - v_0) \tag{28}$$

3.4. *Analysis on the inaccuracy of Equation (1) to Equation (4)*

If $\mu_2 = 0$ and $\delta = 0$, then $a = g\mu_1$, $l_{b,\min} = \frac{v^2 - v_0^2}{2a} = \frac{v^2 - v_0^2}{2g\mu_1}$. Equation (4) is obtained, however, $\mu_2$ is unable to be zero, that is, Equation (4) is incorrect.

If the number of baffle plate is considered as 1, the Equation (20) should be changed into Equation (29) as follow.

$$F_{f0} = P_x \mu_2 = \tau\mu_2 h_p \sigma_y \tag{29}$$

Let us suppose $\delta$ and $\mu_2$ equal to zero to calculate $\sigma_y$, then suppose $\mu_2$ equal to a const (though it could not happen) to calculate $F_f$. Then both sides of Equation (29) are multiplied by $l_b$, the Equation (30) is obtained by

$$F_f = \zeta\mu_2 q_p l_b g \tag{30}$$

Equation (30) is a particular example of the horizontal conveyor and is the original equation of Equation (1). It can be changed into Equation (1) by using the given parameters.

$$F_f = \zeta\mu_2 q_p l_b g = \zeta\mu_2 q_p l_b g \cdot \frac{q_p l_b v_p^2}{q_p l_b v_p^2} = g\zeta\mu_2 \frac{(q_p l_b)^2 \frac{l_b^2}{t_b^2}}{q_p l_b (\frac{v+v_0}{2})^2}$$

$$= \frac{\zeta\mu_2 (I_v \rho)^2 l_b^2 g}{h_p b_1 l_b \rho (\frac{v+v_0}{2})^2} = \frac{\zeta\mu_2 I_v^2 \rho g l_b}{h_p b_1 (\frac{v+v_0}{2})^2} = \frac{\mu_2 I_v^2 \rho g l_b}{(\frac{v+v_0}{2})^2 b_1^2} \tag{31}$$

The derivation of Equation (3) is similar with that of Equation (1). The result of derivation is obtained as following.

$$F_{gL} = 2\tau\zeta\mu_2 q_p g l \frac{\cos\delta}{1+2\tau\zeta\mu_2} \tag{32}$$

## 4. CONCLUSIONS

(1) Equation (1) to Equation (4) are all incorrect because of the following reasons: the number of baffle plate used in calculations is 1, not 2; it is impossible that the value of $\tau$ is 1; it seems contradictory that $\mu_2$ equals to zero to calculate $\sigma_v$ while $\mu_2$ equals to a const to calculate $F_f$.

(2) The derivation of Equation (1) to Equation (4) neglects the effect of slope angle $\delta$ so the four formulas are only particular examples on the basis of $\delta$ equaling to zero.

(3) The value of the Equation (2) only represents the inertial resistance of the material handled in accelerated area.

(4) According to Equation (24), the only driving force $F_v$ that the belt exerts on the material can be calculated by Equation (25) with some related formulae. It is unnecessary to calculate other forces.

(5) The frictional resistance $F_{Af}$ resulting from the material's impulse to the conveyor belt must be considered and can be calculated from Equation (12).

## REFERENCES

ISO 5048: 1989 (E). 1989. Continuous mechanical handling equipment-Belt conveyors with carrying idlers-Calculation of operating power and tensile forces

Xiao Linjing, Wang Qifeng, Sun Kewen. 2006. Discussion on Idler Tilting Resistance and Trough Factor. Bulk Solids Handling 26(5): 322-325

International Mining Forum 2010, Liu et al. (eds) © 2010 Taylor & Francis Group, London, UK. ISBN 978-0-415-59896-5

# Research on moisture emendation formula for coalbed gas content by indirect methods

Zhongyou Tian
*China University of Mining and Technology, Beijing, China*

ABSTRACT: Using coal samples at different metamorphic stages, isothermal adsorption experiments at dryness and different moistures were carried out. The date of dryness coal emended by moisture experiential formulas, and comparing it with the date of practical determination of moisture coal. The results show that the formula of Fushun and Zhangzhancun is better than Eichenberger. When the volatiles is in the range of 33.92%~36.2%, the emendation adsorption isotherm by three moisture emendation equation is all closed to the practical value. The paper proposes to adopt Zhang zhancun formula to emend the moisture effect.

KEYWORDS: Indirect method, gas content, isothermal adsorption, moisture emendation

## 1. INTRODUCTION

Gas content of coal seams refers to the gas volume contained per unit weight of coal. It is one of the basic parameters of mine gas control and utilization. The accuracy of the measured value is significant for the prediction of gas emission, formulation of gas geological map, evaluation of gas extraction, gas reserves calculation and gas utilization (Zhou, Yu 1992). At present, the measurement of gas content of coal seams includes the indirect and direct methods (Lin 1998, Yu 2005). Among them, the indirect method is widely used in scientific research, and applied in the verification of the direct method (Zhao et al. 2006). Currently, the indirect measurement formula of gas content on the moisture correction is still controversial and worth further research.

## 2. MOISTURE EMENDATION FORMULA

The indirect measurement of gas content of coal is based on the research of the physical, mechanical and chemical properties of coal and its interaction with methane, and is calculated by adding the amounts of coal gas adsorption (Q1) and free gas volume (Q2), which are worked out through the Langmuir equation and equation of state separately. Calculated as follows:

$$Q = Q_1 + Q_2 = \frac{abp}{1+bp} + \frac{VpT_0}{Tp_0k} \quad (1)$$

In the formula: V represents the pore volume per unit mass of coal, $m^3/t$; p - the gas pressure, MPa; T0, P0 - the absolute temperature and the absolute pressure of the standard state, 273K, 0.101325 MPa; T - the absolute temperature of the coal seam, K; a, b - the adsorption constant, $cm^3/g$, $MPa^{-1}$; T0 - the experimental temperature of laboratory determination of the adsorption constants, K; k - the compression coefficient of methane.

Moisture in coal gas impacts the adsorption capacity of coal greatly (Chen, Xian 2006), so the presence of moisture in coal mainly affects the calculation of adsorbed gas quantity Q1. At present, the Russian coal chemists Yiqingeer's empirical formula is mainly used (Xie et al. 2007) (referred to as Ai Qin formula) to determine the impact of the natural moisture of coal on methane adsorption. As shown below:

$$Q_{1ch} = Q_{1g} \cdot \frac{1}{1+0.31M} \tag{2}$$

In the formula, $Q_{1ch}$ refers to the adsorption capacity in water conditions, $cm^3/g$, $Q_{1g}$ the gas adsorption capacity of dry coal, $cm^3/g$, M coal moisture, %.

Some scholars from Fushun Branch of China Coal Research Institute (Hu 2007, Zhang 2008) determine the adsorption isotherm of three coal samples of adsorption isotherm various moisture content, and put forward the correction formula of water content (the Fushun-formula) for the consideration of the degree of coal metamorphism. As shown below:

$$Q_{1ch} = Q_{1g} \cdot \frac{1}{1+(0.10+0.0058V)M} \tag{3}$$

In the formula: V refers to the volatile content of coal, %.

Zhang Zhancun and Piliang Ma obtain a new moisture correction factor (Abbreviation as Zhang Zhancun-formula) by selecting coal samples of different ranks and doing adsorption isotherm experiments under different water conditions. As shown below:

$$Q_{1ch} = Q_{1g} \cdot \frac{1}{1+(0.147e^{0.022V})M} \tag{4}$$

In the formula: V refers to the volatile content of coal, %.

## 3. THE VALIDATION AND COMPARISON OF WATER CORRECTION FORMULA

### 3.1. *Coal Samples and Experimental Methods*

In order to observe the applicability of the three water emendation formulas, this paper selects four groups of coal samples of different ranks, which are Zhong Ma Village Coal Mine of Jiaozuo, Da Shu Coal Mine of Feng Feng Group, Wangjiahe Coal Mine of Tong Chuan, Kailuan Coal Mine in Tangshan, and conducts respectively industrial analysis of the samples and the study of true relative density. See table 1.

Table 1. The proximate analysis and true relative density of coal samples.

| Sample Number | Coal Samples | Industrial Analysis | | | True Relative Density |
|---|---|---|---|---|---|
| | | $M_{ad}$/% | Aad/% | $V_{ad}$/% | |
| 1# | Zhongmacun coal | 1.20 | 10.64 | 5.52 | 1.35 |
| 2# | Dashucun coal | 1.53 | 12.29 | 11.94 | 1.41 |
| 3# | Wangjiahe coal | 1.04 | 7.62 | 18.24 | 1.36 |
| 4# | Tangshan coal | 4.24 | 14.58 | 34.68 | 1.23 |

In order to get coal samples with water content under certain water conditions, artificial humidification is used in the laboratory. Steam passes into the samples and the adsorption equilibrium forms after a certain time. The samples with different moisture content are obtained after successive vacuum. The time of filling steam and the duration of vacuum determine the amount of water content. To avoid the loss of water, the adsorption capacity parameter analyzer with the pressurized volumetric method is used for the adsorption isotherm experiment. The experimental temperature is $30^0$C. The drying method of weighing is used to determine the amount of the water content of samples. That is to say, the moisture content should be measured when the samples are removed right after the high-pressure adsorption tank is removed, after high-pressure assay is finished.

The experimental methods stated above are used on the coal samples of four different ranks under dry conditions and certain water conditions.

### 3.2. *Results and Analysis*

First, the adsorption isotherm of the coal samples of the four different ranks is obtained under certain moisture conditions. And then, in the light of the adsorption isotherm of the four samples under dry conditions, and according to each of the three moisture correction formulas, methane adsorption and the related adsorption isotherm are calculated, shown in Figure 1, where, A is the adsorption isotherm curve of dry samples, B is the adsorption isotherm curve of the coal samples of water content corrected according to the Ai Qin formula, C is the adsorption isotherm curve of the coal samples of water content corrected according to the Fushun-formula, D is the adsorption isotherm curve of the coal samples of water content corrected according to the Zhang Zhancun-formula, and E is the adsorption isotherm curve of the measured coal samples of water content.

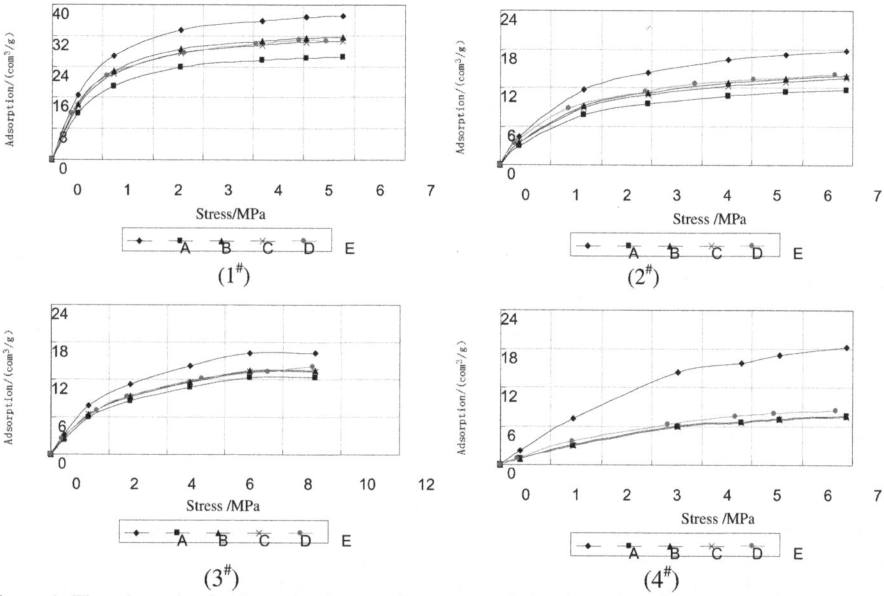

Figure 1. The adsorption isotherm by three moisture emendation formulas of 4 coal samples

Figure 1 shows that $1^{\#}$ refers to the correction result of the methane adsorption isotherms of Zhongmacun coal, the water content of which is 1.29%. The pattern shows that, the adsorption isotherm curves of the coal samples of water content corrected according to the Fushun-formula and Zhang Zhancun-formula are consistent with the measured isotherms, while the adsorption isotherm curve of the coal samples of water content corrected according to the Ai Qin formula is significantly lower. $2^{\#}$ refers to the correction result of the methane adsorption isotherms of Dashucun coal, the water content of which is 1.66%. The pattern shows the similar situation as shown in $1^{\#}$. $3^{\#}$ refers to the correction result of the methane adsorption isotherms of Wangjiahe coal, the water content of which is 1.02%. The pattern shows the similar situation as shown in $1^{\#}$. $4^{\#}$ refers to the correction result of the methane adsorption isotherms of Tangshan coal, the water content of which is 4.55%. The pattern shows that the adsorption isotherm curves of the coal samples of water content obtained according to the three correction formulas correction coincide and make little difference with the measured isotherms. The reasons are that based on the Ai Qin formula, Fushun-formula and Zhang Zhancun-formula consider the difference made by the impact of coal samples containing water of different ranks on gas adsorption capacity. When the volatile content reaches 36.2%, the Fushun-formula becomes the same with Ai Qin formula and when the volatile content reaches 33.92%, Zhang Zhancun-formula becomes the same with Ai Qin formula.

## 4. CONCLUSIONS

Ai Qin formula does not take into account the difference made by the impact of coal samples containing water of different ranks on gas adsorption capacity.

When using the current water correction formulas, the correction of Fushun-type and Zhang Zhancun-formula are closer to the actual, while the correction of Ai Qin formula has big errors. Only when the coal volatile content is between 33.92% and 36.2%, the correction of all the three formulas is close to the actual. This paper recommends Zhang Zhancun-formula for the moisture emendation, to obtain an accurate value of gas content.

## ACKNOWLEDGEMENT

The research is sponsored by National Basic Research Program of China (2005cb221504)

## REFERENCES

Chen Xuexi, Chen Shaojie, Ma Shangquan. 2006. Investigation of the Mode of the Methane Adsorption under different water conditions. Mine Safety 22(5): 89-91. (In Chinese)

Hu Qianting, Zou Yinhui, Wen Guangcai et al. 2007. New technologies of Gas Outburst Prediction with Gas Content Methods. Journal of China Coal Society 32 (3):276-280. (In Chinese)

Lin Baiquan. 1998. Theory and Technology of Mine Gas Control. Xuzhou: China University of Mining and Technology Press. (In Chinese)

Xian Xuefu, Gu Min. 2006. The discussion of The Indirect Measurement of Gas Content. Engineering Sciences 8 (8): 15-22. (In Chinese)

Xie Zhenhua, Chen Shaojie. 2007. Impacts of Moisture and Temperature on the Methane Adsorption of Coal. Journal of University of Science and Technology Beijing 29 (Supplement 2): 22-24. (in Chinese)

Yu Bufan. 2005. A Technical Manual of the Prevention and Use of Mine Gas Disaster Prevention (revised edition). Beijing: Coal Industry Press. (In Chinese)

Yu Qixiang. 1992. Mine Gas Control. Xuzhou: China University of Mining and Technology Press. (In Chinese)

Zhang Zhancun, Ma Piliang. 2008.The Experimental Study of the Impact of Moisture on Gas Adsorption Characteristics of Different Coal Samples. Coal 33 (2): 144-147. (In Chinese)

Zhao Jizhan, Han Baoshan, Chen Zhisheng et al. 2006. Investigation of Estimation Methods of the Seam Gas Content. Coal Geology of China 18 (5): 22-24. (In Chinese)

Zhou Shining, Lin Baiquan. 1992. The Theory of the Occurrence and Flow of the Coal Seam Gas. Beijing: China Coal Industry Publishing House. (In Chinese)

International Mining Forum 2010, Liu et al. (eds) © 2010 Taylor & Francis Group, London, UK. ISBN 978-0-415-59896-5

# Selection of reasonable level of floor roadway in deep mines

Dezhong Li, Xiaohu Xi, Zunyu Xu, Xing Wang
*Anhui University of Science and Technology, Huainan, Anhui Province, China*

ABSTRACT: Choosing horizon of floor roadway scientifically and reasonably is the fundamental guarantee for low maintenance cost of roadway. In this paper, the option of reasonable horizon of floor roadway of deep mine was made by theoretical calculations and numerical simulation, combining with the geological conditions of Jianxin colliery, ultimately to determine the selecting sequence of reasonable normal distance of floor roadway of deep mine, it provides a scientific basis for the option and design of horizon of floor roadway of deep mine in the future.

KEYWORDS: Floor roadway, reasonable level, normal direction; FLAC simulation

## 1. INTRODUCTION

The floor roadways influence tunneling excavation and mining to discard early or late. Especially during the mining, the roadway affected by the dynamic pressure of mining, the dynamic pressure of mining (support pressure) transferring to the floor by a certain form that made the surrounding rock stress of floor redistribute and surrounding rock of the roadway produce deformation persistently. Reasonable selection of horizon of floor roadway can decrease the maintenance cost of roadways.

## 2. OUTLINE OF THE MINE

Mines capacity is about 80 0000 Mg/a in Jianxin colliery, primary recoverable coal seam is B4, which is medium thickness seam, coal thickness ranges from 1.2 to 4.0 m and average is 2.4 m, the distance between coal seams vary from 8 to 11, occurrence of coal seam is more stable. The mining level is -600 m, with two main entry and -600 m east-west service roadway is the belt haulage roadway, which is divided into east mining area, central mining area, west auxiliary mining area and auxiliary mining area on floor 2#.The floor roadway and the main return airway were arranged in the coal floor at the design of district roadway.

In mining district design, the dip head of the floor and dip head of the main return airway are arranged in the floor of coal seam, and the dip head of the floor is used for coal transportation dip head of the main return airway is ready for return air of mining district.

## 3. CALCULATION OF FAILURE DEPTH OF THE FLOOR ROCK BY PLASTICITY THEORY

As the abutment pressure acts on the floor rock mass of the coal body edge, which reach or over the critical value, the rock mass will produce plastic deformation and form plastic zone. When the abutment pressure reach the maximum which lead to failure of some rock, the rock plasticity zone affected by abutment pressure band together, which created the gob floor uplift and the floor rock mass (the plastic deformation) moved to the gob, and formed a continuous slip line field, together with rock mass did not occur the plasticity failure form the slip plane .At this time, the failure of floor rock in the slip plane was the most serious.

The slip line field of floor rock mass, that is, the boundary of plastic zone, is shown in Figure 1. There are three zones which are active limit zone $aa'b$, transient zone $abc$ and passive limit zone $acd$.The slip lines of active limit zone and passive limit zone are made up of two group of lines which are logarithmic spiral and radiation beginning from the point $a$ in the slip line of transient zone. The logarithmic spiral equation is given by

$$r = r_0 e^{\theta \tan \varphi_0} \tag{1}$$

where: $\varphi_0$ is the average internal friction angle for the floor rocks.

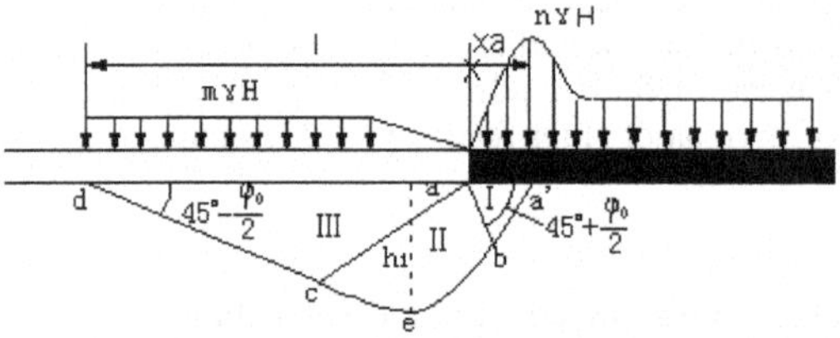


Figure 1. Plastic damage zone of the floor under limit state

After the coal seam is mined, (Fig. 1 I zone) the abutment pressure occur in the floor rocks around the gob, its ultimate strength is over the stress, the rock will occur plastic deformation, and those rocks are compressed in vertical direction, so the expansion certainly happen in horizontal direction, and the expanded rock compress the rock of transition zone (Fig. 1 II zone) and transferred the stress to this zone. The transition zone continued to compress the passive zone (Fig. 1 III zone).Because there was the free face of gob in the passive zone and the stress on the rock of collapse was lower than the original stress in the passive zone, the rock of transition zone and passive zone expanded to the gob under the force transferred from active zone, and the heave of floor rock formed in gob.

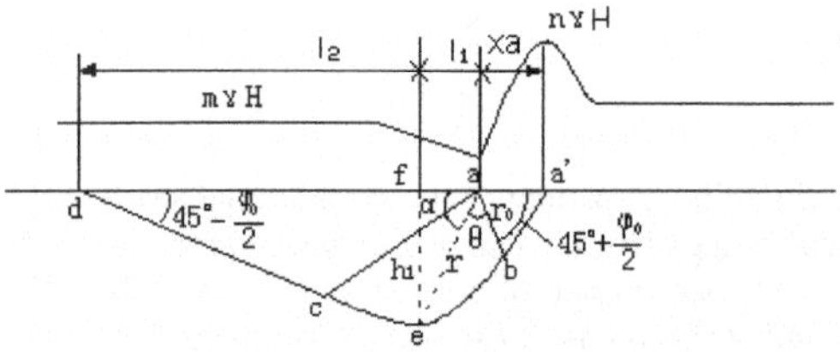


Figure 2. Calculation diagram of depth of maximum damage in the floor

The maximum depth of ultimate plastic failure zone affected by the abutment pressure can be determined on the basis of geometry size of ultimate plastic zone in Figure 2.

In $\Delta aba'$,

$$ab = r_0 = x_a / 2\cos\left(\frac{\pi}{4} + \frac{\varphi}{2}\right) \tag{2}$$

where: $x_a$ is width of coal plastic zone.

In $\Delta aef$ ,

$$h = r\sin\alpha \tag{3}$$

and

$$\alpha = \pi - \left(\frac{\pi}{4} + \frac{\varphi_0}{2}\right) - \theta = \frac{\pi}{2} + \left(\frac{\pi}{4} - \frac{\varphi_0}{2}\right) - \theta \tag{4}$$

Therefore,

$$h = r_0 e^{\theta\tan\varphi_0}\cos\left(\theta + \frac{\varphi_0}{2} - \frac{\pi}{4}\right) \tag{5}$$

The maximum depth of floor failure zone, $h_1$, can be obtained by $\dfrac{dh}{d\theta} = 0$.

$$\frac{dh}{d\theta} = r_0 e^{\theta\tan\varphi_0}\cos\left(\theta + \frac{\varphi_0}{2} - \frac{\pi}{4}\right)\tan\varphi_0 - r_0 e^{\theta\tan\varphi_0}\sin\left(\theta + \frac{\varphi_0}{2} - \frac{\pi}{4}\right) = 0 \tag{6}$$

Therefore,

$$\theta = \frac{\pi}{4} + \frac{\varphi_0}{2} \tag{7}$$

where: $\theta$ is equal to $\dfrac{\pi}{4} + \dfrac{\varphi_0}{2}$, $h$ is the maximum failure depth of floor rocks.

Substitute formula (7) and formula (2) into formula (5), the maximum failure depth of floor rocks $h_1$, is given by

$$h_1 = \frac{x_a \cos\varphi_0}{2\cos\left(\frac{\pi}{4} + \frac{\varphi_0}{2}\right)} e^{\left(\frac{\pi}{4} + \frac{\varphi_0}{2}\right)\tan\varphi_0} \tag{8}$$

According to $x_a = 4.28\,m$, $\varphi_0 = 45^\circ$, the maximum failure depth of floor rocks of B4 coal seam face in Jianxin colliery, $h_1$:

$$h_1 = 12.9m \tag{9}$$

The maximum failure depth of floor rocks caused by mining failure is about 12.9 m during the normal mining in B4 coal seam of Jianxin colliery. In the choice of horizon of the floor roadway, the maximum of floor rocks under the face should be considered first.

## 4. REASONABLE HORIZON AND NORMAL DISTANCE OF FLOOR ROADWAY

### 4.1. *Text and indenting*

Because of the difference of floor lithology and normal distance, mining pressure in face and stability of surrounding rock of the roadways. The basic situation of surrounding rock of the roadways $S$ is determined by the ratio of the environment of roadway stress $K\gamma H$ to the strength of surrounding rock $\sigma_c$, that is:

$$S = \frac{K\gamma H}{\sigma_c} \tag{9}$$

where: $K$ is the attenuation coefficient of floor stress, generally $K = 2.981 - 0.886\log z$;

$z$ is normal distance; $\sigma_c$ is strength of surrounding rock.

Because the maximum failure depth of floor rocks caused by mining failure is about 12.9 m during the normal mining in B4 coal seam, five relatively better horizons are chosen by comparison at where the normal distance of 12.9 m below the coal seam floor .

Using the formula (9) make the comparative analysis of stability of each horizon and floor roadway of layout of normal distance, and to determine the reasonable horizons and normal distance, as shown in Table 1.

Table 1. Option of Reasonable Horizon of Floor Roadways

| Lithology | Thickness (m) | Normal distance (m) | $\sigma_c$ (MPa) | $S$ | Selecting sequence | Description |
|---|---|---|---|---|---|---|
| Siliceous sandstone | 7~12 | 15 | 38 | 0.765388 | 1 | Brown grey and thin and medium thick layered |
| Siltstone | 4~8 | 22 | 30 | 0.914144 | 5 | Charcal grey |
| Rough siltstone | 2~3 | 30 | 40 | 0.783877 | 3 | Dark gray thick layered |
| Fine sandstone | 2~4 | | 32 | | | Dark gray |
| Fine sandstone | 7~10 | 40 | 30 | 0.780787 | 2 | Charcoal grey |
| Siltstone | About 20 | 50 | 25 | 0.885428 | 4 | Shallow–Dark gray |

The floor roadway is laid out in siliceous sandstone where the normal distance $z$ is 15 m, in fine sandstone where the normal distance $z$ is 40 m, in siltstone and fine sandstone where the normal distance $z$ is 30 m, in siltstone where the normal distance $z$ is 50 m and siltstone where $z$ is 22 m. We can see that properties of surrounding rock have a big effect on the stability of floor rock roadways. The effect of surrounding rock property on roadway is greater than that of normal distance $z$ between roadway and upper coal seam. Therefore, when the position of floor rock roadway is determined, the horizon that the relatively large strength of surrounding rock should be selected first. Even the normal distance between the horizon has been selected and upper coal seam is relatively small, the effect of the selected horizon on roadway maintenance will not too much.

Conversely, if the rock roadway locates in the horizon of soft surrounding rock, even the normal distance between the horizon and the upper coal seam is very large, mining the upper coal seam has an effect on the roadway maintenance.

Relationship between the stability of floor rock and the normal distance $z$ (between roadway and the upper coal seam) is: the larger the normal distance is, the better the stability of rock is.

## 5. SIMULATION OF SELECTING THE REASONABLY HORIZON OF FLOOR ROADWAY

The numerical simulation were carried out by nonlinear large deformation program FLAC$^{2D}$ (Fast Lagrangian Analysis of Continua in 2 Dimensions).This program was developed for geotechnical engineering by Itasca Consulting Group, Inc in America, and established on the basis of algorithm of Fast Lagrangian Analysis of Continua in 2 Dimensions. Numerical model of this simulation is shown in Figure 3.

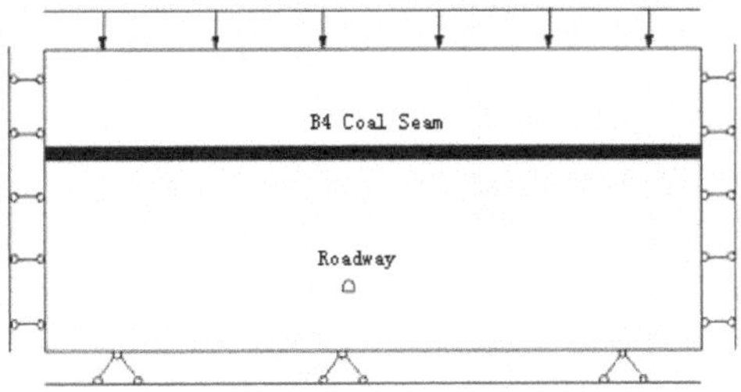


Figure 3. Numerical model

## 4.1. *Simulation Process*

The floor rock roadway model are designed in different horizons, its the normal distance between the horizon and the coal seam, and the normal distances are 15 m, 22 m, 30 m, 40 m, 50 m, respectively. The optimization of roadway horizon layout is carried out by simulating roadway deformation features affected by mining in the B4 coal seam face in different horizons.

The lithology and strength of these five horizons are shown in Table 1, and the other required parameters are determined by the geological instruction in Jianxin colliery. The working face is strike longwall and mined from right to left. Based on the position of floor roadway, the horizontal distance of simulation is from 100 m to -100 m, and footage is 10 m in working face.

In order to further understand the characteristic of dynamic change in floor roadway during the face advancing, the measuring points of the stress were arranged in roadway's sides.

## 4.2. *Simulation Results*

Simulation Results of surrounding rock stress of floor rock of five different horizons is shown in Figure 4 during the face mining. We can see that the surrounding stress of roadway is determined from small to large, the normal distance $z$ is 15 m, 40 m, 30 m, 50 m, 22 m respectively before cross mining; and after cross mining, each roadway placed in the distressed zone, the surrounding rock stress of roadway decreased gradually and tended to be stable.

The surrounding rock stress of roadway where the normal distance $z$ is 15 m decreased to stable firstly, and the roadway where the normal distance $z$ is 15 m tended to be stable last, and the stress satisfies the sequence from shallow horizon to deep horizon. The Vertical Stress of surrounding rock of roadway in each horizons is near after pressure relief; the Horizontal Stress of surrounding rock of roadway where the normal distance $z$ is 15 m is the minimum after pressure relief and keep stable, and the other Horizontal Stress of surrounding rock of roadway is near.

Integrated these, the situation of stability of floor roadway sort is laid out in siliceous sandstone where the normal distance $z$ is 15 m, fine sandstone in where the normal distance $z$ is 40 m, siltstone and fine sandstone in where the normal distance $z$ is 30 m, siltstone in where the normal distance $z$ is 50 m and siltstone where $z$ is 22 m. And these are the same results as the theoretical calculations.

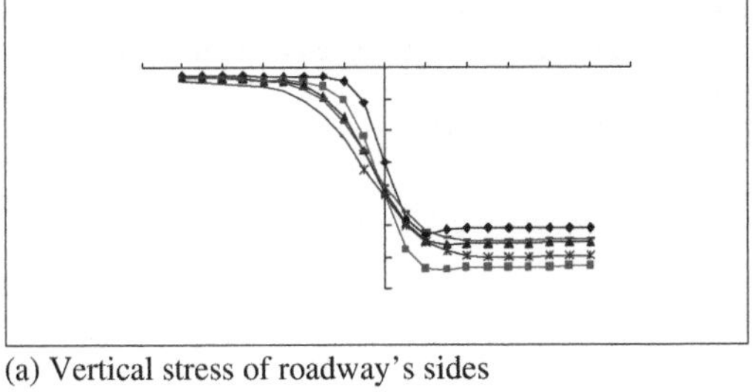

(a) Vertical stress of roadway's sides

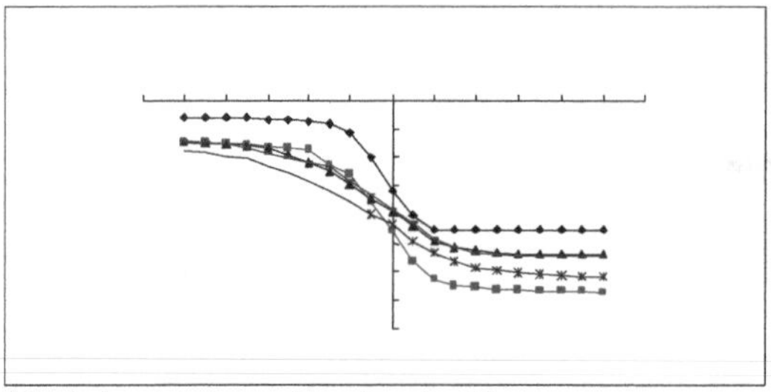

(b) Horizontal stress of roadway's sides

Figure 4. Stress Change Curves of Rock Roadway Under the Different Horizon During Cross-mining

## 6. CONCLUSIONS

1) The maximum failure depth of floor rocks caused by mining failure is about 12.9 m during the normal mining in B4 coal seam of Jianxin colliery.
2) The stability of floor roadway is determined by two leading factors that is, strength of surrounding rock and normal distance of floor roadways. The bigger normal distance is not always better. The selection principle includes: firstly, relative stability of the rock and relative higher strength are selected; secondly, the normal distance should be considered.
3) By theoretical calculations and numerical simulation, the selecting sequence of reasonable normal direction of B4 coal seam floor roadway is laid out in siliceous sandstone where the normal distance $z$ is 15 m, in fine sandstone where the normal distance $z$ is 40 m, in siltstone and fine sandstone where the normal distance $z$ is 30 m, in siltstone where the normal distance $z$ is 50 m and siltstone where $z$ is 22 m in Jianxin colliery. This selecting sequence can be used for designing the selecting horizons of floor roadway, then, the horizon is determined reasonably by the space need and the connecting project quantity of roadway system.

## REFERENCES


Dong Zhao-xing, Wu Shi-liang. 2004. Roadway Engineering [M]. Xuzhou: China University of Mining And Technology Press. (In Chinese)

Dai Chang-qing. 2005. Study on the Floor Water Invasion Regularity Mining on the Top of Water-pressured. Anhui University of Science and Technology. (In Chinese)

Gong Ji-wen. 2002. Numerical Model Method of Stress and Strain - Introduce to Numerical Model Software FLAC. Journal of East China Geological Institute 25(3): 220-227. (In Chinese)

Jiang Fu-xing. 2004. Mining Pressure and Strata Control [M]. Beijing: China Coal Industry Publishing House. (In Chinese)

Jiang Jin-quan. 1993. Stress and Movement of Stop Wall Rock [M]. Beijing: China Coal Industry Publishing House. (In Chinese)

Li De-zhong, Xia Xin-chuan, Han Jia-gen. 2005. Mining Technology for Deep Mine [M]. Xuzhou: China University of Mining and Technology Press. (In Chinese)

Liu Shi-chun. 2008. Research on Destroyed Depth of Stope Floor and Defending Water Technology in Yonggu Colliery. Coal Technology 17(1): 78-80. (In Chinese)

Ma Nian-jie, Hou Zhao-jiong. 1995. Theory and Application of Mining Pressure for Preparation Roadway [M].Beijing: China Coal Industry Publishing House. (In Chinese)

Ni Xi-min. 2001. Discussion on the significance of reasonable selecting stone-bed of mining area system gateway. Coal Technology 20(1): 55-56. (In Chinese)

Qian Ming-gao, Shi Ping-wu. 2003. Mining Pressure and Strata Control [M]. Xuzhou: China University of Mining and Technology Press. (In Chinese)

International Mining Forum 2010, Liu et al. (eds) © 2010 Taylor & Francis Group, London, UK. ISBN 978-0-415-59896-5

# MC simulation research on adsorption and desorption of methane in coal seam

Xijian Li
*School of Safety Engineering, China University of Mining & Technology, Xuzhou, Jiangsu, China; School of Mining, Guizhou University, Guiyang, Guizhou, China; Guizhou Key Laboratory of Comprehensive Utilization of Non-metallic Mineral Resources, Guiyang, Guizhou, China*

Baiquan Lin
*School of Safety Engineering, China University of Mining & Technology, Xuzhou, Jiangsu, China*

Hao Xu
*School of Mining, Guizhou University, Guiyang, Guizhou, China*

ABSTRACT: At present, the study of methane adsorption and desorption inside coal only stay in the level of its mechanism and invertible experiment. Study about methane molecules' adsorption and desorption characteristic in coal pores in microcosmic level is few. In this study, relationship of macrofluid parameters and micromolecules physical quantity which represent methane adsorption and desorption characteristic was established applying thermodynamics, statistical mechanics and coal chemistry theories. Random movement of methane molecules in coal pores was simulated adopting Monte Carlo molecular simulation technology. Action of methane molecules in adsorption and desorption process was studied. Research methods of this paper have some unique characteristics. Firstly, it is an innovative method that theory of methane molecules random movement in the microcosmic was combined with thermodynamics, statistical mechanics theories to study methane adsorption and desorption characteristic. Secondly, model of methane molecules' adsorption and desorption process was established. Computer simulation and laboratory experiment were both used to inspect and verify each other. Thirdly, simulation program was developed adopting VC++6.0 development environment. Fourthly, relationship graph can be drawn between different parameters including porosity, pressure, temperature etc. and methane adsorption quantity adopting computer graphics visualization technology.

KEYWORDS: Methane, adsorption and desorption, Monte Carlo, molecular simulation

## 1. INTRODUCTION

Coal is natural adsorbent, and has great capacity to adsorb methane ($CH_4$). Generally, there are two ways of which the methane exists in coal: Free State and adsorbed state. On the macro view, the major factors which affect coal methane molecular adsorption and desorption are coal's porosity, pressure, temperature, the coal metamorphism level, coal moisture, coal gas component, etc. On the micro view, the result of coal methane molecular adsorption is mainly the in-iterative result of coal molecular structure and methane molecular. Researches of this area are carried mostly in macroscopic level, but less in microscopic level. In this paper, with the help of research achievement of coal molecular structure, molecular thermodynamics and surface physicochemical theories, combined with computer simulating technology, innate character of coal methane molecular adsorption was discussed in microscopic level. The study is of great significance in guiding the coal gas ($CH_4$) exploitation and mine gas control in mining practice.

## 2. POTENTIAL ENERGY MODEL

As porous medium, coal has a lot of pores and crannies. So, in this study slit pores model, which are widely used currently, was adopted to study coal pores.

In this paper, Truncation Drift Lenard-Jones potential was used to represent intermolecular potential among flowing molecules (Chen, Xu 2007):

$$\phi_{ff} = \begin{cases} \phi_{LJ}(r) - \phi_{LJ}(r_c) & r < r_c \\ 0 & r > r_c \end{cases} \tag{1}$$

where: $\phi_{ff}$ denotes intermolecular potential, r is intermolecular radius, $r_c$ is truncation radius, $r_c$ generally evaluates $2.5\sigma - 3.5\sigma$ here evaluates $r_c = 2.5\sigma$.

$$\phi_{LJ}(r) = 4\varepsilon_{ff}\left[\left(\frac{\sigma_{ff}}{r}\right)^{12} - \left(\frac{\sigma_{ff}}{r}\right)^{6}\right] \tag{2}$$

$\varepsilon_{ff}$ and $\sigma_{ff}$ respectively represent energy parameters and dimensions parameters of flowing molecules, which can be looked up in corresponding references.

Potential energy between flowing molecules and single slit pores was modeled adopting mean field potential model, if define the direction which is perpendicular to slit pores wall as Z axis, then potential energy can be expressed as:

$$\phi_{fw}(z) = 2\pi\rho_w\varepsilon_{fw}\sigma_{fw}^{\ 2}\Delta\left[0.4\left(\frac{\sigma_{fw}}{z}\right)^{10} - \left(\frac{\sigma_{fw}}{z}\right)^{4} - \left(\frac{\sigma_{fw}^{\ 4}}{3\Delta(0.61\Delta + z)^3}\right)\right] \tag{3}$$

where: $\rho_w$ denotes number density of the carbon pores walls, here it is 114 nm-$^3$; $\Delta$=0.335 nm; z is distance between flowing molecules and coal pores walls, $\varepsilon_{fw}$ and $\sigma_{fw}$ are interaction parameter. LJ interaction constant among different atoms usually determined by the following formula:

$$\begin{cases} \sigma_{fw} = (\sigma_{ff} + \sigma_{ww})/2 \\ \varepsilon_{fw} = (\varepsilon_{ff} \bullet \varepsilon_{ww})^{1/2} \end{cases} \tag{4}$$

As to fixed pore diameter, total potential energy of flowing molecules in coal pores can be express as sum of potential energy among flowing molecules and potential energy between fluid and coal pores walls: $\phi_T = \phi_{ff} + \phi_{fw}(z) + \phi_{fw}(H - z)$.

## 3. SIMULATION PROCESS OF GRAND CANONICAL ENSEMBLE

Grand canonical ensemble (μVT) requires chemical potential (μ), volume (V) and temperature (T) of the system to be constant in simulation process. When slit aperture is fixed, periodic boundary condition can be set in and axis direction.

During implementation steps of simulating, every step of circling includes the following three kind of attempt (Steele 1974): (1) to insert a molecule into the simulation box; (2) to delete a molecule from the simulation box; (3) to move a molecule from one position to another in the simulation box. In the simulation, equal weight molecule selection algorithm was adopted; making attempt of insert, delete and move of equal probability as 1/3. During simulating, variable quantities was converted to dimensionless ones:

$$\begin{aligned} &\mu^* = \mu/\varepsilon_{ff}, V^* = V/\sigma_{ff}^3, T^* = k_B T/\varepsilon_{ff} \\ &P^* = P\sigma_{ff}^3/\varepsilon_{ff}, H^* = H/\sigma_{ff}, \rho^* = \rho\sigma_{ff}^3 \end{aligned} \tag{5}$$

Where ρ denotes number density, H is slit wide, P is pressure,* is contrast quantity.

In the simulations, as to each slit aperture, under certain temperature and pressure, process of flowing molecules adsorption by slit aperture will go through many configurations till it is balanced. Generally there need to be $10^5$- $10^7$ equilibrium configurations. Part of them is pre-blance configurations to rejected, another part were post-blance configurations to be averaged. In this simulation we take $2\times10^5$ as step number, the front $1\times10^5$ times are used to balance simulation. The after $1\times10^5$ times are used to conduct adsorption quantity statistics. In order to make sure the program can be operated in personal computer, number of simulations must be limited in certain scope. In this study, the model can simulate 50 times configuration changes per second, so the time that needed to conduct 200 000 times simulation was $t=2\times10^5/(50*3600)\approx$ $\approx$ 1.11 (hours), there is 7 balance points, and the total time of simulation time is 7.77 (hours), flow chart of the simulation is demonstrated as Figure 1.

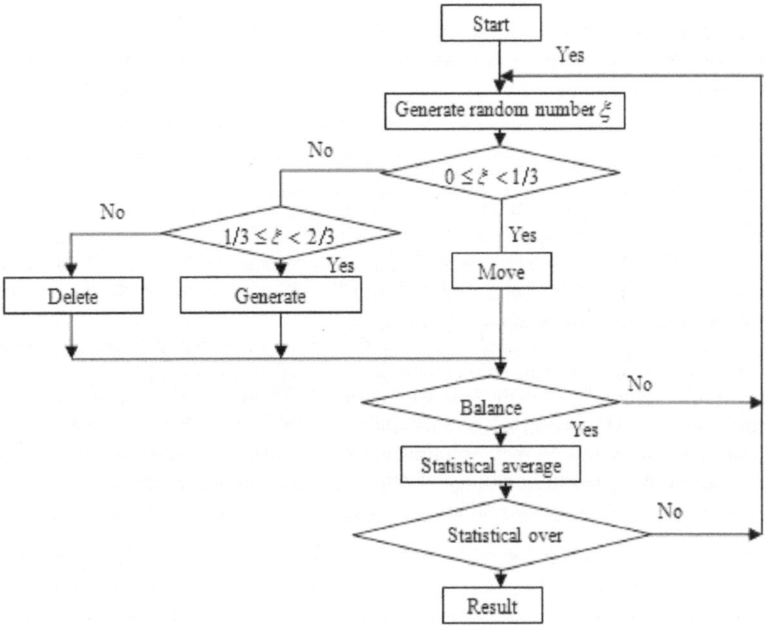


Figure 1. Flow chart of GCMC simulation

## 4. DISCUSSION ON INFLUENCING FACTORS OF ADSORPTION

On the macro view, many factors influence adsorption and desorption of methane gas in coal, such as coal-seam gas-pressure, temperature, proximate analysis of coal(including moisture content, ash content), coal's porosity, specific surface area, type of coal (metamorphic grade) etc. How to connect those influencing factors to microcosmic molecule simulation, and convert those macroscopic influencing factors to microcosmic simulation factors is the key question the simulation study need to resolve.

Indicated from definition of adsorption quantity $\Gamma = \rho_T / (N_A \cdot \sigma_{ff}^{\ 3} \cdot \rho_c)$ (Frenkel 1996) ($mmol \cdot g^{-1}$), adsorption quantity of a unit mass of coal is related to average fluid number density:

$$\rho_T = \frac{1}{H}\int_0^H \rho(z)dz \tag{6}$$

where: $\rho_c$ is parameter related to coal density, through this microscopic influence factors can be related to micro-simulation.

The influence factors can be divided into two types: (1) influence factors that is determined by adsorption formula; (2) factors that influence boundary conditions of the model.

### 4.1. *Influence factors that is determined by adsorption formula*

In the study of adsorption, pressure can not be defined within porous medium (coal) but molecular number in connected to acceptance probability of the insert, delete and move process.

Acceptance probability is related to chemical potential, so pressure of particle source that is related to chemical potential is:

$$\beta\mu^{B} = \beta\mu^{0}_{id.gas} + \ln(\beta p / \varphi) \tag{7}$$

$$\beta\mu^{0}_{id.gas} = k_B T \ln \Lambda^3 \tag{8}$$

Macro pressure *P*, temperature *T* and pore diameter *H*, influence acceptance probability of insert, delete and move movement of methane fluid molecule in the adsorption process. They effect molecular number absorbed, and can be converted to micro-adsorption parameters.

### 4.2. *Factors that influence boundary conditions of the model*

Water in coal generally coal seam contain some water, coal pores are occupied by water, so to determine content of methane in the coal, Effective porosity should be a basis to calculate the content of methane (Li 2006). In this study, it is suggested that because of the exist of moisture, micro pores volume were occupied by water molecules in the primitive cell. Make the adsorption volume reduce into $V_{absorption} = V_{primitive\ cell} - V_{water}$.

Ash content adsorption of methane in coal is mainly the action of Van der Waals' force between macromolecule of coal and methane molecules. Ash of coal is converted by mineral substance in coal. Its origin is minerals of plant, and detained in coal forming process, the ash content is usually small, and its influence to adsorption ability is little (Xu 2007). Compared with adsorption ability of coal molecules to methane molecules, the minerals' adsorption ability is much smaller, so in this study, adsorption ability of minerals was ignored. Minerals only occupied some space of coal; make carbon atom relatively fewer in each unit volume (adsorbing material reduced).

$V_{daf}$ is generated when organic matter of coal are heated in the absence of air, it includes $CH_4$, $C_2H_6$, $H_2$, CO, $H_2S$, $NH_3$, COS, $H_2O$, $CnH_2n$, $CnH_{2n-2}$ and benzene, naphthaline, phenol etc. it also includes hydrocarbon of $C_5$~$C_{16}$, pyridine, pyrrole, thiofuran and other likewise compound (Zhang 2003). Study shows, principal Components of $V_{daf}$ are carbon-containing compound, so it can be classified as adsorbate.

So the final adsorption quantity formula is:

$$\Gamma_m = \frac{\langle N(H) \rangle}{N_A} \cdot \frac{(1 - V_{water} / V_{primitive\ \ cell} - y)}{(\rho_c) \cdot \sigma_{ff}^{\ 3}} \tag{9}$$

where: $\langle N(H) \rangle$ denotes molecular number in primitive cell, $N_A$ is avogadro's number, $\rho_c$ is apparent density of coal, $\sigma_{ff}$ is dimensions parameters of methane molecule. $V_{water}$ is water volume occupied in primitive cell, it is obtained through transformational calculation.

## 5. SIMULATION RESULTS AND ITS DISCUSSION

In this study, based in theoretical analysis and adsorption simulation model established, simulation program of MC methane molecule adsorption was developed. Simulation results compared with laboratory test results obtained using WY-98 Adsorption Constant Tester, the laboratory test results include: the adsorption constant *a* and *b*, the relative coefficient $\gamma$, and adsorption curve. By comparison, accuracy of the program was tested and verified; reasons of inaccuracy that existed are discussed.

### 5.1. *Comparison of simulation and experiment results*

In this paper, data of adsorption pore volume and specific surface are quoted from references (Fu, Qin 2003).

### 5.1.1. *Comparison of simulation and experiment results of Qinglong coal sample*

Data of experiment and simulation results of Qinglong coal sample are shown in Table 1.

Table 1. Experiment and simulation result of Qinglong coal sample.

| Data type | Proximate analysis | | | Apparent density | Actual density | Pore volume | Specific surface | Temperature |
|---|---|---|---|---|---|---|---|---|
| | Mad | Aad | Vdaf | | | | | |
| | % | % | % | $g/cm^3$ | $g/cm^3$ | $10^{-4}cm^3 \cdot g^{-1}$ | $m^2 \cdot g^{-1}$ | °C |
| Experiment | 1.05 | 8.22 | 6.61 | 1.49 | 1.41 | | | 30 |
| Simulation | 1.05 | 8.22 | 6.61 | 1.49 | 1.41 | 278 | 8.276 | 30 |

Here the pore volume capacity and specific surface are of adsorption pores.

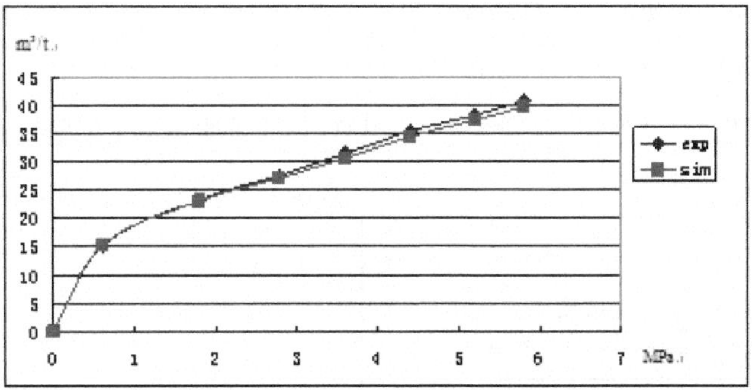


Figure 2. The simulation and experiment data contrast diagram of Qinglong coal sample (30°).

When the adsorption is balanced, data of adsorption quantity of each point are shown in Table 2.

Table 2. Data of adsorption quantity of each points of Qinglong coal sample

| Pressure (MPa) / Adsorption quantity ($m^3/t$) / Type | 0.6 | 1.8 | 2.8 | 3.6 | 4.4 | 5.2 | 5.8 |
|---|---|---|---|---|---|---|---|
| Experiment | 15 | 23.2 | 27.5 | 31.5 | 35.5 | 38.1 | 40.5 |
| Simulation | 15.2695 | 23.109 | 27.1137 | 30.6767 | 34.2589 | 37.2685 | 39.5023 |
| Absolute error ($m^3/t$) | 0.2695 | 0.091 | 0.3863 | 0.8233 | 1.2411 | 1.0223 | 0.9977 |
| Relative error to experiment (%) | 1.8 | 0.39 | 1.4 | 2.61 | 3.5 | 2.68 | 2.46 |

Comparison table of experiment and simulation adsorption constant *a*, *b* are indicated as Table 3.

Table 3. Comparison table of experiment and simulation adsorption constant *a*, *b*

| | Adsorption constant *a* | Adsorption constant *b* | Correlation $\gamma$ |
|---|---|---|---|
| Experiment | 52.710 | 0.467 | 0.983 |
| Simulation | 54.083 | 0.396 | 0.984 |
| Absolute error | 1.373 | | |
| Relative error (%) | 2.6 | | |

As shown in Figure 2 and error analysis in Table 2 and Table 3. Adsorption curve of simulation are in close agreement with experiment adsorption curve. It certified the simulation method and program developed are of good practicability and accuracy. Results of experiment and simulation is closed, it is of great possibility that Qinglong coal sample is similar to coal sample quoted from the references.

### 5.1.2. *Comparison of simulation and experiment results of Xiaotun coal sample*

Data of experiment and simulation results of Xiaotun coal sample are shown in Table 4.

Table 4. Data of Xiaotun coal sample experiment and simulation

| data type | proximate analysis | | | apparent density | actual density | pore volume | specific surface | temperature |
|---|---|---|---|---|---|---|---|---|
| | Mad | Aad | Vdaf | | | | | |
| | % | % | % | $g/cm^3$ | $g/cm^3$ | $10^{-4} cm^3 \cdot g^{-1}$ | $m^2 \cdot g^{-1}$ | °C |
| experiment | 1.67 | 13.22 | 6.00 | 1.58 | 1.48 | | | 30 |
| simulation | 1.67 | 13.22 | 6.00 | 1.58 | 1.48 | 185 | 4.805 | 30 |

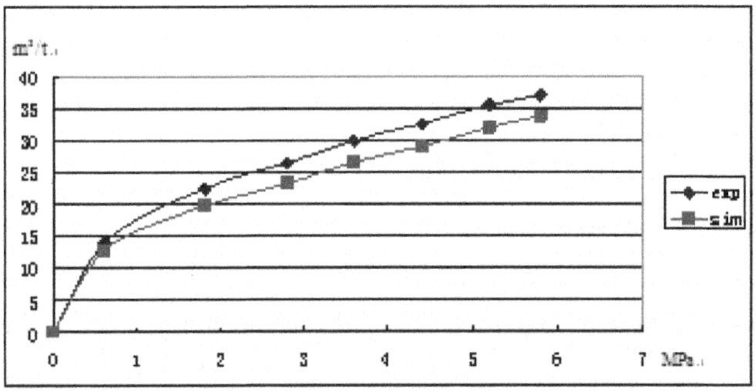


Figure 3. The simulation and experiment data contrast diagram of Xiaotun Mine (30°C)

When the adsorption is balanced, data of adsorption quantity of each point are shown in Table 5.

Comparison table of experiment and simulation adsorption constant *a*, *b* are indicated as Table 6.

As shown in Figure 3 and error analysis in Table 5 and Table 6. Results of experiment and simulation of Xiaotun coal sample is relatively closed. Because the ash content of Xiaotun coal sample is comparatively large, so there exists certain error in the simulation.

Table 5. Data of adsorption quantity of each points of Xiaotun coal sample

| Pressure (MPa) / Adsorption quantity ($m^3/t$) / Type | 0.6 | 1.8 | 2.8 | 3.6 | 4.4 | 5.2 | 5.8 |
|---|---|---|---|---|---|---|---|
| Experiment | 14 | 22.5 | 26.5 | 30 | 32.6 | 35.5 | 37 |
| Simulation | 12.7895 | 19.6993 | 23.3059 | 26.6685 | 28.9808 | 31.8759 | 33.6341 |
| absolute error ($m^3/t$) | 1.2105 | 2.8007 | 3.1941 | 8.1129 | 3.6192 | 3.6241 | 3.3659 |
| relative error to experiment (%) | 8.65 | 12.45 | 12.05 | 11.10 | 11.10 | 10.21 | 9.10 |

Table 6. Comparison table of experiment and simulation adsorption constant *a*, *b*

| | Adsorption constant *a* | Adsorption constant *b* | Correlation $\gamma$ |
|---|---|---|---|
| Experiment | 47.060 | 0.541 | 0.990 |
| Simulation | 46.306 | 0.396 | 0.988 |
| Absolute error | 0.754 | | |
| Relative error(%) | 1.6 | | |

## 5.2. *Adsorption curve under different temperature*

Temperature is one of the major factors which influence adsorption. In this study, it is an important input parameter. As shown in Figure 4, it demonstrates adsorption curves under different temperature. Feasibility of simulation method was tested as well as accuracy of simulation program.

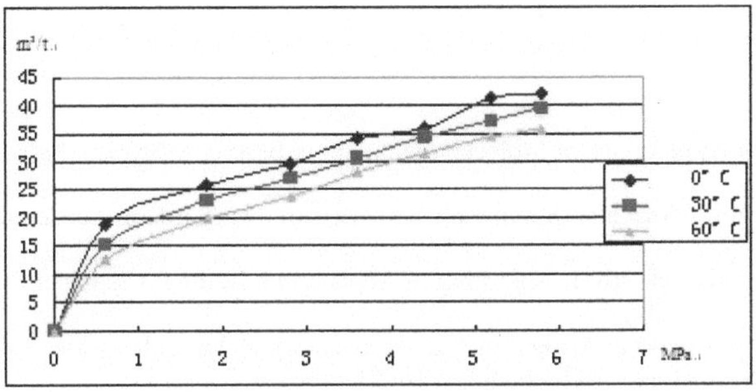


Figure 4. The isothermal adsorption curve in different temperature of Qinglong coal sample

Adsorption constant of Qinglong coal sample under different temperature are shown in Table 7.

Table 7. Adsorption constant of Qinglong coal sample under different temperature

| Temperature | Adsorption constant *a* | Adsorption constant *b* | Correlation $\gamma$ |
|---|---|---|---|
| °C | m³/t | MPa⁻¹ | |
| 0 | 55.466 | 0.461 | 0.976 |
| 30 | 54.083 | 0.396 | 0.984 |
| 60 | 54.223 | 0.31 | 0.987 |

As shown in Figure 4 and Table 7, as temperature rises, adsorption isotherm line drops. On the micro view, as temperature rises, methane's activity is increased; methane molecules' energy is enhanced. It is easier for methane molecules to get rid of the intermolecular attraction and be desorbed. The isothermal adsorption curves of different temperature also confirm the principle that with the rise of temperature, adsorption quantity and adsorption constant a will drops, maximal adsorption ability will decline finally. The conclusion above confirms the feasibility of simulation method and accuracy of simulation program.

Tests show the adsorption isotherm line does not exactly satisfy experience formula: $X=X_o e^{n(to-t)}$, $n = 0.02/(0.993+0.07p)$, but it satisfy formula $X=X_o e^{n1(to-t)}$, ($n \neq n1$), given $n1 = A/(B+Cp)$, through linear regression process, we got $n1 = 0.0067/(0.993+0.274p)$, the reason why the modulus changed might be the following. Parameters used to calculate potential energy between coal molecule and methane molecule are substituted by Parameters of carbon atom and methane molecule. Take parameters of carbon atom as the substitution of coal molecule will cause changes of chemical potential, thus the calculation probability changes.

## 6. CONCLUSIONS

This work did provide that results of simulation and experiment are in general agreement. Unavoidably there exists certain error, one reason is that the model is idealized; other reasons are from calculation of boundary conditions. Which include the following ones: attractive force between methane molecule and coal is substituted by attractive force between methane molecule and carbon atom; coal pore diameter is supposed to be equal, dispersity of coal pore diameter is ignored; ash content is treated as mineral of no adsorption ability; and all components of $V_{daf}$ are treated as mineral of full adsorption ability because most components are carbon-containing compound.

So concentration of future study should be research on the influence of micro-simulation system, which caused by physical and chemical feature of coal, including macromolecular structure of coal, pore configuration, and composition of coal (including ash content, $V_{daf}$, etc.). In the next step study, based on direction of laboratory experiments (including tests of adsorption constant *a*, *b*, coal pore volume, adsorption specific area, dispersity of adsorption pores), we will improve and perfect the simulation system, thus simulation error may be reduced.

## REFERENCES

Chen Zheng-long, Xu Wei. 2007. Theory and practice of Molecular simulation. Beijing: Chemical Industry Press. (In Chinese)

Frenkel D., Smit B. 1996. Understanding Molecular Simulation. Amsterdam: Amsterdam Academic Press: 99-260

Frenkel D., Smit B. 2002. Molecular Simulation – from Algorithm to Application. Beijing: Chemical Industry Press.

Fu Xue-hai, Qin Yong. 2003. Prediction Theory and Method of Gas Permeability in Multiphase Medium Coal Seam. Xuzhou: China University of Mining & Technology. (In Chinese)

He ke-rong. 2007. Monte Carlo simulation of adsorption process in cellular carbon materials. Shanghai: TongJi University. (In Chinese)

Li Jian-ming. 2006. Prevention Technique of Coal and Gas Outburst. Xuzhou: China University of Mining & Technology. (In Chinese)

Steele W. A. 1974. The Interactions of Gases with Solid Surfaces. Oxford: Pergamon

Xu Fang-de. 2007. Practical Technique of Coal Gas Drainage. Xuzhou: China University of Mining & Technology. (In Chinese)

Zhang Shuang-quan, Wu Guo-guang. 2004. Coal Chemistry. Xuzhou: China University of Mining & Technology. (In Chinese)

International Mining Forum 2010, Liu et al. (eds) © 2010 Taylor & Francis Group, London, UK. ISBN 978-0-415-59896-5

# Research on coal mine gas sensor systems based on two-wavelength

Mengran Zhou, Haiqing Zhang, Hongwei Wu
*Anhui University of Science and Technology, Huainan, Anhui Province, China*

ABSTRACT: Gas explosion coal mine is the major security accident, for the past shortcomings of gas detection and the detection of gas concentration measurements, which cause wrong measurement data, so two-wavelength gas sensing system is designed to reduce the error greatly. Selecting the right wavelength, the system uses the same light source and the same absorption pool to eliminate the impact of the light source changing, the errors caused by different absorption parameters and interference absorption caused by background. The result shows that this method effectively improves the detection sensitivity, precision and stability.

KEYWORDS: Two-wavelength, K coefficient, error interference

## 1. INTRODUCTION

Coal mine methane gas mainly consists of methane gas. In China coal mine accidents, the casualties caused by methane gas explosion accounted for more than 50% of all the significant accident casualties, real time monitoring of methane gas content and preventing its explosion have a important significance (Jin, Stewart 1993; Zhou, Li 2007; Zhou, Li et al. 2007).

The traditional spectroscopy absorption technique of methane detection system is usually affected by cross-interference of other gas composition, it is difficult to precise correction of data errors caused by background gas cross-interference, the problem becomes more and more severely when detecting content is very low (Jin, Stewart 1997; Jin, Stewart 1993; Inaba, Kobayasi 1979; Dakin, Wade 1987). The two-wavelength K coefficient fiber-optic methane sensor system designed in this paper is combined by optical spectroscopy techniques and modern technology, it makes spectroscopy analysis which used to laboratory gas analysis develop to gas online monitoring devices. Two-wavelength uses the same light source to achieve two beams of monochromatic light, so the impact caused by light source change is very small, and low-noise signal can be recorded (Fried 2008). The characteristics of optical fiber are that fiber materials have stable performance, and remains under the environment of high temperature and pressure, low temperature, high corrosive and other harsh environment, which make a leap of spectroscopy analysis technology in detection sensitivity and remote control, etc.

## 2. PRINCIPLES OF TWO-WAVELENGTH K COEFFICIENT

### 2.1. *Principles of two-wavelength selection*

The sensor system based on two-wavelength absorption spectroscopy technique has many advantages, but how to choose the right two-wavelength is the key to design. This paper uses the K coefficient method in two components detection to select the measurement wavelength and reference wavelength of methane. Two-wavelength K coefficient is also known as K-ratio method.

In the two-wavelength selection adding a K-ratio device can make $A_{\lambda_1}$ or $A_{\lambda_2}$ signal amplify, after further processing data, putting the signal into the difference calculation circuit and obtaining the general relationship:

$$K=\frac{A_{\lambda_2}}{A_{\lambda_1}} \tag{1}$$

$$\Delta A = A_{\lambda_2} - KA_{\lambda_1} = n_2 A_{\lambda_2} - n_1 A_{\lambda_1} \tag{2}$$

Application of K coefficient method can make the general form of differential absorbance of interfere components as formula (1), also can be expressed as formula (2). If it coexists two components, we can suppose absorbance spectral of interference components, testing components and mixtures at two wavelengths respectively as $A_{\lambda_1}$, $A_{\lambda_2}$, $A'_{\lambda_1}$, $A'_{\lambda_2}$, $A''_{\lambda_1}$, $A''_{\lambda_2}$, shown in Figure 2.

$$\begin{aligned}\Delta A &= n_2 A''_{\lambda_2} - A''_{\lambda_2} = n_2 (A_{\lambda_2} + A'_{\lambda_2}) - (A_{\lambda_1} + A'_{\lambda_1}) \\ &= (n_2 A_{\lambda_2} - A_{\lambda_1}) + (n_2 A'_{\lambda_2} - A'_{\lambda_1})\end{aligned} \tag{3}$$

Adjusting the coefficients makes $n_2 A_{\lambda_2} = A_{\lambda_1}$, then $\Delta A = n_2 A'_{\lambda_2} - A'_{\lambda_2}$, it eliminates influence of interference components and the mixture and the processing of multi-component interference is similar.

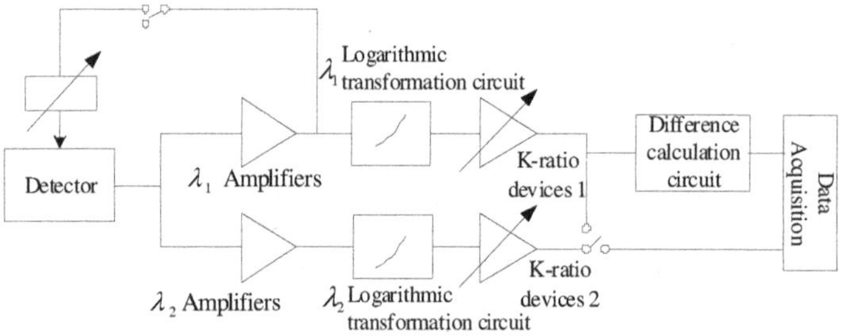


Figure 1. K coefficient figure

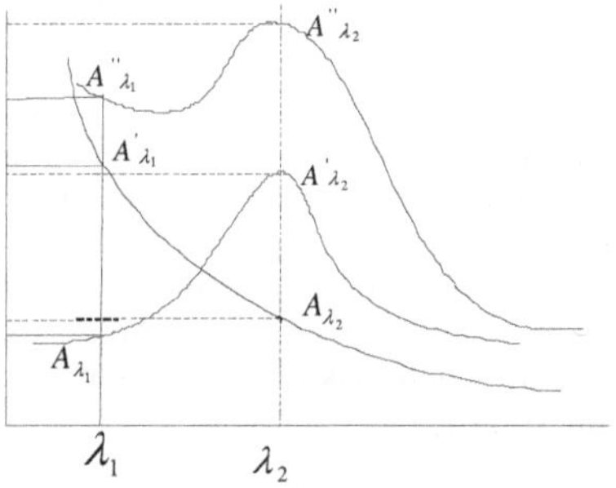


Figure 2. Component spectrum interference

## 2.2. *Principles of two-wavelength measurement*

Two-wavelength detection method is developed upon traditional basis, the theoretical basis is the differentiate technology and equal absorption wavelength. The difference of tradition method is that it uses two different wavelengths (measure wavelength and reference wavelength) to determinate a sample on the same time, to overcome the shortages of a single wavelength determination and improve the determination results precisely and accurately.

Light source basically uses broadband light source, it selected by two monochromators of different wavelengths, the measurement wavelength is $\lambda_1$, the reference wavelength is $\lambda_2$, after the two wavelengths light going through the gas chamber the output are:

$$P(\lambda_1) = P_0(\lambda_1)K(\lambda_1)\exp((-a_{\lambda 1}cl) + \beta) \tag{3}$$

$$P(\lambda_2) = P_0(\lambda_2)K(\lambda_2)\exp((-a_{\lambda 2}cl) + \beta) \tag{4}$$

In the formula, $P(\lambda_1)$, $P(\lambda_2)$ respectively means the output light intensity when light detector selects the wavelength of $\lambda_1, \lambda_2$; $P_0$ means luminous intensity of light source; $K(\lambda_1)$, $K(\lambda_2)$ respectively means photoelectric sensitivity of light detector at $\lambda_1, \lambda_2$; $\beta$ means the background interference. After adjusting the optical path makes $K(\lambda_2)P_0(\lambda_2) = K(\lambda_1)P_0(\lambda_1)$, according to formula (3) / (4) we can get:

$$c = \frac{1}{(a_{\lambda_2} - a_{\lambda_1})l} In \frac{P(\lambda_1)}{P(\lambda_2)} \quad \rightarrow \quad \frac{P(\lambda_1)}{P(\lambda_2)} = \exp(-a_{\lambda_1}cl + a_{\lambda_2}cl) \tag{5}$$

Analyzing theory we can get that light intensity output value of measurement wavelength is much less than light intensity output value of the reference wavelength, that is $P(\lambda_1)/P(\lambda_2) < 1$, after further processing of formula (5) we get:

$$c = \frac{1}{(a_{\lambda_2} - a_{\lambda_1})l} \frac{P(\lambda_1) - P(\lambda_2)}{P(\lambda_1)} \tag{6}$$

## 3. EXPERIMENT

Selecting the light source is mainly based on the absorption spectroscopy range of methane. Methane gas has four inherent vibration: $V_1 = 2913.0\ cm^{-1}$, $V_2 = 1533.3\ cm^{-1}$, $V_3 = 3018.9\ cm^{-1}$, $V_4 = 1305.9\ cm^{-1}$, corresponding wavelength respectively is $\lambda_1 = 3.43\ \mu m$, $\lambda_2 = 6.53\ \mu m$, $\lambda_3 = 3.31\ \mu m$, $\lambda_4 = 7.66\ \mu m$. There are many overtone band and combination band in the near infrared band, and relative absorption spectrum is maximum when $2V_3 + V_2$ is $1.3312\ \mu m$ and $2V_3$ is $1.6654\ \mu m$ (Wang, Xu 1986). In the experiment we select near $1.3312\ \mu m$, light source technology in the region is comparatively mature, broadband LED light source which the center wavelength is $1.3312\ \mu m$ is designed to two-wavelength of methane gas concentration detection system.

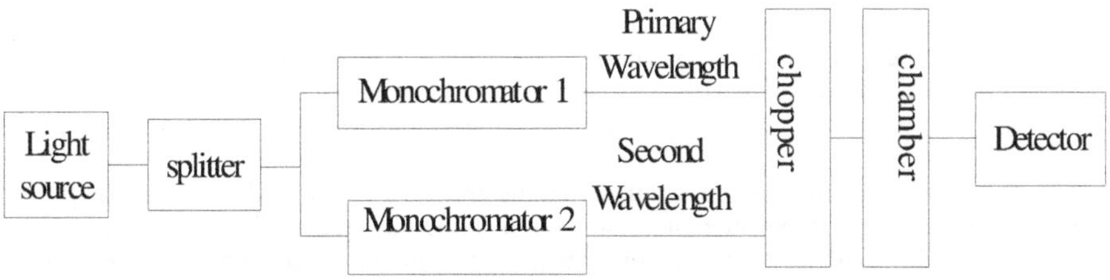


Figure 3. Two-wavelength K coefficient fiber-optic methane sensor system diagram

The corresponding band of laser is splitted into two beams by splitter. In the two-wavelength selection, the measurement wavelength is usually selected absorption peak of 1.3312 $\mu$m. Using K coefficient to calculate, analyze and select reference wavelength in 1.3731 $\mu$m. Because wavelength is fixing, so using optical filters obtain monochromatic light in order to simplify the equipment installation and save cost. Two-wavelength detection need to separately measure monochromatic light of different wavelengths, so in the system using the same detector chopper it needs chopper device to inject the two light beams of different wavelengths into the detector alternately, and then the two signals are separated by synchronization signal amplifier, then the purpose of separate detection is achieved.

The design of two-wavelength K coefficient fiber-optic methane sensor system don't use chamber, but two beams of measure wavelength and reference wavelength pass the same absorption pool in the same position, in order to eliminate the errors caused by different absorption pool parameters, it is convenient to select absorption pool. At room temperature, methane at 1.3312 $\mu$m has the largest absorption coefficient of a = 83 $atm^{-1}cm^{-1}$ (1 atm = 101.325 kPa) (Dubaniewicz, Chilton 1993) when methane measurement absorption pool length is L = 50 cm. Detectors use the window type photomultiplier tube of large area to accept light to transform the pulse light signals of two wavelengths of different phase in the same light path into corresponding pulse electrical signals. Because of the same optical path, the errors of two beams measurement are eliminated. In addition, the larger light area can effectively capture the scattering signal, therefore the sensitivity and precision of detection can be improved.

Table 1. Experimental data

| Serial number | Input values (%) | Output values (%) | Relative error (%) | Serial number | Input values (%) | Output values (%) | Relative error (%) |
|---|---|---|---|---|---|---|---|
| 1 | 0.000 | 0.000 | 0.00 | 11 | 2.000 | 2.042 | 2.10 |
| 2 | 0.200 | 0.202 | 1.00 | 12 | 2.200 | 2.186 | -0.64 |
| 3 | 0.400 | 0.393 | -1.75 | 13 | 2.400 | 2.437 | 1.57 |
| 4 | 0.600 | 0.589 | -1.83 | 14 | 2.600 | 2.623 | 0.87 |
| 5 | 0.800 | 0.809 | 1.11 | 15 | 3.000 | 2.930 | -2.34 |
| 6 | 1.000 | 1.021 | 2.10 | 16 | 3.400 | 3.365 | -1.02 |
| 7 | 1.200 | 1.189 | -0.92 | 17 | 3.700 | 3.757 | 1.53 |
| 8 | 1.400 | 1.428 | 2.00 | 18 | 4.000 | 4.094 | 2.36 |
| 9 | 1.600 | 1.631 | 1.93 | 19 | 4.300 | 4.276 | -0.57 |
| 10 | 1.800 | 1.781 | -1.05 | 20 | 4.500 | 4.489 | -2.54 |

Before the experiment, firstly calibrating the system, that is to say, when the concentration of methane gas in measuring chamber is 0, the output value of the detection of methane concentration is also 0 by adjusting the device to ensure that the two monochromators output intensity are equality. After calibrating we use different concentrations of standard methane gas to compare the system, we can get the relative error by comparing measurement recording values to the standard gas value. From table 1 we can get that when the methane concentration is less than 5%, the relative error is at ± 2.54%, in the works of Dubaniewicz (Dubaniewicz, Chilton 1993) when the methane volume fraction C is less than 5%, the relative error is ± 7%, in comparison, the project in this paper reduces a lot of measurement error ,this method can overcome the defects caused by interference from system, and provide reliable guarantee for the following signal analysis; the lowest detectable concentration of this system can reach to 0.001%, in the woks of Wang (Wang]) the system sets with two path and two wavelength, the minimum detection concentration of the system is 0.02%, by comparing we can see that the detection sensitivity and accuracy of two-wavelength single optical methane sensor in this system are also greatly improved.

Testing the stability of the system, firstly injecting the 1.13% standard methane gas into the chamber and adjusting the experimental value, if it is stable then recording its value, the system continuously runs 11 hours for system stability test. Experimental results are shown in Figure 5, from the figure we can see that the fluctuations area of the recorded value is minimal. Through calculating the real-time records data, the detection maximum value is 1.138%, the minimum is 1.129%, and the average value is 1.133%.

Above experimental data show that the design of the project in this paper greatly reduces the measurement errors, improve measurement accuracy and system stability and sensitivity.

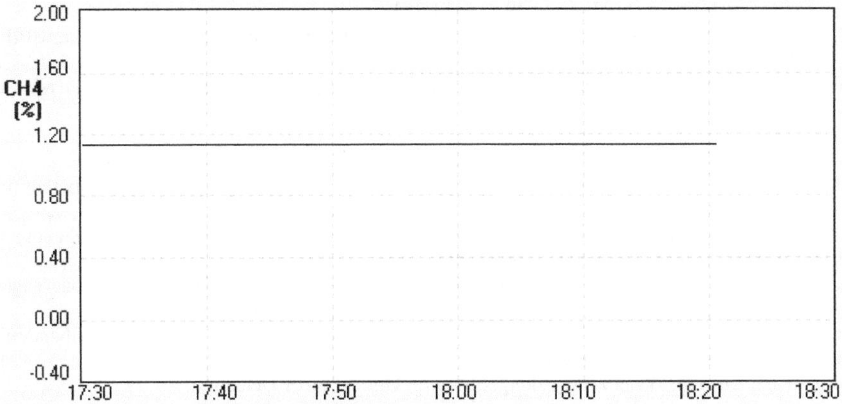


Figure 4. A real-time record of trend analysis

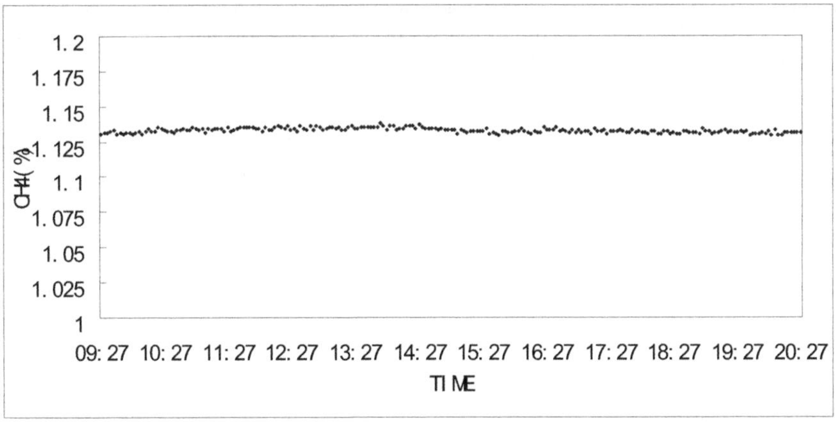


Figure 5. System stability test

## 4. CONCLUSIONS

Using two-wavelength absorption spectroscopy technique designs a two-wavelength single optical fiber methane sensor based on K coefficient method, in the experiments light of the same source will be splitted into two beams, so that two-wavelength monochromatic light intensity is relatively stable, the system can be measured separately monochromatic light intensity of methane gas. Using an absorption cell to eliminate background error, and greatly reduces measurement error from the measurement environment and the background interference. The results show that using this method can eliminate background interference well, effectively improve the accuracy, and have a universal reference value for the other two-wavelength fiber measurement system.

## ACKNOWLEDGEMENT

The project is supported by National Natural Science Foundation of China (50574005) and Natural Science Foundation of Education Department of Anhui, China (KJ2009A023).

## REFERENCES


Dubaniewicz T.H., Chilton J.E., Dobroski H. Jr. 1993. Fiber optical atmospheric mine monitoring [J]. IEEE Transaction on Industry Application 29(4): 740-753

Dakin J.P., Wade C.A., Pinchbeck D. 1987. A Novel Optical Fiber Methane Sensor [J]. Proc. SPIE, 734(5): 187-190

Fried A. 2008. Tunable infrared laser instruments for airborne atmospheric studies [J]. Applied Physics B 92: 409-417

Inaba H., Kobayasi T., Hirama M. 1979. Optical-Fiber Network System for Air-Pollution Monitoring over a Wide Area by Optical Absorption Method [J]. Electronics Letters (23): 749-751

Jin W., Stewart G., Philp W. 1997. Limitation of absorption-based fiber optic gas sensors by coherent reflections [J]. Appl. Opt., 36(25): 6251-6255

Jin W., Stewart G., Culshaw B. 1993. Absorption measurement of methane gas using a broad light source and interferometric signal processing [J]. Optics letters 18(6): 1364-1366

Jin W., Stewart G., Culshaw B. 1993. Absorption Measurement of Methane Gas Using a Broad Light Source and Interferometric Signal Processing [J]. Optics Letters 18(16): 337-339

Wang Shu-ren, Xu Guang-ren, Mai Guang-xin. 1986. Two-wavelength Spectrometer [M]. Shandong Technology Book Concern. (In Chinese)

Wang Yan-ju, Wang Yu-tian, Wang Zhong-dong. Study on Optical Fiber Detection

Zhou Meng-ran, Li Zhen-bi. 2007. Research on coal-mine gas monitoring system controlled by annealing simulating algorithm [J]. Proceedings of SPIE 3rd International Symposium on Advanced Optical Manufacturing and Testing Technologies 6723: 271-278. (In Chinese)

Zhou Meng-ran, Li Zhen-bi, Zhu Zongjiu. 2007. Research of Examination Technology of the Second-harmonic in Coal Mine Gas Detector [J]. Proceedings of the 8th International Conference on Electronic Measurement & Instruments: 587-591. (In Chinese)

International Mining Forum 2010, Liu et al. (eds) © 2010 Taylor & Francis Group, London, UK. ISBN 978-0-415-59896-5

# Numerical simulation analysis of coal-gas outburst during excavation in coal seams

Feng Cai, Zegong Liu
*Anhui University of Science and Technology, Huainan, Anhui Province, China;*

*Key Laboratory of Coal Mine Safety and Efficient Exploitation of Ministry of Education, Huainan, Anhui Province, China;*

*Key Laboratory of Integrated Coal Exploitation and Gas Extraction, AUST, Huainan, Anhui Province, China*

ABSTRACT: Coal and gas outburst is a kind of complex engineering-evoked disaster. According to the foundational theory of coal-rock deformation and gas seepage, taking consideration of nonuniform material mechanics characteristic of coal-rock medium as well as nonlinear permeability characteristic during cracking, and taking advantage of SPH method of LS-DYNA, coal-gas outburst evoked by excavation was numerically simulated. The simulation results reappeared the process of outburst: under the interaction of gas pressure, strata stress and mechanical property of coal-rock, the whole process of outburst including cracks in coal were produced, grew, penetrated each other as well as the coal were ultimately thrown out. The nonlinear essence of the process, that the gradual failure evoked sudden change under the condition of disturbance of mining activity, was revealed. And the evolution of stress field during the cracking process of gassy coal-rock was revealed. This provided theoretic foundation and scientific reference for further research on the mechanism and controlling technology of outbust.

KEYWORDS: Coal-gas outburst; SPH; solid-gas coupling; numerical simulation

## 1. GENERAL INSTRUCTIONS

Coal and gas outburst is a kind of complex engineering-evoked disaster, and it can destroy facilities and ventilation system, suffocate and bury workers, and even cause the accident of gas blasting and coal fire, and ultimately interrupt production. This disaster brings not only enormous economic loss, but also the tragedy of dead workers and closing mine. From 1843, after the first outburst broken in Issac Coal Mine in France, there are more than 30 thousands outburst accident in the world (Cai et al. 2007). But in which, the number of outburst coal mine, the sum of outburst per year, as well as the average intensity ranks first of the world, and the total number of outburst is the 1/3 of all (Wei, Xiao 2005). So, it is one of the most serious disasters in Chinese coal mines, and it has been the important problem which is needed to be solved. So, it has important significance to deeply carry on the foundational research on the outburst mechanism in gassy coal seam under the condition of mining, that can impel the development of high-production-and-high-efficiency technology, and prevent and control this disaster evoked by mining.

To research the outburst mechanism, there were many fruitful research works made by domestic and international scholars (Wang 2004, Long 2008, Thabet 2001), many theories and models were put forward, and this theories and models impelled the development of the research on outburst mechanism greatly. However, though the theoretic foundation was provided by fracture mechanics and damage mechanics, both fracture mechanics and damage mechanics were hard to explain the whole process, because of complicated interaction between rock-coal and gas.

Based on the foundational theory of deformation and seeping flow of gas, coupled compressible gas with deformed rock-coal, and took consideration of nonlinear characteristic of rock-coal in material mechanism as well as nonlinear variation of permeability of rock-coal during deformation and crack, the SPH description, which can describe the couple interaction between solid and gas, was built, and the whole process, that of deformation and crack and ultimately outburst, is numerically simulated, taking advantage of this SPH description.

## 2. FOUNDATION THEORY OF SPH PARTICLE

### 2.1. *SPH (Smoothed Particle Hydrodynamics) description*

As the first mesh-free algorithm of LS-DYNA, SPH was applied widely in the analysis of smash and separating of continuous body (Liu 2003, Xu 2001, Wang 2002).

During the using of SPH, the material was replaced by particles, and every particle has their own mass and fluidity, and conversation equation controls the calculating process. The properties of each particle are: mass ($m$), velocity ($v$), and other properties depending on different problems. The mass conservation equation is:

$$\frac{d\rho}{dt} = -\rho div(v) \tag{1}$$

This equation can be approximated by SPH described in equation (2):

$$\frac{d\rho}{dt}(x_i) = \sum_{j=1}^{N} m_j \left(v(x_j) - v(x_i)\right) A_{ij} \tag{2}$$

Momentum equation can be described into:

$$\frac{dv^a}{dt}(x_i) = \sum_{j=1}^{N} m_j \left( \frac{\sigma^{\alpha,\beta}(x_j)}{\rho_i^2} A_{ij} - \frac{\sigma^{\alpha,\beta}(x_i)}{\rho_j^2} A_{ji} \right) \tag{3}$$

Energy equation is:

$$\frac{dE}{dt}(x_i) = -\frac{P_i}{\rho_i} \sum_{j=1}^{N} m_j \left(v(x_j) - v(x_j)\right) A_{ij} \tag{4}$$

The time integral cycle can be described by Figure 1.

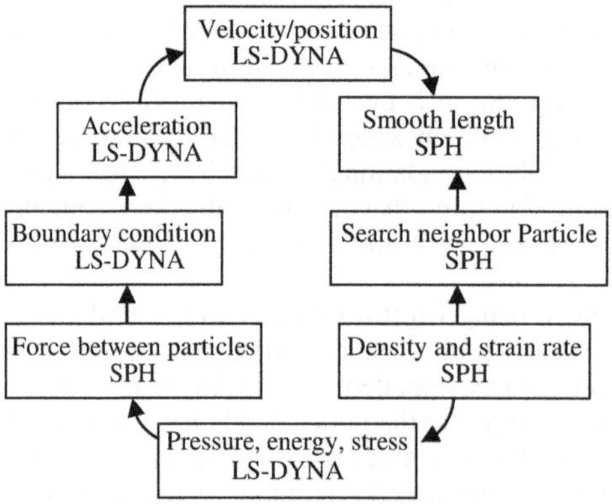


Figure 1. One SPH circle

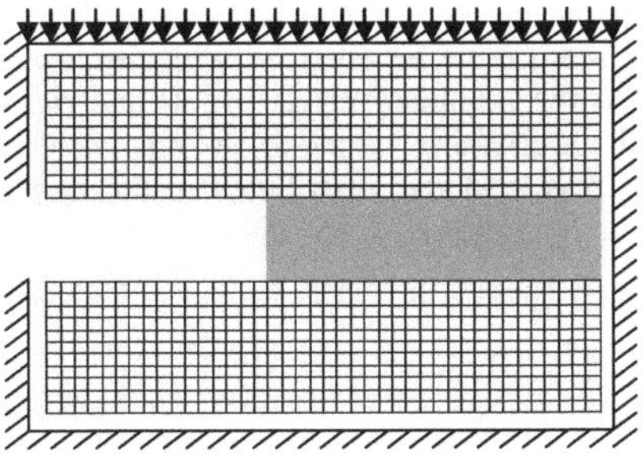

Figure 2. Numerical model of coal and gas outburst

### 2.2. *Shock wave and artificial viscosity*

Shock wave was generated from rock-coal seams during outburst, and this made pressure, density, velocity of mass point and energy jump, and discontinuity made the resolution difficult. Generally, artificial viscosity was joined into pressure to deal with the discontinuity generated by shock wave.

Artificial viscosity used the following algorithm (Yang 2005, Tsukasa 2002):

$$q = \begin{cases} \rho l\left(c_0 l\left|\dot{\varepsilon}_{kk}\right|^2 - c_1 a\left|\dot{\varepsilon}_{kk}\right|\right) & \dot{\varepsilon}_{kk} < 0 \\ 0 & \dot{\varepsilon}_{kk} > 0 \end{cases} \tag{5}$$

In which, $l$ is cube root of the volume of element, $a$ is local sound velocity, $\rho$ is density, $\left|\dot{\varepsilon}_{kk}\right|$ is the trace of strain rate tensor, that is $\left|\dot{\varepsilon}_{11}+\dot{\varepsilon}_{22}+\dot{\varepsilon}_{33}\right|$, $c_0$ (its value is 1.5) and $c_1$ (its value is 0.06) is dimensionless constant.

After introducing artificial viscosity ($q$), the stress's calculation equation changes into:

$$\sigma_{ij} = S_{ij} + (p+q)\delta_{ij}$$

In which: $p$ is pressure, $S_{ij}$ is deviatoric stress tensor.

## 3. CALCULATING MODEL AND BOUNDARY CONDITION

### 3.1. *Calculating model*

Figure 2 is the explicit calculation model for outburst. The elements in Lagrangian mesh in the upper and bottom of the model were rock in roof and floor of coal, and the middle coal elements were SPH particles. The dimension of the whole model was 20×15×0.5 *m*, the height of coal was 3 *m*, the length of excavated part was 8 *m*, and the length of unexcavated part is 12 *m*.

Rock elements used Lagrangian description, taking consideration of lower demand in computational accuracy of roof and floor rock during outburst, the mesh was relatively large, and the dimension of element was 0.5×0.5 *m*, and the sum was 920; coal elements used SPH description, and the mesh was very small, and there were 18850 elements, which diameter is 0.1 *m*.

### 3.2. *Boundary condition*

In explicit calculation model as shown in figure 2, particular constraint of LS-DYNA was used in the model.

When shock wave met boundaries, cracks, weak faces or damage, boundary effect, which generates stress concentration or magnification (Z. 2002, Jiang 2000). When the degree of stress concentration or magnification was over the breaking point, the coal would be cracked. In the process of numerical simulation, infinite practical coal seam must be replaced with finite numerical model because of limit of computer's ability, so the boundary effect in left and right boundary must be prevented. This present paper used unreflecting boundary to resolve this question effectively, as shown in figure 1. The strata-stress load was distributed in the upper boundary, according to the bury depth and Haym's hypothesis (Cao 2010, Yin 2009), $\sigma_y = \bar{\gamma} H$ , the strata stress loaded on the upper boundary of the model could be calculated.

## 4. RESULT AND ANALYSIS OF NUMERICAL SIMULATION

### 4.1. *Distribution and analysis of stress during outburst*

Figure 3 shows the whole process of outburst of coal and gas. It was shown in Figure 3 that the time from the coal wall been cracked to fierce outburst of coal and gas was very short. In this ultra-short time, fierce physical change and energy transfer of the gas contained in the soft and hard coal happened.

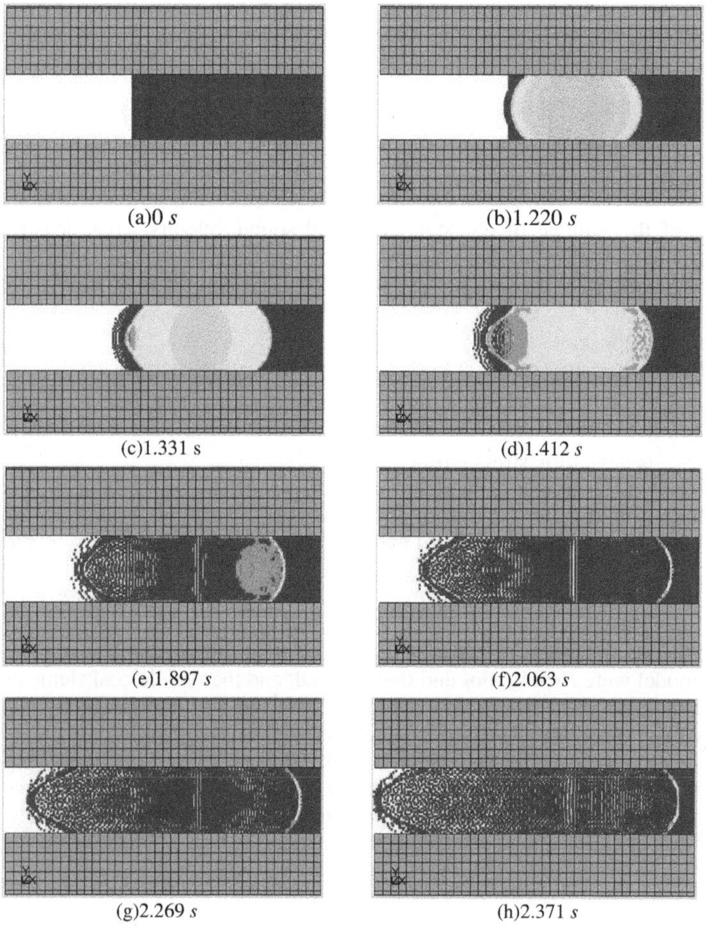


Figure 3. Stress distribution while outburst during excavation in coal

From Figure 3(a) to Figure 3(b), it probably needed 1.22 *s*. In this period, hard coal part in coal seam broke suddenly and lost the capacity of support because of action of external force or enormous internal energy, and soft coal part inside coal seam got the task to support. But because the coal is a kind of brittle material, when it was buried in deep strata, because it was under the condition of three dimensional stress, the compressive strength was very large and had the capacity to support; when it was revealed, because it was under the condition of single dimensional stress, the compressive strength was very small, and it was very easy to crack. Because of the crack of outside hard coal, inside high-pressure gas gushed out in the direction of roadway, at the same time, the adsorbed gas desorbed quickly, high-pressure gas carried cracked coal particles gushed into roadway space. Outside hard coal broke suddenly, and revealed soft coal support the upper load, but it was smashed quickly, so, the area of stress concentration was always near the revealed coal. Because of big hardness, dense and low permeability of the outside hard coal, it prevented inside gas from migrating into roadway space, so the gradient of gas pressure increased. After break of outside hard coal, inside gas gushed to the roadway space quickly. So, the break of outside hard coal in headings leaded to the redistribution of support stress, gas pressure and its gradient, and further leaded to the redistribution of stress in the inside soft coal, which would take effect of support, and plenty of gas gushed out, and ultimately high-pressure gas carried cracked coal particles gushed into roadway space. And this signified the start of outburst.

At 1.331 *s*, inside soft coal overcame the impeding effect of outside hard coal and started to gush out, as shown in figure 3(c). In this moment, the soft coal contacted with atmosphere, and the gas pressure decreased sharply in the earliest place of contact to atmosphere (middle of figure 3(c)), and the gas pressure approached atmosphere pressure in some place which has contacted to atmosphere and the pressure gradient increased sharply between inside and outside of coal wall. The particle, under the action of elastic energy and pressure gradient, of soft coal was on the flying state. In the area of boundary coal wall, a stress concentration shell formed. At the same time, because the support stress moved into the deep of the coal constantly, inside soft coal support the upper load, and a new stress concentration shell formed in the inside of the old shell. And then the old shell gushed to the roadway space. So, as the process of outburst, stress concentration shell formed and gushed out constantly.

At 1.897 *s*, because the move velocity of earlier outburst coal particle, “head of outburst” was formed, and the distance to further outburst coal particle was enlarged, as shown in figu-re 3(e).

As the exposing face enlarged, inside gas contacted with mining atmosphere, the capacity to carry coal particles decreased. So, the number of outburst coal particles began to decrease after summit, then old stress concentration shell could be saved, and new formed stress concentration shell could move into the deep coal gradually. In this time, the strength of outburst decreased gradually, and formed a series of shells, as shown in Figure 3(f), 3(g), 3(h).

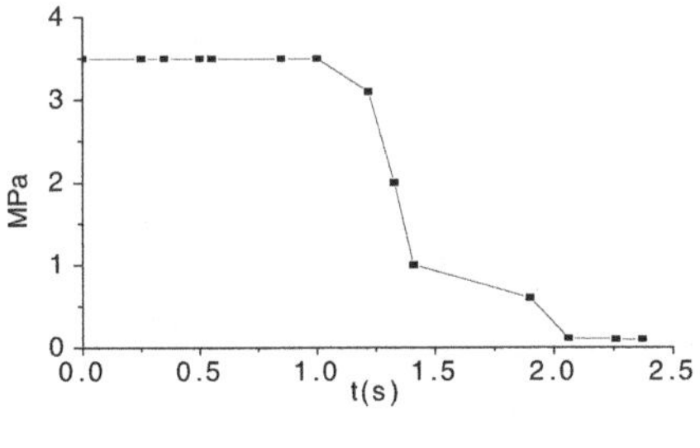


Figure 4. The change of horizontal stress of coal unit in the center of outburst shell with time

## 4.2. *Shell-shaped stripping out of coal*

Stress concentration shell existed in the whole process of outburst, as shown in Figure 3, and it was generated by the interaction between upper pressure on revealed soft coal and inside high-pressure gas. Because of the effect of stress concentration shell, the revealed soft coal was compressed densely, and blocked the transferring path of gas, so inside gas pressure increased and gas pressure gradient increased sharply. At the same time, the revealed and compressed soft coal could bear the enormous pressure and be crushed. So the just blocked path of gas opened suddenly, and gas carried crushed coal gushed to the roadway space, and formed the shell-shaped stripping out of coal. And then, the stress concentration moved to inside new revealed coal. And this process recycled again and again, leaded to constantly outburst of gas and coal.

### 4.3. *Force analysis of mass point of outburst coal*

From Figure 4, horizontal stress of outburst shell was generated by inside gas pressure. After the start of outburst, horizontal stress of mass point of outburst coal decrease to zero sharply. And this was complete identity with released process of practical outburst.

### 4.4. *Kinematic analysis of mass point of outburst coal*

Figures 5 and 6 described the kinematic state of mass point of outburst coal in the process of outburst. The three curves of A, B, C respectively represented the kinematic state of mass point in the middle of outburst shell, the position of distance of 0.3 m and 0.6 m to the middle of outburst shell.

From Figures 5 and 6, mass point of outburst coal began to accelerate, and reached the summit at 1.6 s and stopped accelerate. From 1.1 to 1.6 s, because of high gas pressure inside coal, the velocity of mass point of outburst coal increased sharply, and this could be seen steep curve in velocity curves and absolute value increasing sharply in acceleration curves.

After 1.6 *s*, gas pressure decreased sharply, the velocity of mass point of outburst coal maintained its value for a short time and began to decrease. At this time, external force on mass point of outburst coal was very small, the reason of movement of mass point was that the mass point had inertia. So velocity curves went gently.

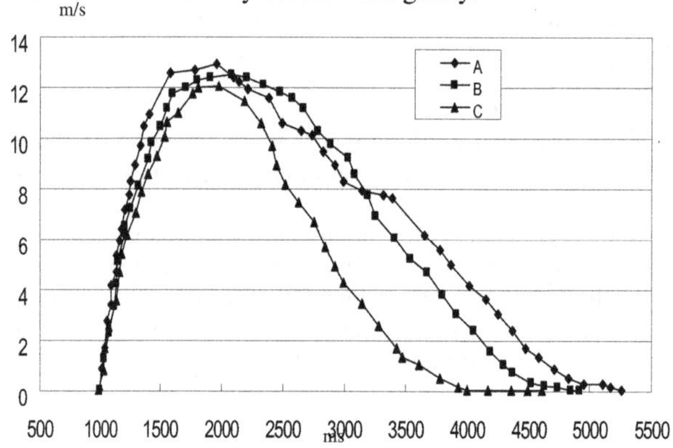


Figure 5. The change of horizontal velocity of coal unit with time

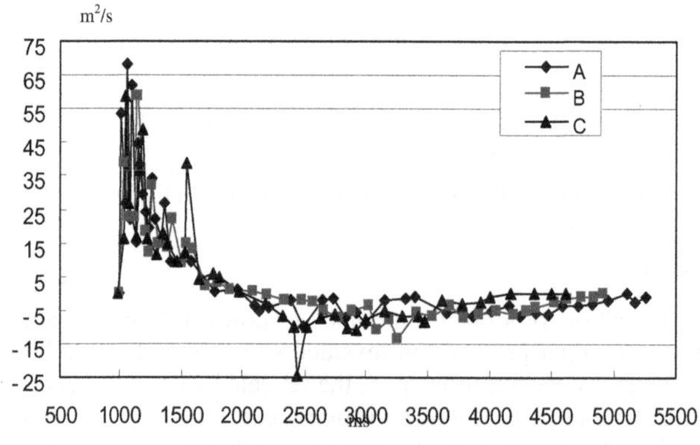


Figure 6. The change of horizontal acceleration of coal unit with time

## 5. CONCLUSIONS

1) This numerical simulation reappeared the whole process, that is, under the comprehensive interaction of gas pressure, strata stress and mechanical property of rock-coal, cracks in gassy coal generated, grew, interacted and penetrated into each other, and crushed coal ultimately lost stability and was thrown out.
2) Through numerical simulation, there always was stress concentration shell. And this stress concentration shell was formed by newly revealed soft coal supporting upper rock strata. As the process of outburst, outside shell gushed out, and stress concentration shell went inside and formed a new shell. In the process of outburst intensity decreased and stopped, outside shells didn't gush out, and stress concentration shell went inside and formed a new shell. Stress concentration shell was the main force of shell-shaped stripping out of coal.
3) Numerical simulation results provided theoretic foundation and scientific grounds for further research on the mechanism and preventing and control technology of outburst of coal and gas.

## ACKNOWLEDGEMENTS

Parts of the research were performed under grant of National Science Support Plan Project (2007BAK28B01) and Scientific Creative Team Plan Project for High School of Anhui Province (2006KJ005TD).

## REFERENCES

Cai Feng, Liu Zegong. 2007. Numerical Simulation and Analysis of Improving Permeability by Deep-hole Presplitting Explosion in loose-soft and Low Permeability Coal Sea. Journal of China Coal Society 32(5): 499-503. (In Chinese)

Cao Shugang, Guo Ping, Li yong et al. 2010. Effect of gas pressure on gas seepage of outburst coal. Journal of China Coal Society 35(4): 595-599. (In Chinese)

Jiang Chenglin. 2000. Linear Discriminatory Analysis on Coal-Gas Outburst Danger and Its Critical Value. Journal of china university of mining & technology 29(1): 63-66. (In Chinese)

Liu Wentao. 2003. Study on the damage constitutive model for rock and the effects of underground explosion. Chinese Journal of Rock Mechanics and Engineering 22(2): 342-346. (In Chinese)

Long Qingming, Zhao Xusheng, Sun Dongling et al. 2008. Experimental study on coal permeability by adsorp tion. Journal of China Coal Society 33 (9): 1131-1134. (In Chinese)

Thabet, A., Haldane, D. 2001. Three-dimensional numerical simulation of the behavior of standard concete test specimens when subjected to impact loading. Computers and Structures 79(1): 21-31. (In Chinese)

Tsukasa, Ohba, Hiromitsu, Taniguchi. 2002. Effect of explosion energy and depth on the nature of explosion cloud A field experimental study. Journal of Volcanology and Geothermal Research 115(1): 33-42.

Wang Jinhua. 2004. China mechanized roadheader status and bolt support technology in mine seam roadway. Coal Science and Technology 32(1): 6-10. (In Chinese)

Wang Zhongqi, Zhang Qi, Bai Chunhua. 2002. Numerical simulation on influence of hole depth on explosion effect. Chinese Journal of Rock Mechanics and Engineering 21(4): 550-553. (In Chinese)

Wei Guoying, Zhang Shujun, Xin Xinping. 2005. Study on Technology of Preventing Coal and Gas Outburst during Excavating in Outburst Coal Bed. China Safety Science Journal 15(6): 100-104. (In Chinese)

Xiao Hongfei, He Xueqiu, Feng Tao et al. 2005. Research on coupling laws between eme and stress fields during deformation and fracture of mine tunnel excavation by flac3d simulation. Chinese Journal of Rock Mechanics and Engineering 24(13): 2304-2309. (In Chinese)

Xu Ying, Zong Qi. 2001. Theory and application of Underground engineer blasting. Xuzhou: Press of China University of Mining Technology. (In Chinese)

Yang Renshu, Niu Xuchao, Shang Housheng, et al. 2005. Dynamic caustics analysis of crack in samdwich materials under blasting stress wave. Journal of China Coal Society 30(1): 36-39. (In Chinese)

Yin Guangzhi, Li Xiaoshuang, Zhao Hongbao et al. 2009. Experimental study of effect of gas pressure on gas seepage of outburst coal. Chinese Journal of Rock Mechanics and Engineering 28(4): 697-702. (In Chinese)

Z. Fang, J. P. Harrison. 2002. Application of a local degradation model to the analysis of brittle fracture of laboratory scale rock specimens under triaxial conditions. International Journal of Rock Mechanics & Mining Sciences 39(6): 459-476

International Mining Forum 2010, Liu et al. (eds) © 2010 Taylor & Francis Group, London, UK. ISBN 978-0-415-59896-5

# Application of multi-step sectional sealing technology in gas pressure measurement in coal seam by a downward borehole

Qilin He
*Key Laboratory of Coal Mine Safety and Effecient Exploitation of Ministry of Education, Huainan, Anhui Province, China*

Wei Peng
*School of Energy and Safety, Anhui University of Science and Technology, Huainan, Anhui Province, China*

ABSTRACT: In order to get the exact value of gas pressure, high-pressure grouting method is used to sealing fracture around the borehole in rock, which can block the connection between the borehole and the fracture in rock. This article introduces a method that using multi-step sectional sealing technology to mesure the gas pressure in broken coal seam. An advanced method to measure the gas pressure in coal seam comes up according to a large number of actual measurements and analysis of the borehole sealing technology, which has certain value of extension and application.

## 1. INTRUDUCTION

Gas pressure in coal seam is the most fundamental parameter to forecast coal seam gas emission and flow, which is also one of the coal and gas outburst motivity. It is very important to get the exact value of gas pressure for forecasting coal and gas outburst, and establising effective measure about preventing coal and gas outburst.

At present, the common method to measure the gas pressure is called direct method, that is drilling from the rock roadway to the coal seam, then sealing the borehole and measuring the coal seam gas pressure. It can be divided into vary methods according to different materials in sealing borehole, such as capsule and cement slurry. While in actual measurement process, grouting cement slurry into the fracture in rock is the most popular method, which is not only very simple in operating, very cheap , but also has a good measurement result. Many researchers think sealing the downward borehole is very difficult in gas pressure measurement technology, because the water and cement slurry often flow into the bottom pressure measurement chamber due to the gravity. Through a lot of pressure measurement practice, a new multi-step sectional sealing technology with cement slurry, polyurethane and some other materials is gradually improved and mature, the new technology was used in many coal seam pressure measurement of North area of Zhangji Mine and achieved good results.

## 2. PRINCIPLE OF COAL SEAM GAS PRESSURE MEASUREMENT

Drill a borehole from the rock roadway into a coal seam, and then connect a serial of tube from the coal seam to the outside through the borehole with a gas gauge. At last, grout cement slurry into the borehole to block the connection between coal seam and outside. As the original gas in the borehole has got out during the drilling, it leaves a low gas pressure chamber. The gas in the coal seam around borehole flows into the borehole, and the pressure in it increases gradually. Since the volume of coal around the borehole is much larger than the volume of coal borehole, and the gas absorption amount of coal seam is much greater than the free gas amount, the gas pressure in the borehole will gradually approach the original coal one after a period of gas permeation. The pressure can be read from the gauge outside the borehole.

In the point of pressure measurement principle, once measuring a coal seam pressure, just drilling a borehole to the right coal seam, then sealing the borehole, a steady flow of gas will go to the measureing chamber from the coal seam,, the pressure in the chamber will finally reach the value of original coal seam.

## 3. MULTI-STEP SECTIONAL SEALING TECHNOLOGY

### 3.1. *Significance of multi-step sectional sealing technology*

Because of the specific occurrence conditions of coal seam and the influence of mining, there are many fractures in coal and rock which causes the borehole is connected with groundwater or air. When we measure the coal seam gas pressure, the borehole is often filled by groundwater, or the borehole conducts to the air, so we can't get the exact value of gas pressure. Because the high cost of drilling borehole underground, once a borehole cant't be used to measure gas pressure, it will not only bring a trouble for gas controlling, but also cause large loss of money, people and time. The technology of multi-step sectional sealing can solve the problem.

### 3.2. *Principle of multi-step sectional sealing*

When the drilling of a pressure measurement borehole is completed, put a serial tube into it. There are additional baffles at both end of the borehole.Sealing both end with polyurethane block. Besides the tube with gas gauge, there are two pipes(a grouting pipe and a slurry spilling pipe) in the borehole, and each of them is connected with a high-pressure valve. When sealing the borehole, after the polyurethane at both end of borehole solidify, fill cement slurry with pump through the grouting pipe. The grounting procedure is in two steps. First, grouting 3–5 m in length with cement. A second grouting operates when the former cement solidify (24 hours after the first grouting).until the cement slurry gets out from the spilling pipe, and then close the valve, keep grouting on with pressurized cement again , to improve the impermeability of sealing.

After the solidification of cement which grouting in the first step, it can take the pressure of the second step, to prevent the flow of pressurized cement slurry grouting to the pressure measurement chamber, and the cement can flow into the fractures to improve the sealing quality.

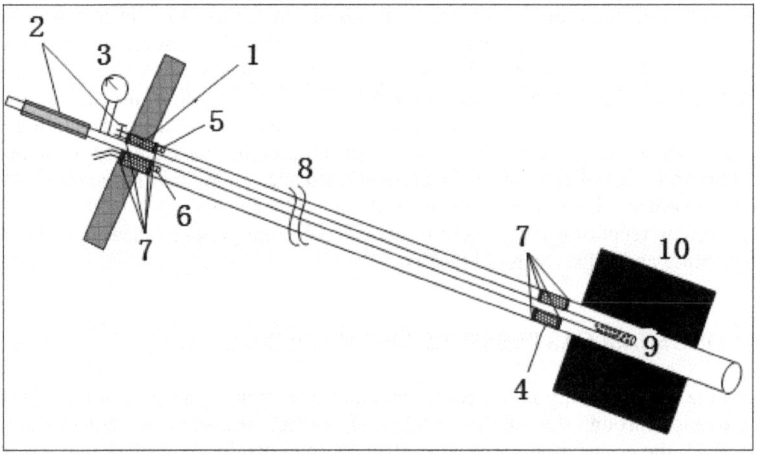


Figure 1. Sketch map of downward borehole sealing and pressure measurement
1 - Cotton and polyurethane; 2 - Ball Valve; 3 - Pressure Gauge;
4 - Cotton and polyurethane; 5 - Slurry spilling pipe; 6 - Grouting pipe;
7 - Baffle; 8 - Cement with coagulant; 9 - Tube with holes; 10 - Coal seam

### 3.3. *Site operation*

(1) Drill the borehole according to the criterion, and then clean it with pressured air, the borehole must go across the coal seam floor more than 0.5 m.

(2) Bandage cotton with polyurethane (set the fermentation time of polyurethane to 5 minutes) between the two baffles at the end of sealing device. Push the sealing device into the borehole quickly. At last, install the front sealing devices, bandage cotton with polyurethane between the two baffles which is the same as the end l,

(3) After the polyurethane fully fermenting, inject the cement slurry into the middle of borehole through the grouting tube with pump, the grouting length is 3–5 m. It's best to use thick slurry to avoid the slurry goes across the cotton-polyurethane barrier to the pressure measurement chamber.

(4) After the solidification of first step cement slurry ,the second grouting goes on until the cement slurry gets out from the spilling pipe, and then close the valve, keep grouting on. When the grouting pressure reaches 4 Mpa, stop the second grouting, then close the valve of grouting pipe.

(5) When the sealing cement solidified (24 hours after grouting), install the pressure measurement devices, and fill high pressure nitrogen into the measurement chamber through a T-branch tube until the pressure reach the estimated value of coal seam gas pressure. Monitor the pressure changes of measurment chamber by pressure gauge. If the pressure doesn't change within 24 hours, the value of pressure gauge is the actual coal seam gas presure. The period of measurement is about 2 or 3 days, so it is also a fast measurement method.

## 4. APPLICATION

The 6th and 8th coal seam in North area of Zhangji Mine were identified as seam with potential outburst danger, and the coal mine as mine with potential outburst danger by qualified evaluation company. According to relevant laws and regulations, if the seam with potential outburst danger can mining protection seam, it must do it first; If there is no protection seam for mining, the tunnelling in the coal seam must go after gas drainage through boreholes in the roof and floor of coal seam, and elimination outburst measure about coal working face is draining gas from its roof and floor through boreholes. The 8th coal seam of the 2nd West mining area is a seam with potential outburst danger due to its depth. We designed 4 pressure measurement bo-rehole in the coal seam, and get the gas pressure successsfully.

Table 1. Gas pressure measured in No.8 coal seam in No.2 West Panel

| Time for measurement/ /days | Gas pressure (MPa) | | | |
|---|---|---|---|---|
| | Elevation -468.7 m | Elevation -492.0 m | Elevation -543.7 m | Elevation -586.2 m |
| 1 | 0.02 | 0.08 | 0.05 | 0.12 |
| 2 | 0.05 | 0.17 | 0.21 | 0.39 |
| 3 | 0.25 | 0.28 | 0.37 | 0.57 |
| 4 | 0.31 | 0.39 | 0.49 | 0.75 |
| 5 | 0.33 | 0.47 | 0.71 | 0.89 |
| 6 | 0.33 | 0.48 | 0.84 | 1.25 |
| 7 | 0.33 | 0.48 | 0.86 | 1.25 |

According to Figure 2, gas pressure changes with coal seam buried depth, and there is a linear relation between them.

$$P = 0.0078H-3.349 \quad (1)$$

where: P - gas pressure, MPa; H - coal seam buried depth, m.

From the result, we can get a conclusion that the gas pressure increases with coal seam buried depth inearly, the linear relation accords with coal mine actual gas emission situation and coal seam gas pressure gradient variation rules in methane area. It indicates that the application of multi-step sectional sealing and pressure measurement technology in measuring gas pressure is practicable, reliable and successful.

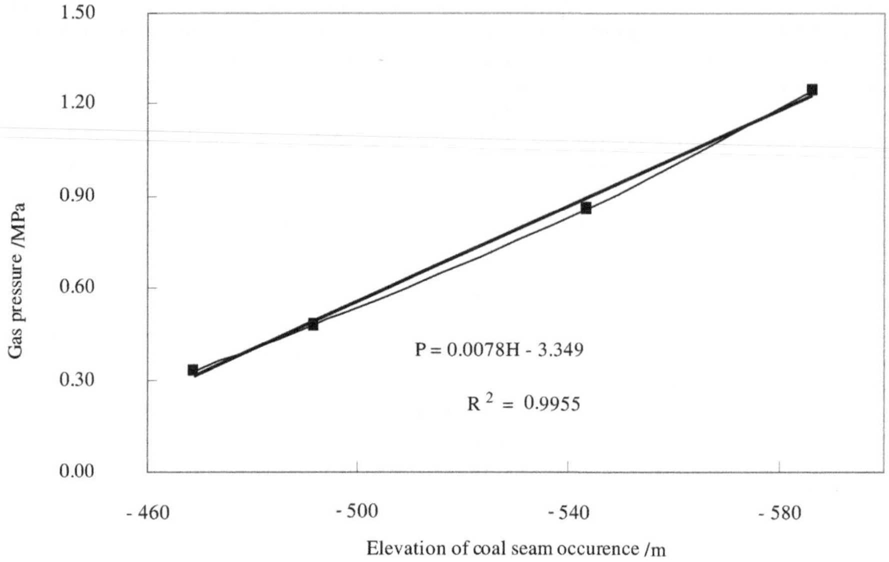


Figure 2. Relationship between gas pressure and elevation of coal seam occurrence

## 5. CONCLUSIONS

Multi-step sectional sealing and pressure measurement technology can effectively solve the problem that rock fracture water goes into the pressure measurement chamber or weakly grouting pressure causes the pressure measurement chamber conducting to the air. Those problems will lead the failure of gas pressure measurement. It appeares that the technology of sealing and measurment can get the exact gas pressure in different type of coal seam, which is also very simple, practicable, and has certain value of extension and application.

## REFERENCE


Chen Daxin 2007. Second grouting reinforcement and water sealing technology in mine auxiliary shaft of Renlou Mine (1): 72. (In Chinese)

Ge Junling, Wang Baishun 2009. Application of Sealing Technology of Downhole for Measurement of Gas Pressure. Safety in Coal Mines (3): 28-29. (In Chinese)

Zhang Tianen 1990. Elementary Knowledge for Preventing Coal and Gas Outburst. Beijing: Coal Industry Press. (In Chinese)

International Mining Forum 2010, Liu et al. (eds) © 2010 Taylor & Francis Group, London, UK. ISBN 978-0-415-59896-5

# Deduction analysis on development characteristics of mining induced stress shell and mining induced fracture in deep longwall mining

Ke Yang
*Key Laboratory of Integrated Coal Exploitation and Gas Extraction, Anhui University of Science and Technology, Huainan, Anhui Province, China*

Guangxiang Xie
*Key Laboratory of Coal Mine Safety and Efficient Exploitation of Ministry of Education; Anhui University of Science and Technology, Huainan, Anhui Province, China*

ABSTRACT: key issue in underground mining is to understand and master distributing and evolving patterns of stress and fissure induced by mining profoundly, and to control and utilize the action of rock pressure effectively. Moving coordinate system and ellipsoid equation have been carried to construct the 3D model of MISS analysis based on rock pressure and strata behaviors of deep longwall mining in Huainan and Huaibei coal mines. Developing configuration equations of mining induced stress shell (MISS) and evolving extension of mining induced fissure (MIF) were obtained under the conditions of subcritical extraction. And the relational expressions of factors during the developing of MISS-MIF were gained by analyzing factors affected on development characteristics of MISS-MIF. Constructed model has been applied to investigate into calculating and deducing analysis on MISS and MIF with longwall mining on the strike in special geological and mining conditions. In which, development characteristics of MISS-MIF were obtained that are in conformity with experimental results. Integrated analyzing shows that 3D model is feasible in calculating and deducing development characteristics of MISS-MIF.

KEYWORDS: Longwall mining, mining-induced stress shell, mining-induced fissure, subcritical extraction, half- ellipsoid equation

## 1. INTRODUCTION

Longwall mining has been a popular safety and high effective mining method in mining deep coal resources. Engineering practices show that disaster-causing mechanism and prevention, which induced by mining induced stress (MIS) and MIF developing such as disability of surrounding rock, rock burst etc., has been a science difficult problem that all engaged mining engineering should face(Song 2003, Qian et al. 2003, Xie et al. 2006, Xie 2007, Peng 2008, Yuan 2008).

And development characteristics of MIS and MIF, deformation and displacement patterns of surrounding rock, and rock or coal dynamic disasters are influenced by 3-D rock pressure development. Especially, some strata behaviors happened at working face frontage and rear, near gates are all controlled by the MISS development in longwall mining (Wang et al. 2003; Xie 2005a,b; Xie et al. 2006, 2007; Yang 2007; Yang, Xie 2008; Shen et al. 2008; Yang, Xie 2009).

Thus according to typical geological and mining conditions, to establish the MISS development model and deduce the calculating procedure of MISS and MIF developing characteristics is important and essential to build theoretical basis that will play an key role in safety and high effectively mining deep coal resources.

## 2. MODELING AND ANALYZING OF MISS DEVELOPMENT

Engineering practice and research show that the MISS composed of high stress bundles exists in the surrounding rock of a longwall mining face. MISS located unbroken strata and original coal seam is the main load bearing structure that bear and transfer the load of overlying strata. The development configuration of MISS is changing and to be a relative stable state with the longwall face advancing and its development configuration is approximate to a half space ellipsoid shell. It is essential the development process of MISS and MIF are simultaneous. Namely, the development extent of MIF extends with the mining induced stress (MIS) transferring while strata are broken when the MIS value is better than the critical loading strength. There is an unloading and MIF evolving area in inner of MISS and some MISS structure is located in plastic MIF area especially plastically. Consequently, MISS and MIF development are interactional and there is some inevitable relationship between MISS and MIF in geometric characteristics and development configuration.

### 2.1. *MISS development model*

Based on longwall mining analysis, the MISS development (Fig. 1) was been established within fundamental assumptions as follows (Yang, Xie 2010). Firstly, the heights of MISS and MIF are all to the maximum of given geological and technological conditions when advanced distance is approximate equal the face length and it is assumed a subcritical extraction. Secondly, the inside rock of caving zone and fractured zone (CFZ) are regarded as loose body, other rock in bedrock are regarded as elastic-plastic bodies, and the overburden layers are also regarded as loose body which those load are equably applied on bedrock. Thirdly, the fractured and broken path lines are all beeline and MISS and MIF development characteristics are isotropic along coal seam without considering special geological structure and gates' excavation.

Analysis model of MISS and MIF development is shown as Figure 1.

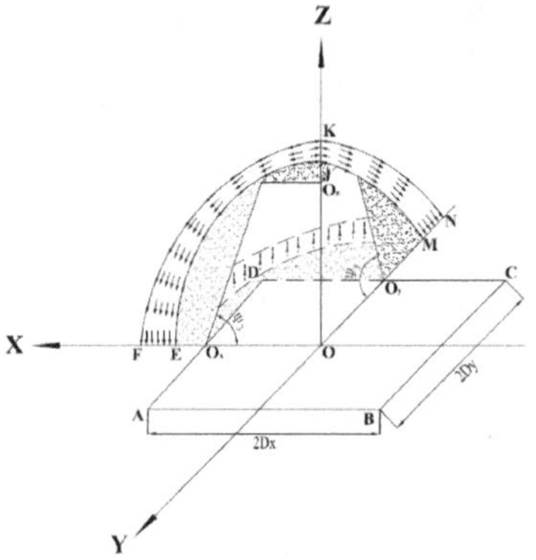


a) comprehensive model (2Dx = 2Dy)

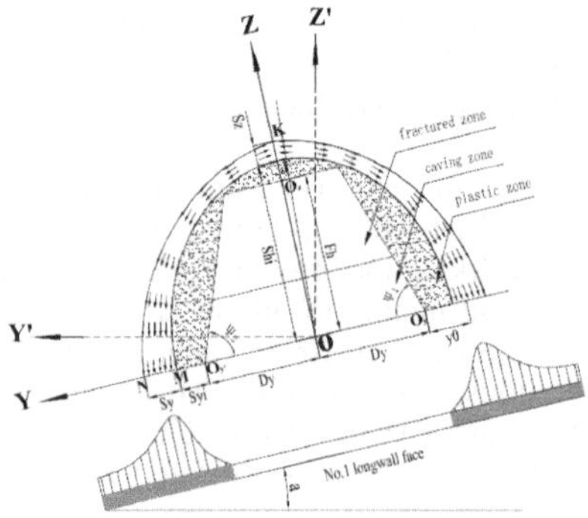


b) dip section (2Dx > 2Dy)

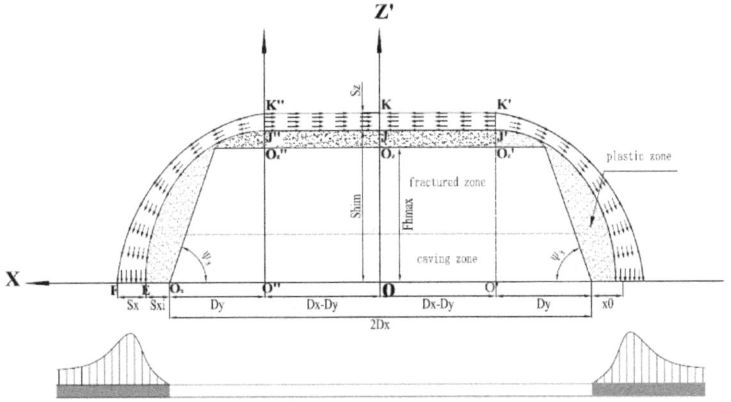


c) strike section (2Dx > 2Dy)

Figure 1. Analysis model of MISS and MIF development

In the model: $2D_x$ (AB segment) and $2D_y$ (BC segment) is the advanced length on the strike and face length to the dip respectively. Positive directions of X, Y, and Z are face advancing direction, dip direction, and overlying strata direction respectively. The origin of coordinate is located at the geometric centre open-off cut and is removed with advancing, such as origin of coordinate will be removed $D_x$ distance along the advancing direction while face has been advanced $2D_x$ distance. The OxE segment and OxM segment is distance of inside MISS base to wall on the strike and to the dip remarked $S_{xi}$ and $S_{yi}$ respectively. The EF segment and MN segment is the MISS base width on the strike and to the dip remarked $S_x$ and $S_y$ respectively. The OOz segment is the height of caving zone and fractured zone remarked $F_{hm}$. The OJ (O'J' or O"J") and OK (O'K' or O"K") segment is the inside height and exterior MISS remarked $S_{hi}$ and She respectively. The JK segment is the thickness of MISS remarked $S_z$. The Ψ1, Ψ2, Ψ3, Ψ3' is the broken angle along the dip, rise, and rear-frontage strike direction respectively.

## 2.2. *MISS development equation deduction*

According to half-ellipsoid equation (Equation 1), the ellipsoid equation can be deduced if a, b, and c is only known.

$$\frac{x^2}{a^2}+\frac{y^2}{b^2}+\frac{z^2}{c^2}=1 \quad z>0 \tag{1}$$

So, while advancing distance is less than or equal to face length ($2D_x \leq 2D_y$), where $a=D_x+S_{xi}$, $b=D_y+S_{yi}$ , $c=S_{hi}$ $a=D_x+S_{xi}+S_x$, $b=D_y+S_{yi}+S_y$, and $c=S_{hi}+S_z$ are only introduced into Equation 1 and inside and exterior MISS development equations (I-MISS and E-MISS) can be deduced shown as Equation 2, which their enveloped part is as well as MISS body exists in strata. As the same as above, I-MISS and E-MISS development equations can be deduced when advancing distance is more than face length ($2D_x>2D_y$) shown as Equation 3.

$$2D_x \leq 2D_y: \begin{cases} I-MISS: \dfrac{x^2}{(D_x+S_{xi})^2}+\dfrac{y^2}{(D_y+S_{yi})^2}+\dfrac{z^2}{(S_{hi})^2}=1 \\ E-MISS: \dfrac{x^2}{(D_x+S_{xi}+S_x)^2}+\dfrac{y^2}{(D_y+S_{yi}+S_y)^2}+\dfrac{z^2}{(S_{hi}+S_z)^2}=1 \end{cases} \tag{2}$$

$$
2D_x > 2D_y : \begin{cases} x \in [(D_x - D_y), D_x] \cup [-D_x, -(D_x - D_y)] : \begin{cases} I-MISS: \dfrac{x^2}{(D_y + S_{xi})^2} + \dfrac{y^2}{(D_y + S_{yi})^2} + \dfrac{z^2}{(S_{hi})^2} = 1 \\ E-MISS: \dfrac{x^2}{(D_y + S_{xi} + S_x)^2} + \dfrac{y^2}{(D_y + S_{yi} + S_y)^2} + \dfrac{z^2}{(S_{hi} + S_z)^2} = 1 \end{cases} \\ x \in [(D_x - D_y), D_x] \cup [-D_x, -(D_x - D_y)] : \begin{cases} I-MISS: \dfrac{y^2}{(D_y + S_{yi})^2} + \dfrac{z^2}{(S_{him})^2} = 1 \\ E-MISS: \dfrac{y^2}{(D_y + S_{yi} + S_y)^2} + \dfrac{z^2}{(S_{him} + S_z)^2} = 1 \end{cases} \end{cases} \tag{3}
$$

where: $2D_x$ is face advancing distance (m); $2D_y$ is face length to the dip (m); $S_{xi}$ is distance of inside MISS base to face wall on the strike (m); $S_{yi}$ is distance of inside MISS base to roadway wall to the dip (m); $S_{hi}$ and $S_{him}$ is height and the max height of inside MISS (m) respectively; $S_x$ is width of MISS base on the strike (m); $S_y$ is width of MISS base to the dip, $S_z$ is thickness of MISS on the $Z$ direction (m).

According to Equation 2 and Equation 3, MISS development is dynamically evolving process and some parameters such as height and width of MISS, will be to the maximum of special geological and mining condition dynamic with face advancing.

### 2.3. *Analysis on factors affected MISS-MIF development*

In summary, MIF evolvement extension zone can be deduced and obtained if mined dimension ($2D_x$, $2D_y$), height of caved and fractured zones ($F_h$, $F_{hmax}$), broken angle of overlying strata ($\Psi_1$, $\Psi_2$, $\Psi_3$), and coal seam plastic zones ($x_0$, $y_0$) are known. And MISS development equations can be deduced and obtained if other MISS heights ($S_{hi}$, $S_{he}$, $S_h$, $S_{him}$, $S_{hem}$, $S_{hm}$), distance of inside MISS base to wall ($S_{xi}$, $S_{xe}$, $S_{yi}$, $S_{ye}$), and width or thickness of MISS are known. Each determination parameter of functional relationship between MISS and MIF can be deduced shown as Equation 4, Equation 5 and Equation 6 with integrated analyzing on physical-mechanical properties of coal-rock, geological conditions and mining method.

$$
(S_{hi}, S_{he}, S_h, S_{him}, S_{hem}, S_{hm}) = f(F_h, 2D_y, (x_0, y_0), \psi) \tag{4}
$$

$$
(S_x, S_{xi}, S_{xe}) = f(x_0) \tag{5}
$$

$$
(S_y, S_{yi}, S_{ye}) = f(y_0) \tag{6}
$$

Where, the S and F remark stands for MISS and MIF respectively. And the subscript *h*, *i*, *e*, *m*, *x*, *y*, and *0* stands for height, inside of MISS base, exterior of MISS base, face advancing direction, dip direction, and plastic zone respectively.

Because MISS is a high stress zone formed mining induced fissure, rock broken, and stress redistribution, above parameters are the major acting factors on stress redistribution rules and will lead different value of MISS-MIF development.

## 3. DEDUCTION AND BACK ANALYSIS ON MISS-MIF DEVELOPMENT

### 3.1. *Geological conditions and mining method*

In order to deduce MISS-MIF characteristics and prove applicability of MISS analysis model, special engineering of longwall mining with fully mechanized top-coal mining (FMTC) and without filling was introduced into derivation process based on theory and experiment analyses, which main geological and mining parameters shown as Table 1. Consequently, MISS-MIF development configuration parameters can be synthetically obtained by rhetorical and experimental deduction as fellows.

Table 1. Main geological and mining parameter

| Items | Mining depth (m) | Overburden layer thickness (m) | Main roof thickness (m) | Coal seam thickness (m) | Coal seam dip angle (°) |
|---|---|---|---|---|---|
| Value | 640 | 380 | 260 | 5.4 | 13 |
| Items | Coal specific gravity (KN/ $m^3$ ) | Rock specific gravity (KN/ $m^3$ ) | Coal shear strength (MPa) | Coal hardness coefficient | Coal cohesion (MPa) |
| Value | 14 | 25 | 2 | 0.6 | 1.25 |
| Items | Coal internal friction angle (°) | Mining height (m) | Face length to the dip (m) | Face length on the strike (m) | Pillar width (m) |
| Value | 32 | 5.4 | 240 | 1600 | 5 |

## 3.2. *MIF evolvement extension deduction*

By similar material experiment analyzing, the CFZ height can be calculated with regression formula shown as Equation 7. Thus the CFZ height is 118 m to 136 m and the maximum height is 124 m within special engineering as above.

$$F_h = \frac{100\sum M}{0.5\sum M + 1.6} \pm 8.9 \tag{7}$$

And coal seam plastic zones and overlying strata broken angle can be synthetically deduced and obtained by theory, experiment, and experience shown as Table 2.

Table 2. Value of coal seam plastic zones and overlying strata broken angle*

| Items | | | Theory | Physical simulation | Numerical simulation | In-situ observation | Engineering experience | Synthesis |
|---|---|---|---|---|---|---|---|---|
| Plastic zones (m) | $x_0$ /X$^+$ | top | 18.9 | | 18~20 | 20 | 10~24 | 18~24 |
| | | middle | 12.2 | 20 | 10~12 | | 10~24 | 10~24 |
| | | bottom | 14 | | 15~17 | 15.6 | 10~24 | 14~24 |
| | $x_0$ /X$^-$ | middle | 10.7 | | | | 8~18 | 8~18 |
| | $y_0$ | dip | 5.3 | | | | | 5 |
| | | rise | 5.6 | 5~7 | 3~10 | | | 3~10 |
| Broken angle (°) | $\Psi_1$ | | | | | | 56.8 | 57 |
| | $\Psi_2$ | | | | | | 60.2 | 60 |
| | $\Psi_3$ | | | 66 | 58 | | 58 | 58~66 |
| | $\Psi_{3'}$ | | | 59 | 58 | | 58 | 58~59 |

*(Chen, Qian 1994, Yang 2007)

## 3.3. *MISS development equation deduction*

By synthetically investigating into stress redistribution and if stress concentration coefficient is selected as 1.05, characteristics of MISS base and functional relations between MISS and MIF are deduced shown as Table 3, Equation 8 and Equation 9.

Table 3. Characteristic value of MSS-Base width

| | Face advancing | | | | Open-off cut | Coal seam along the dip | | |
|---|---|---|---|---|---|---|---|---|
| | Items | top | middle | bottom | | Items | dip | rise |
| Value | $x_0$ | 19 | 14 | 15 | 12 | $y_0$ | 5 | 5.5 |
| | $S_{xi}$ | 3.6 | 4.2 | 4.7 | 4.2 | $S_{yi}$ | 3.0 | 3.2 |
| | $S_{xe}$ | 43.5 | 26.8 | 29.8 | 22.4 | $S_{ye}$ | 15.4 | 15.9 |
| | $S_x$ | 39.9 | 22.6 | 25 | 18.2 | $S_y$ | 12.4 | 12.7 |
| Ratio | $S_{xi}/x_0$ | 0.19 | 0.30 | 0.32 | 0.35 | $S_{yi}/y_0$ | 0.61 | 0.59 |
| | $S_{xe}/x_0$ | 2.29 | 1.91 | 1.98 | 1.87 | $S_{ye}/y_0$ | 3.08 | 2.89 |
| | $S_x/x_0$ | 2.10 | 1.61 | 1.67 | 1.52 | $S_y/y_0$ | 2.47 | 2.3 |
| Synthesis | AVE($S_{xi}/x_0$) | | 0.29(1/3) | | | AVE($S_{yi}/y_0$) | 0.6(3/5) | |
| | AVE($S_{xe}/x_0$) | | 2.01(2) | | | AVE($S_{ye}/y_0$) | 2.99(3) | |
| | AVE($S_x/x_0$) | | 1.73(5/3) | | | AVE($S_y/y_0$) | 2.4(12/5) | |

On the strike: $$\begin{cases} S_{xi} = 1/3x_0 \\ S_{xe} = 2x_0 \\ S_x = 5/3x_0 \end{cases} \tag{8}$$

To the dip: $$\begin{cases} S_{yi} = 3/5\,y_0 \\ S_{ye} = 3y_0 \\ S_y = 12/5\,y_0 \end{cases} \tag{9}$$

If just one longwall mining is only considered under the above special geological conditions and mining method, the MISS development characteristics and some dynamic configuration equations can be deduced shown as Table 4 and Equation 10, in which MISS developing is to stability when face advancing distance is equal to face length.

Table 4. Development characteristics of MISS (m)

| Items | X direction | | | Y direction | | Z direction | |
|---|---|---|---|---|---|---|---|
| | | Face advancing | Open-off cut | | | | |
| value | $S_{xi}$ | 4 | 4 | $S_{yi}$ | 3 | $S_{zi}$ | 150 |
| | $S_{xe}$ | 27 | 24 | $S_{ye}$ | 16 | $S_{ze}$ | 162 |
| | $S_x$ | 23 | 20 | $S_y$ | 13 | $S_z$ | 12 |

$$\begin{cases} X^+YZ^+: \begin{cases} I-MISS: \dfrac{x^2}{124^2}+\dfrac{(y-123)^2}{123^2}+\dfrac{z^2}{150^2}=1 \\ E-MISS: \dfrac{x^2}{147^2}+\dfrac{(y-123)^2}{136^2}+\dfrac{z^2}{162^2}=1 \end{cases} \\ X^-YZ^+: \begin{cases} I-MISS: \dfrac{x^2}{124^2}+\dfrac{(y-123)^2}{123^2}+\dfrac{z^2}{150^2}=1 \\ E-MISS: \dfrac{x^2}{144^2}+\dfrac{(y-123)^2}{136^2}+\dfrac{z^2}{162^2}=1 \end{cases} \end{cases} \tag{10}$$

where: $X^+YZ^+$ stands for the advancing quadrant and $X^+YZ^+$ stands for the open-off cut quadrant.

In summary, because of different geological conditions and mining method, some determination parameters are different that result in MISS is geometrically asymmetrical and non-uniform as far as stress distribution is concerned. Its span does not agree with its thickness. All show that the MISS shape is related to the structure of the working face.

By synthetically and comparatively analysis, deduction results are in conformity with experimental results shown as Table 5. Consequently, the model is feasible and reasonable.

Table 5. Comparative results of MISS in above engineering case (m)

| Location | Numerical simulation results | Theory deduction results from MISS model |
|---|---|---|
| Working face | 72 | 36~91 |
| Rear 15m to working face | 82.2 | 77~110 |
| Rear 100m to working face | 118.2 | 120~144 |
| Rear 100m to working face | 133.6 | 146~173 |

## 4. CONCLUSIONS

Based on investigation into characteristics of MISS-MIF, the MISS analysis model was built up and development equations in face advancing were deduced. Factors affected upon MISS-MIF development are obtained according to integrated analysis on interaction between MISS and MIF evolving with longwall mining. Research results were applied into engineering practice show that the model is feasible and reasonable, that are in conformity with experimental results, and can be used to deduce some development characteristics and equations of MISS-MIF.

## ACKNOWLEDGEMENT

This study was financially supported by the Science and Technological Fund of Anhui Province for Outstanding Youth under Grant No. 08040106839, the National Basic Research Program of China under Grant No. 2005cb221503 and No. 2010CB226806, the Natural Science Research Project of Anhui Province for Colleges and Universities under Grant KJ2009A139, and the Outstanding Innovation Group Program of Anhui University of Science and Technology

## REFERENCES

Chen Yanguang, Qian Minggao. 1994. Strata Control around Coal Face in China. Xuzhou: China University of Mining and Technology Press

Peng Syd S. 2008. Coal Mine Ground Control. Third Edition, Published by Syd S. Peng, Qian Minggao

Miao Xiexing, Xu Jialin. 2003. Theory of key strata control. Xuzhou: China University of Mining and Technology Press

Shen B., King A., Guo H. 2008. Displacement, stress and seismicity in roadway roofs during mining-induced failure. International Journal of Rock Mechanics and Mining Sciences 45(5): 672-688

Song Zhenqi. 2003. Study on the Basic Information of Stratum Movement and Mechanics on Predicting and Controlling Heavy Accidents in Coal Mine. Beijing: China Coal Industry Publishing Press

Wang Yuehan, Deng Kazhong, Wu Kan et al. 2003. On the dynamic mechanics model of mining subsidence. Chinese journal of rock mechanics and engineering 22(3): 352-357

Xie Guangxiang. 2005a. Influence of mining thickness on mechanical characteristics of working face and surrounding rock stress shell. Journal of China Coal Society 30(3): 6–11

Xie Guangxiang. 2005b. Study on mechanical characteristics of fully mechanized top-coal caving face and surrounding rock stress shell. Journal of China Coal Society 30(3): 309–313

Xie Guangxiang, Yang Ke, Chang Jucai, et al. 2006. Surrounding rock abutment pressure distribution and thickness effect of dynamic catastrophic in fully mechanized sublevel mining stope. Journal of China Coal Society 31(6): 731-735

Xie Guangxiang, Yang Ke, Chang Jucai. 2007. Study on distribution characteristics of 3D stress and thickness effects of coal-seam in unsymmetrical disposal and fully-mechanized top-coal caving. Journal of China Coal Society 26(4): 775-779

Xie Guangxiang. 2007. Three-dimensional Mechanical Characteristics of Rocks Surrounding the Fully Mechanized Top-coal Caving Mining Faces. Beijing: China Coal Industry Publishing Press

Xie Heping, Peng Suping, He Manchao. 2006. Basic Theory and Engineering Practice of Deep Mining. Beijing: Science Press

Yuan Liang. 2008. Theory and Practice of Integrated Pillarless Coal Production and Methane Extraction in Multiseams of Low Permeability. Beijing: China Coal Industry Publishing Press

Yang Ke. 2007. Study on the Evolving Characteristics and the Dynamic Affecting of Surrounding Rock Macro Stress Shell and Mining Induced Fracture. Doctoral dissertation. Huainan: Anhui University of Science and Technology: 83-110

Yang Ke, Xie Guangxiang. 2008. Caving thickness effects on distribution and evolution characteristics of mining induced fracture. Journal of China Coal Society 33(10): 1092-1096

Yang Ke, Xie Guangxiang. 2009. Distribution of mining fissures induced by fully mechanized caving mining and analysis on their evolvement characteristics. Mining Safety & Environmental Protection 36(4): 1-4

Yang Ke, Xie Guangxiang. 2010. Modeling and analyzing on the development of mining induced stress shell in deep longwall mining. Journal of China Coal Society 35(7): 1066-1071

*International Mining Forum 2010, Liu et al. (eds) © 2010 Taylor & Francis Group, London, UK. ISBN 978-0-415-59896-5*

# Study on geothermal temperature distribution characteristics and it's affecting factors in an Huainan Mining Area

Guangquan Xu, Zegong Liu, Weining Wang
*School of Earth Science and Environmental Engineering, Anhui University of Science and Technology, Huainan, Anhui Province, China*

Peiquan Li, Xiangjin Huang
*Huainan Mining Group, Huainan, Anhui Province, China*

ABSTRACT: Geothermal, which brings about coal mining heat-hazard, has influence on environment of mining and at the same coal mining area, there are large different temperatures in vertical and horizontal direction. Taken Huainan coal mine as an example, after collected the hydrological drilling and geothermal data of the new and old Huainan mining area systematically, the horizontal and vertical geothermal distribution characteristics have been studied. And also the affecting factors of abnormal geothermal were analyzed. Study results show that owing to the terrain and the rock thermal conductivity, the geothermal and the geothermal gradient at the same level of the mine site are higher than the syncline anticline geothermal sites, and Quaternary loose layer is not conducive to the diffusion of geothermal; the intrusion of igneous rock causes high temperature; groundwater flowing also makes the geothermal abnormal.

KEYWORDS: geothermal hazard, geothermal distribution, affecting factor

## 1. INTRODUCTION

Huainan is a coal-based industrial city. In recent years, with the increasing depth of coal mining and affected geological factors, many mining section temperatures have exceeded that of coal mine safety regulations which is under the 30°C, so mining workers have suffered from different degrees of thermal hazard (Hu 2004). The high underground temperature, which has harm on the healthy of the workers, makes a threat to the workers safety. With deteriorative work environment, work efficiency has been reduced gradually. Geothermal has become a problem which can not be ignored during mining. So, exploring the geological temperature, studying geothermal distribution characteristics of mining area, and analyzing affecting factors have an important significance in how to exploit coal mining.

## 2. BACKGROUND

Huainan is situated near the Huaihe River, in north of Anhui Province, and the general elevation is above sea level 20 to 30 m. It belongs to warm temperate and semi-humid monsoon climate; average annual temperature is 15.3°C; average relative humidity was 74%; average annual amount of precipitation is 926 mm, but the average annual evaporation is 1642.2 mm, so, evaporation is greater than precipitation (Su, Zhang 2000).

Huainan mining area is located in the southern margin of North China plate. To east is Tan-Lu fault; to west is Fuyang-Macheng fault; to north is Bengbu uplift, and to south is Laorencang-Shouxian faults which is adjacent to Hefei Mesozoic sinking. Width of East-west is 180 kilometers, and width of north-south is from 15 to 20 kilometers. Nearly east-west structure exists in the form of a ramp faults basin, north and south sides are the overlapped fans, among which was a synclinorium structure.

Huainan regional geology structure and the geometric array configuration reflect that extruded action from south and north forms structure framework of two ramp folds. There are some strikes over faults, while normal faults of northeastward also exist, which is roughly parallel to the Tan-Lu faults, and is west slope of the ladder-type structure. The igneous rocks are mainly distributed at the exploration area of Panji, Shangyao and Dinjing, and rock isotope age is about 110 million (Li, Zhu 2007).

Special geological condition results in different grade distribution of geothermal which is involved in zonal region and its complexity.

## 3. MINING GEOTHERMAL DISTRIBUTION

### 3.1. *Geothermal data*

Geothermal field reflects the distribution of temperature from the Earth's interior different depths at a moment .If ground temperature remains constant perennially; the constant temperature zone is called. A lot of perennial observation materials show that in this area a constant depth is 30 meter, and its corresponding constant temperature is 16.8°C. Temperature of -1000 meter depth ranges from 24°C to 50.6°C, and the average value is 37.2°C.

Geothermal gradient are affected by rock physical properties and thermal structure, so it can not only reflect the temperature change of stratigraphic section, but also shows regional characteristics. Geothermal gradient in different mines changes greatly, and even in the same mine at different mining sections, the temperature gradient is not the same value. The ground temperature in different conditions sees Table 1. The table reflects that geothermal gradient (from 1.4 to 4.1°C/hm) in Panji area, is higher than one in Wangfenggang mine (from 0.8 to 2.1°C/hm). One of mining temperature such as Dingji mine which is higher than other geothermal gradient ranges from 1.4 to 4.1°C/hm. When mined at -800 meter, temperature is at 46.5°C, already belonging to three high-temperature zone.

Table 1. Geothermal parameters in Huainan Mining Area

| Mine Name | Mine production /$Mt\cdot a^{-1}$ | Borehole number | Mine depth range /m | Temperature range at bottom /°C | Temperature gradients /$°C\cdot hm^{-1}$ | Abnormal rate of temperature gradient /% | Geothermal gradient anomaly rate /% |
|---|---|---|---|---|---|---|---|
| Wangfenggang | 3.0 | 21 | -957~-1260 | 24~42.2 | 0.8~2.1 | — | 66 |
| Dingji | 5.0 | 56 | -716~-1075 | 37.4~52.7 | 1.4~4.1 | 58 | 2 |
| Guqiao | 5.0 | 10 | -819~-996 | 38.5~43.8 | 2.9~3.6 | 30 | — |
| Liuzhuang | 3.6 | 15 | -750~-873 | 43.7~45.2 | 2.2~2.6 | — | — |
| Panbei | 4.0 | 19 | -770~-860 | 30.5~43.6 | 2.0~3.3 | 42 | — |
| Xinzhuangzi | 2.7 | 16 | -565~-870 | 27.5~32.5 | 0.8~1.8 | — | 25 |
| Panyi | 3.0 | 12 | -610~-653 | 29.5~49.2 | 2.5~3.9 | 34 | — |
| Pansan | 3.0 | 26 | -586~-682 | 31.4~54.8 | 2.6~3.8 | 61 | — |

In normal area, geothermal gradients range from 1.6 to 3.0°C/hm. If the geothermal gradient is more than 3.0°C/hm, it is regarded as abnormal, otherwise, it is negative anomalies (Wang 2005). Exploited results show that there are different values of geothermal gradients in different mines (Table 1).

### 3.2. *Horizontal distribution of temperature*

Taken Dingji mine as an example, which is more thermal hazard, the different temperature contours can be drawn from borehole data of different depths. Detail study indicates that horizontal distribution of temperature: (1) there is higher temperature in the north and lower in south, and on the different levels, such as -300 m, -500 m, -700 m and -900 m; (2) with the increasing of mining depth, the geothermal gradient also increases and abnormal zones exist in the north and in the center. Especially, at the north of the 12th borehole, which depth is -900 m, temperature, is up to 50.2°C, higher than normal temperature.

### 3.3. *Vertical distribution of temperature*

According to statistic data about temperature (T) and burial depth (H) from all the different coal mines, statistic analysis finds that the increase of temperature is accompanied with depth increase, and temperature and burial depth of coal seam are a positive correlation. After collected data systematically, analyzed results show that temperature in northwest zone is higher than in southeast zone, and geothermal gradient is up to 4.1°C/hm. When mining depth comes to 1000 meter, temperature gets to 50.6°C.

Table 2. Regressive analysis of results between geothermy and depth in different mine areas

| Mine name | Borehole number | Regression equation | Correlation coefficient |
|---|---|---|---|
| Wangfeng-gang | 21 | $T$(°C)=16.384-0.012$H$(m) | 0.965 |
| Dingji | 56 | $T$(°C)=18.260-0.037$H$(m) | 0.995 |
| Guqiao | 10 | $T$(°C)=17.308-0.031$H$(m) | 0.997 |
| Liuzhuang | 15 | $T$(°C)=21.045-0.023$H$(m) | 0.908 |
| Panbei | 19 | $T$(°C)=19.450-0.027$H$(m) | 0.992 |
| Xinzhuangzi | 16 | $T$(°C)=17.223-0.012$H$(m) | 0.989 |
| Panyi | 12 | $T$(°C)=22.730-0.032$H$(m) | 0.964 |
| Pansan | 26 | $T$(°C)=21.450-0.038$H$(m) | 0.956 |

## 4. AFFECTING FACTORS OF TEMPERATURE ANOMALY

Mine temperature status reflects comprehensive factors, including geological structure, litho logy, magmatic activity, and hydro geological conditions. So, based on comprehensive analysis and related geothermal features, study on the regional geothermal field discovers that the relative high and low temperature are very important, and distribution of temperature in Huainan mining mainly is affected from the thermal heat inside the earth, that is, the deeper depth is, the higher temperature is. Abnormal temperature in Huainan mining area is caused as follows.

### 4.1. *Influence of basement fluctuations and rock conductivity*

Basement fluctuations have some positive correlation on lateral changes of temperature field. For the fold structure and morphology, at the same level, the temperature of anticline site is higher than syncline site. For example, Chengqiao-Panji Anticline geothermal gradient generally ranges from 3.5 to 4.5°C/hm, but due to impact of syncline terrain, Wangfenggang mine temperature gradient is lower than 2.0°C/hm.

Rocks physical properties are one of the main reasons, because dense crystalline rock is harder and more conductive, and increasing temperature rate is lower, therefore, its temperature becomes abnormal lower in the section which conductive heat is better. However, the soft non-crystalline rocks, which has lower conductivity, slower heat transfer and temperature improves rapidly. So, there is higher abnormal temperature in sections which thermal conductivity is better. Part of rocks thermal conductivity shows in Table 3.

Table 3. Thermal conductivity of partial rocks

| Rock kinds | Thermal conductivity $(W \cdot m°C)^{-1}$ | Rock kinds | Thermal conductivity $(W \cdot m°C)^{-1}$ |
|---|---|---|---|
| Granite | 2.51~3.76 | Limestone | 1.67~2.93 |
| Basalt | 2.09 | Dolomite | 3.76~5.85 |
| Gneiss | 2.09~4.60 | Sandstone | 1.67~4.60 |
| Ultra basic rock | 3.76~4.39 | Shale | 1.25~2.51 |
| Schist | 2.93~4.60 | Coal | 0.21~0.58 |
| Quartzite | 2.93~7.96 | Coke | 0.12~0.17 |

It is because of differences in physical properties of hot rock, leading to higher thermal conductivity along the consequent layer direction, but lower along vertical direction .Such results make a mount of heat concentrate on the site of basement uplift and Anticline axis.

### 4.2. *Influence of soil layer thickness and coal seams*

Mining coal strata are covered with more than 400 meter of Quaternary loose sedimentary, forming pot cover effect which is low heat dissipation, but poly thermal effect is obvious. Under the condition of same depth and in the same geological conditions, the thicker its overlying Quaternary is, the higher the ground temperature is. In northern mines of Huaihe River, thickness covered with Cenozoic sedimentary prevents heat from spreading around and temperature, accumulated in the coal strata, causes high temperature. But, in southern mines of Huaihe River, thickness of covered lay is relatively thinner, and its temperature is relatively lower.

An amount of heat produces when coal strata and sulfide in coal has been ox dated. Compared with other sedimentary rocks; coal has a much lower thermal conductivity. Thicker coal seam possesses higher geothermal gradient; whereas more coals strata hinder heat from dissipation.

### 4.3. *Intrusion of magmatic rocks*

The magmatic rock, intruded along rock strata, was discovered at Panji and Dingji exploration region, making geothermal in these areas anomaly (Li, Li 2008). Pansan mine geothermal illustrates that magmatic activity has an important impact on temperature. According to survey data from geological reports, parts of Pansan mine temperature is influenced by the magmatic activity, and the age of magmatic activity is at the Yanshan time (dated back to 110 million), and the distribution of the residual heat almost diffuses. However due to thick covering, a small part of the residual heat may remain so that the local ground temperature becomes anomaly.

The magmatic rock is distributed at the northern part of Pansan mine, near the anticline axis. And its thickness gradually becomes thin from north to south, indicating that the direction of magma intrusion also is from north to south, and at the same time, intrusion channel also is at the anticline axis which temperature is higher.

### 4.4. *Groundwater activity impact on thermal temperature*

Groundwater activity is the most active factor, which has flowing ness and high capacity. In general, the vertical movement of groundwater has more influent than horizontal movement. In the absence of groundwater flowing area, surrounding rock temperature is only controlled by conduction, and the vertical cure of temperature with depth is almost straight, whereas in groundwater activities region, the surrounding rock temperature suffers interaction from conduction and groundwater flowing ness (Wang, Gui 2002).

When the groundwater flows down vertically along the original rock fissures or cracks, and the curve of the surrounding rock temperature will become concave from straight, unless groundwater flows up, the temperature curve become convex line. When the groundwater flows horizontally along rock layer, owing to groundwater flowing along the isotherm or near isothermal surface, the temperature will make surrounding rock a balance status, so, the surrounding temperature has less influent than groundwater flowing ness.

## 5. CONCLUSIONS

1) Research show that special hydro geological conditions make the distribution of geothermal temperature with regional property and complexity.
2) The distribution of temperature indicates that geothermal gradient increases with depth and due to the influence of main four factors, abnormal points exist in some area such as Panji anticline.
3) Regularities of distribution of geothermal temperature provide an important direction for how to forecast the varying tendecy of different mines thermal temperature and what some practicable measure is taken.

## ACKNOWLEDGEMENT

The research is sponsored by Eleventh Five-Year Scientific Research Projects of Ministry of Science and Technology, China.

## REFERENCES

Hu Shao-long. 2004. Distributed characteristic of geothermal and analysis of its factors in Liuzhuang Mine [J]. Mining Safety & Environmental Protection 31(5): 26-28. (In Chinese)

Li De-zhong, Li Bing-bing. 2008. Preliminary discussion on deep heat hazard prevention and control technique about Huainan and Huaibei mines [J].China Coal 34 (4): 64-66. (In Chinese)

Li Hong-yang, Zhu Yao-wu, Yi Ji-cheng. 2007. Geothermal variation and analysis of abnormal factor in Huainan Mining areas [J]. Safety of Coal Mines 396 (11): 68-71. (In Chinese)

Su Yong-rong, Zhang Qi-guo. 2000. Preliminary analysis on the geothermal situation of the Panxie mine in the Huainan coal field [J]. Geology of Anhui 10 (2): 124-129. (In Chinese)

Wang Wen, Gui Xiang-you, Wang Guo-jun. 2002. The prevention of thermal hazard [J]. Mining Safety & Environmental Protection 29 (3): 31-33. (In Chinese)

Wang Xing. 2005. Occurrence of the Weihe River basin geothermal resources and its exploitation [M]. Xian: Shanxi Science and Technology Press: 59. (In Chinese)

International Mining Forum 2010, Liu et al. (eds) © 2010 Taylor & Francis Group, London, UK. ISBN 978-0-415-59896-5

# Experimental studies on oxidation pattern of coal by TG-DSC-GC technology

Wei Peng
*University of Science and Technology of China, Hefei, Anhui, China*

*Key Laboratory of Coal Mine Safety and Efficient Exploitation of Education Ministry of Education, Huainan, Anhui Province, China*

*Anhui University of Science and Technology, Huainan, Anhui Province, China*

Qi-lin He
*Anhui University of Science and Technology, Huainan, Anhui Province, China*

Xin-yu Ge
*Zhangji Coal Mine of Huainan Mining Group, Huainan, Anhui Province, China*

ABSTRACT: Coal mine fire is a common accident, and spontaneous combustion is the major cause of it. Thermo-gravimetry (TG) and Differential Scanning Calorimetry (DSC) technologies were used to study the coal oxidation process. The whole process can be divided in to 5 stages: they are moisture evaporation stage, oxygen absorption stage, thermal decomposition stage, combustion stage and burnout stage. The oxidation characteristic values of each stage were analyzed in order to find out the oxidation and spontaneous combustion pattern; TG and DSC combined with Gas Chromatography (GC) technologies were used to study the gas production pattern in the whole oxidation process of different coal. According to the study results, carbon monoxide (CO) can be used as the main index gas for the prediction of coal spontaneous combustion in the low temperature stage, and ethene ($C_2H_4$) can be used as the subsidiary one. Finally, the pattern of oxygen demand at oxidation process was proposed.

KEYWORDS: Thermal analysis, gas chromatography, coal oxidation, spontaneous combustion, index gas

## 1. INTRODUCTION

Coal is a mixture of complex microstructure. The active constituent of it can be easily oxidized in the air, forming peroxide and generate heat. The peroxide is then decomposed into CO, $CO_2$ and other gas products, and generates more heat. The accumulation of heat can cause the spontaneous combustion of coal. In the low temperature stage of coal oxidation, some of the gas products can be used to predict the spontaneous combustion. The oxidation process is related to the properties of coal, the environment of oxidation, the rank of coal and so on, so different species of coal have different oxidation gas products. It is very important to find a suitable gas product as the index gas for the prediction of spontaneous combustion.

## 2. EXPERIMENTS

### 2.1. *The preparation of materials*

Four species of coal with different rank were picked out for the experiments: they are lignite from Ning-xia, gas coal from Gu-bei, gas fat coal from Li-yi, anthracite from Bai-shan. At first, a series of industrial analysis were performed to all the coal, the results are shown in Table 1.

Strip the outer layer of the fresh coal sample (without drying process), grind its core into a particle size of 120 mesh (0.1 mm), put the powders into the glass bottle, keep the bottle tightly sealed for further experiment. 20 Mg coal powder was used for each experiment in air atmosphere.

Table 1. Industry analysis of experimental coal samples

| No. | Coal Species | Moisture (%) | Volatile Compound (%) | Ash (%) | Fixed carbon (%) |
|---|---|---|---|---|---|
| 1. | lignite from Ning-xia | 13.17 | 28.9 | 4.33 | 53.6 |
| 2. | gas coal from Gu-bei | 3.72 | 28.06 | 6.81 | 61.41 |
| 3. | gas fat coal from Li-yi | 0.7 | 19.12 | 24.79 | 55.39 |
| 4. | anthracite from Bai-shan | 1.56 | 5.87 | 27.56 | 65.01 |

### 2.2. *Experimental equipments and methods*

In the experiments, a thermal analysis instrument with model number of TGA/SDTA851e from METTLER-TOLEDO Company was used to determine the TG-DSC curve, and a GC instrument with model number of GC-4085 was used to determine the gas species and concentration emitted from coal during the oxidation process.

In order to study the coal oxidation pattern at low temperature, the temperature rising rate was variable during the whole heating process of thermal analysis. From 30 to 300°C, the temperature rising rate was 2°C/min; from 300 to 800°C, the temperature rising rate was 10°C/min. Once the temperature increased 10°C, collect the gas production from the thermal analysis instrument for GC analyzing.

## 3. EXPERIMENT RESULTS AND DISCUSSION

### 3.1. *The thermal analysis curve of coal sample*

Figures 1-4 are the thermal analysis curve of four species of coal sample, the trend of them were the same.

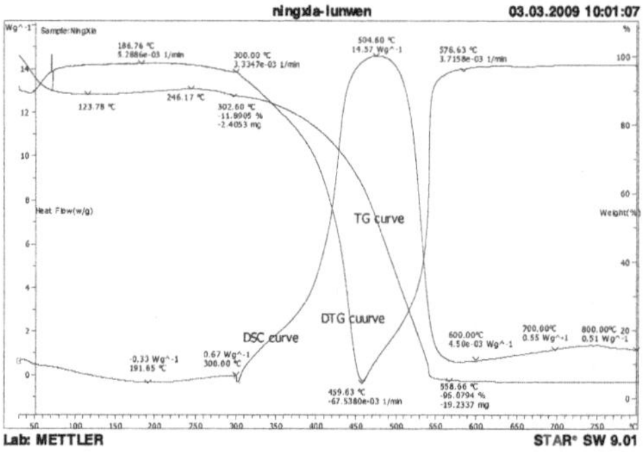


Figure 1. Thermal analysis curves of lignite coal sample from Ning-xia

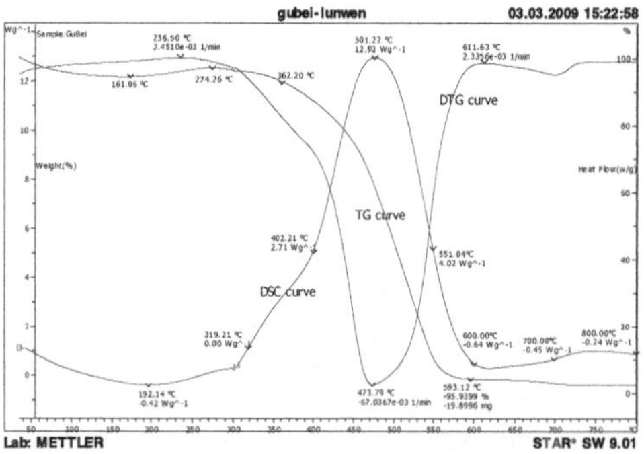


Figure 2. Thermal analysis curves of gas coal sample from Gu-bei

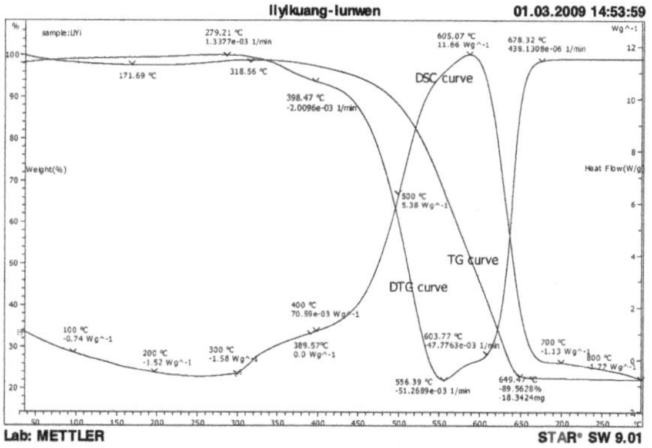


Figure 3. Thermal analysis curves of gas fat coal sample from Li-yi

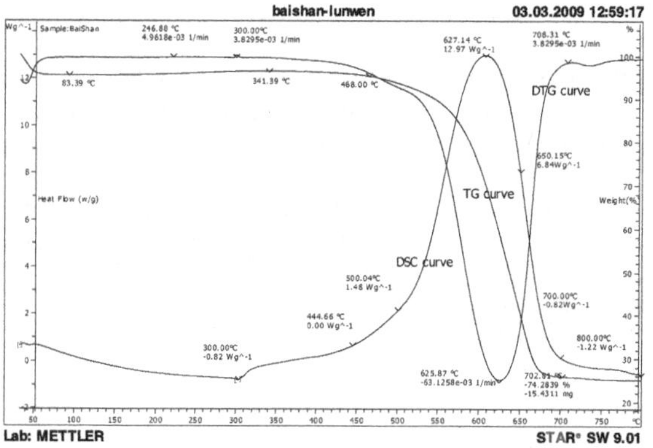


Figure 4. Thermal analysis curves of anthracite coal sample from Bai-shan

Take lignite coal sample from Ning-xia (Figure 1) for example, the oxidation process could be divided into 5 stages:

(1) Moisture evaporation stage (from 30 to 123.78°C). Since the evaporation of the water in the coal absorbed heat, the DSC curve was below zero, the absorbed heat was decided by the moisture content. Since the moisture content of coal sample was low, the DSC curve had no sharp endothermic peak. The coal sample adsorbed oxygen, which accompanied the water evaporation (He, Wang 2006). In the stage, water evaporation lost weight, while oxygen adsorption gained weight. The weight lost by water evaporation was much heavier than the weight gained by oxygen adsorption, so the total effect was losing weight, the TG curve descended.

(2) Oxygen absorption stage (from 123.78 to 246.17°C). Since the coal sample absorbed oxygen and was oxidized during the heating process, the weight of coal sample gained and the heat production increased when the temperature rising. Both the TG curve and DSC curve rose with the temperature increasing.

(3) Thermal decomposition stage (from 246.17 to 302.60°C). The oxidation of coal continued, the volatile component appeared, the gas production emitted, the weight of coal sample decreased, heat production increased. The TG curve descended, while the DSC curve rose.

(4) Combustion stage (from 302.60 to 558.66°C). The coal sample was ignited, the weight decreased sharply and the heat production increased rapidly. The TG curve descended sharply, while the DSC curve rose steeply and reached the maximum value at 504.66°C.

(5) Burnout stage (from 558.66 to 800°C). Since the combustible component burnt out at previous stage, the weight decreased rarely, the heat production approached to 0. The TG curve descended slowly and the DSC curve remain 0 above.

Other species of coal could be analyzed the same way, all the results are shown in Table 2.

Table 2. The oxidation characteristic values of different species of coal

| No. | Moisture evaporation stage | | Oxygen absorption stage | | Thermal decomposition stage | | Combustion stage |
|---|---|---|---|---|---|---|---|
| | Temperature range /°C | Weight change /% | Temperature range /°C | Weight change /% | Temperature range /°C | Weight change /% | Temperature range /°C |
| 1. | 30-124 | -3.24 | 124-246 | 1.39 | 246-303 | -13.57 | 303-559 |
| 2. | 30-161 | -1.68 | 161-274 | 2.75 | 274-362 | -8.65 | 362-593 |
| 3. | 30-172 | -2.52 | 172-319 | 1.93 | 319-398 | -9.48 | 374-649 |
| 4. | 30-83 | -1.66 | 83-341 | 0.996 | 341-468 | -4.57 | 469-703 |

At the oxygen absorption stage, the oxygen was mainly consumed by chemical absorption and reactions, the weight change was the amount of absorbed oxygen. Because every species of coal had different ignition temperature, the oxygen consumption at the stage was different. According to the data listed in Table 2, the oxygen consumption amount of the bituminous coal (No. 2 and 3) was large, the lignite's (No. 1) was moderate and anthracite's (No. 4) was small.

At the thermal decomposition stage, the alkyl side chains and the oxygen-containing functional groups of condensed aromatics split, the gas production emitted (Xie 2002, Saurabh, Pradeep 1996, Zhang et al. 1993). According to the data listed in Table 2, the weight change of lignite (No. 1) was large, the bituminous coal's (No. 2 and 3) was moderate and anthracite's (No. 4) was small. It showed that lignite had more alkyl side chains and the oxygen-containing functional groups, the porosity and the specific surface area was big. While the coal rank became higher, the number of alkyl side chains and the oxygen-containing functional groups reduced. Further more, the initial temperature and temperature range of coal decomposition increased when the coal rank became higher due to the structure differences.

At the coal combustion stage, the ignition point of anthracite (No. 4) was the high, the bituminous coal's (No. 2 and 3) was moderate and the lignite's (No. 1) was low. The lignite's ignition point was 170 lower than anthracite's.

### 3.2. *The GC curve of different coal sample during the oxidation process*

The gas production release pattern of different coal during the oxidation process is given in Figure 5.

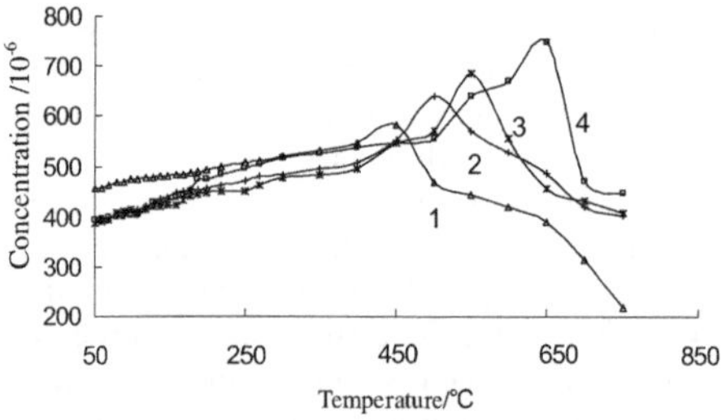


a. $CO_2$

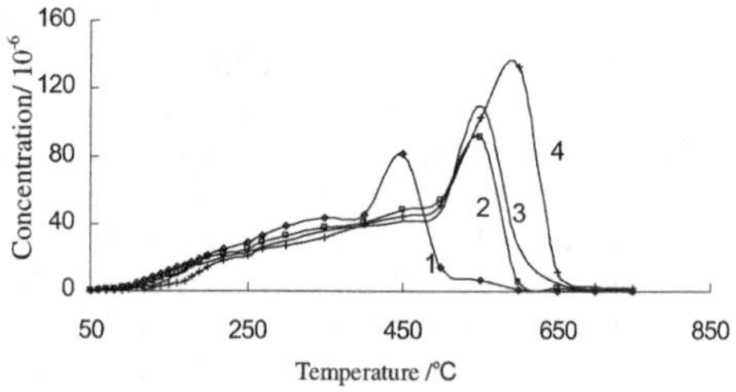


b. CO

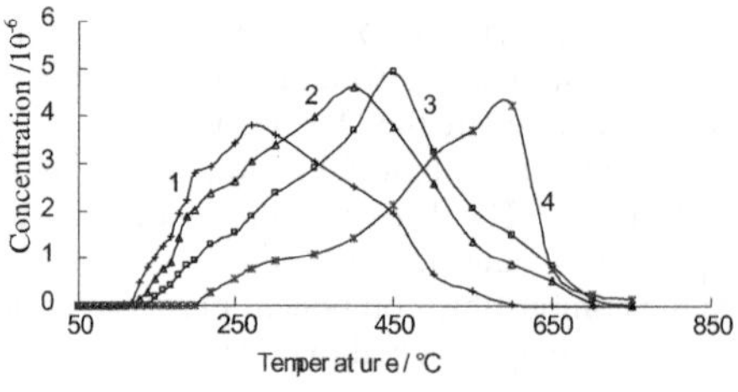


c. $C_2H_4$

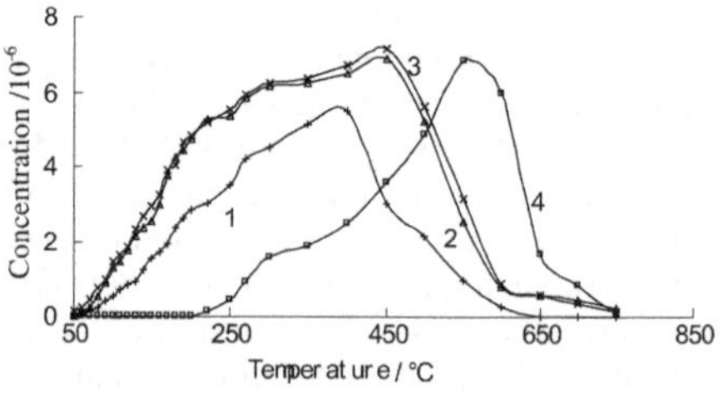


d. $C_2H_6$

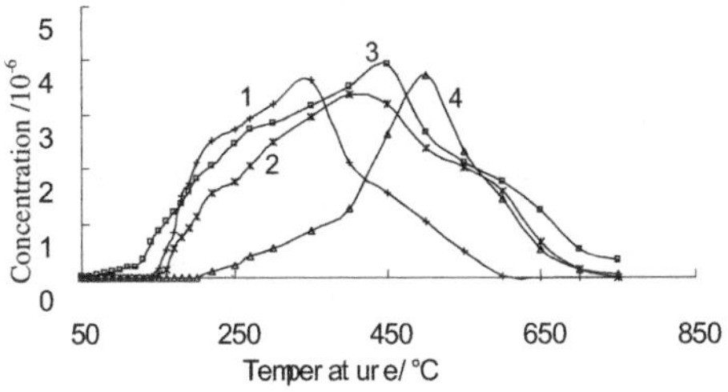


e. $C_3H_8$

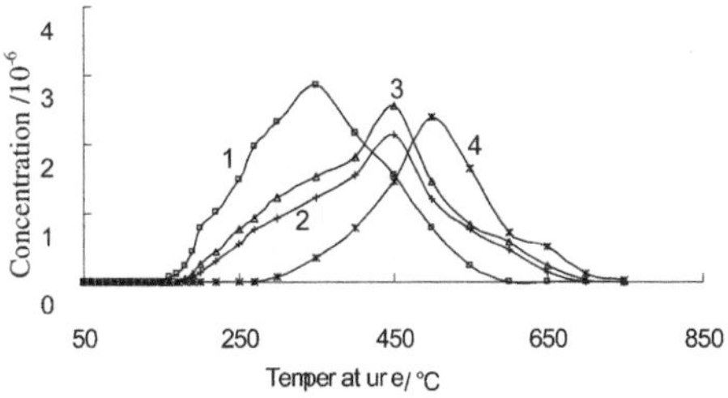


f. $C_2H_2$

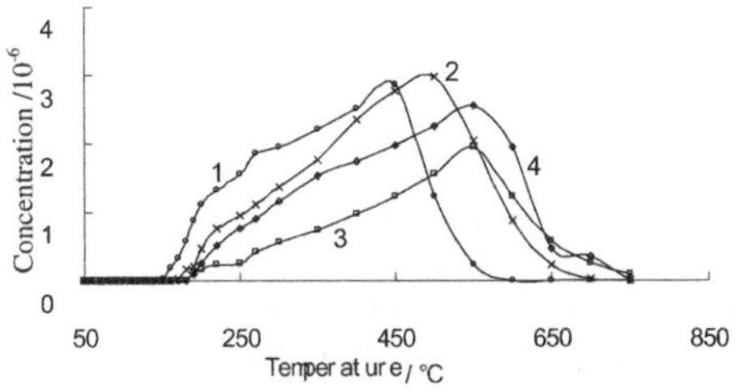


g. $H_2$

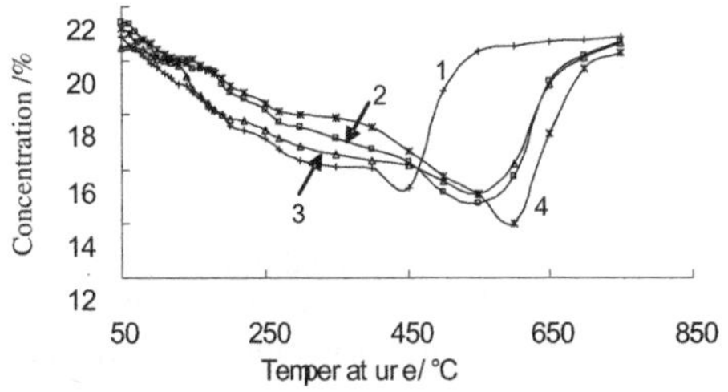


h.$O_2$

Figure 5. The concentration of specific gas during the oxidation process
*1 - lignite, 2 - gas coal, 3 - gas fat coal, 4 - anthracite

The concentrations of 8 gases were measured by the GC instrument, as shown in Figure 5. After the analyzing, we found that:

(1) At the low temperature oxidation stage (from 50 to 300°C), the production of CO and $CO_2$ increased with the temperature. In the same air flow rate and same temperature rising rate, the sequence of CO's concentration in the gas production from different coal was: lignite's > gas coal's > gas fat coal's > anthracite's. The temperature of starting producing CO: lignite's was 50°C, gas coal's was 60°C, gas fat coal's was 70°C and anthracite's was above 80°C. Therefore, CO can be used as main index gas for prediction the spontaneous combustion of coal.

(2) At the beginning of oxidation process (about 50°C), there was $C_2H_6$ emitted from the lignite, gas coal and gas fat coal. The production of $C_2H_6$ increased with the temperature, which were caused by the desorption of $C_2H_6$ from the coal sample.

(3) At 130°C, $C_3H_8$ emitted from lignite coal sample, and the temperatures for emission were 150°C and 250°C from gas coal and anthracite coal samples.

(4)The emission pattern for $C_2H_4$ was similar with CO. The temperature of starting producing $C_2H_4$: lignite's was 100°C, gas coal's and gas fat coal's was above 140°C, and anthracite's was above 250°C. Therefore, $C_2H_4$ can be used as subsidiary index gas for prediction the spontaneous combustion of coal.

(5) $C_2H_2$ usually appeared at high temperature: lignite's was above 160°C, gas coal's and gas fat coal's was above 180°C, and anthracite's was above 300°C. The concentration of $C_2H_2$ in the gas production was very low. $H_2$ was also found from the gas production, which may come from the hydrocarbon splitting.

(6) The oxidation of coal consumed oxygen, the oxygen demand was concerned to the coal rank. In the low temperature oxidation stage, the oxygen demand of lignite is large, and the oxygen concentration in the gas production was low. The oxygen concentration in the gas production from anthracite was the highest.

## 4. THE GAS PRODUCTION PATTERN AT LOW TEMPERATURE OXIDATION STAGE

Many researchers proposed different index gases for predicting the spontaneous combustion of coal (Wang 2008, Xie et al. 2003, He, Wang 2005, Liu 2003, Lu et al 2004), due to different coal rank, experiment equipments and methods. The macro oxidation pattern of the same coal rank was similar. That is:

(1) The index gas means a gas that can indicate the coal's oxidation. For a certain rank of coal, different gas productions appear at different temperature. According to the experimental results, at the low temperature oxidation stage, when the temperature rises, the sequence of gas production appearance is: $CO_2/CH_4 \rightarrow CO \rightarrow C_2H_6/C_3H_8 \rightarrow C_2H_4 \rightarrow C_2H_2/H_2CO_2$ appears at low temperature, the alkyne ($C_2H_2$, etc.) appears at high temperature. When $CO_2$ appear, the coal is at slow oxidation stage. When CO and $C_2H_4$ appear, the coal is at rapid oxidation stage. When $C_2H_2$ appear, the coal is at fully oxidized stage.

(2) The appearance temperature and the amount of gas production are different for different species of coal. If the rank of coal is low (such as lignite), the coal is easy to burn and the amount of gas production is large.

## 5. CONCLUSIONS

1) At experiment temperature range (from 30 to 800°C), the oxidation process of different species of coal can be divided into 5 stages, according to the TG-DSC curve.
2) Different ranks of coal are different in oxidation characteristic value, such as moisture lose ratio, oxygen absorption ratio, temperature range of each oxidation stage, coal decomposition ratio and so on.
3) At low temperature oxidation stage, carbon monoxide (CO) can be used as the main index gas for the prediction of coal spontaneous combustion, and ethene ($C_2H_4$) can be used as the subsidiary one.

## ACKNOWLEDGEMENTS

This work is supported by Opening Project of State Key Laboratory of Fire Science, under Grant No. HZ2009-KF13.

## REFERENCES

Chen Qin-mei, Huang Ying-hua, Ren De-qing, Cai Gen-cai. 1997. Evaluation of the Combustion Characteristic of Pulverized Coal by Using TG-DTA-T-DTG and EGD-GC. Journal of East China University of Science and Technology 23(3): 286-291. (In Chinese)

He Qi-lin, Wang De-ming. 2005. Comprehensive study on the rule of spontaneous combustion coal in oxidation process by TG- DTA - FTIR technology. Journal Of China Coal Society 30(1):53-57. (In Chinese)

He Qi-lin, Wang De-ming. 2006. Kinetics of oxidation and thermal degradation reaction of coal. Journal of University of Science and Technology Beijing 28(1): 1-5. (In Chinese)

Liu Fei 2003. Study on Index Gases Released by Coal's Low Temperature Reacting. Nanjing: Nanjing University of Technology. (In Chinese)

Lu P., Liao G.X., Sun J.H., Li P.D. 2004. Experimental research on index gas of the coal spontaneous at low-temperature stage. Journal of Loss Prevention in the Process Industries 17: 243-247

Wang De-ming 2008. Coal Mine Fire. Xuzhou: China University of Mining and Technology Press. (In Chinese)

Xie Ke-chang. 2002. The structure and reactivity of coal. Beijing: Science Press. (In Chinese)

Saurabh Bhat, Pradeep K. Agarwal. 1996. The effect of moisture condensation on the spontaneous combustibility of coal. Fuel 75(13): 1523-1532

Xie Zhen-hua, Jin Long-zhe, Song Cun-yi. 2003. Coal Spontaneous Combustion Characteristics at Programmed Temperatures. Journal of University of Science and Technology Beijing 25(1): 12-14. (In Chinese)

Zhang Peng-zhou, Li Li-yun, Ye Chao-hui. 1993. Solid state $^{13}C$-NMR study of Chinese coal. Journal of Fuel Chemistry and Technology 21(3): 310-316. (In Chinese)

*International Mining Forum 2010, Liu et al. (eds) © 2010 Taylor & Francis Group, London, UK. ISBN 978-0-415-59896-5*

# Research on gas sensor fault diagnosis based on wavelet analysis and VLBP neural network

Junhao Wang, Xiangrui Meng
*Anhui University of Science and Technology, Huainan, Anhui Provincee, China*

ABSTRACT: For four types of common faults of gas sensor, namely offset, impact, drift and periodic types, on the basis of wavelet analysis and VLBP neural network, a method of the gas sensor fault diagnosis is proposed based on the pattern matching and classification of characteristic energy spectrum extracted by wavelet packet decomposition and VLBP neural network. Characteristic energy spectrum of each singularity can be obtained through the wavelet packet decomposition of output signal of gas sensor. After processing, it, as the characteristic vector for training VLBP neural network, adopts the momentum VLBP algorithm for updating weight values at each sample point; and, through learning a variety of characteristics data of fault model, it is used to determine the fault type of sensor. The experiment result shows that, this method can be effectively applied to the fault diagnosis of gas sensor.

KEYWORDS: Gas sensor, wavelet packet, VLBP neural network, fault diagnosis

## 1. INTRODUCTION

As the environment for coal mine underground is poor, long-term high temperature, high pressure and/or high dust may cause damage to sensitivity and linearity of black-white component of gas sensor, critical zero shift, output distortion and serious false alarm (Wang 2007). Once gas sensor sends out a false alarm, it may cause the entire colliery security monitoring system to perform practically no function, produce potential hazards, and even result in extremely serious consequences. Therefore, the research on fault diagnosis of gas sensor has very important practical significance.

Sensor faults are divided into abrupt fault (Abrupt) and incipient fault (Incipient) (Ramanathan, Nithya 2009), and only the abrupt fault is analyzed. Typical manifestations of abrupt sensor faults mainly include offset, impact, drift and periodicity. For abrupt fault signal output by sensor, when the signal frequency traits are relatively abundant, the time domain analysis cannot achieve the desired results. Therefore, method of the gas sensor fault diagnosis is proposed based on wavelet packet decomposition and VLBP neural network. The diagnosis uses wavelet packet decomposition of each node as a characteristic energy vector to more directly reflect transient changes of sensor signal, and adopts momentum VLBP neural network to classify and determine, so as to obtain corresponding fault message.

## 2. SIGNAL ANALYSIS AND FEATURE EXTRACTION

### 2.1. *Wavelet packet decomposition*

Wavelet packet decomposition is a meticulous signal decomposition and reconstruction method based on wavelet analysis. In the wavelet analysis, it is required to only redecompose the previously decomposed low-frequency part every time, and the high-frequency part remains as it has been in the past; therefore, the resolution of the high frequency band is relatively poor. The wavelet packet decomposition decomposes scale-space and wavelet space simultaneously, and gets the spectrum to be further split into fine segments; and, it has a higher time-frequency resolution, and strengthens the ability to analyze signal (Li 2001).

Definition 1: Assumed that $h_k, g_k \ (k \in Z)$ is the filter coefficient of orthogonal wavelet the orthogonal wavelet function corresponding to orthogonal scaling function $\phi(x)$ is $\psi(x)$; and, the two-scale equations that they respectively meet include:

$$\begin{cases} \phi(x) = \sqrt{2}\sum_k h_k \phi(2x-k) \\ \psi(x) = \sqrt{2}\sum_k g_k \phi(2x-k) \end{cases} \tag{1}$$

Where, $u_0(x) = \phi(x)$, $u_1(x) = \psi(x)$, getting $u_n(x)$ meets the following two-scale equations:

$$\begin{cases} u_{2n}(x) = \sqrt{2}\sum_k h_k u_n(2x-k) \\ u_{2n+1}(x) = \sqrt{2}\sum_k g_k u_n(2x-k) \end{cases} \tag{2}$$

Then, $\{u_n(x) : n \in Z\}$ is defined as the wavelet packet corresponding to $\phi(x)$, and $u_n(x-k)$ constitutes the orthogonal basis of $L^2(R)$ space.

Theorem 1: Assume that $V_j = \overline{span\{\phi_{j,k} : k \in Z\}}$, $W_j = \overline{span\{\psi_{j,k} : k \in Z\}}$, $U_j = \overline{span\{u_{j,k} : k \in Z\}}$ and $V_{j+1} \subset V_j \subset L^2(R)$ where in, $W_{j+1}$ are the orthogonal compliment of $V_{j+1}$ in $V_j$, obtaining the orthogonal wavelet packet value and decomposition formula of wavelet space as follows:

$$\begin{cases} V_j = U^0_{j+k} \oplus U^1_{j+k} \oplus \cdots \oplus U^{2^k-1}_{j+k} = \bigoplus_{m=0}^{2^k-1} U^m_{j+k} \\ W_j = U^{2^k}_{j+k} \oplus U^{2^k+1}_{j+k} \oplus \cdots \oplus U^{2^{k+1}-1}_{j+k} = \sum_{m=0}^{2^k-1} U^{2^k+m}_{j+k} \end{cases} \tag{3}$$

On the basis of Theorem 1, and in accordance with the band feature of gas sensor output signal, the signal is decomposed into three layers of wavelet packets; also the decomposition tree of wavelet packets is shown in Figure 1.

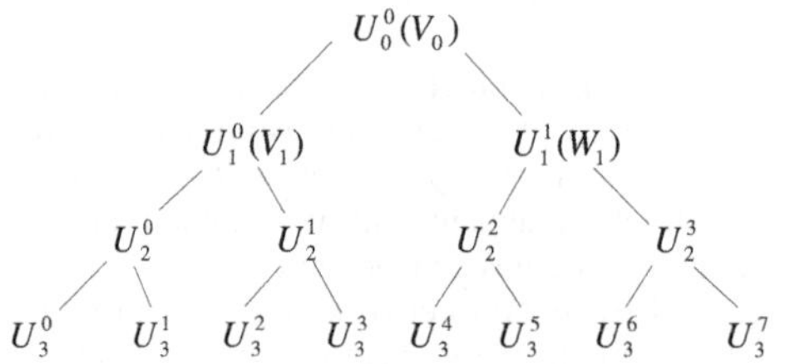


Figure 1. Decomposition tree of wavelet packets of gas sensor signal

## 2.2. *Feature extraction of signal*

In order to obtain the frequency domain characteristics of signal, it is required to select high-resolution mother wavelet from frequency domain. Shannon wavelet is selected, because, among all the orthogonal wavelets, it has the highest resolution (Mallats 1999). Through the wavelet packet decomposition of signal within the whole frequency range, the wavelet packet decomposition coefficient of each sub-band frequency produced from the uniform division of the whole frequency range is obtained.

The decomposition coefficient vectors of the nodes at the third layer are respectively expressed by $\mu_{30}, \mu_{31}, \mu_{32}, \mu_{33}, \mu_{34}, \mu_{35}, \mu_{36}, \mu_{37}$ from bottom to top.

To describe the energy changes more accurately in the signal, certain subtraction method (Xu, Wang 2006) is selected for $\mu_{30}, \mu_{31}, \mu_{32}, \mu_{33}, \mu_{34}, \mu_{35}, \mu_{36}, \mu_{37}$:

(1) Calculate the reduction threshold of decomposition coefficient of each node,

$$Thr\mu_{3i} = \sqrt{\frac{1}{k}\sum_{j=1}^{k} cfs_{3i_j}^2} \tag{4}$$

where: $k$ refers to the length of $\mu_{3i}$,

$$\mu_{3i} = \left[cfs_{3i_1}\ cfs_{3i_2}\ \cdots\ cfs_{3i_k}\right].$$

(2) The coefficient in $\mu_{3i}$ is treated as follows:

$$cfs_{3i_j} = \begin{cases} cfs_{3i_j} & |cfs_{3i_j}| \geq Thr\mu_{3i} \\ 0 & |cfs_{3i_j}| < Thr\mu_{3i} \end{cases} \tag{5}$$

And then, through the reconstruction of wavelet packet decomposition coefficient, the signal of each sub-band frequency can be obtained, with the signal length of each sub-band frequency unchanged, relatively narrow frequency bandwidth and relatively high noise-signal ratio.

It is known by the Parseval Theorem that the signal reconstructed stores all the energy of original signal and can achieve the characterization of all the characteristics of the original signal. Assumed that, the reconstruction signal of the frequency band (i) at the third layer after wavelet packet decomposition is $S_{3i}$, and corresponding signal energy is $E_{3i}$:

$$E_{3i} = \int |S_{3i}|^2\, dt = \sum_{m=1}^{N} |x_{im}|^2 \tag{6}$$

Where, N refers to data length, and $x_{im}$ refers to amplitude of reconstruction signal at discrete point ($S_{3i}$). As sensor fault has great influences on the signal energy within each frequency band, the signal energy within the frequency band decomposed can be normalized.

$$\overline{E_{3i}} = \frac{E_{3i}}{\sum_{i=0}^{7} E_{3i}} \tag{7}$$

As a result, the characteristic vector of signal, namely $V = \left[\overline{E_{30}}\ \overline{E_{31}}\ \overline{E_{32}}\ \overline{E_{33}}\ \overline{E_{34}}\ \overline{E_{35}}\ \overline{E_{36}}\ \overline{E_{37}}\right]^T$, is constructed by a group of signal energy values $E_{3i}$ after the normalization; and, it, as a sample, is input into the classifier in VLBP neural network for fault identification and classification.

Through analyzing characteristics of four types of abrupt gas sensor faults (namely offset type, impact type, drift type and periodic type), it is can be seen that, the output signal energy have a significant step change in case of offset fault; impact in wavelet transformation is mainly manifested as a relatively large peak value; drift has output signal based on the drift of the original signal at certain speed rate; periodic has output signal energy mainly centralized within the frequency domain around fault point (Tian 2007).

Through the analysis and induction, and with references to relevant literatures, the standardized characteristic energy spectrum of wavelet packet decomposition frequency band for training neural network (including normal sensor) is obtained, as shown in Figure 2.

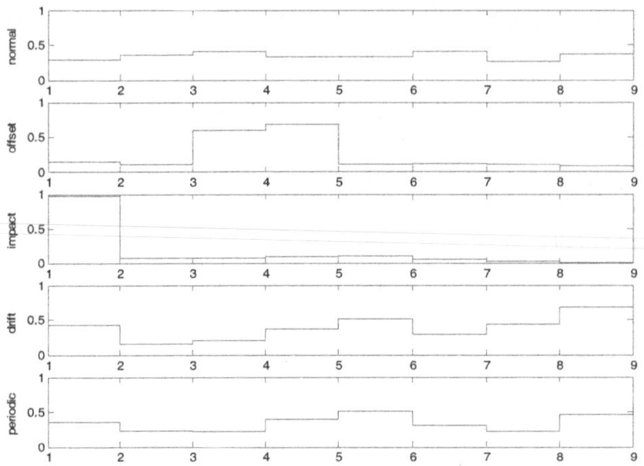


Figure 2. Characteristic energy spectrum under five modes of gas sensor

## 3. ESTABLISHMENT OF VLBP NEURAL NETWORK MODEL

### 3.1. *Neural network structure*

VLBP (Variable Learning Rate Back Propagation) neural network is a kind of unidirectional multi-layer feed forward neural network, including an input layer, an output layer and one or more hidden layer. Neurons in the same layer are not unrelated, with inter-neuronal forward connection. Through selecting appropriate network structure according to the complexity of object, the mapping of any nonlinear function between input space and output space can be achieved (Wei 2005).

First of all, the mapping relations between sensor fault type and neural network output space are established; and, because four types of abrupt faults of gas sensor is mainly researched, it is allowed to establish the mapping relations shown in Table 1 (including normal type sensor). And then, construct a neural network with single hidden layer, and the number of the nodes in the output layer is 3. It is calculated according to the length of characteristic vector provided through 3-layer wavelet packet decomposition that, the number of nodes in the input layer is 8. The number of nodes in the hidden layer obtained through the subtractive clustering algorithm (Wang, Yang 2002) is 7. Therefore, the neural network adopts 8-7-3 structure, as shown in Figure 3, and the transfer function of nodes in the hidden layer is set by $f(x)=1/(1+e^{-x})$.

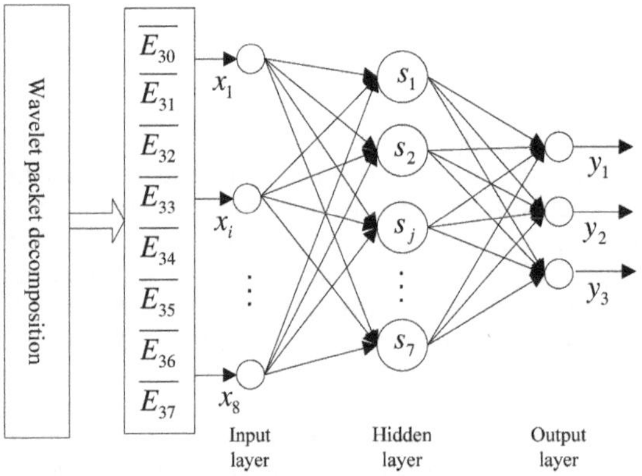


Figure 3. Neural network structure

Table 1. Mapping relations between sensor status and neural network output value

| Sensor status | NN output value |
|---|---|
| I ( normal ) | 0 0 0 |
| II ( offset ) | 0 0 1 |
| III ( impact ) | 0 1 0 |
| IV ( drift ) | 0 1 1 |
| V ( periodic ) | 1 0 0 |

### 3.2. *Learning algorithm*

The basic BP algorithm is easy to fall into local minimum and low convergence speed during the iterative computation; therefore, it is necessary to improve the basic BP algorithm. To improve training time and precision of training, it is required to adopt momentum VLBP algorithm of update weight value at each sample point; and, this algorithm can effectively increase the convergence speed and recognition accuracy of neural network. The algorithm is described as follows:

Definition 2: Assumed that the desired output of the unit (j) on the output terminal in the iteration (n) is $d_j$ (n) and the actual output is $y_j$ (n), the error signal $e_j$ (n) of this unit is defined as follows:

$$e_j(n) = d_j(n) - y_j(n) \tag{8}$$

As shown in Def. 2, the total squared error instantaneous value $\xi(n)$ of output terminal is:

$$\xi(n) = \frac{1}{2}\sum_{j\in c} e_j^2(n) \tag{9}$$

where: c includes all output units.

Then, the modified formula of weight value of network $w_{ji}$ is given by:

$$\Delta w_{ji}(n) = (1-\alpha)\eta \frac{\partial \xi(n)}{\partial w_{ji}(n)} + \alpha \Delta w_{ji}(n-1) \tag{10}$$

where: $\alpha$ refers to momentum factor,
$0 \le \alpha \le 1$ and $\eta$ refers to network learning rate.

The adjustment rules between learning rate $\eta$ and momentum factor $\alpha$ in Formula (10) is shown as follows (Dai 2002):

(1) Mean square error after updating increases, and exceeds or equals to certain value (approximately between 0.01~0.06), with learning rate $\eta$ multiplying by $\theta$ $(0<\theta<1)$, and momentum factor $\alpha$ is set to zero.

(2) Mean square error after updating increases and is less than, with learning rate $\eta$ unchanged, and momentum factor $\alpha$ is set as the initial value.

(3) Mean square error reduces or remains unchanged after updating, with learning rate $\eta$ multiplying by $\varepsilon$ $(1 < \varepsilon)$, and momentum factor $\alpha$ is set as the initial value.

## 4. EXPERIMENTAL VERIFICATION

This experiment is performed at Zhangbei Mine of Huainan Mining Industry Co. Ltd. First of all, extract the characteristic vector (V) from the characteristic energy spectrum of frequency band of standard wavelet packet decomposition described in Item 2.2 of this article. Each status has 15 groups of samples, which are used for training VLBP neural network. The minimum allowable error in this experience is 0.0001, and through 1400-step training, the network is converged into the desired precision.

And then, 20 groups of observed data of sensor are obtained from the gas database of Zhangbei Mine under each of the five modes. Corresponding characteristic vector samples are extracted after the wavelet packet decomposition, and used for verifying the trained neural network, with a group of output results shown in Table 2, and the classification of all the sample models identified as Figure 4. The experiment shows that, the method of design in this article can effectively identify and diagnose fault type of gas sensor. It has already achieved the expected design effects, and it is practically feasible for the fault diagnosis of gas sensor.

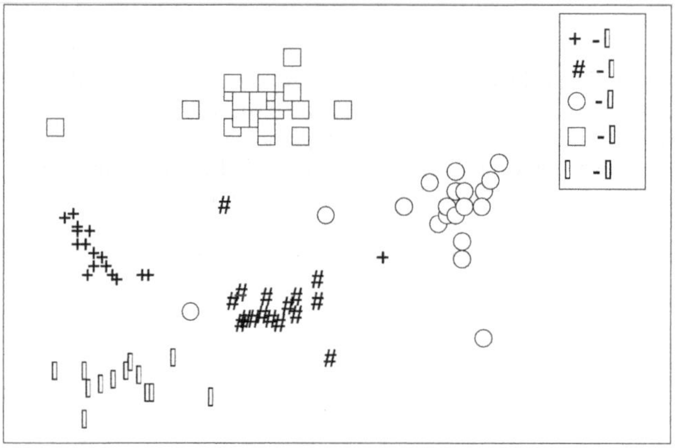

Figure 4. Sample recognition and classification results

Table 2. Experimental Output Results

| Test samples | Output | Diagnosis | Recognition Rate (%) |
|---|---|---|---|
| 0.2213 0.3136 0.3820 0.3289 0.3229 0.4118 0.2554 0.3648 | 0.0002 0.0001 0.0004 | I | 99.9 |
| 0.1256 0.0912 0.5942 0.6436 0.1211 0.1112 0.0975 0.0845 | 0.0004 0.0006 0.9983 | II | 99.9 |
| 0.9865 0.1684 0.0727 0.0900 0.1021 0.0323 0.0214 0.0125 | 0.0010 0.9881 0.0009 | III | 99.8 |
| 0.4241 0.1415 0.2418 0.3139 0.5123 0.2813 0.2205 0.4462 | 0.0001 1.0000 1.0000 | IV | 100.0 |
| 0.3120 0.2864 0.2364 0.3123 0.4100 0.3146 0.2320 0.4564 | 0.9999 0.0001 0.0002 | V | 100.0 |

## 5. CONCLUSIONS

The gas sensor fault diagnosis method based on wavelet packet decomposition VLBP neural network is proposed. Wavelet packet decomposition can troubleshoot signals encountered the nonstationary signal, and effectively get the characteristics of the fault information.

Adopted the momentum VLBP algorithm for updating weight values at each sample point, with shorter training time and higher diagnostic accuracy, shows better robustness and stronger ability of classification. Experimental results point that this method can effectively increase the effectiveness and accuracy of the gas sensor fault diagnosis method, thus greatly improve the stability and reliability of the monitoring system for coal mine safety in production.

## ACKNOWLEDGEMENT

The research is sponsored by the Key Program of the Natural Science Funds of Anhui province (KJ2010A084).

## REFERENCES

Dai Kui. 2002. Neural network design. Beijing: Mechanical Industry Press. (In Chinese)
Li Jian-ping. 2001. The wavelet analysis and signal processing. Cambridge: Cambridge university press.
Mallats. 1999. A wavelet tour of signal processing. New York, USA: Academic Press: 67-126
Ni Kevin, Ramanathan, Nithya. 2009. Sensor network data fault types. ACM Transactions on Sensor Networks 5(3): 1-29
Tian Yu-peng. 2007. The sensor principle. Beijing: Science Press. (In Chinese)
Wang Hong-bin, Yang Xiang-lan, Wang Hong-rui. 2002. An Improved Learning Algorithm for RBF Neural Networks. Systems Engineering and Electronics 24(6): 103-105. (In Chinese)
Wei Hai-kun. 2005. Neural network theory and method of structure design. Beijing: National Defense Industry Press. (In Chinese)
Wang Ji-jun. 2007. Gas monitoring system of intelligent fault diagnosis technology. Jinan: Shandong University of Science and Technology: 32-33. (In Chinese)
Xu Tao, Wang Qi. 2006. Based on wavelet packet neural network fault diagnosis method of sensor. Chinese Journal of Sensors and Actuators 19(4): 1060-1064. (In Chinese)

International Mining Forum 2010, Liu et al. (eds) © 2010 Taylor & Francis Group, London, UK. ISBN 978-0-415-59896-5

# Technology of interactive collecting and processing mine ventilation network information

Yingdi Yang, Guoshu Zhang
*School of Energy and Safety, Anhui university of science and technology; Key Laboratory of Coal Mine Safety and Efficient Exploitation of Ministry of Education, Huainan, Anhui, China*

ABSTRACT: Aimed at the problems existing in collecting and treatment ventilation network information, the paper puts forward a method about the mine ventilation system diagram's combination of manual analysis and automatic identification, which can collect network information interactively and draw a simple ventilation system graph with GIS function and a network with face centralized layout in center, inlet air down and outlet air up; Associational branches are moved with the nodes. A simplified fixed Q method is used in network calculation, which is complete to meet the requirement. The result is shown in the network graphs. Through in situ test, the result shows that the interactive collection means is feasible; simple ventilation system is simple and intuitive; the efficiency of drawing and editing network graphs is very high; the range of searching cotree's branches and independent loop is greatly reduced, and the velocity is improved.

KEYWORDS: Network information, interactive collection, simulation mode

## 1. INTRODUCTION

Mine ventilation network information is extracted from the topological relation of roadways (lines) and crosses (points) in mine ventilation system diagram. It's the premise and basis of researching ventilation system. At present, there are three main methods to get original information of ventilation system:

The first method builds network information instructive files with manual analysis on different CAD diagrams' topological relation, but it is of low efficiency and needs long period.

The second one uses self-developed mine-used CAD software, and when mine ventilation system diagram is being drawn, topological relation will be built and managed automatically, such as the MineCAD developed by Huannai Mining Institute (Li 1996, Li 1997) MVSS developed by Liaoning Technical University (Jia 2004, Li 2004, Ni 2004, Su 1996) and so on, but it is time-consuming to redraw and difficult to promote with impact of operation habit.

The third one is based on the automatic identification technology for mine ventilation network information of AutoCAD: DXF file will be automatically identified and transformed into mine ventilation network information, some professors like Liu Jian from Liaoning Technical University and Lin Zaikang (Lin 2000) from China University of Mining and technology have detailed study on this aspect, but the accuracy rate and efficiency are both low.

Taking the previous study into consideration, the authors develop mine-used CAD software, through mine ventilation system diagram (AutoCAD Electronics Version), combines manual operation with automatic identification organically, collects ventilation network information interactively, and realizes automatic drawing network diagram and calculation function. The software mainly has the following characteristics:

1) Collecting data correctly and rapidly with querying, timely amending to the accuracy of the information collected.

2) Drawing simple ventilation system diagram with single and dual-line automatic conversion display and GIS quick search function.

3) Automatic conversion between ventilation system and network chart; dividing into intake roads faces and return roads; faces are centralized in the middle of network diagram, which has a good visual.

4) Use simplified fixed air quantity method to make network calculation, which lessens the searching scope of cotree chords and independent loops before network calculation with Scott-Hinsley, and improves the network calculation speed

## 2. DESIGN IDEA

The software is based on the first (traditional) mine ventilation network information steps to design, as is shown in Figure1. It is divided into three modules: collecting data, processing network and calculating network.

## 3. MAIN FUNCTIONS

### 3.1. *Collecting data*

1) Collection

With the combination technology of manual analysis and automatic identification, data collection refers to the gather the topological relationship between roadways and crosses. If collecting the physical parameters of roadways like length, position etc., it must be divided into single stage line (straight line) and multi-stage line.

Collecting single stage line needs to click the start and end point position; the software will automatically add the start node and end node. And adding a record to save parameters: start, start number, end, end number, length etc. into database. Meanwhile, inputs the other parameters: name, category, support, shape, area and air quantity and so on.

Collecting multi-stage line needs to add the position of each turn point, and only the start and end are saved, turning point's location will be recorded and temporally numbered in AutoCAD, but does not appear in the capture software or be saved.

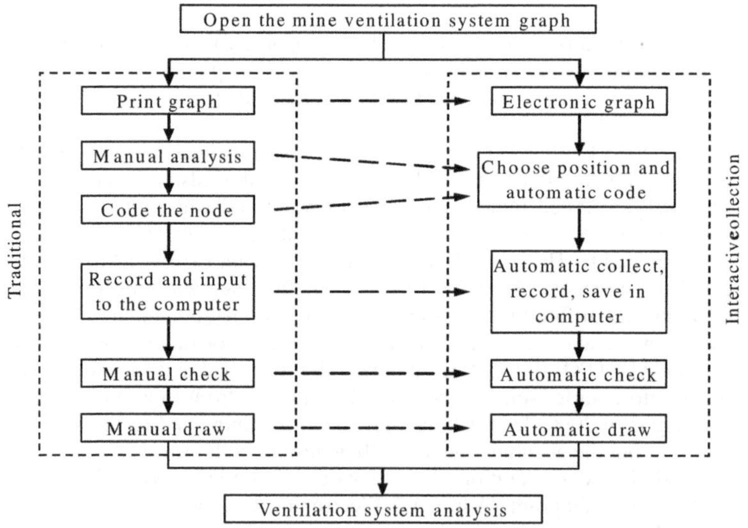


Figure 1. The flow chart of interactive collection

2) Query path

Whether the result is correct or not can be calibrated by query path which collects data quickly. For a complete ventilation system, path is from the intake through faces to the return. So path must meet the requirements:

Inlet from the intake shaft, through a face (no series air generally), return from the return-air. So any path that can not meet the condition is either not complete (roads were not finished) or wrong (endless loop or series air path), and necessary to modify. However, doing path query in regional (mining area) roadways is not necessary to meet the condition, and the major work is to inspect if there is series air between the inlet and return air in the given network .

3) Loading data and verifying

Collection may not be completed by one time, and needs to switch continuously between AutoCAD and collection software. When reloading the collected data after it's saved (txt format), open the appropriate ventilation system diagram at the same time, and check the loaded data and primitive in "Newlayer" layer, the collected data will not be collected again but directly displayed. Process the no match data. After processing, collecting will be continued along the last operation.

### 3.2. *Process graph*

1) Drawing simple ventilation system graph

With fast inquiry the roadway information, simple ventilation system graph reflects the location of roadway on locating the nodes, and implies the topology relationship of network, roadway parameters. There're some problems to handle during drawing (Wang 1996).

(1) Process to dual lines that joint in the surface

Steps: ① Determine two line segments intersect or not. ② If intersect, calculate the coordinates of four intersected nodes: JD1, JD2, JD3 and JD4 ; ③ According to the spatial relation, determine the upper and lower positions to break. Figure 2a and Figure 2b are the results before and after processing to dual lines that joint in the surface and disjoint in the space.

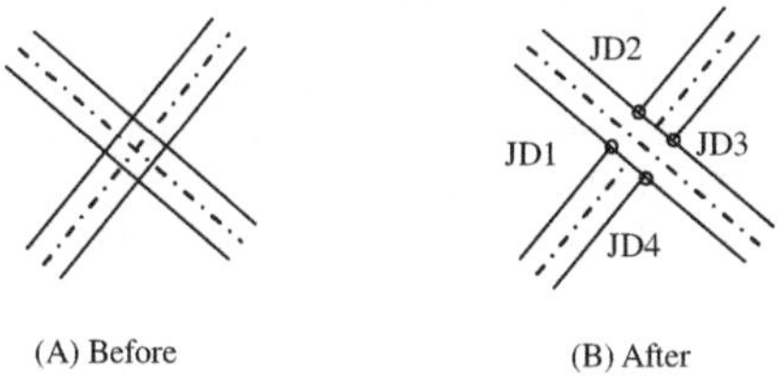


Figure 2. The process of the surface joint and the space disjoint

(2) Process to dual lines joint in the surface

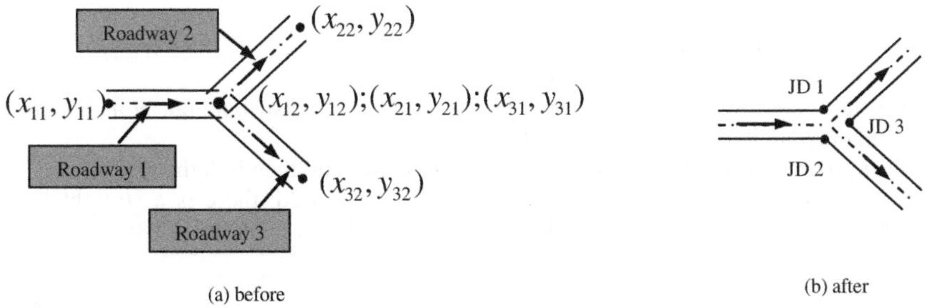


Figure 3. The process of the surface joint

Mathematical processing model of dual lines intersection node automatic hiding

$$\begin{cases} y = k_i x + b_{i1} \quad or \quad y = k_i x + b_{i2} \; i, j = 1,2,3 \cdots n; i \neq j \\ y = k_j x + b_{j1} \quad or \quad y = k_j x + b_{j2} \end{cases} \qquad (1)$$

where: $k_i$, $k_j$ are the slope of roadway $i, j$;

$b_{i1}$, $b_{j2}$ are the larger intercepts of straight lines both sides of the roadway;

$i, j, b_{i1}, b_{j2}$ are the smaller data;

$n$ is the number of roadways in and out of the node.

According to the above system of equations and quadrant which each roadway straight line in, with combination of two respectively, and take Figure 3 as an example, three solving equations can be established:

$$\begin{cases} y = k_1 x + b_{11} \\ y = k_2 x + b_{21} \end{cases} \qquad (2)$$

$$\begin{cases} y = k_1 x + b_{12} \\ y = k_3 x + b_{32} \end{cases} \qquad (3)$$

$$\begin{cases} y = k_2 x + b_{22} \\ y = k_3 x + b_{31} \end{cases} \qquad (4)$$

Solve it; get the coordinates of four nodes: JD1, JD2, JD3.Then based on the coordinates of the intersecting points and the beginning and ending nodes of roadway, do roadway drawing.

2) Faces centralized arrangement network automatically generated

The current methods of drawing network can not comprehensively and clearly demonstrate some information like the mine ventilation type, the number of ventilation system and intake air shaft, the number and type of faces etc. (Tian 2007, Lin 2006, Liu 2003).

Based on the ventilation network information collected, the author puts forward a new drawing network method, whose characteristics are:

- the network is divided into three areas: inlet area(the inlet branches), faces area (the face branches), return area(return branches);
- all the face branches are centralized in the middle of the network, which visually and directly shows the number of system. Also add a mark to a face branch (like there is a local fan mark, then there is the driving branch), which can express the type of face branches (like coal mining branches, driving branches, chamber branches etc);
- distribute the nodes with the longest path method. Put the path including the most nodes to the leftmost (or rightmost) boundary, and according to the mining area and crossing situation, distribute the other face while avoiding crossing as possible;
- show the number of intake shaft and return shaft visually and directly. The intake shaft is in the lower, and return shaft is in the upper;
- treat multi-fan: According to geographic location, successively distribute fan-branch from left to right, and draw an icon on the fan-branch;
- treat multi-system return linked: First draw the subsystems, and then link the return;
- treat parallel branch: According to parity, if the number is odd, there is a straight line in middle, the others successively distribute left and right; if it is even, just distribute left and right.

As Figure 4, it is a network of the exhaust ventilation system with 1 intake shaft, 6 face branches (including a driving branch) and the number of system is 1.

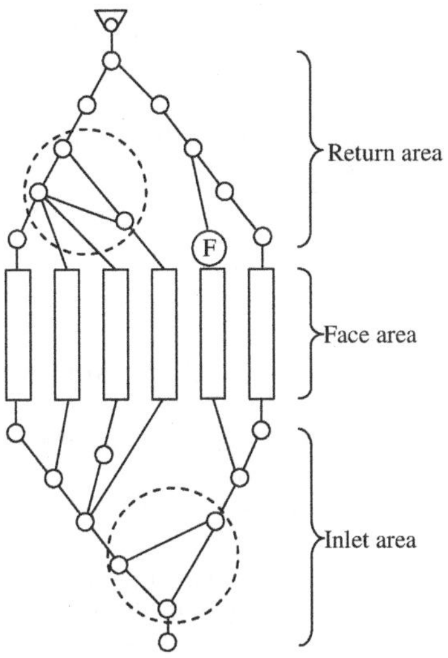


Figure 4. The network graphs of the exhaust ventilation system

### 3.3. *Simplified fixed Q method to make network calculation*

The substance of this method takes a branch when the Q is known as a regulation point. During network calculation, don't allow the fixed Q branch to participate in the process of iteration with Q unchanged. When network calculation ends, the Q of other branches is calculated, the regulation parameters could be calculated. The simplified fixed Q method removes the branches whose Q are known and the branches whose Q can be calculated through the branches given, and makes network calculation to the surplus branches.

During the process of analysis, only some Q of the branches (mostly face branches) is known, the number is generally given by experience or through the actual measurement. Then make network calculation through the fixed Q and Scott-Hensley method. Before network calculation, firstly search cotree chords and independent loops from all the branches, and then establish independent loops equations. As Figure 4, the number of cotree chords $f$ (also the number of independent loops) is:

$$f = n - m + 1 = 27 - 20 + 1 = 8 \qquad (5)$$

where: $n$ – the number of branches, including virtual branches; $m$ – the number of nodes.

Through Q balance to the branches in Figure 5, find that the Q of only 6 branches are unknown in two dashed circles, the other 21 branches can be regarded as fixed Q branches, and do not participate, and the number of independent loops should be 2. So, after removing the branches not participating in calculation, the scope of searching cotree chords and selecting independent loops will lessen greatly and the calculation speed will also improve greatly. When doing ventilation system analysis, especially for large and complex mine, the simplified fixed Q method is more effective.

## 4. EXAMPLE APPLICATIONS

For example, a coal mine ventilation system optimization network data in 2006 in Huaibei Mining Group, since the mine ventilation system graph (AutoCAD) does exist, it's troublesome to redraw. On the basis of opened AutoCAD, with interactive ventilation network data collection, through selecting nodes by hand, we can automatically save ventilation network data and simultaneously draw the network.

Ventilation network data: a system, 80 branches, 46 nodes, 31 face branches (including air doors), the number of independent loops is: $f = n - m + 1 = 80 - 46 + 1 = 35$. After Q is balanced, remove all fixed Q branches, only 6 independent loops, including 22 branches, using simplified fixed Q method to make a calculation, the searching scope will lessen nearly 3/4, calculation time will lessen 2/3,and the accuracy is unchanged.

## 5. CONCLUSIONS

Through on field test, it can be found out:

- realize mine ventilation network information's accurate, and rapid acquisition; meanwhile draw the simple ventilation system graph with GIS;
- automatic generation Network and adjustment time decreases greatly, and the information covered in network is very rich, and the visual effect is quite good;
- using the simplified fixed Q method, calculation time is reduced greatly, calculation results conform to actual situation completely.
- this technology can realize collecting information, drawing and analyzing ventilation system integration, which is higher in the overall efficiency than manual analysis and automatic recognition, and better than self-developed CAD software in popularization and application. If the technology can achieve network interconnection and multi-person simultaneously acquisition, the workload will be reduced.


## ACKNOWLEDGEMENT

The study is supported by Anhui Province College Young Teachers Scientific Research "Allotment Planning" Key Project (2009SQRZ067) and National Natural Foundation (No. 50874005)


## REFERENCES


Li Husheng. 1996. Mine ventilation network diagram CAD software development [J]. Safety in Coal Mines 10: 1-4. (In Chinese)

Li Hu-sheng. 1997. Automatic generation of ventilation network diagram from mine ventilation system diagram [J]. Xi'An Mining Institute Journal 6: 127-130. (In Chinese)

Li qi,Liu Jian. 2004. Auto-plot of Layer Ventilation Network Equilibrium Diagram [J]. China Safety Science Journal 8: 25-27. (In Chinese)

Lin Jian-guang, Jiang Zhong-an. 2006. Mine Ventilation Network Solution System with Functions of Network Chart Plotting and Fan Selecting [J]. Mining Technology 6(3): 388-391. (In Chinese)

Lin Zai-kang, Yan Xue-feng, Tao Wei zhong et al. 2000. Module of CAD Software System in Mining [J]. Journal of China University of Mining & Technology 29(1): 67-69. (In Chinese)

Liu Jian, Jia Jin-zhang, Zheng Dan. 2003. Study of mathematical model of plotting of equilibrium of ventilation network based on independent path idea [J]. Journal of China Coal Society28 (2): 153-156. (In Chinese)

Jia Jin-zhang, Wei Shi-chuan, Liu Jian. 2004. Network simplification technology on mine ventilation simulation sysytem[J]. Journal of Liaoning Technical University 8: 433-435. (In Chinese)

Ni Jing-feng, Liu Jian, Li Yu-cheng. 2004. Establishment of topology relationships and maintenance of visualization of mine ventilation network [J]. Journal of Liaoning Technical University12: 724-726. (In Chinese)

Su Qing-zheng, Liu Jian. 1996. Mine Ventilation Simulation System Theory and Practice [M]. Beijing: Coal Industry Press. (In Chinese)

Tian Wen-ming, Du Cui-feng, Zhang Hao. 2007. Study of Ventilation Network Solution Software with Function of Network Graph Drawing [J]. Express Information of Mining Industry 23 (5): 35-39. (In Chinese)

Wei Chong-guang. 2005. AutoCAD and AutoCAD development [M]. Beijing: Chemical Industry Press. (In Chinese)

Wang De-ming, Li Yong-sheng. 1996. Decision Support System for Rescuing Disaster during Mine Fire Period [M]. Beijing: Coal Industry Press. (In Chinese)

International Mining Forum 2010, Liu et al. (eds) © 2010 Taylor & Francis Group, London, UK. ISBN 978-0-415-59896-5

# Design of coal pillar with roadway driving along goaf in fully mechanized top-coal caving face

Jucai Chang, Guangxiang Xie, Ke Yang
*Key Laboratory of Coal Mine Safety and Efficient Exploitation of Ministry of Education; Anhui University of Science and Technology, Huainan, Anhui Province, China*

ABSTRACT: Small coal pillar with reasonable width plays important roles in not only improving recovery ratio and eliminating dynamic disasters, but also in keeping the stability of entry surrounding rock. In order to analyze reasonable layout of gob-side entry driving in fully mechanized top-coal caving face (FMTC face), numerical simulation, in-situ measurement and theoretical calculation are carried out. The stress redistribution rules of coal mass in dip direction are put forward by metering the stress after coal extraction. At the same time, mechanical characteristics of coal pillars with different widths, 3 m, 5 m, 7 m, 10 m, 15 m and 20 m, respectively, are obtained by analyzing stress field, displacement field and fractured field characteristics. Reasonable width of district sublevel coal pillar is determined by calculating the width of limit equilibrium region and the cracked range of the entry surrounding rock due to mining of adjacent face based on the results of in-situ measurement and numerical simulation. The engineering practice proves that the width of coal pillar is scientific and reliable. It provides a scientific basis for reasonable layout of gates and determining parameters of the coal pillar in FMTC face. It can be favorable to improve maintenance states of entry, and also benefit to improve the recovery in FMTC face, and it provides beneficial lessons for determining reasonable width of gob-side entry driving of other similar conditions.

KEYWORDS: Fully mechanized top-coal caving face, roadway driving along goaf, coal pillar width

## 1. INTRODUCTION

Small coal pillar with reasonable width plays important roles in not only improving recovery ratio and eliminating dynamic disasters, but also in keeping the stability of entry surrounding rock. Lots of facts prove that the stability of the entry is determined to a great extent by the width of coal pillar (Yang, Jiang 2001, Bai, Wang 2000, Chen, Jin 1998), especially gob-side entry of FMTC face. Because of the unreasonable coal pillar width, there are probably roof falling and large deformation existing in many entries, which effects production safety badly. Presently, experience analogy method is usually used to determine the width of small pillar, which are with limitations and blindness to some extent in many engineering practices under different complicated geological condition. Therefore, how to determine the width of coal pillar reasonably and scientifically is attached great importance upon safety in production of FMTC face. In this paper, the compositive method, which consists of in-situ measurement, numerical simulation and theoretical calculation, is adopted based on engineering geology and exploitation technology of 1151(3) FMTC face in Xieqiao colliery. It is expected to be of guidance for engineering practice.

## 2. GEOLOGICAL CONDITIONS

Xieqiao FMTC Face 1151(3), from the switchyard below Section E-C westward to the cut connecting road eastward, in which works Seam C13-1 averages 5.4m thick, dipping 12°~15°, averaging 13°. The rate of mining height to caving height is 1:1~1.08. The strike length is 1674 m and the dip length is 231.8 m. The face is at an elevation of -588 ~ -662 m, of which the seam floor of the air return is at an elevation of -558.0~-602.8 m and the seam floor of the haulage drift is at -636.5 ~ -662.0 m. The coal seam in the face is in a state of stability and equilibrium before extraction, and develops generally two-layer carbonaceous mudstone or tonstein. The roof and floor of coal seam distributes as follows: The main roof is powdery packsand, which thickness is about 6.2 m; the immediate roof is comprised of mudstone and coal, which thickness is 3.3 m; the immediate floor is mudstone, which thickness is 1.5 m; the main floor is siltstone, which thickness is 2.8 m.

## 3. ANALYSIS OF IN-SITU MEASUREMENT

### 3.1. *Text and indenting*

Because of coal extraction, the equilibrium state of the primary stress around face will be damaged and stress filed in surrounding rock will redistribute accordingly. The distribution laws of side abutment pressure are the premises for choosing the entry position and determining the coal pillar width. To obtain the distribution rules of side abutment pressure of 1151(3) FMTC face, stress meters (KSE-II -1) are installed in coal entity of the headentry underlier to monitor the stress after coal extraction. The results are shown in Figure 1.

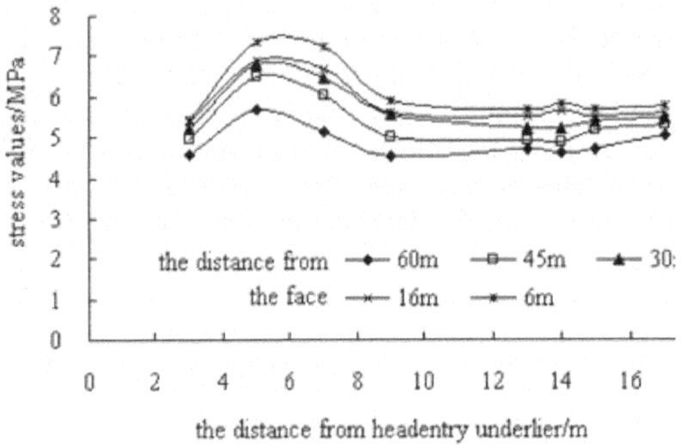


Figure 1. The variation of coal mass' stress in the head entry along the inclination

As shown in Figure 1, the stress of sideward coal mass increases gradually with the face advancing, and the peak value of abutment pressure appears at a distance of 7 m from the entry side in dip direction, then decreases gradually to a certain stable value, so it can conclude that the stress-decrease zone is at the distance of 3 m~7 m from the underlier, which is the optimum position for driving the gob-side entry.

## 4. ANALYSIS OF NUMERICAL SIMULATION

The numerical simulation software FLAC3D for rock mechanics is adopted to analyze the distribution and evolvement rules of displacement field, damage field and stress field with different width of coal pillar. According to the rule of FLAC3D, the Mohr-Coulomb yield criterion is taken to judge the damage of surrounding rock, and strain-softening model shows the properties of residual strength decreasing gradually with deformation development after coal mass' destroy. The simulation model is 500 m long in the strike direction, its inclinational width is 600 m and height is 314.17 m, which is divided into 95332 3D units, 112739 knits. Displacement of the model are restricted on boundaries, horizontal ones by the four sides and the vertical ones by the bottom, with vertical loadings exerted from above, in simulation of the weight of overlying strata. The simulation width of coal pillar is 3 m, 5 m, 7 m, 10 m, 15 m and 20 m, respectively. The results are as follows.

### 4.1. *Stress field characteristics of the entry with different wide coal pillars*

The distribution of peak value characteristics of vertical stress with different coal pillars is shown in Figure 2.

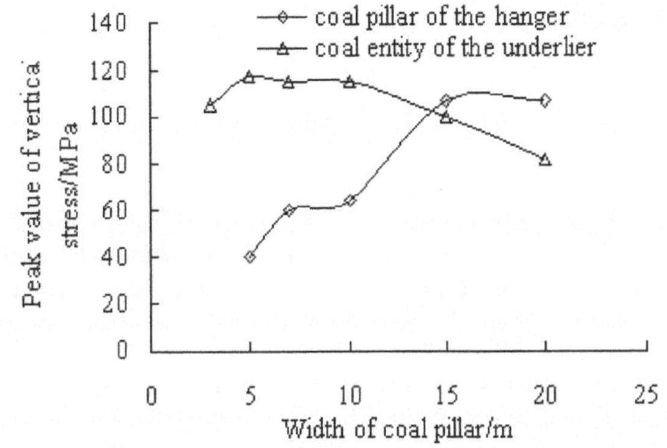


Figure 2. Vertical stress peak values with different pillars

It is observed that the stresses both in coal pillar and coal entity are variable when width of entry protecting coal pillar is different. With the increase of coal pillar width, the peak value of vertical stress in coal pillar increases gradually and that in coal entity decreases gradually. It shows that the bearing capacity of coal pillar improves. With the increase of coal pillar width, the stress transfers from coal entity to coal pillar gradually. When the width of coal pillar increase to a certain value, for example 15 m, the stress peak value in coal pillar is equal to that in coal entity. Here, both coal pillar and coal entity bear biggish stress peak value. However, when the widths of coal pillar increases further, for example 20 m, the stress peak value of coal pillar is larger than that of coal entity; the coal pillar mainly bears the loading.

### 4.2. *Displacement field characteristics of the entry with different wide coal pillars*

Relative displacement curves of the roof-floor with different coal pillars are illustrated by Figure 4.

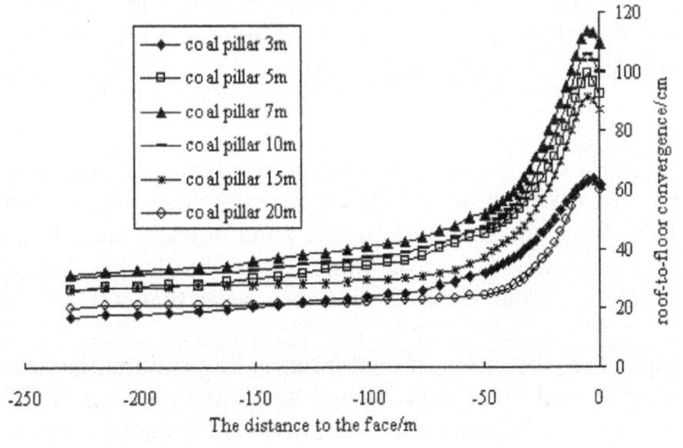


Figure 3. Relative displacement curves of the roof-floor

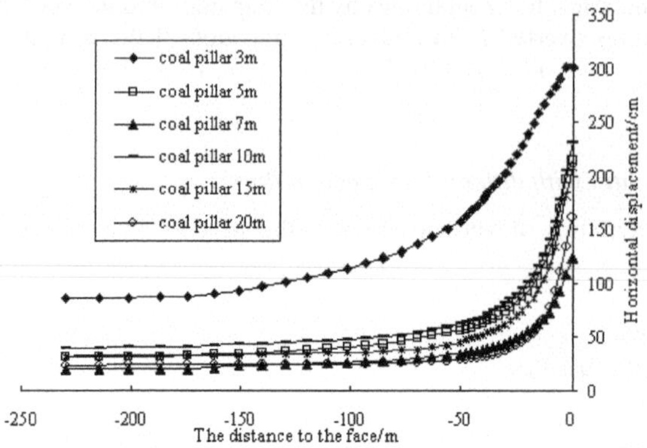


Figure 4. Relative horizontal displacement of the ribs

As shown in Figure 3, when the width of coal pillar increases from 3 m to 20 m, the roof-to-floor convergence firstly increases gradually, and then decreases. Concretely, when the width increases from 3 m to 7 m, the roof-to-floor convergence increase gradually, and then decreases gradually when the width of coal pillar increases from 7 m to 20 m, namely, the displacement value reaches the maximal when the width of coal pillar is 7 m.

Relative horizontal displacement curves of the ribs with different wide coal pillars are shown in Figure 4. It can be seen that horizontal displacement of the ribs is great near the face along the strike, and as the distance to the face increase, the displacement decrease gradually. When the width of coal pillar increases from 3 m to 7 m, the horizontal displacement of the ribs decreases gradually, however, when the width of coal pillar increases from 7 m to 10 m, the displacement increases suddenly, and then the horizontal displacement decreases gradually when the width of coal pillar increases from 10 m to 20 m.

### 4.3. *Fractured field characteristics of the entry with different wide coal pillars*

Distribution of fractured field in the entry surrounding rocks with different wide pillars is shown in Figure 5.

As shown in Figure 6: Plastic failure zones in coal pillars have interpenetrated within 15 m wide coal pillars. When coal pillar is 20 m wide, plastic failure zones have not interpenetrated, and there is the elastic core zone with the width of 5~8 m in coal pillar.

When coal pillar is 3 m wide, there are 10 m wide plastic failure zones in coal entity of the underlier. There are about 5 m wide plastic failure zones in coal entity of the underlier for 5~15 m wide coal pillars. When coal pillar is 20 m wide, the coal mass of the underlier is not destroyed over 50 m ahead of the face, however, there are about 5 m wide plastic failure zones within 50 m, which shows that the coal mass of the underlier is destroyed badly when coal pillar is very small, and with moderate wide coal pillars, there are certain failure zones in coal entity ahead of face, but with large wide coal pillar, there are no failure zones in coal entity ahead of the face, except in the mining influenced area.

### 4.4. *Analysis of reasonable width of coal pillar*

For a large coal pillar, for example 20 m, the distribution of vertical stress in coal pillar is double peak at a certain distance ahead of the face, and there are plastic zone in each side of coal pillar, and there is an elastic core zone between two stress peak values, so coal pillar has enough supporting capacity to keep its stability. However, the loss of coal is very large, that is to say the recovering ratio is badly affected.

For a very small coal pillar, for example 3 m, the stress of coal pillar is low and the maximal displacement area mostly lies in the center of coal pillar buttock, and horizontal displacement is much larger than that of other coal pillars. The coal pillar has been destroyed completely, and its supporting capacity is deteriorated and the stability of the entry is affected badly.

For the coal pillar with moderate width, for example 15 m, on one hand, the plastic zone in coal pillar have interpenetrated, the coal mass of coal pillar generate plastic yield as a while, and there is not elastic core zone in the middle of coal pillar. On the other hand, there is large stress peak value in coal pillar acts on coal pillar all along. It keeps coal pillar bearing large compression stress and doesn't transfer. So it is the worst width for the stability of the entry and coal pillar.

For the small wide coal pillars, for example 5~10 m, the plastic zone in coal pillar have interpenetrated, the coal mass of coal pillar generate plastic yield as a while, and there is not elastic core zone in the middle of coal pillar. But the maximal displacement of the entry surrounding rock center at the roof, the coal pillar have not destroyed completely, and the stress of the entry surrounding rock can transfer rapidly from coal pillar to coal entity of the face, there is no region of high stress in coal pillar, and the stress of coal pillar is small. So the coal pillar can keep preferable stability if reasonable supporting measures are adopted in time to maintain the coal pillar integrity and improve the residual strength of coal pillar. But it is well known that the width, from 7 m to 10 m, is the boundary point between softening and strengthening in coal pillar. So the width of coal pillar should be smaller than the width of 7 m, and the width of 5 m is better within the range.

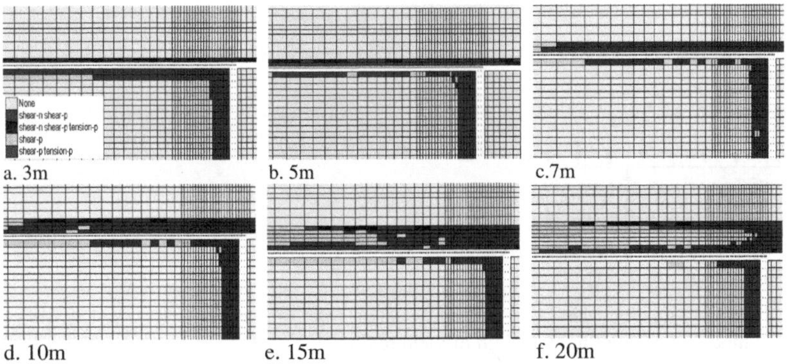


Figure 5. Distribution of fractured field in the entry surrounding rocks with different wide pillars

## 5. THEORETICAL CALCULATION OF REASONABLE WIDTH OF COAL PILLLAR

### 5.1. *Determination to reasonable width of coal pillar*

To avoid the influence of fixed abutment pressure and residual abutment pressure to the entry, and to decrease convergence rate of the surrounding rock, keep the entry stable, reduce the loss of district sublevel coal pillar, the width of coal pillar should be as small as possible. If the width of coal pillar is very small, the stability and bearing capacity is very low, because there are fractured zone and plastic zone in the coal pillar. Besides, all bolts are in fractured surrounding rock, with no footing of reinforcing, the anchoring force is very low, and don't keep the entry stable. So there is a reasonable width to ensure the supporting load of the entry is enough and the loss of coal resource is small, as shown in Figure 6. The reasonable width of coal pillar can be calculated by the following expression:

$$\begin{cases} B = x_1 + x_2 + x_3 \\ x_3 = 0.15 \sim 0.35(x_1 + x_2) \end{cases} \tag{1}$$

where: B, $x_1$, $x_2$ and $x_3$ are the width of small pillar, the width of yield region due to mining of the previous district sublevel, the effective width of bolt in small coal pillar, respectively.

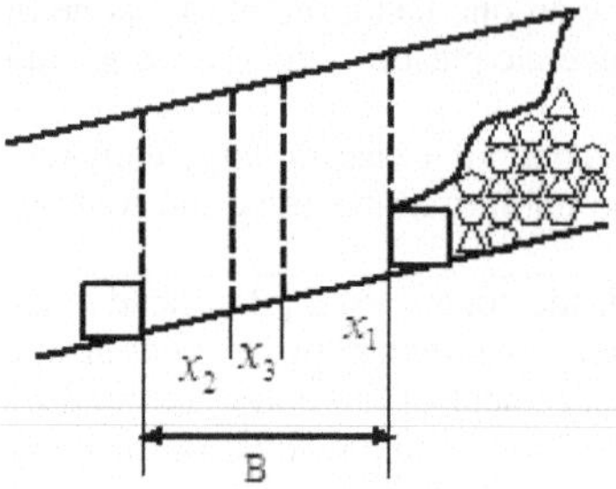


Figure 6. Determination to reasonable width of small pillar

## 5.2. *Determination to the width of yield region x1*

On the Figure 7 $x_1$ is the width of yield region, m; M is the height of the entry of the previous district, m; $p_x$ is lateral restriction of the gob applying to coal pillar, MPa; $\sigma_y$ is the vertical stress at the interface of the coal-rock mass, MPa; $\alpha$ is dip angle of the coal seam, degree; $\sigma_{yl}$ is ultimate strength of coal pillar, MPa.

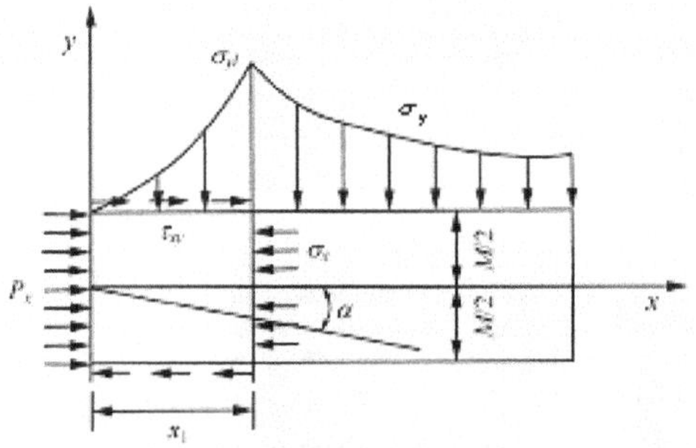

Figure 7. Mechanical model of calculating the width of yield region

The mechanical model of calculating the width of yield region is founded based on the theory of ultimate equilibrium, considering the influence of coal seam dip angle, as shown in Figure 7. The equilibrium equations to solving the stress at the interface of yield region are as follows.

$$\begin{cases} X - direction: \dfrac{\partial \sigma_x}{\partial x} + \dfrac{\partial \tau_{xy}}{\partial y} + X = 0 \\ Y - direction: \dfrac{\partial \sigma_y}{\partial y} + \dfrac{\partial \tau_{xy}}{\partial x} + Y = 0 \\ \tau_{xy} = -(C_0 + \sigma_y \tan \phi_0) \end{cases} \tag{2}$$

So the width of yield region is solved, as shown in Formula (3).

$$x_1 = \frac{M\beta}{2\tan\varphi_0} \ln\left[\frac{\beta(\sigma_{yl}\cos\alpha\tan\varphi_0 + 2C_0 + M\gamma_0\sin\alpha)}{\beta(2C_0 + M\gamma_0\sin\alpha) + 2P_x\tan\varphi_0}\right] \tag{3}$$

Where, $\beta$ is coefficient of lateral pressure at the interface between yield region and core zone, $\beta$= μ/(1-μ), μ is Poisson's ratio; $\varphi_0$ is angle of friction between coal mass and roof – floor, degree; $C_0$ is cohesion at the interface of coal seam roof and floor, MPa; $\gamma_0$ is the average body force of rock mass, MPa.

For 1151(3) FMTC face, the parameters are as follows: $\sigma_{yl}$ = 9.891 MPa, $M$ =3 m, $\alpha$ =13°, μ = 0.32, $\beta$ = 0.47, $\varphi_0$=28, $C_0$=2 MPa, $\gamma_0$ =0.025 MPa, $p_x = 0$, $x_2 = 2\,m$.

So $x_1$ can be obtained by Formula (3) referring to the parameters mentioned above.

$$x_1 = 1.093\,m \tag{4}$$

The yield region of coal pillar will expand to the deep and the width of plastic zone will be enlarged due to mining disturbance in practice, which is generally described by a so-called coefficient of disturbance. By analyzing lots of in-situ data and characteristics of FMTC mining (Xie, Yang 2006), here, the coefficient of disturbance is 1.8. So the width of yield region $x_1$ is determined as 1.97 m.

### 5.3. *Reasonable width of small coal pillar*

The width of small coal pillar, B, can be obtained considering $X_1$ = 1.97 m, $x_2 = 2m$. The results are as follows:

$$B = 4.57 \sim 5.36\,m \tag{5}$$

In sum, reasonable width of small pillar is 5m by in-situ measurement, numerical simulation and theoretical calculation. In practice when the next face was mined in Xieqiao Colliery, 5 m wide coal pillar was adopted to protect the air-return entry. Facts have proved that the maintained effect of the entry is better, which ensured the normal coal winning, and improved the recovery ratio.

## 6. CONCLUSIONS

Reasonable width of district sublevel coal pillar is determined, and reasonable layout of FMTC face gob-side entry driving is analyzed by calculating the width of limited equilibrium region and the cracked range of the entry surrounding rock due to mining of adjacent face based on the results of in-situ measurement and numerical simulation.

The engineering practices prove that the designed width of coal pillar is scientific and reliable by this method. It provides a scientific basis for reasonable layout of gates and determining parameters of the coal pillar in FMTC face. It can be favorable to improve maintenance states of entry, and also benefit to improve the recovery in FMTC face. At the same time it provides beneficial lessons for determining reasonable width of gob-side entry driving of other similar conditions.

## ACKNOWLEDGEMENT

This work was supported by Science and Technological Fund of Anhui Province (No.KJ2010A 090), Science and Technological Fund of Huainan city (No. 2009A05009) and Outstanding Academic Innovation Team of AUST in China.

## REFERENCES

Bai Jian-biao, Wang Wei-jun, Hou Chao-jiong et al. 2000. Control mechanism and support technique about gateway driven along goaf in fully mechanized top coal caving face. Journal of China Coal Society 25(5): 478-481. (In Chinese)

Chen Qing-min, Chen Xue-wei, Jin Tai et al. 1998. Ground Behavior and Support Technique of a Roadway Driven Along Gob in a Fully Mechanized Sub-level Caving Face. Journal of China Coal Society 23(4): 382-3854. (In Chinese)

Xie Guang-xiang, Yang Ke, Liu Quan-ming. 2006. Study on Distribution Laws of Stress in Inclined Coal Pillar for Fully-mechanized Top-coal Caving Face. Chinese Journal of Rock Mechanics and Engineering 25(3): 545-544. (In Chinese)

Yang Yong-Jie, Jiang Fu-xing, Ning Jian-guo et al. 2001. Method of determining rational width of small coal pillar protecting roadways driving along next goaf supported by bolting and meshing of fully mechanized sublevel face. The Chinese Journal of Geological Hazard and Control 12(4): 81-84. (In Chinese)

International Mining Forum 2010, Liu et al. (eds) © 2010 Taylor & Francis Group, London, UK. ISBN 978-0-415-59896-5

# Roadside supporting design for gob-side entry retaining and engineering practice

Yingfu Li
*Key Laboratory of Coal Mine Safety and Efficient Exploitation of Ministry of Education; Anhui University of Science and Technology, Huainan, Anhui Province, China*

Xinzhu Hua
*Anhui University of Science and Technology, Huainan, Anhui Province, China*

ABSTRACT: In order to determine the width of roadside backfilling body, new mechanical model of roadside supporting resistance is established based on taking the dip angle of coal seam and supporting force of roadside coal seam into account. Based on this model, computational formulas of roadside supporting resistance and supporting force of roadside coal seam are derived, and function mechanism of roadway-in bolting support is analyzed. Moreover, the technique of advancing roadside combined support is proposed to solve the problem of hysteretic support of roadside backfilling body, to provide early stronger roadside supporting force, and to improve self-bearing capacity of roof strata. The technique is that a gap is made in advance at the end of the working face and combined support of bolting, cable, steel strip, and wire netting is adopted to support the gap roof. According to the theoretical analysis, a roadside supporting design is made for haulage entry of No. 512(5) working face in Xieyi mine of Huainan, and the function relationships are explored between roadside supporting resistance and the width of roadside backfilling body, the dip angle of coal seam, roadway width, feature size of fractured roof strata, supporting resistance of roadway roof.

KEYWORDS: Mechanical model, roadside supporting resistance, roadside supporting design

## 1. INTRODUCTION

The technique of the gob-side entry retaining can well realize protection of roadway with no pillar, increase the recovery rate of coal resource, decrease workload of roadway excavation, realize Y-type ventilation, and solve effectively the problem of gas accumulation and gas overflow at the upper-corner.

Roadside backfilling technique of high water materials is a main developing trend. However, there exist some problems of this technique as follows:

- most of former mechanical models don't involve the dip angle of coal seam and supporting force of roadside coal seam, which leads to inaccurate calculation of roadside supporting resistance and causes waste of roadside backfilling materials or lack of roadside supporting resistance.
- roadside backfilling body is of certain width, but roadside supporting resistance is simplified for concentrated force, which can hardly reflect the relationship between the width of roadside backfilling body and roadside supporting resistance.

In view of the above-mentioned problems, new mechanical model of roadside supporting resistance will be built to design roadside support for haulage entry of No. 512(5) working face in Xieyi mine of Huainan.

## 2. MECHANICAL MODEL OF ROADSIDE SUPPORTING RESISTANCE AND CALCULATION

### 2.1. *Formula derivation of roadside supporting resistance*

After coal is mined, the goaf forms unloading space, when unloading space is up to a certain extent, roof strata of the goaf would cause active caving under weight, this type of roof caving belongs to layered caving from the bottom up. If timely active support is adopted to support roadway roof before roof separation occurs, roadside support of stronger setting load is provided early, roof strata will fracture along the outside of roadside backfilling body, and form cantilever remaining border in the goaf, the remaining border and roadside backfilling body will bear together partial load of roof strata. Now, roadway roof is stable, bearing load of roadway is mainly caused by deformation pressure, roadway support bears smaller load. Otherwise, roof strata will fracture along the roadside of the coal seam, roadway roof is vulnerable to disturbance and is instable, bearing load of roadway is mainly caused by deformation pressure and caving pressure together, roadway support bears larger load.

Roof strata of the gob-side entry retaining is simplified to "superimposed continuous layered model" which the cohesion between layers is neglected, then roof strata is divided hypothetically into several segments by "strip method", the most dangerous cross-section of roadway support can determine roadside supporting resistance under different boundary conditions, and the model involves the dip angle of coal seam and supporting force of roadside coal seam.

Calculate the force of every section by using balanced system of force, analyzing from No. m layer, as is shown in Figure 1.

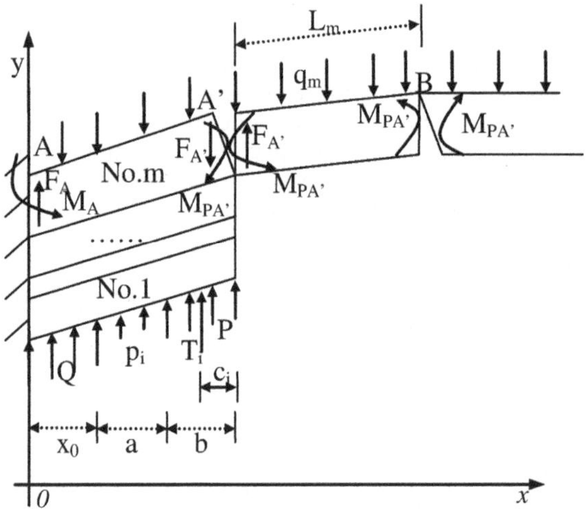


Figure 1. Mechanical model of roadside supporting resistance

At the section of A'B, equilibrium equations of forces in Y-direction are as

$$\sum y = 0, \sum M_{A'} = 0$$

Then

$$F_{A'} - q_m L_m \cos\alpha - \gamma H_m L_m = 0\,, 2M_{PA'} - \int_0^{L_m \cos\alpha} q_m \cdot x \cdot dx + \gamma H_m L_m \frac{1}{2} L_m \cos\alpha = 0$$

At the section of AA', equilibrium equation of moment at the point of A is as

$$\sum M_A = 0$$

So

$$
\begin{aligned}
& P[b(x_0+a)+\frac{b^2}{2}] = \frac{1}{2}\sum_{i=1}^{m}(q_i+\gamma H_i)(x_0+a+b+\sum_{j=0}^{i-1}\cos\alpha H_j\tan\beta_j)^2 \\
& + L_m(q_m\cos\alpha+\gamma H_m)(x_0+a+b+\sum_{j=0}^{m-1}\cos\alpha H_j\tan\beta_j)+M_{PA'} \\
& + \sum_{i=1}^{m-1}F_i(x_0+a+b+\sum_{j=0}^{i-1}\cos\alpha H_j\tan\beta_j)-M_Q-\sum_{i=1}^{m}M_{Ai} \\
& - p_i(\frac{a^2}{2}+x_0 a)-\sum_{i=1}^{t}T_i(x_0+a+b-c_i)
\end{aligned}
\qquad (1)
$$

Where P is roadside supporting resistance; $q_i$ is vertical external load of No. I layer; $q_m$ is vertical external load of No. m layer; $H_j$ is the thickness of No. J layer; $\beta_j$ is rupture angle of No. J layer, $H_0=0$, $\beta_0=0$; $L_m$ is feature size of fractured roof strata; $M_{Ai}$ is bending moment of No. I layer; $F_i$ is shearing force of the remaining border for No. I layer; $F_{A'}$ is shearing force of the remaining border for No. m layer; $\alpha$ is the dip angle of coal seam; $\gamma$ is average bulk weight of overlying strata; x0 is plastic zone width of roadside coal seam; a is roadway width; b is the width of roadside backfilling body; $M_Q$ is bending moment which is caused by supporting force of roadside coal seam; $p_i$ is supporting resistance of roadway roof; m is total number of roof caving; $T_i$ is active supporting force of No. I cable; t is the number of cable; $c_i$ is the distance from No. I cable to the outside of roadside backfilling body; $M_{PA'}$ is the ultimate bending moment of No. m layer.

At free boundary of two ends, it is possible to express the ultimate bending moment ($M_{PA'}$) as

$$M_{PA'} = \frac{L_m^2}{4}(q_m\cdot\cos^2\alpha+\gamma H_m\cdot\cos\alpha)$$

At free boundary of one end and fixed boundary of other end, it is possible to express the ultimate bending moment ($M_{PA'}$) as

$$M_{PA'} = \frac{L_m^2}{2}(q_m\cdot\cos^2\alpha+\gamma H_m\cdot\cos\alpha)$$

Based upon the principle of combination beam, it is possible to express $q_i$ as

$$q_i = \frac{E_i H_i^3\gamma(H_1+H_2+\ldots+H_{i+1})}{E_1 H_1^3+E_2 H_2^3+\ldots+E_{i+1}H_{i+1}^3} \qquad (2)$$

where: $E_i$ (i=1, 2… m) is elasticity modulus of No. I layer; Hi is the thickness of No. I layer.

## 2.2. *Calculating supporting force of roadside coal seam*

Based upon the theory of limit equilibrium, if bearing load of roadside coal seam exceeds the limit of strength ($\sigma_c$), surrounding rock of roadway will cause plastic failure, the strength of yielded coal seam is equal to residual strength ($\sigma^*_c$). With the development of plastic failure towards the interior of roadside coal seam, compressive strength of roadside coal seam gradually increases, and roadside coal seam will lie in an elastic stage again when compressive strength is equal to $\sigma_c$.

A lot of practices show that yielded roadside coal seam is of certain bearing capacity, and that roadside coal seam can be taken as ideal elastic-plastic softening strain model, as is shown is in Figure 2.

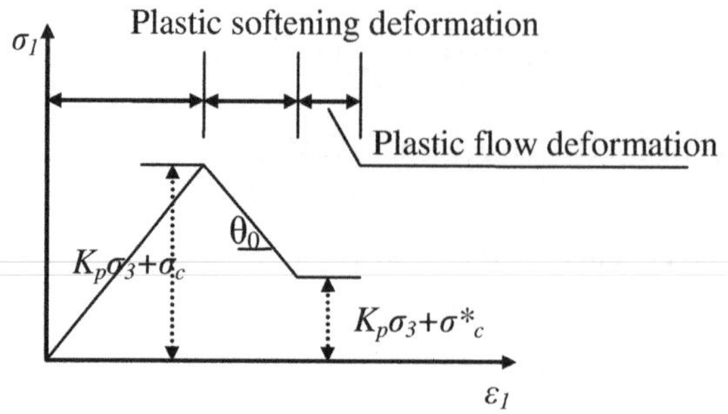


Figure 2. Ideal elastic-plastic softening strain model

Based upon strength conditions of Mohr-Coulomb model, under the condition of triaxial stress, when roadside coal seam is in the stage of plastic softening deformation, compressive strength condition can be expressed as

$$\sigma_1 = K_P\sigma_3 + \sigma_C - M_0\varepsilon_1^{\ P} \ , \ K_p = \frac{1+\sin\varphi}{1-\sin\varphi}$$

Where $\sigma_1$ is major principal stress; $\sigma_3$ is minor principal stress; $K_p$ is triaxial factor; $\varphi$ is internal frictional angle of coal seam; $\varepsilon_{1p}$ is plastic deformation of coal seam; $\sigma_c$ is uniaxial compressive strength; $\theta_0$ is softening angle; $M_0$ is softening modulus, $M_0=\tan(\theta_0)$.

In the stage of plastic flow deformation, uniaxial compressive strength of coal seam ($\sigma_c$) reduces to residual strength ($\sigma^*_c$). Now, coal is almost completely broken. In a similar way, when roadside coal seam is in the stage of plastic flow deformation under the condition of triaxial stress, compressive strength condition can be expressed as

$$\sigma_1 = K_P\sigma_3 + \sigma^*{}_c$$

Within the range of limit equilibrium of roadside coal seam, one side of roadside coal seam is free, and stress is released, thus vertical stress ($\sigma_y$) is more than horizontal stress ($\sigma_x$), and the angle between $\sigma_y$ and σ1 is very small, the angle between $\sigma_y$ and $\sigma_1$ is also very small, it is possible to express as $\sigma_y=\sigma_1$, $\sigma_x=\sigma_3$.

When x=0, coal is in the state of limit equilibrium, the stress at the junction of non-elastic region and elastic region is the peak of abutment pressure, so, $\sigma_y=k_\gamma H$.

When $x=x_0$, $\sigma_x=p_x$, so, $\sigma_y=K_p p_x + \sigma^*_c$.

For $x \in [0, x_0]$, suppose supporting force of roadside coal seam (Q) is of linear distribution, its value is

$$Q = \frac{-k\gamma H + (K_p p_x + \sigma^*{}_c)}{x_0} x + k\gamma H \qquad (3)$$

Bending moment ($M_Q$) which is caused by supporting force of roadside coal seam is expressed as

$$M_Q = \int_0^{x_0} Q \cdot x \cdot dx = \frac{[k\gamma H + 2(K_p p_x + \sigma^*{}_c)]x_0^{\ 2}}{6(a+b+x_0)}$$

Plastic zone width of roadside coal seam ($x_0$) is expressed as

$$x_0 = \frac{AM}{2\tan\varphi} \ln\left[\frac{k\gamma H + \dfrac{C}{\tan\varphi}}{\dfrac{C}{\tan\varphi} + \dfrac{p_x}{A}}\right] \qquad (4)$$

Where: C is cohesion of roadside coal seam; A is coefficient of horizontal pressure, A=u/(1-u); u is Poisson's ratio of roadway coal seam; M is roadway height; $p_x$ is supporting resistance of roadway coal side; k is stress concentration factor; H is mining depth.

## 2.3. *Feature size of fractured roof strata*

Suppose roof strata can be taken as ideal Rigid-Plastic model before roof strata fractures, plastic limit analysis method is used to calculate $L_m$. Roof strata is inclined to form the failure mechanism which is of three fixed boundary and one free boundary when roof strata caves, as is shown in Figure 3.

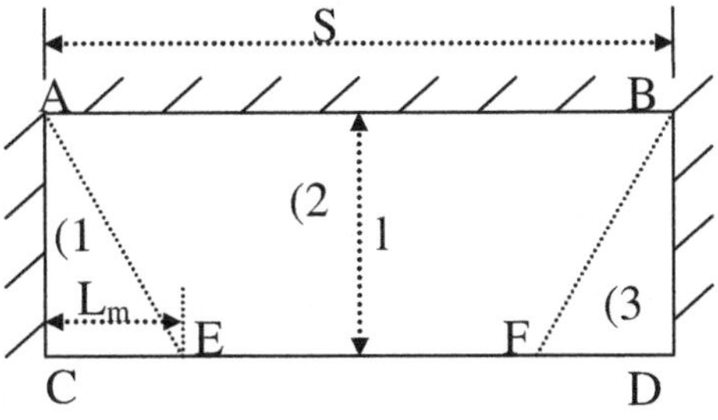


Figure 3. Failure mechanism of rectangular roof strata

Yielding line is AB, AC, BD, AE, and BF. Total virtual dissipation work (U) in yielding line and total virtual work (W) which is caused by external force are expressed as

$$U = M_P\delta(\frac{4l}{L_m} + \frac{2L_m + S}{l}) \quad W = \frac{2}{3}lL_m q_s\delta + \frac{1}{2}lq_s(S - 2L_m)\delta$$

Where δ is virtual displacement; $M_P$ is the ultimate bending moment of roof strata; l is caving step of roof strata; S is the length of working face; $q_s$ is the ultimate load.

According to the equation of W=U, $q_s$ is expressed as

$$q_s = \frac{6(4l^2 + 2L_m^{\ 2} + SL_m)}{l^2 L_m(3S - 2L_m)} M_P$$

Moreover, the value of qs meets the equation below.

$$\frac{dq_s}{dL_m} = 0$$

So

$$L_m = l(-\frac{l}{S} + \sqrt{\frac{l^2}{S^2} + \frac{3}{2}}) \qquad (5)$$

## 3. FUNCTION MECHANISM OF ROADWAY-IN BOLTING SUPPORT

In Formula (1), $\sum M_{Ai}$ is total bending moment of caving layers, and is self-bearing capacity of roof strata. Bolting support can make partial layered caving roof strata be anchored and form a whole, bending resistance at A point can be greatly improved, as is shown is in Formula (6). Meanwhile, bolting can provide certain active supporting force which can made bearing load of roadside support diminish.

$$\sum_{i=1}^{m} M_{Ai} = \sum_{i=1}^{m} \frac{H_i^2}{6}\sigma_{ti} < \frac{(\sum_{i=1}^{n} H_i)^2}{6}\sigma_{ti} + \sum_{i=n+1}^{m} \frac{H_i^2}{6}\sigma_{ti} \tag{6}$$

where: $\sigma_{ti}$ is tensile strength of No. I layer; n is the number of anchored layers.

In Formula (3) and Formula (4), bolting support can reinforce roadside coal seam, improve residual strength and supporting force of roadside coal seam, and diminish roadside supporting resistance.

In order to provide early stronger roadside supporting force, avoid premature fracture of the remaining border along the roadside of the coal seam and hysteretic support of roadside back-filling body, and improve the self-bearing capacity of roof strata, a gap is made in advance at the end of the working face and combined support of bolting, cable, steel strip, and wire netting is adopted to support the gap roof.

## 4. WIDTH CALCULATION FOR ROADSIDE BACKFILLING BODY

The function relationship between mechanical property of roadside backfilling materials and roadside supporting resistance (P) is as follows.

$$P = \frac{p_t}{k_s} \tag{7}$$

Where $p_t$ is compressive strength of roadside backfilling body at the age of 1d; $k_s$ is safety factor.

Based upon Formula (1) and Formula (7), roadside supporting resistance and the width of roadside backfilling body can be determined.

## 5. ROADSIDE SUPPORTING DESIGN AND ENGINEERING PRACTICE

Ground elevation of No. 512(5) working face in Xieyi mine of Huainan is +19.0 m~+25.5 m, according to the distribution of coal seam in the section, the length is 1340 m along the strike of working face, the length of working face is 150 m~195 m; The level of haulage entry is -780 m, The mine belongs to deep mine; Average thickness of No. C15 coal seam is 0.39~1.31 m, the dip angle of coal seam is 19°~22°, and average value is 20°. Immediate roof is mudstone, average thickness is 2.3 m, brittle, broken; main roof is fine sandstone, average thickness is 2.5 m, crack is growing, hardness is very large; immediate floor is shale, average thickness is 1.5 m. In order to adopt Y-type ventilation in No. 512(5) working face, the technique of the gob-side entry retaining is applied to haulage entry. Haulage entry is of vertical wall and a half arch cross-section, clear width of 5.0 m, and clear height of 4.0 m.

### 5.1. *Width calculation of roadside backfilling body*

Backfilling system of pumping-over high water rapid hardening material is applied to roadside backfilling. Chief ingredients of roadside backfilling material are as follows: silicate, sand, fly ash, water and additive. Experimental compressive strength of roadside backfilling body is shown in Table 1.

Table 1. Compressive strength of roadside backfilling body

| day | 1 | 3 | 5 | 7 | 28 |
|---|---|---|---|---|---|
| compressive strength ($10^6$ Pa) | 3.5 | 9 | 10 | 12 | 14 |

According to occurrence conditions of roof strata in No. 512(5) working face, the parameters are selected as follows.

$\alpha = 20°$, $\gamma = 23$ KN/m$^3$, C = 1.56 MPa, H = 802 m, $\varphi = 35°$, k = 2.8, M = 2.8 m, u = 0.223, $p_x = 0.13$ MPa, $\sigma_0 = 1.5$ MPa, $\sigma_c = 13.8$ MPa, $p_t = 3.5$ MPa, a = 5.0 m, $k_s = 1.2$, $F_i = 0$ KN, S = 175 m, l = 10 m, $H_i = 2.3$ m, $\beta_i = 43°$, $\sigma_{ti} = 1.6$ MPa, $p_i = 0.15$ MPa, m = 6.

If safety factor is given to 1.2, it is suitable that the width of roadside backfilling body takes 2.5 m. Required roadside supporting resistance is $2.907\times10^6$ Pa.

### 5.2. *Design for advancing roadside combined support*

With the mining of working face, a gap is made in advance at the end of the working face, the length of the gap is 5 m, the width of the gap is 3.5 m, combined support of bolting, cable, steel strip and wire netting is used to support the gap roof.

(1) Bolting, which is attached to steel strip and wire netting, is adopted to support the gap roof, support parameters of bolting are as follows: the diameter of 22 mm, the length of 2500 mm, steel type of 20MnSi, lengthened anchoring, line-row space of 800×800 mm.

(2) Cables between bolting are installed according to the forms of "2-1-2", "1" indicates that one cable is installed to the middle of the gap, "2" indicates that row spacing between two cables is 1600 mm; the diameter of cable is 22 mm, its length is 6500 mm. The area of cable pallet is 300×300 mm. Cables are installed on the heels of roadway excavation.

(3) If deformation and pressure of the gap roof is both very large, individual hydraulic prop of DZ series and metal articulated roof beam of HDJA-1000 series should be used to reinforce the gap roof.

(4) Wire netting or thick wooden panel is used to prevent the gap from spalling rib, as is shown in the Figure 4.

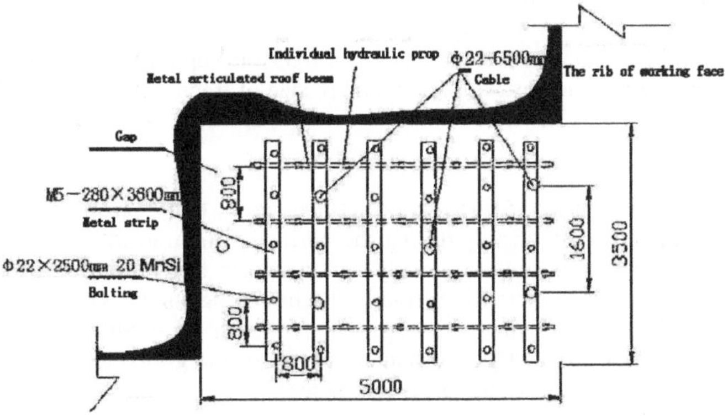


Figure 4. Supporting parameters of advancing roadside combined support

### 5.3. *The function relationships between roadside supporting resistance and influencing factors*

Based upon mechanical parameters of surrounding rock in No. 512(5) working face, the function relationships between P (Pa) and b (m), $\alpha$ (Radian), a (m), $L_m$ (m), $p_i$ (Pa) are expressed as

$$P = 14313b^2 - 365984b + 5\times10^6 \quad , \quad R^2 = 0.9865 \tag{8}$$

$$P = 3\times10^6 e^{-0.0193\alpha} \quad , \quad R^2 = 0.8376 \tag{9}$$

$$P = 3066195 - 1.059186p_i \tag{10}$$

$$P = 29878a + 3\times10^6 \quad , \quad R^2 = 0.9922 \tag{11}$$

$$P = 1698937 + 89236.86L_m + 1210.531L_m^2 \tag{12}$$

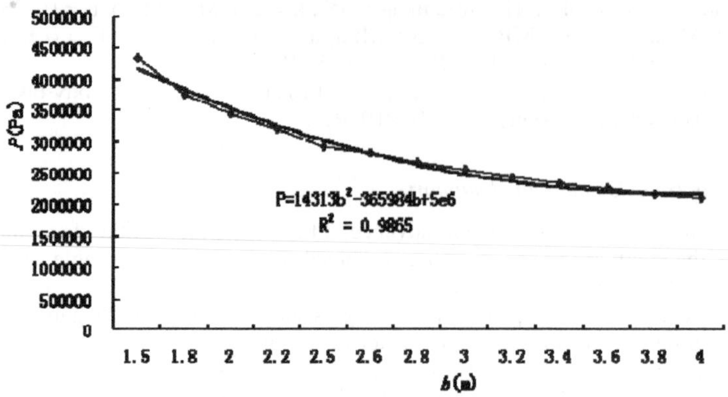


Figure 5. Relationship between the widths of roadside backfilling body and roadside supporting resistance

As can be seen from Figure 5, roadside supporting resistance gradually decreases with the increasing of the width of roadside backfilling body, but after the width of backfilling body exceeds 3.5 m, curve gradually inclines to flatten; it indicates that supporting effect of roadside backfilling body is not obvious only through increasing the width of roadside backfilling body.

According to Formula (9) and Formula (10), it indicates that roadside supporting resistance tends to decrease with the increasing of the dip angle of coal seam and supporting resistance of roadway roof.

According to Formula (11) and Formula (12), it indicates that roadside supporting resistance tends to increase with the increasing of roadway width and feature size of fractured roof strata.

### 5.4. *Analysis for supporting effect*

Pressure observation of over 2 months in No. 512(5) working face shows that bearing load of roadside backfilling body is of increasing tendency during the period of mining, in beginning period bearing load of roadside backfilling body is quickly increasing, in later period bearing load of roadside backfilling body is slowly increasing, finally it tends to be stable. The largest roof-to-floor convergence of roadway is 473 mm; the largest convergence of roadway-side is 551 mm. The success of industrial test proves that mechanical model of the gob-side entry retaining is suitable and width design for roadside support is reasonable.

## CONCLUSIONS

1) New mechanical model of roadside supporting resistance is established, it involves the dip angle of coal seam and supporting force of roadside coal seam. Moreover, the formula of roadside supporting resistance is derived, and function mechanism of roadway-in bolting support is analyzed.
2) Calculating formula of supporting force of roadside coal seam and feature size of fractured roof strata are derived by using the theory of limit equilibrium and plastic limit analysis method.
3) Roadside supporting design is made for haulage entry of No. 512(5) working face in Xieyi mine of Huainan, and the function relationships are explored between roadside supporting resistance and the width of roadside backfilling body, the dip angle of coal seam, roadway width, feature size of fractured roof strata, supporting resistance of roadway roof.
4) Pressure observation of over 2 months in No. 512(5) working face shows that mechanical model of roadside supporting resistance is suitable, and that width design for roadside backfilling body is reasonable.

ACKNOWLEDGEMENT

The project is sponsored by National Natural Science Foundation of China (No. 50774001). Part of the research is supported by fund of Outstanding Academic Innovation Team of Anhui University of Science and Technology.

REFERENCES

Gao Wei, Jian Xue-yun. 1997. Analysis of width plastic zone in left strip coal pillars by strip mining [J]. Shanxi Mining Institute Learned Journal 15(2): 142-147. (In Chinese)

Hua Xin-zhu. 2002. Discussion on Technology of Backfill along Goaf Side of Gateway in Fully Mechanized Caving Face [J]. Mine Construction Technology 19(2): 31-34. (In Chinese)

Hua Xin-zhu, Ma Jun-feng, Xu Ting-jiao. 2005. Study on controlling mechanism of surrounding rocks of gob-side entry with combination of roadside reinforced cable supporting and roadway bolt supporting and its application[J]. Chinese Journal of Rock Mechanics and Engineering 24(12): 2107-2112. (In Chinese)

Li Xue-hua. 2008. Control theory and technology of surrounding rock stability of gob-side entry diving in fully-mechanized sublevel caving mining faces [M]. Xuzhou: China University of Mining and Technology Press: 114-119. (In Chinese)

Qian Ming-gao, Shi Ping-wu. 2003. Mine stress and ground control [M]. Xuzhou: China University of Mining and Technology Press: 59-80. (In Chinese)

Sun Heng-hu, Zhao Bing-li. 1993. Theory and practice of the gob-side entry retaining [M]. Beijing: China Coal Industry Publishing House: 57-77. (In Chinese)

Wu Shi-yue, Guo Yong-yi. 2001. Gas control of mechanical mining coal face of high yield using Y model ventilation manner [J]. Xi'an University of Science & Technology Journal 21(3): 205-208. (In Chinese)

*International Mining Forum 2010, Liu et al. (eds) © 2010 Taylor & Francis Group, London, UK. ISBN 978-0-415-59896-5*

# Research on a new structure of bolt-end based on broken mechanism of FRP bolt

Yingming Li
*Key Laboratory of Coal Mine Safety and Efficient Exploitation of Ministry of Education; Anhui University of Science and Technology, Huainan, Anhui Province, China*

Nianjie Ma
*School of Resources and Safety, China University of Mining and Technology, Beijing, China*

ABSTRACT: The failure of bolt-end often occurs when roadway rib is supported by bolt. The broken mechanism of FRP bolt is revealed by mechanics analysis. Its broken mechanism can be summarized that bolt-end stress increases rapidly under an eccentric load, which leads to a break at this point and then spreads to the entire cross-section. A new structure of FRP bolt-end called metal sleeve-indentation is presented and manufactured based on its broken mechanism and by studying the structure of FRP bolt-end. These experimental results under normal and eccentric load show that the new FRP bolt with new bolt-end structure not only has perfect connection feature but also has proofed eccentric breaking capacity.

KEYWORDS: Coal rib supporting, FRP bolt, bolt-end, eccentric load, round indentation

## 1. INTRODUCTION

After gateway was constructed, concentrated stress comes into being in two sides of gateway and original three-dimensional stress of coal rib changes into biaxial stress. The abutment position of coal transfers to deep part of coal under concentrated stress (Ma 1995).

The rib spalling will happen in different degree since coal is very soft. Furthermore, the rib spalling of coal widens the gateway support span, which makes it get more difficult to support gateway.

Therefore, people lay more and more emphasis on coal rib supporting. It is very popular to strengthen coal rib with bolt. The coal rib bolt should meet the two pieces of demands. Firstly, coal rib could not be destroyed in order to keep the stability of coal rib and prevent rib spalling of coal wall when coal rib is supported. Secondly, the bolt could be cut by mining cutter. In other words, spark should be avoided when cutter comes across rib bolt (Li 2008).

In order to meet the first demand, the bolt needs to possess very big tension and certain shear intension; the shear intension needs to be as little as possible for the second demand. It is impossible for present bolt that the bolt has the two contradictory performances at the same time. For example, shear intension of metal bolt is so much that it could not be cut; as for bamboo and wooden bolt, their shear intension is so little that their anchoring forces are not enough to satisfy supporting demand.

FRP (Fiberglass-Reinforced Plastics) bolt makes it possible to solve the problem discussed above. At present, The FRP bolt produced by outside/in china could be cut and its tension is very high. It is found by a great deal of practice in mine that almost all bolt broke in bolt-end such as FRP, bamboo and wooden bolt but anchor top and body are rarely broken. So, it is necessary to study broken mechanics and FRP bolt structure.

2. FRP BOLT BROKEN MECHANISM

After the bolt is fixed on the rib, bolt-end often is affected under eccentric load. There are two conditions: one is because the surface of the rib is not perpendicular to the drilling hole, as shown in Figure 1 (a), the other is because rib surface is not smooth shown as Figure 1(b). So, the FRP bolt-end under this eccentric load is analyzed theoretically as follows.

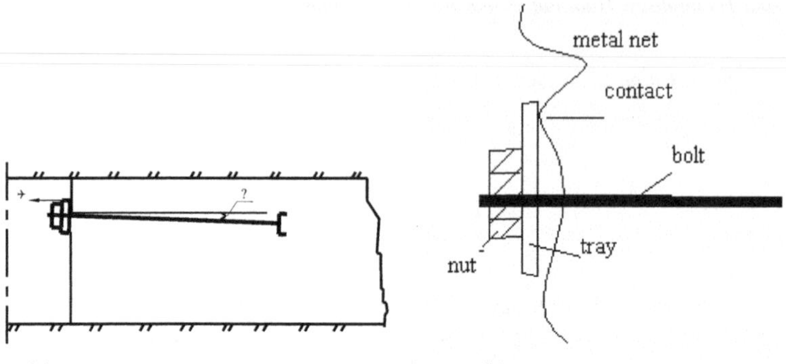


(a) (b)
Figure 1. The bolt construction with the eccentric load

Similar to that stated by Kong et al., we postulated the hypothesis that the bolt-end is fully affected by the eccentric load which is a concentrated force at the inscribed circle of the bolt nut (Kong 2003). The calculation model is shown as Figure 2.

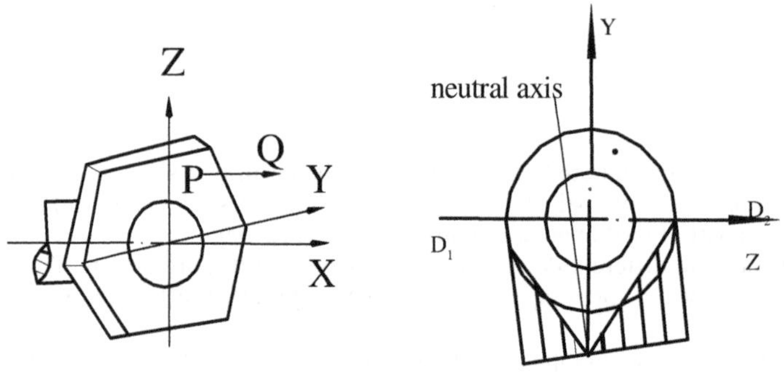


(a) (b)
Figure 2. Mechanics model of bolt-end under eccentric load

The axis of the bolt serves as the X axis, while the two main inertial axes centered in Figure 2 (a) and (c) are the Y and Z axes. It is assumed that the eccentric tension Q is parallel to the axis of the bolt body, whose working point is in the first quadrant and its coordinates are $y_Q$ and $z_Q$, as shown in Figure 2 (a). The eccentric tension Q is simplified from the point ($y_Q$, $z_Q$) to ($y_Q$, 0) and then to the centre of the nut (0, 0). After that the eccentric tension Q is transformed into a couple of bending moments ($M_y^0$ and $M_z^0$) which act on the XZ and XY planes respectively.

Given the mechanisms of material, the stress at any point (y, z) in the cross section of the bolt is as follows:

$$\sigma = \frac{Q}{A} + \frac{M_y z}{I_y} + \frac{M_Z y}{I_z} = \frac{Q}{A}(1 + \frac{z_Q z}{i_y^2} + \frac{y_Q y}{i_z^2}) \tag{1}$$

Where A is the applicable sectional area of the bolt-end and $i_y$ and $i_z$ are the inertial radii for the Y and Z axes.

For any point ($Y_0$, $Z_0$) on the neutral axis, the equation of the neutral axis is derived as:

$$1+\frac{z_Q z}{i_y^2}+\frac{y_Q y}{i_z^2}=0 \tag{2}$$

From Equation(2), the neutral axis is a beeline which does not cross the section centre O. Providing $y_0 = 0$ or $z_0 = 0$, the intercept of the neutral axis with the Y and Z axes is easily obtained as follows:

$$a_y=-\frac{i_z^2}{y_Q}, a_z=-\frac{i_y^2}{z_Q} \tag{3}$$

In order to determine the magnitude of direct stress at the most dangerous point on the bolt-end cross section, we assumed that Q acts on point P/ which is in Z axis and in the inscribed circle of nut. All points on the circle of the bolt-end cross section are dangerous points of which point D, where Y is 0, is the most dangerous point.

From Equation (1), the direct stress of point D is

$$\sigma_D=\frac{Q}{A}+\frac{Q\cdot\frac{s}{2}\cdot\frac{d}{2}}{\frac{\pi d^4}{64}}=\frac{Q}{A}\cdot(1+4\frac{s}{d}) \tag{4}$$

Where A is the diameter of the inscribed circle of the nut and d the nominal diameter of the bolt-end screw thread.

To nuts M16, M18, M20 in which each number refers to d in Eq (4), their corresponding inscribed diameters(s) are 24, 27 and 30, where s/d is usually 1.5. We obtained $\sigma_D$=7σt where $\sigma_t$ is stress on the bolt-end. If we ignore the effect of s, then $\sigma_D$=5$\sigma_t$. It can be shown that, if the eccentric load of the bolt-end is considered as a concentrated load, the magnitude of direct stress ranges from 5 to 7 times the general load without the eccentric load. Obviously, the eccentric load is worse at the bolt-end.

## 3. THE NEW FRP BOLT FEATURE

Based on the FRP bolt-end failure mechanism, there are two technological ways to prevent the bolt from eccentric failure. The first is to avoid or reduce the eccentric load; the second is to enhance the anti-eccentric-load capacity of the bolt.

Now the structure of FRP bolt-end mainly includes injection molding and direct-right-hand thread molding. Because technical demand of the injection molding tail is so strict that before injecting, it need peel about 1mm firstly, cut a thread with 1mm in depth then, inject last, which makes bolt-end area of section reduce 15% in compare with bolt's and lead to bolt tail intensity reduced, the bolt tail intension which should be strengthened declines greatly in practice; The diameter of direct-right-hand thread is small, shear strength is low and it is not proper (Li 2007).

Under this condition, the new bolt-end structure, which is metal bushing and indentation (as shown in Figure 3), puts forward. By profound research and fully considering the FRP bolt-end should have both very high tensile strength and much shear strength. Its principal structure characteristics include: bolt-end has metal bushing with thread, the bushing was connected with bolt body by one or several fillisters with taper, the bushing is pressed in body of FRP bolt and they scarf each other and make it as a whole, which not only make full use of the high tensile strength of the FRP bolt body, but also make use of the high shear strength of metal bushing. This kind of bolt tail structure meet the coal-side supporting requirements, it will not be broken due to large shear strength of bushing in normal operating conditions, at the same time, it will not bring about spark because the shear strength of FRP bolt body is low and it is cut more easily in process of the cut mining machine work meeting bolt in coal rib.

Figure 3. Bolt-end Structure with metal bushing and round indentation

The FRP bolt with this kind of structure does not fail in practical application. At the same time, this structure has following characteristics due to indentation joint. Firstly, it adopts taper joint face and makes full use of high intensity of mechanical connection. It also overcomes shortage of stress concentration due to normal mechanical connection, because force transfer along the bigger transfer surface. Secondly, working load changes direction by taper face and makes much circumferential normal pressure on the combining face, so it increases shear strength of bone line. Thirdly, the working load and radial normal pressure and utmost shear intension increase in linearity with working load. When working load reaches a certain value, the combining face generates shear and sliding and bone line failure, but the force transferring in bolt body and bushing further increase with enlarge of the working load.

The relationship of bolt extension and drawing force is shown as Figure 4. The new structure not only provides certain working resistance, but adapts to the enclosing rock distortion by sleeking relatively between bolt body and bushing. It makes up the defect of low extension and is very well fit for coal rib support. It can be seen that the bolt intension can be kept because of good connection between metal bushing tail and bolt body when drawing load descends.

The bushing covers uncured soft bolt-end at once after producing bolt body. The bushing is suppressed to groove by machine for bolt-end pressing. The bolt is laid in solidifying box to make bolt-end solidified. So the new structure is realized.

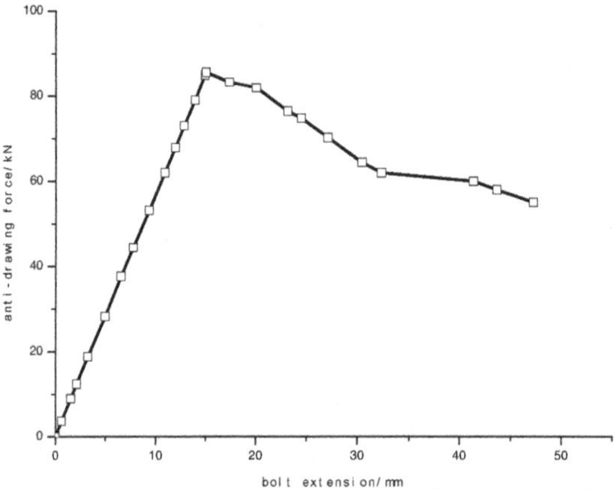


Figure 4. The relationship between bolt extension and drawing force

In order to determine anti-eccentric-load capacity of the new bolt, the eccentric load is simulated with different eccentric angle. Relationship curve between anti-drawing-force and extension under different eccentric angle curve is shown as Figure 5. The metal sleeve-indentation bolt under eccentric load have similar trend with under normal load and still there are rapid increasing resistance and decreasing resistance stage. The maximum anti-drawing-force goes down with eccentric angle improving. The maximum anti-drawing-force is 76 kN when eccentric angle is 3°. The maximum anti-drawing-force is 64 kN when eccentric angle is 15°. The experiment results show the new FRP bolt under eccentric load still meets coal rib support demand.

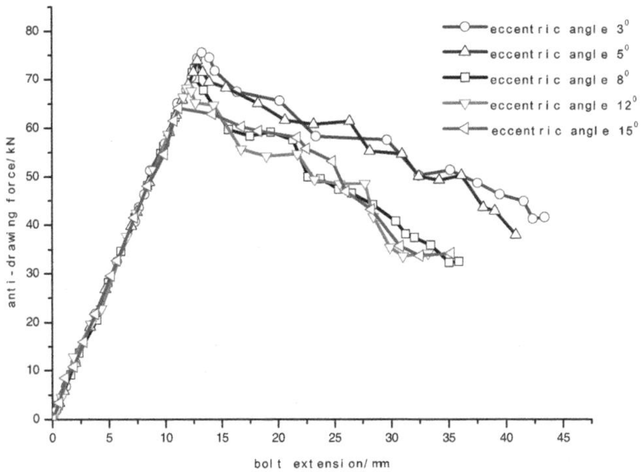

Figure 5. Relationship between drawing strength and extension of different taper indentation blot

4. CONCLUSIONS

(1) Eccentric load in the course of rib support is unavoidable. FRP bolt rods have great extension strength which is generally 600~700 MPa. However, FRP bolts have no capacity of plastic deformation and extensibility is just 1%~1.5% low. Bolt-end stress increases rapidly under an eccentric load, which leads to a break at this point and then spreads to the entire cross-section.

(2)A new structure of bolt-end is presented and its principal structure characteristics are as below: bolt-end has metal bushing with thread, the bushing was connected with bolt body by one or several fillisters with taper, the bushing is pressed in body of FRP bolt and they scarf each other and make it as a whole, which not only make full use of the high tensile strength of the FRP bolt body, but also make use of the high shear strength of metal bushing.

(3)The contact feature of the new FRP bolt-end is good under normal load. The experiment results show the new FRP bolt under eccentric load still meets coal rib support demand. The maximum anti-drawing-force is 76 kN when eccentric angle is 3°. The maximum anti-drawing-force is 64 kN when eccentric angle is 15°.

ACKNOWLEDGEMENT

The research is sponsored by fund of Anhui Province College Natural Science Key Projects (No. KJ2010A100).

REFERENCES

Kong H, Wang M S, Ma N J. 2003. Study on breaking mechanism of rock bolt-end. Chinese Journal of Rock Mechanics and Engineering 22(3): 383-386

Li Ying-ming. 2007. Study on Failure Mechanism and New Structure of FRP Bolt-end [Ph.D. dissertation]. Beijing: China University of Mining & Technology. (In Chinese)

Li Ying-ming, Ma Nian-jie. 2008. Research on the new FRP bolt & its application [J], Proceeding of The Third International Symposium on Modern Mining and Safety Technology: 1042-1045. (In Chinese)

Ma Nian-jie, Hou Chao-jiong. 1995. Gateway and preparation roadway rock pressure theory and application [M]. Beijing: Coal industry press. (In Chinese)

International Mining Forum 2010, Liu et al. (eds) © 2010 Taylor & Francis Group, London, UK. ISBN 978-0-415-59896-5

# Research on deformation and failure laws of aquifuge of coal seam floor during pressurized working

Zhaoning Gao
*School of Energy and Safety, Anhui University of Science and Technology, Huainan, Anhui Province, China*

Xiangrui Meng, Xiangqian Wang
*Key Laboratory of Coal Mine Safety and Efficient Exploitation of Ministry of Education, Huainan, Anhui Province, China*

ABSTRACT: Taking into consideration of the geological conditions of Face 1028 of Suntuan Coal Mine of Huaibei Group Limited of Mining Industry, the present research on the simulation of the dynamic process of the coal seam under working is conducted by turning to numerical analysis, with the aim of observing the laws governing the floor stress and displacement variations as the working face advances. It is concluded that the floor stress and displacement are always dynamic, leading to the difference between the displacements of the front and back of working face. Through analysis, the difference is attributed to the fact that before working, the stress is accumulated; after working, the stress is relieved and then restored, hence the added stress. Then theoretical analysis is turned to in order to analyze the damage factors of the floor under working, leading to the conclusion that the depth of floor failure serves as a function of the internal friction angle $\varphi$, the uniaxial compression strength $\sigma_c$, the maximum compression strength coefficient n and the working depth H. Based on the results, measures are put forward for the reduction of the floor failure. For the sake of earliest possible monitoring, prediction and prevention of the floor water burst from the working face and goaf, investigations are conducted into the floor failure depth and the working fissures for Face 1028 of Suntuan Mine by turning to direct current conductivity CT.

KEYWORDS: pressurized working, displacement difference, floor deformation and failure

## 1. INTRODUCTION

The distribution of the stress, and the deformation and failure of the roof and floor of the coal face under working have been the major concern of the safety production of coal mines. Therefore, the correct determination of the failure depth of the floor under the impacts of working plays a vital role in the prediction of the water resistance capacity of the floor (Zhang 2006). When it comes to the coal seams being threatened more seriously by water burst from floor, during the working, attention should be more focused on the study of the floor failure laws. In recent years, the mechanic methods shed much light on the recognition of floor deformation failure of the face under working. However, pure mechanic analysis cannot effectively bring us to the knowledge of the laws (Wang 2000, 2004, Bai 1997).

In order to study the distribution of the stress, the deformation and failure of the floor, we have conducted numerical simulation tests and based on the test results, come to the clear conclusion in this regard by turning to the integration of theoretical analysis and on-spot testing.

## 2. THE GEOLOGICAL CONDITIONS OF THE WORKING FACE

Face 1028 is situated in Suntuan Mine with a strike of 1742 m and an inclination of 180 m. The coal seam ranges from 1.91 m ~ 4.11 m in thickness with an average of 3.42 m. The coal seam is simply structured with a layer of carbonaceous mudstone partings in between; of the thickness of about 0.3 m. The coal seam angle is 17° with a hardness of 2. In the roof and floor, there are mudstone, sandstone, silt and fine sandstone. The average distance between Seam 10 and Seam 82 is 76.4 m and the downward distance from the limestone water ranges between 1.69 m ~ 68.31 m with an average of 58.38 m. The long wall mechanized working is adopted with the roof fully collapsing.

With the data obtained from the limestone surface boreholes, Limestone 3 and Limestone 4 and aqueous, with a water head of 20 m in height. The underground inflow for boreholes 1-1, 1-2 and 2-1 reaches its maximum of 8 $m^3/h$ but with a pressure of 4.0 MPa.

## 3. THE SIMULATION ANALYSIS OF THE FLOOR WATER BURST VALUES

### 3.1 *Model Establishment and Parameter Selection*

According to the geological conditions of Suntuan Mine, the strike length, the inclination width and the height of the simplified model are 360 m, 360 m and 250 m respectively. In the calculation, the immediate strata with similar physical characteristics will be regarded as a single stratum. According to the mechanic tests of the samples, the model for calculation is simplified into a structure composed of 18 strata with the mechanic parameters shown in Table 1.

The whole model is divided into 95480 three-dimensional units with 103194 nodes. Limited by the calculation capacity, the model is supposed under a vertical load or the gravity of the rock bo-dy up to the surface. The pressure resistant water acts on the bottom of the aquefuge of the floor and the rock body is only under its own gravity. The boundary conditions for the geological model are as follows. Four sliding bearings are placed on the four sides for the control of the horizontal displacement, but with vertical displacement allowable, and a sliding bearing is placed at the bottom for the control of the vertical displacement. However, the top is free of displacement control. And in the experiment, the working face floor pressure is assumed as 4.0 MPa.

Table 1. Mechanical parameters of rock mass

| Lithology | Bulk Modulus /GPa | Shearing Modulus/GPa | Density /kg/$m^3$ | Cohesion /MPa | Uniaxial tensile strength /MPa |
|---|---|---|---|---|---|
| Siltstone | 8.798 | 6.218 | 2664.22 | 17.504 | 1.29 |
| Mudstone | 8.842 | 6.061 | 2747.198 | 18.9 | 0.61 |
| Seam 10 | 1.189 | 1.118 | 1574.237 | 1.25 | 0.345 |
| Mudstone | 7.808 | 6.632 | 2619.3 | 18.9 | 0.61 |
| Siltstone | 8.798 | 6.218 | 2664.22 | 17.504 | 1.29 |
| Fine Sandstone | 8.798 | 6.218 | 2646.739 | 17.504 | 1.85 |
| Medium-sized sandstone | 11.033 | 8.834 | 2663.517 | 26.9 | 3.46 |
| Siltstone | 8.798 | 6.218 | 2664.22 | 17.504 | 1.29 |
| Mudstone | 7.808 | 6.632 | 2619.3 | 18.9 | 0.61 |
| Limestone | 22.619 | 11.047 | 2090 | 6.72 | 1.58 |

### 3.2 *Analysis of the results*

As the working face advances from the open cut, the rock stratum stress field is thrown out of balance, leading to the redistribution of the stress. Before the breakage, the rock in roof of the goaf is combined with the front and rear supports of the face to a systematic structure to bear the load of the overburden and convey the load to the surroundings, resulting in an abutment pressure. As the working face advances, the overburden will go through the stages of breakage, stabilization and destabilization. Then there will be a free space in the goaf for rock displacement, leading to the redistribution of the vertical stress in the surrounding areas of the working face. And consequently, in front of the working face, there will be a supporting pressure formed in advance. As the working face advances to a substantially distant point, the goaf behind it will be stabilized due to the compression and the supporting stress will gradually be restored to the original state. In this case, the pressure will contribute to the impacts on the stability of the surrounding spaces. During the advance of the working face, the floor of the seam being worked will be under the supporting pressure integrated with water pressure, leading to the formation in advance of the loaded compression zone horizontal to the floor, the formation of the pressure relief swelling zone near the working face and the formation of the re-stabilization zone in the goaf. Under the supporting pressure, the floor in depth can be divided into the severely affected, the affected and the slightly affected zones as shown in Figure 1 and Figure 2.

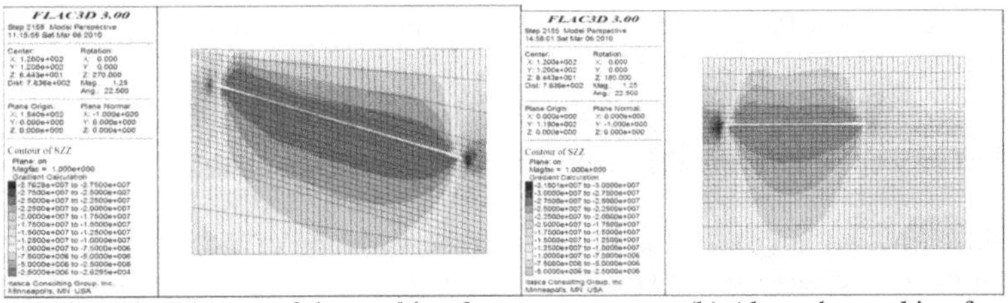

(a) Along the advance of the working face (b) Along the working face

Figure 1. The vertical stress distribution of the floor

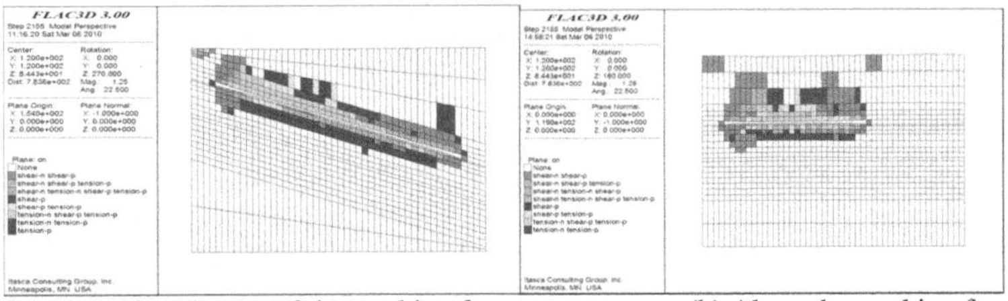

(a) Along the advance of the working face (b) Along the working face

Figure 2. Features of the fractured zone of the floor

(1) In case of normal advance, 50 m in front of the working face, the rock strata of the floor begin to change in stress under an added load. 20 m in front of the face, the stress increases drastically and the peak value appears 5 m in front of the face. The point serves as a watershed, behind which the stress drastically decreases. About 5 m behind the working face, the rock stress in the goaf reaches its minimum, indicating that the broken floor contributes to the stress decrease. With the collapse of the roof, about 10 m in the goaf behind the working face gradually increases but can never reach the value for the original rock, indicating the floor stress is still being relieved. In the place about 60 m behind the working face, the stress is basically stabilized, indicating that the collapsed waste is compressed to support the roof. Generally speaking, in places 2~20 m in front of the working face, the floor stress is apparently higher than the stress for the original rock strata. In the area, the floor rock strata are under the integrated pressure of the concentrated stress and water pressure. The front part is subject to the horizontal extrusion while the rear part is subject to the horizontal tension, leading to the vertical cracks and tensile fissures in the rock body. The stress is distributed in the area 2 m in front of and 10 m behind the working face as follows. In the area 2 m in front of the working face, under the supporting pressure, the coal body is in a relaxed failure, unable to pass on the supporting stress to the floor rock. Consequently, the stress for the floor is decreasing. In the area 10m behind the working face, the collapsed waste in the goaf and the rock strata in the roof have not formed an inte-

grated whole, unable to pass on the stress to the floor. Consequently, the floor stress is also decreasing.

(2) Differences in stress concentration and relief are witnessed at different floor depths. In front of the working face, the nearer to the floor, the greater the stress concentration; but behind the working face, the nearer to the floor, the greater the stress relief.

(3) Water bursts are related to the impacts of floor failure on the aquifer. The simulation results indicate that before the first pressure concentration, the floor failure is confined in range. With the advance of the working face and the maximization of overhang span of the main roof, the roof breaks, resulting in a balance of the Voussoir Beam and a bed separation in the roof. Consequently, the stress on the coal wall in front of the working face increases drastically and the first pressure on the roof appears. At this juncture, the failure reaches 12.0m in depth behind the open cut and it reaches about 10.0 m in depth near the working face, a failure of a moderate degree. However, as the first periodic pressure for the working face appears, the floor failure reaches its maximum in depth of about 13.0 m. As the working face advances to 200 m, the in-depth failure of the floor comes to be stabilized, with the maximum failure depth ranging 15 m~17 m.

## 4. A THEORETIC ANALYSIS OF THE FAILURE DEPTH FOR THE FLOOR

The floor failure induced by working varies in scope with the variation of the working scope and the stress distribution of the surrounding areas in the goaf. Taking into consideration of the stress distribution of the floor rock, the profile along the coal seam inclination is taken as the profile for calculation. According to the Saint Vansant Principle, the static equivalent force is substituted for the bearing stress for the profile along the coal seam inclination. Then combined with the Mohr-Coulomb failure criteria, the maximum floor failure depth h during the working period is obtained (Peng, Wang 2001):

$$h = \frac{(n+1)H}{2\pi}\left(\frac{2\sqrt{k}}{k-1} - \cos^{-1}\frac{k-1}{k+1}\right) - \frac{\sigma_c}{\gamma(k-1)} \tag{1}$$

$$k = \frac{1+\sin\varphi}{1-\sin\varphi} \tag{2}$$

where: n is the maximum stress concentration coefficient; H is the working depth (m); $\sigma_c$ is the uniaxial compression strength (MPa); $\gamma$ is the rock density ($MN/m^3$); and $\varphi$ is the internal friction angle.

For Face 1028 in Suntuan Mine, the maximum working depth reaches 528 m. Taking into account the litho logical characteristics of the rock and the results obtained from physical and mechanic tests, in addition to the scale effects, 1/4 of the test value is converted into the rock strength (Peng, Wang 2001) and the following data are arrived at: the average compression strength of the rock is 14.7 MPa, the internal friction angle is 15.5°, the rock density is 0.026 $MN/m^3$, and the calculated floor failure depth is 15.7 m. According to Formulae (1) and (2), h as a function of H, $\sigma_c$ and $\varphi$ are shown in Figure 3(a) ~ 3(d).

(1) In Figure 3(a) and Figure 3(b), the floor failure depth as a function of the internal friction angle $\varphi$ and the uniaxial compression strength are shown. Apparently, with the increase of the internal friction angle and the uniaxial compression strength, the floor failure depth will decrease during the working operation, pointing in the possibility that by improving the floor of the coal seam, the physical and mechanical characteristics of the rock in floor at various depths can be improved to different extents. In this way, the floor failure development can be effectively controlled.

(2) In Figure 3(c) and Figure 3(d), the floor failure depth as a function of the maximum concentration coefficient n and the working depth H are shown. Apparently, with the increase of the maxi-mum concentration coefficient n and the working depth, the floor failure depth will increase during the working operation. With the gradual decrease and the exhaustion of the coal resources, coal mining will develop at increasing depths. For this reason, the risk of water burst will increase. However, wherever the geological conditions are definite, the working depth will

also be definite. Therefore, any attempt at exercising constraints on working depths to control the floor failure development will apparently of course not be feasible.

The maximum concentration coefficient n is determined by such factors as the dynamic variation of the overhang span, the deformation and failure of the rock strata surrounding the working face and the technological parameters. Therefore, it is possible to inject water into the roof, to force the collapse of the roof, to shorten the working length and so on to reduce the maximum concentration coefficient n and to control the floor failure development.

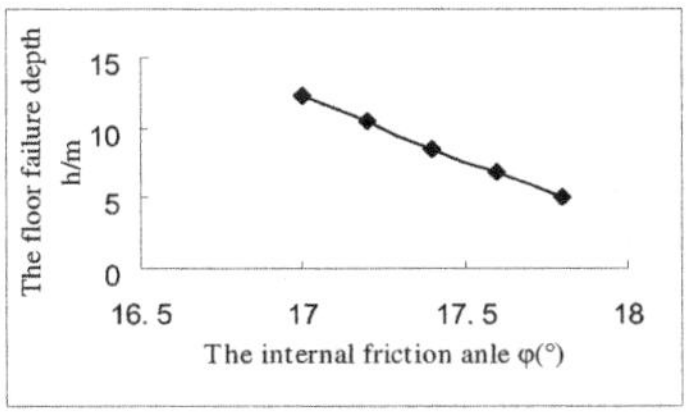


(a)

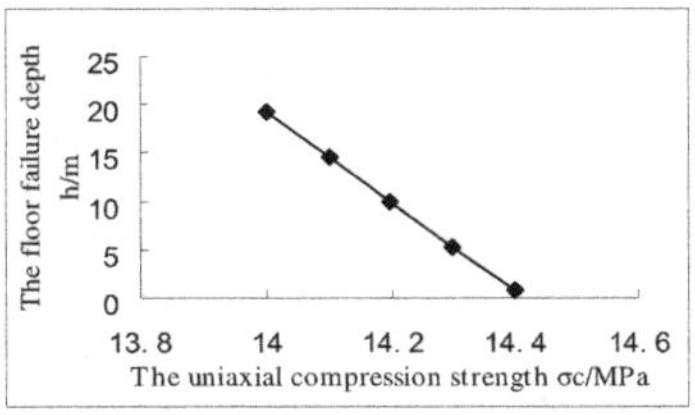


(b)

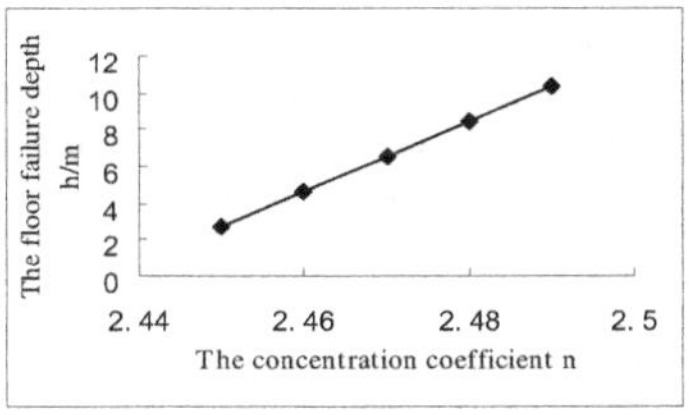


(c)

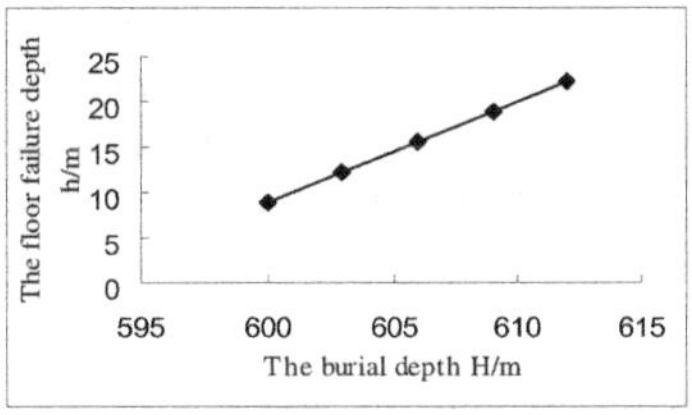


(d)

Figure 3. h as a function of φ H $\sigma_c$ and n

## 5. THE FIELD TEST

In addition to the numerical simulation and theoretic analysis, tests are conducted on spot of the floor failure depth by turning to the conductivity CT method. The test is conducted in the return airway of Face 1028, with two holes bored and the horizontal distance between the holes being 5m. Two copper electrodes are placed in the Hole 1 and Hole 2. In Figure 4, section graphs of the conductivity CT are shown when the holes are placed at the distances of 75 m and 31 m from the working face, when the working face advances near the holes and when the holes are behind the working face.

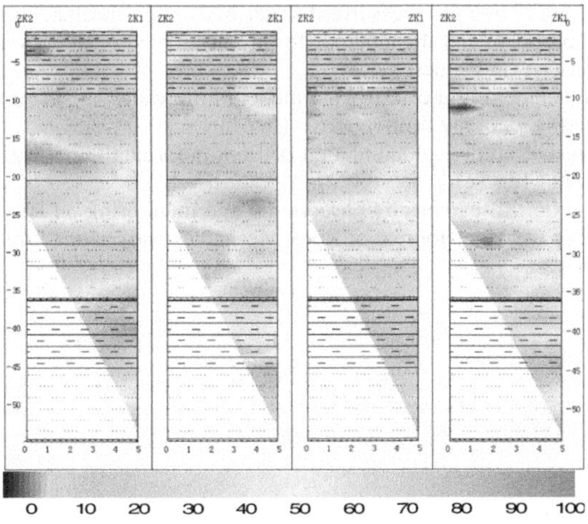


Figure 4. Conductivity contrasts obtained from four tests

As shown in Figure 4, with test results compared, a conclusion can be drawn that for Face 1028 of Suntuan Mine, under the integrated pressure of coal supporting and water confinement, the floor failure ranges between 0 ~ 17 m. 17 m ~ 33 m under the coal seam floor, affected by the working operation, the rock deforms elastically. However, with the advance of the working face, the elastic deformation will be restored to normal.

## 6. CONCLUSIONS

1) The laws governing the variation of the floor stress with the advance of the working face are simulated by turning to numerical analysis, leading to the conclusion that the displacement difference is induced by the stress concentration before working, the stress relief after working and the stress addition resulted from the stress restoration.
2) By applying theoretic analysis to the analysis of impacts of coal working on the floor failure, a conclusion is reached that the floor failure varies in depth with the variation of the internal friction angle φ, uniaxial compression strength σc, the maximum concentration coefficient n and the working depth H. Based on the analysis, ideas are proposed as measures to counter floor failure.
3) By applying the DC conductivity CT, comprehensive dynamic surveys are conducted of the floor failure depth and the working cracks of Face 1028 in Suntuan Mine, with the aim of the earliest possible monitoring, prediction and prevention of the floor water burst in the working face and the goaf.


## ACKNOWLEDGEMENT

The research is sponsored by the Special Research Funds of the Ph. D Program of Institutions of Higher Learning (No.200803610001), the New Instructor Funds of The Special Research Funds of the Ph. D Program of Institutions of Higher Learning (No. 20093415120001) and the Research Funds for Excellent Youngsters of Anhui Province (No.10040606Y31), the Key Project of Science and Technology of Anhui Educational committee (No.KJ2010 A081; No. KJ2009A66)


## REFERENCES


Bai Chen-guang, Li Liang-jie, Yu Xue-fu. 1997. On the establishment of a cusp catastrophic model for in-stability of the key stratum of the water-bearing floor. Journal of China Coal Association 22(2): 149-154. (In Chinese)

Peng Su-ping, Wang Jin-an. 2001. Safety working above the pressurized water body – on mechanisms for the failure of the floor under the tensile coal seam and the methods for predication and prevention of the water burst. Beijing: Press of Coal Industry. (In Chinese)

Wang Lian-guo, Song Yang. 2000. On the establishment of a catastrophic model of the water burst from the floor of the coal seam. Journal of Engineering Geology 8(2):160-163. (In Chinese)

Wang Lian-guo, Bi Shan-jun, Song Yang. 2004. On numerical simulation of floor deformation damage laws. Mine Pressure and Roof Management. (In Chinese)

Zhang Ping-song, Wu Ji-wen, Li Sheng-dong. 2006. Researches by observation and monitoring on laws governing floor failures of the coal seam under working. Journal of Rock Mechanics and Engineering 25 (Additional Issue 1): 3010-3013. (In Chinese)

International Mining Forum 2010, Liu et al. (eds) © 2010 Taylor & Francis Group, London, UK. ISBN 978-0-415-59896-5

# Experimental study on strength development law of filling body in gob-side entry retaining

Xiaoyu Lu
*School of Energy and Safety, Anhui University of Science and Technology, Huainan, Anhui Province, China*

Liang Yuan
*National Engineering Research Center for Coal Mine Gas Controlling, Huainan, Anhui Province, China*

Xinzhu Hua
*Key Laboratory of Coal Mine Safety and Efficient Exploitation of Ministry of Education, Huainan, Anhui Province, China*

ABSTRACT: The uniaxial compressive and splitting tension strengths exhibit two important mechanical properties of filling body in gob-side entry retaining. Therefore, experimental study on the strength development law of filling body was conducted. The specimens were fabricated in Xieyi mine. The uniaxial compressive and splitting tension strengths were tested at different curing aging time of 1, 3, 7 and 28 days. Finally, regressive formulas are given to describe the relationship between the strength and aging time. The strength increases very quickly in the first 3 days, and the strength in 3 to 7 days increases a bit more slowly than in the first 3 days, and the strength in the latter 21 day increases much more slowly than in the first 7 days. The uniaxial compressive and splitting tension strengths of 3 days aging time and 7 days aging time can reach about 35% and 60% of that of 28 days aging time, respectively.

KEYWORDS: Gob-side entry retaining, uniaxial compressive strength, splitting tension strength

## 1. INTRODUCTION

Technology of gob-side entry retaining can realize protection of roadways without coal-pillar (Hua, 2006). Most of the achievements are about the moving rule of surrounding rock during the gob-side entry retaining, the support theory and technology inside the roadway and beside the roadway have been obtained (Sun, Zhao 1993; Yuan 2008). However, little literature about the mechanical properties of the filling body is reported. The uniaxial compressive and splitting tension strengths exhibit two important mechanical properties of filling body in gob-side entry retaining.

The strength development law of filling body plays a very important role in the safety and stability of gob-side entry retaining. Therefore, combined with the engineering practice of Xieyi mine in Huainan, experimental research on the strength development law of filling body is con-ducted. The research has an important instructive meaning on the work of gob-side entry retaining.

## 2. EXPERIMENT DESIGN

The materials of filling body are composed of cement, fly ash, gravel, sand, water and compound admixture (Yuan 2008). By the mixing proportion optiminal design, mass ratios of each component include the following : cement (10 ~ 20%), fly ash (7 ~ 40%), gravel (15 ~ 40%), sand (15 ~ 25%) and water (10 ~ 30%). Compound admixture is 0.5 ~ 2.0% of the total mass of cement and fly ash. The mixing proportion is the same as that of Xieyi mine in Huainan city.

## 3. UNIAXIAL COMPRESSIVE STRENGTH DEVELOPMENT LAW OF FILLING BODY

Uniaxial compressive strength is one of the most basic and important mechanical parameters of filling materials. And it determines other mechanical properties of filling body, such as elastic modulus, peak strain, and ductility index. Therefore, it is important to research on the compressive properties of filling body.

### 3.1. *Specimen size and test method*

After the materials of filling body have been mixed with water, they would be fed into the mould with a size of 100×100×100 mm. Before the formwork been removed, the specimens must be cured in underground mine for 1 ~ 2 days. One day ahead of the uniaxial compressive test, specimens were drilled into cylinder specimen with a size of 50×100 mm by using core-drilling machine, and the two end faces of each specimen must be grinded smoothly. Each group has 4 specimens. According to the Chauvenet Criterion (Xiao 1985), suspicious data can be rejected. Then the average value of each group is used as the uniaxial compressive strength. With the rock mechanics testing system (Figure 1), the uniaxial compressive test was done by displacement control method at a rate of 0.02 mm/s.

### 3.2. *Test results and analysis*

Test results of uniaxial compressive strength at different curing aging time of 1d, 3d, 7d and 28d are shown in table 1.

Table 1. Test results of compressive strength of filling body

| Aging time | Numbering | | Diameter | Height | Compressive strength | | Notes |
|---|---|---|---|---|---|---|---|
| day | | | mm | mm | MPa | MPa | |
| 1 | A | AD-1 | 47.36 | 92.30 | 4.201 | 5.28 | Actual aging time is 2d. |
| | | AD-2 | 48.02 | 97.96 | 5.941 | | |
| | | AD-3 | 47.74 | 95.92 | 5.698 | | |
| 3 | B | BD-1 | 48.24 | 96.38 | 11.337 | 11.229 | Actual aging time is 4d. |
| | | BD-2 | 48.56 | 96.82 | 12.332 | | |
| | | BD-3 | 48.86 | 97.38 | 8.832 | | |
| | | BD-4 | 48.08 | 97.24 | 12.415 | | |
| 7 | C | CD-1 | 47.44 | 97.82 | 21.906 | 17.397 | Actual aging time is 8d. |
| | | CD-2 | 48.72 | 97.84 | 19.139 | | |
| | | CD-3 | 48.16 | 98.12 | 17.797 | | |
| | | CD-4 | 48.24 | 96.80 | 10.746 | | |
| 28 | D | DD-1 | 48.60 | 94.08 | 16.096 | 24.392 | - |
| | | DD-2 | 48.74 | 93.54 | 29.114 | | |
| | | DD-3 | 48.72 | 98.58 | 28.473 | | |
| | | DD-4 | 48.74 | 97.32 | 23.883 | | |

From the Figure 2, it can be seen that there is a very marked logarithmic relationship between compressive strength of filling body and curing aging time. The formula of compressive strength and curing aging time can be obtained by regression.

$$f_c = 7.2358\ln(x) + 1.0233,\ R^2 = 0.9855 \quad (1)$$

Where $f_c$= compressive strength of filling body, MPa; $x$ =aging time of the specimen, d。

The uniaxial compressive strength, calculated by the regressive formula, at the aging time of 1, 3, 7 and 28 days are 1.02 MPa, 8.97 MPa, 15.10 MPa, and 25.13 MPa, respectively. It can be found that the compressive strength increases very quickly in the first 3 days, and the strength of 3 days aging time can reach 35.7% of that of 28 days aging time. From 3 days to 7 days, the strength increases a bit more slowly than the first 3 days, and the strength of 7 days aging time can reach 60.0% of that of 28 days aging time. After the aging time of 7 days, the strength increases much more slowly than the first 7 days, and the strength in the latter 21 day only increases 40.0% of that of 28 days aging time.

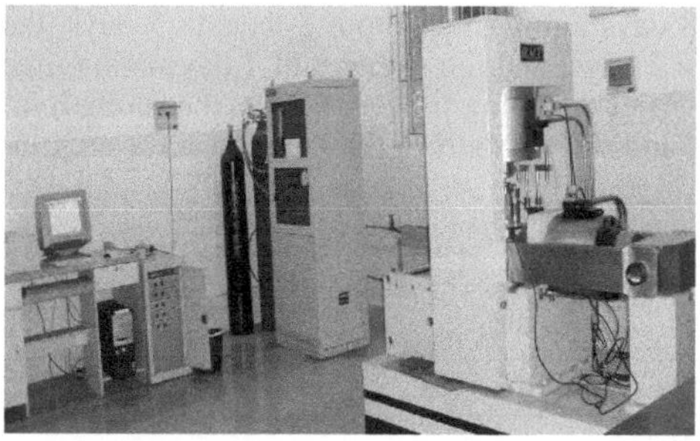

Figure 1. The rock mechanics testing system

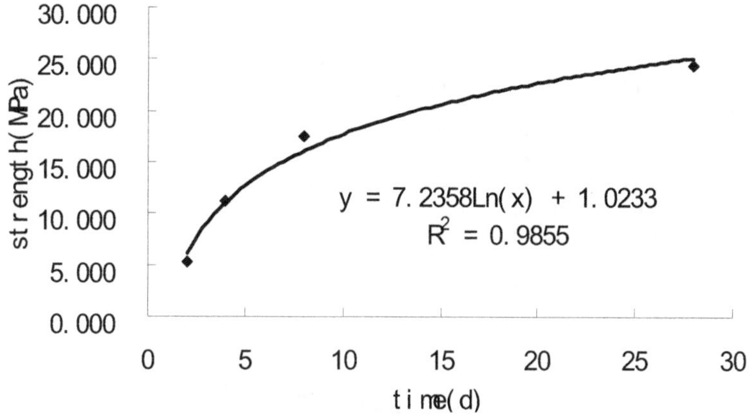


Figure 2. Relationship between uniaxial compressive strength and aging time

## 4. SPLITTING TENSION STRENGTH DEVELOPMENT LAW OF FILLING BODY

Damage, crack and their development always have great relationship with tension strength. So tension strength is an important mechanical parameter. Therefore, splitting tension method was adopted to test the tension strength of filling body.

### 4.1. *Specimen size and test method*

Specimen size is 50×40 mm. According to the Chauvenet Criterion, suspicious data can be rejected. The average value of each group is used as the tension strength. The splitting tension strength test was done by displacement control method at a rate of 0.01 mm/s.

### 4.2. *Test results and analysis*

Test results of splitting tension strength at different curing aging time of 1d, 3d, 7d and 28d are shown in table 2.

In the Figure 3, it can be seen that there is a very marked logarithmic relationship between splitting tension strength of filling body and aging time. The formula of splitting tension strength and aging time can be obtained by regression.

$$f_t = 0.5241\ln(x) + 0.0458, \quad R^2 = 0.9513 \tag{2}$$

where: $f_t$ = splitting tension strength of filling body, MPa ; $x$ = aging time of the specimen, d.

The splitting tension strength, calculated by the regressive formula, at the aging time of 1, 3, 7 and 28 days are 0.046 MPa, 0.621 MPa, 1.065 MPa, and 1.792 MPa, respectively. It can be found that the splitting strength increases very quickly in the first 3 days, and the strength of 3 days aging time can reach 34.68% of that of 28 days aging time. From 3 days to 7 days, the strength increases a bit more slowly than the first 3 days, and the strength of 7 days aging time can reach 59.46% of that of 28 days aging time. After the aging time of 7 days, the strength increases much more slowly than the first 7 days, and the strength in the latter 21 days only increases 40.54% of that of 28 days aging time.

Table 2. Test results of splitting tension strength of filling body

| Aging time | Numbering | | Diameter | Height | Tension strength | | Notes |
|---|---|---|---|---|---|---|---|
| day | | | mm | mm | MPa | MPa | |
| | | AP-1 | 47.62 | 41.24 | 0.198 | | |
| 1 | A | AP-2 | 47.50 | 43.54 | 0.263 | 0.165 | |
| | | AP-3 | 47.60 | 41.36 | 0.098 | | |
| | | AP-4 | 47.60 | 43.60 | 0.102 | | |
| | | BP-1 | 48.40 | 45.18 | - | | Data file destroyed. |
| 3 | B | BP-2 | 48.14 | 41.04 | 0.362 | 0.380 | |
| | | BP-3 | 48.12 | 39.18 | 0.398 | | |
| | | CP-1 | 48.34 | 37.24 | 0.989 | | |
| 7 | C | CP-2 | 48.52 | 44.20 | 1.223 | 1.168 | |
| | | CP-3 | 48.68 | 41.88 | 1.079 | | |
| | | CP-4 | 48.48 | 41.36 | 1.379 | | |
| | | DP-1 | 48.64 | 43.34 | 1.996 | | |
| 28 | D | DP-2 | 48.48 | 40.52 | 1.403 | 1.812 | |
| | | DP-3 | 48.60 | 42.56 | 2.123 | | |
| | | DP-4 | 48.30 | 42.28 | 1.727 | | |

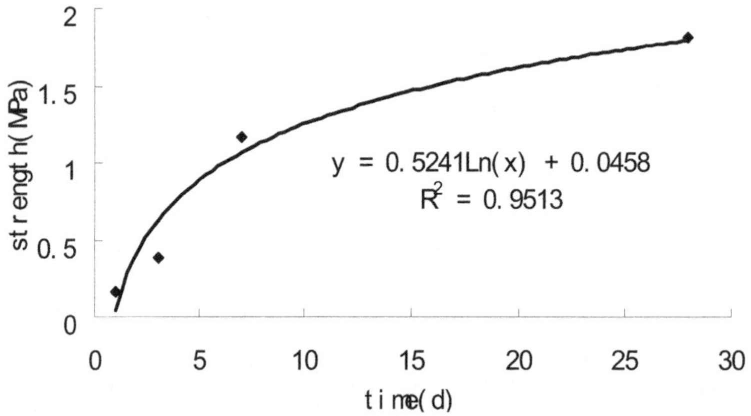


Figure 3. Relationship between splitting tension strength and aging time

## 5. CONCLUSIONS

The uniaxial compressive and splitting strengths were tested at different curing aging time of 1, 3, 7 and 28 days. The strength development laws of uniaxial compressive and splitting tension strengths were experimented and analyzed, and fitting formulas are proposed to describe the relationship between the strength and the curing aging time. Experimental results show that the strength increases very quickly in the first 3 days, and the strength in 3 to 7 days increases a bit more slowly than in the first 3 days, and the strength in the latter 21 days increases much more slowly than in the first 7 days. The uniaxial compressive and splitting tension strength of 3 days aging time and 7 days aging time can reach about 35% and 60% of that of 28 days aging time, respectively. The results of this research have an important instructive meaning on the work of gob-side entry retaining.

## ACKNOWLEDGEMENT

The study is sponsored by National Natural Science Fund of China (50774001).

## REFERENCES

Hua Xin-zhu. 2006. Development status and improved proposals on gob-side entry retaining support technology in China. Coal Science and Technology 34(12): 78-81. (In Chinese)

Sun Heng-hu, Zhao Bing-li. 1993. Theory and coal practice of gob-side entry retaining. Beijing: China Industry Publishing House. (In Chinese)

Xiao Ming-yao. 1985. Theory and application of errors. Beijing: China measure Publishing House. (In Chinese)

Yuan Liang. 2008. Theory and practice of coal mining and gas extraction without coal-pillar. Beijing: China Industry Publishing House. (In Chinese)

International Mining Forum 2010, Liu et al. (eds) © 2010 Taylor & Francis Group, London, UK. ISBN 978-0-415-59896-5

# Wet shotcreting dual-mode design and dynamic zoning support mechanism and technology for deep shaft

Yunhai Cheng
*Key Laboratory of Coal Mine Safety and Efficient Exploitation of Ministry of Education, Anhui University of Science and Technology, Huainan, Anhui Province, China*

Guandong Wang
*Technical Center of Xinwen Mining Industry Group, Xintai, Shandong Province, China*

Shanxin Guo
*Technical Center of Xinwen Mining Industry Group, Xintai, Shandong Province, China*

ABSTRACT: A dual-mode system suitable for wet shotcreting of deep shaft workings is proposed. Dynamic zoning support principle and technology has been proposed based on the mechanism of deep shaft concrete, making the wet shotcreting design can adapt to the rock surrounding conditions and stresses variation. Through system mechanics experiments, the result shows that the wet shotcreting and dry shotcreting have the same support effect while the design thickness of wet shotcreting is 83% of dry shotcreting, and the wet shotcreting saves 17% material. Application shows that the dust concentration of wet shotcreting is about 1/7 of the national standard (dust concentration of dry shotcreting is nearly 10 times of the national standard), realizing non-respirable dust basically and avoiding the hazard of workers' breathing respirable dust, which completely changes the phenomenon of dust diffusing situation in the past. Wet shotcreting reduces rebound ratio approximately 12%. The uniaxial compressive strength of wet shotcreting is about 1.48 times of dry shotcreting, which enhances the support effect. The above study shows that deep shaft wet shotcreting achieves an ideal effect.

## 1. INTRODUCTION

With the increasing coal mining depth every year, more and more Chinese mines enter into the deep mining (with mining depth greater than 800 m). As the deep shaft workings is under difficult conditions such as crushed zone and stress concentration zones, concrete body of half wet shotcreting cracks and spalls quickly, it is difficult to achieve the role of effective closure of rock workings.

In addition, ventilation system of deep shaft is complex and with long line, dry shotcreting produces a lot of dust which is difficult to remove out in time, resulting in poor working conditions which is a major hazard source for the occurrence of coal mine pneumoconiosis. So the current situation of deep shaft dry shotcreting must be changed quickly.

Wet shotcreting can solve the above-mentioned drawbacks of dry shotcreting. Dust concentration and rebound volume of wet shotcreting is greatly reduced and workings condition gets fundamental improvement. It can accurately control the water-cement ratio and improve the quality of shotcreting; production efficiency is improved. Therefore, wet shotcreting is an inevitable trend of mining development. But so far there is little successful application of coal mine wet shotcreting in China (Holzer 2003, Ding 2003, RMEM 2002, Dimmock 2003).

The mining depth of Juye Coal Mine in Shandong Province is near 1000 m. The main stress is about 1.8 times of the vertical stress. The concrete cracks soon after dry shotcreting, which isnot applicable for the crashed surrounding rock support of stress regions of deep shaft highland, which restrict the wet spraying of coal mine in China are: the equipment coordination is not therefore, the research team carried out wet shotcreting under the background of this project.

## 2. DUAL-MODE DESIGN OF WET SHOTCRETING OF DEEP SHAFT

Appropriate coordination of sets of equipment is a prerequisite for wet shotcreting. Key factors strong, which is difficult to form support mechanized production line. In abroad, wet shotcreting units consisting of adding, mixing, transporting and wet shotcreting operation comprehensive system, is a complex and bulky system formed by ground concrete adding station, concrete mixing station, concrete conveying equipment and concrete loading equipment, so it is difficult to use under the complex space of coal mine in China. Therefore, the research group proposed dual-mode design suitable for wet shotcreting of China's deep shaft operation.

### 2.1. *Mode 1: ground proportioning system + main workings centralized concrete mixing system +long-distance concrete supply wet shotcreting mode*

The test site is at the secondary auxiliary workings in the North Area. This workings is a development workings (5‰ uphill), with a net sectional area of 21.42 $m^2$.

Therefore, the research team designed the ground proportioning system + main workings centralized concrete mixing system + long-distance concrete supply wet shotcreting mode. Its characteristic is ground proportioning. The material is conveyed to underground centralized concrete mixing system. A number of rock excavation working surface use a common concrete mixing system, using a concrete tank specialized for coal mine use to convey mixture (adding hydrated agent) to each shotcreting locations. A number of large development open workings excavation (usually rock workings) at the same horizontal section is suitable for this model.

Figure 1 is the system schematic diagram of this model, of which 1-stone; 2-sand; 3-standard sieve; 4-proportioning machine; 5-dump truck; 6-loading machine; 7-mixing; 8-high efficiency slushing agent; 9-water; 10-hydration control agent; 11-mine concrete transport vehicle; 12-small wet shotcreting machine; 13-pressure fan; 14-compressed air; 15-dosing pump; 16-alkali-free liquid accelerating agent.

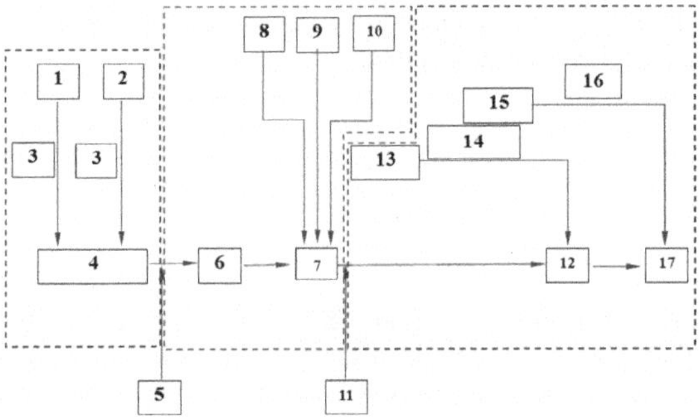


Figure 1. System schematic diagram of mode 1

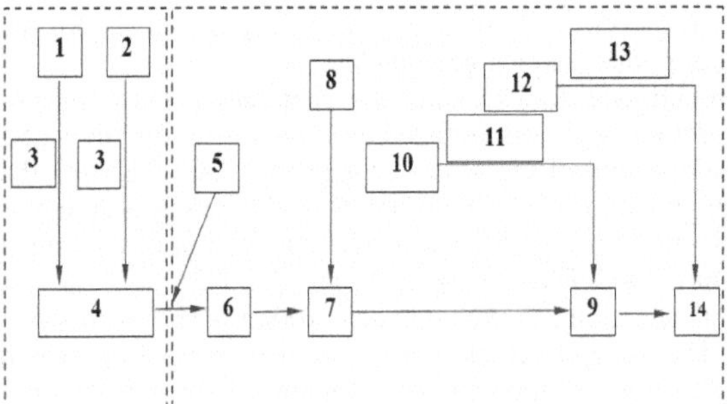


Figure 2. System schematic diagram of mode 2

The system operates as follows: use special proportioning machine for proportioning; load concrete material with dump truck; the dump truck carried the material to -810 underground mixing station through auxiliary vertical shaft; and then dump the material into the lifting hopper of mixing station; and then lift into the mixing bucket with adding cement; add water and proportional additive; mix 2–3 min; load into the concrete transport vehicle; transport to the construction site; carry out wet shotcreting.

### 2.2. *Mode 2: ground proportioning system + mixing, loading and wet shotcreting-integration model*

The test site is on an uphill at north wing of mine, more than 3000 m from pit bottom, with a net sectional area of 16.82 $m^2$, preparing for workings. Limited by the small space, the mode 1 uses centralized concrete mixing system and concrete transport vehicle specialized for coal mine use. It is difficult to set up such large mixing equipment in such a small area and conduct tank transport. Therefore, there is no domestic suitable wet shotcreting system under such condition.

Therefore, the double-helix mixing + double-helix piston pump integrated system has been developed for loading and spiral mixing, with a small size, which ensures the maximum wet shotcreting effect. An integrated mode formed by ground proportioning system + mixing, loading and wet shotcreting system, which makes the wet spray shotcreting system more simple and is an important supplement to model 1. It can adapt to the narrow space environment of coal mine, creating condition for the general promotion of wet shotcreting of deep shaft workings.

Figure 2 is the schematic diagram of wet shotcreting mode 2, of which, 1-stone; 2-sand; 3-standard sieve; 4-proportioning machine; 5-dump truck for loading; 6-mixing machine and loading machine; 7-mixing; 8-high efficiency slushing agent; 9-water; 10-hydration agent; 11-mine concrete transport vehicle; 12-small wet shotcreting machine; 13-pressure fan; 14-compressed air; 15-dosing pump; 16-alkali-free liquid accelerating agent.

The system operates as follows: use proportioning machine for proportioning; load concrete material with dump truck; the dump truck carried the material to the excavating working surface through auxiliary vertical shaft; and then dump the material into the lower end of loading machine; and then add cement; add water and proportional additive at the double helix mixing station; mix 2–3 min; carry out wet shotcreting.

## 3. SHOTCRETING DYNAMIC ZONING SUPPORT PRINCIPLE AND APPLICATION FOR DEEP SHAFT

### 3.1. *Mechanism of shotcreting for deep shaft*

The deep shaft anchoring and shotcreting support rock workings deformation characteristics can be divided into three stages, namely, deceleration stage (stage I), approximate linear constant speed stage (stage II) and accelerating stage (stage III). The rock workings concrete damage of deep shafts of Juye Coal Mine is concentrated in stage I and III (Yu 2005).

Characteristic of stage I is that the concrete cracks in a very short time after shotcreting. As shotcrete support is used, the shotcrete on the surface of surrounding rock subject to shear failure under the effect of ground pressure and broken rock.

Characteristic of stage III is that cracks often appear at the arch and shoulder spray layer of the workings, followed by falling off and local caving, but other parts of support remain in a relatively stable situation (see Figure 3). Figure 4 is numerical calculation based on project implementation location, which shows the arch and shoulder parts are stress concentration area. See Figure 5, when the workings subjects to pressure, under the pressure of surrounding rock (P), the workings surrounding rock surface bears tensile stress (F), when the tensile stress becomes greater than the tensile strength (T) of concrete, the shotcrete layer destruction occurs. At this point, the shotcrete layer does not work and can not maintain the stability of the workings (Jiang 2004).

Therefore, the concrete destruction of deep shaft rock workings at stage I is mainly shear failure. Tensile stress and shear stress act together at stage III, and the high tensile stress mainly causes tensile damage. The concrete effect of stage I is bonding resistance. The concrete effect of stage II is beam effect. The concrete effect of stage III is thin shell effect.

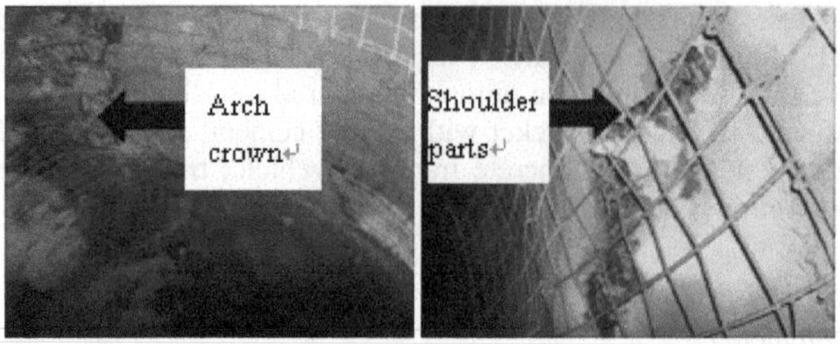


(a) (b)

Figure 3. Spray layer damage of arch and shoulder parts

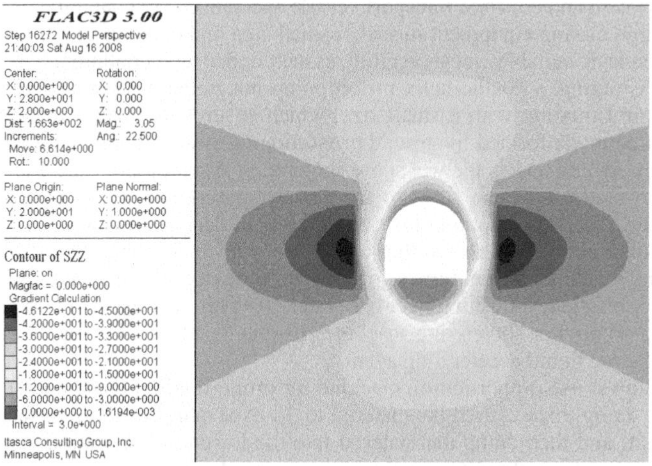


Figure 4. Vertical stress distribution map around workings

Through system mechanical experiments on large size wet shotcrete sample such as horizontal, vertical loading and flexural strength, the research group gets the following conclusions: (1) increase the thickness of the first shotcrete layer aiming at the damage characteristics of stage I, enhancing the initial resistance of surrounding rock and shotcrete layer; reduce the thickness of the second shotcrete layer to reduce the brittleness of shotcrete layer aiming at the damage characteristics of stage III, making that the surrounding rock is not prone to create separate layer and can maintain sufficient bearing capacity; (2) From the analysis of flexural strength, the equivalent thickness of wet shotcrete is 1.18 compared to the dry shotcrete. See Figure 6, it shows that the optimum thickness of wet shotcrete should be between 150–125 mm, which can ensure the flexural strength of wet shotcrete test piece, saving material cost.

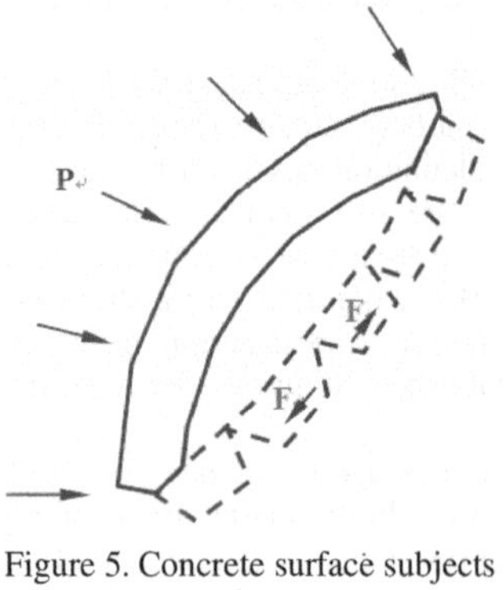


Figure 5. Concrete surface subjects to tensile stress damage

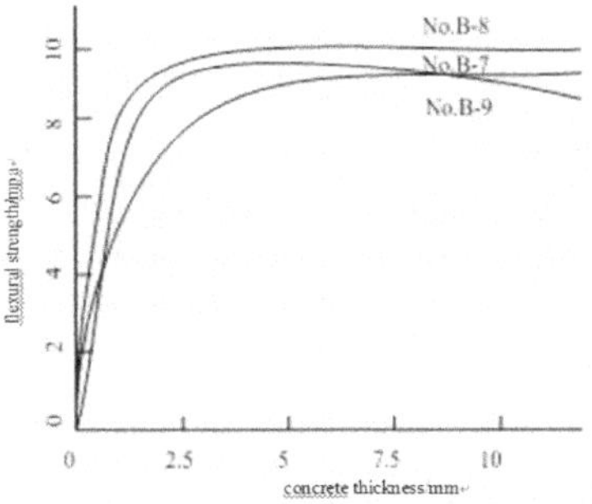


Figure 6. Wet shotcrete beam flexural strength- thickness chart

### 3.2. *Dynamic zoning support technology of shotcrete*

Wet shotcreting dynamic support technology of high stress broken rock workings for deep shaft has been proposed based on the above analysis.

(1) Support intensity classification: determine the support strength of shotcrete layer according to the overall stability of surrounding rock, such as very stable surrounding rock (Ky≤ 0.12). The shotcrete layer design lies close to surrounding rock; there are no high requirements on its flexural strength and compressive strength, and the ordinary concrete can meet the engineering needs. The unstable rock (0.4 <Ky≤ 0.5), the shotcrete layer design is concentrated on high bonding strength to prevent rock loose and falling; If there is high horizontal stress, high bending strength will also be involved, ensuring effective support of beam effects; dosage of side wall accelerating agent is less than crown.

(2) Fractionated wet shotcreting: the shotcreting is divided into two or several times according to the deformation stages of surrounding rock and New Austrian Tunneling Method supporting mechanism;

(3) Different wet shotcreting designs for different zones along the same workings: shotcreting design is different from zone to zone because the variation of surrounding rock structure and hydrological and geological conditions. For example, when there is severe flooding on the rock surface, the concrete formulations need to improve, such as adding internal curing agent to increase bonding strength;

(4) Strengthen support for the key parts of cross-section along the same workings: if the shotcreting layer is located in different rock stratum, for example, the crown plate is very weak rock, or stress concentration of high ground stress downward arch crown and two shoulder, then the shotcreting layer design of very-weak zone and stress concentration parts are different from other parts.

Reference (Jiang 2004): $K_y = \gamma h / f_T m \sigma_c$, where, $K_y$ is stability factor of workings; $r$- the average volume-weight of overlying rock, $r$ = 2.6 × 103; h-workings burial depth; $\sigma_c$- rock uniaxial compressive strength; K- stress concentration factor of surrounding rock, arch workings, k=3; $f_T$ - structure characteristic factor of rock mass, $f_T$ = 0.7-1.0. (1.0 for homogeneous unbroken lots, 0.95 for layered unbroken lots, 0.9 for homogeneous crack lots, 0.85 for homogeneous broken lots, 0.8 for layered crack lots, and 0.7 for layered broken lots). To classify the stability of surrounding rock according to $K_y$, if the workings through layer, the compressive strength shall take a small value.

### 3.3. *Engineering application*

Engineering application sites are auxiliary secondary main workings in the northern zone and north wing track uphill of the first mining area.

Take the north wing track uphill of the first mining area as an example to explain engineering application. Use Mode 2 to conduct wet shotcreting. According to different rock structures passed by workings construction, shotcreting dynamic zoning support design can be divided into A, B, C, D and E zones, see Table 1.

Table 1. Shotcreting zoning support design

| Content | Stability classification | Shotcrete effect | Zoning support design (proportioning of concrete materials) |
|---|---|---|---|
| A zone | From workings crown to shoulder parts: Ky > 0.5, unstable surrounding rock | Muddy fine sandstone (from workings crown to shoulder parts): support and closure. | Muddy fine sandstone from workings crown to shoulder parts: steel fiber (6%)+high efficiency slushing agent (0.7%)+alkali-free liquid accelerating agent (6%)+hydrated agent (0.5%)+curing agent (5%) |
| | Other parts: 0.2 < Ky≤0.3, stable surrounding rock | Fine sandstone (other parts): to prevent weathering. | Other parts: high efficiency slushing agent (0.3%) +alkali-free liquid accelerating agent (4%) +hydrated agent (0.5%) + TCC735 curing agent (5%) |
| B zone | Ky > 0.5, unstable surrounding rock | Muddy fine sandstone (at workings crown section): support and closure. | Steel fiber (6%) + high efficiency slushing agent (0.7%) + alkali-free liquid accelerating agent (6%) +hydrated agent (0.5%) +curing agent (5%) |
| C zone | Ky≤0.12, very stable surrounding rock | To prevent weathering. | Alkali-free liquid accelerating agent (4%)+hydrated agent (0.5%), if steel fiber is added, steel bar net can be omitted. |
| D zone | Ky > 0.5, mudstone section, extremely instable | Mudstone (at workings crown section): support and closure. | Consistent with design of B zone |
| E zone | Ky > 0.5, overall extremely instable | Crown section: support and closure<br>Two wall rock: closure | steel fiber + high efficiency slushing agent + alkali-free liquid accelerating agent + hydrated agent |

### 3.4. *Actual measurement and analysis of wet shotcreting*

Engineering application sites are auxiliary secondary main workings in the northern zone and north wing track uphill of the first mining area. The two sites were measured simultaneously. Figure 7 is wet shotcreting final roadway picture, which shows that the workings keep intact, with compact shotcrete and good finish degree.

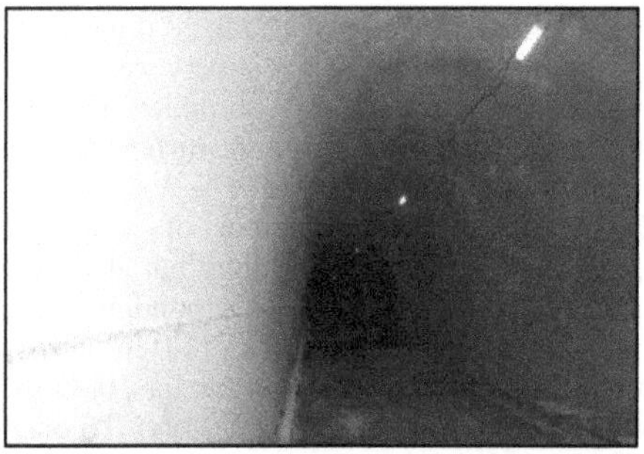

Figure 7. Roadway picture of wet shotcrete

Comprehensive analysis data is shown in Table 2. The data shows that: (1) wet shotcreting achieves respirable dust free basically, effectively avoiding the hazard of workers' breathing respirable dust. (2) Wet shotcreting reduces the total cost. For example, compared to dry shotcreting, the rebound of wet shotcreting reduces significantly. The original design thickness of the dry shotcreting is 150 mm, support effect of wet shotcreting with a thickness of 125 mm is better than dry shotcreting, which makes efficiency increase 17%, saving material costs. Especially dynamic zoning support design, achieves the most reasonable and economical material proportion, which gets the optimal supporting results and brings economic benefits for construction companies. (3) Achieves full mechanization of the system, increases the thickness of one-time shotcrete and the production efficiency.

Table 2. Shotcreting zoning support design

| Contents | Respirable dust concentration | Rebond ratio | Early stage uniaxial compressive strength | Late uniaxial compressive strength | Late flexural strength | Material savings | Construction efficiency |
|---|---|---|---|---|---|---|---|
| Wet shotcreting compared with dry shotcreting | 1/37 | -12% | 2.2 times | 1.48 times | 1.6 times | Saving 17% | 1.7 times |

Table 2 shows that the strength of wet shotcreting is much higher than dry shotcreting. The high strength wet shotcreting with superior mechanical properties effectively maintains the stability of surrounding rock of newly-excavated workings of deep shaft and solves the problems of dry shotcreting, such as quickly cracking and spalling and can not adapt to the high ground stress crushed surrounding rock support, ensuring the construction safety. Its safety benefit is particularly significant.

## 4. CONCLUSIONS

According to the current situation of dry shotcreting for deep shaft, this paper proposed a dual-mode system suitable for wet shotcreting of deep shaft workings, which provides assurance for wet shotcreting. Dynamic zoning support principle and technology has been proposed based on the mechanism of deep shaft concrete, making the wet shotcreting design can adapt to the rock surrounding conditions and stresses variation. The design thickness of wet shotcreting got from system mechanics experiments. Industrial test shows that wet shotcreting reaches the stability of surrounding rock of newly-excavated workings of deep shaft and long-term stability. The test achieved initial successful.

## ACKNOWLEDGEMENTS

The research is sponsored by the National Natural Science Foundation of China (No.40674017, No.50534080), open fund of State Key Laboratory of Coal Resources and Safe Mining (No.2007-04), open doctor innovation fund of Shandong Province (No.200703020), Chinese postdoctoral science foundation (No.20080440304) and Chinese postdoctoral second special financial aid foundation (No.200902048).

## REFERENCES

Holzer L., Winnefeld F., Lothenbach B., Zampini D. 2003. Proceedings of the 11th International Congress on the Chemistry of Cement, Durban, South Africa, Vol. 1: 236

Ding Y. 2003. Eigenschaften von Faserbeton and Faserspritzbeton, ibidem Verlag. Hannover: ibi2 Dem-Verlag

RMEM T. 2002. Test and design methods for steel fiber reinforced concrete, BEN DINGTE ST. Materials and Structuers/Matriauxet Constructions 35(11): 579-582

Dimmock, Ross. 2003. Sprayed concrete-Advanced technologies, Concrete (London) 37(9): 14-18

Yu Y.P. 2005. Research on Bolt-net Support under High Stressed and Fragmentized Rockmass Condition Shenyang: Northeastern University: 56-78. (In Chinese)

Jiang F.X. 2004. Ground Pressure and Strata Control. Beijing: China Coal Industry Publishing House 280-282. (In Chinese)

*International Mining Forum 2010, Liu et al. (eds) © 2010 Taylor & Francis Group, London, UK. ISBN 978-0-415-59896-5*

# Experimental study on the influence of effective confining pressure on liquid permeability in coal seam

Pin Lv, Zhisheng Wang, Li Huang
*Key Laboratory of Coal Mine Safety and Efficient Exploitation of Ministry of Education; Anhui University of Science and Technology, Huainan, Anhui Province, China*

ABSTRACT: There exists obvious effect of fluid-solid coupling during process of coal seam water infusion when water is injected into coal seam, the relationship between coal permeability and confining pressure was studied by self-made coal permeability testing equipment. The fitted equations of coal permeability were obtained, which demonstrated the interaction between coal permeability and confining pressure. On the basis of experimental data, the sensitivity coefficient Ck of permeability under stress was gotten. The results show that the permeability of different coal will drop regularly along with the increasing of the confining pressure, and meet the mathematic relationship of quadratic polynomial; different permeability of coal sample has similar sensitivity to the change of confining pressure.

KEYWORDS: Effective confining pressure, coal permeability, seepage

## 1. INTRODUCTION

In the process of coal seam water infusion, there exists very complicated interaction between fluid and coal. In one hand, the change of coal stress will lead to the change of the characteristic of water seepage. On the other hand, the change of seepage properties has a further effect on coal stress field. This interrelated and interdependent relationship between the coal and seepage is called coupling effect between seepage field and stress field. In the process of coal seam outburst, permeability is an important parameter which reflects the degree of difficulty of coal seam effusion. Under the effect of solid-liquid coupling ,there are many factors which influence on seepage flow in coal, among which the extension of coal pore and fissure have a important influence on seepage and stress will directly cause the changes of pore and fissure. The-refore, it is significant to have a research on the influence of stress on seepage. This text starts from experimental study of the variation laws of coal bed seepage under the stress. Put coal samples into the holder of permeability test system, and then exert different effective con-fining pressure on them. Research the variation laws of coal samples' permeability as confining pressure changing, analyze the coal samples' stress sensitiveness, and provide reliable theoretical basis for the effective application of dustproof technology as coal seam effusion.

## 2. EXPERIMENTAL TEST PRINCIPLES AND COAL SAMPLE PREPARATION

### 2.1. *Experimental test principles*

This experiment mainly tested coal samples' permeability under the various confining pressure and analyses its variation laws. For this, a set of coal and rock permeability test system was developed by us. The system mainly consists of four parts:

1) coal sample holder. 2) water injection regulatory system which consists of constant pressure pump. 3) confining pressure regulating system.4) data acquisition and processing system, just as Figure 1 shows. At first, putted the coal simples (ϕ25 mm x 60 mm) according to certain requirement) into holder (Figure 2). And then connected holder, constant pressure pump and confining pressure regulating system with high pressure resistant and stainless steel pipes, all which consisted of the test system. Injected the common tap water, used as displacing media, into the coal simple in holder, then regulated required different water pressure at the entry of the holder by consistent pressure pump. There was pressure transducers installed at the entry and exit of the holder, so that we could test their changing pressures exactly. Also with flow transducer and flow meter at the exit pipeline, the changes of export flow could be tested and analyzed. Using confining pressure regulating system, we exerted pressure on coal simples and regulated it to constant pressure by hand. At last, analyzed and calculated the data which were collected by transducers with the data acquisition and processing system (Figure 3), and got the permeability parameter of coal samples according to relevant standard.

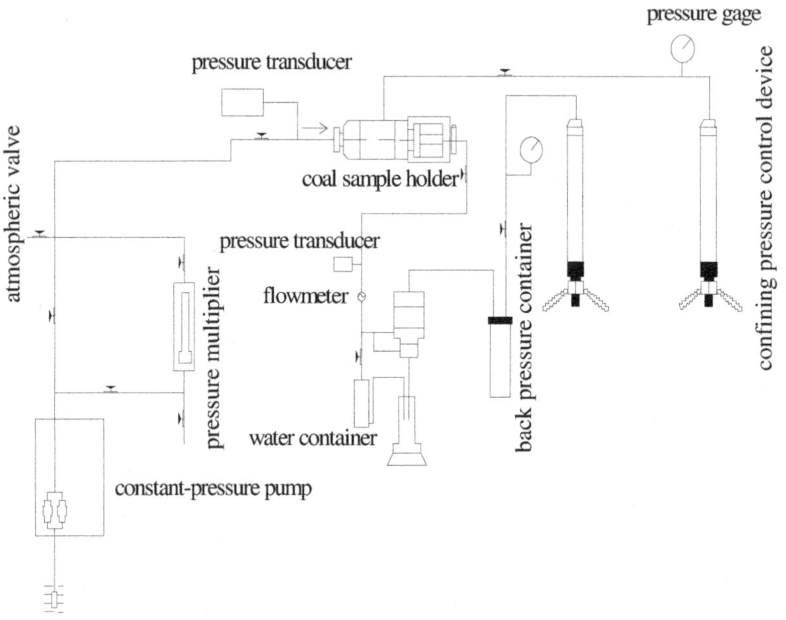


Figure 1. Schematic map of permeability testing system

## 2.2. *Coal sample preparation*

Experimental coal sample preparation could be divided into two types: one was to harder coal, took large coal briquette from the spot, encapsulated them immediately by certain requirements, and then transported them to laboratory, put them on the drilling machine and drilled them into coal cores with different specifications. Another was to the softer coal, stripped some coal from fresh coal surface, got one-third after mixing them, and encapsulated them immediately according to certain requirements, transported them to laboratory, processed them into pulverized coal, and then placed them in press molding machine, compressed into coal cores with different specifications artificially.

The experimental coal samples numbered $s_1$, $s_2$ and $s_3$ were taken from ZhangJi coal mine slot 13 and Panbei coal mine coal face 11213 and 11218 respectively. It was hard to dull them into coal cores since coal from that three place were rather soft. So we made them by artificial compressing. At first, comminuted and processed the coal samples, screened out pulverized coal and then put them into the cylindrical mould of the press molding machine, pressured to 100 MPa, hold two hours with constant pressure. Finally, three group coal samples were made out ($\phi$ 25 mm x 60 mm).

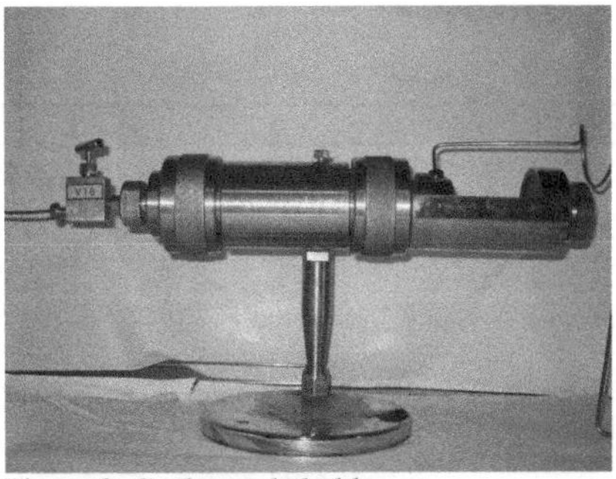

Figure 2. Coal sample holder

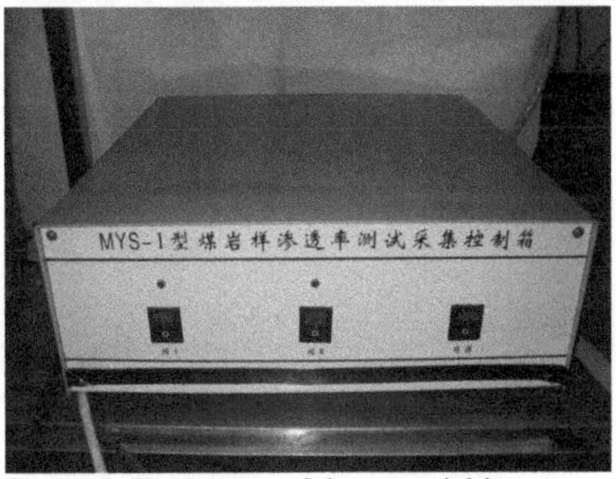


Figure 3. Equipment of data acquisition

## 3. EXPERIMENTAL RESULTS AND ANALYSIS

### 3.1. *The influence of effective confining pressure to permeability of coal samples*

With the coal and rock permeability test system developed by ourselves, we had a test to the changes of permeability of this three coal samples under different confining pressure, the results show as Table 1. By analyzing the data we got three mathematical expressions (Table 2) about nonlinear variation of the three coal samples' permeability as effective confining pressure changing. The results indicate that there are a relationship of quadratic polynomial with one variable between the coal samples' permeability and effective confining pressure, and it have a high fitting accuracy. Its fitted curve shows as Figure 4. Just as the research achievement shows (Huang, Wang 2007, Liu, Liu 2001, Peng, Deng, Qi 2008, Tang, Pan, Li 2006, Zhao 1994), which have demonstrated that there are three forms of relationship (power low, exponent and quadratic polynomial with one variable) between permeability and confining pressure in the experiment of fluid-solid coupling, this experiment have got the third relationship and also have verified above research achievement primely.

In addition, we can make out that all coal samples' permeability present the declining trend but with different amplitude as confining pressure increasing from Figure 4. When confining pressure adds from 4 MPa to 7 MPa, the permeability of s1 reduce nearly 42%, while s2 and s3 reduce 78% and 51% respectively, which fully illustrates that different coal samples have different permeability for their various individual characteristics. Meanwhile, when confining pressure adds to more than 7 MPa in the descending process of permeability, every curve has an inflection point. Before the point, slope of the curves are large, which shows that the permeability have a rapid declining as pressure increasing. After it, the curve tend to be gently, which shows the permeability mainly have no change as pressure increasing. The reasons are as follows: As confining pressure increasing, the coal samples are compressed, their pores turn tiny and their fissures closed, which are bad for fluid infiltration and lead to coal sample permeability decreases. But when confining pressure adds to about 7 MPa, coal samples' pores and fissures have basically been compacted so that the rising of confining pressure has less influence on permeability.

Table 1. Experimental data of coal samples' permeability

| NO. | Confining pressure (MPa) | | | | | | |
|---|---|---|---|---|---|---|---|
| | 4 4.5 5 5.5 6 7 8 | | | | | | |
| S1 | 0.52 | 0.46 | 0.41 | 0.37 | 0.33 | 0.3 | 0.29 |
| S2 | 0.23 | 0.17 | 0.13 | 0.09 | 0.08 | 0.05 | 0.04 |
| S3 | 0.49 | 0.38 | 0.32 | 0.29 | 0.24 | 0.21 | 0.19 |

Table 2. Data about fitted curves of relationship between permeability and confining pressure

| NO. | Equations of the fitted curve | Correlation coefficient |
|---|---|---|
| S1 | $K = 0.0176\sigma^2 - 0.2683\sigma + 1.3113$ | 0.9988 |
| S2 | $K = 0.0153\sigma^2 - 0.2288\sigma + 0.8936$ | 0.9905 |
| S3 | $K = 0.0236\sigma^2 - 0.3521\sigma + 1.5039$ | 0.9834 |

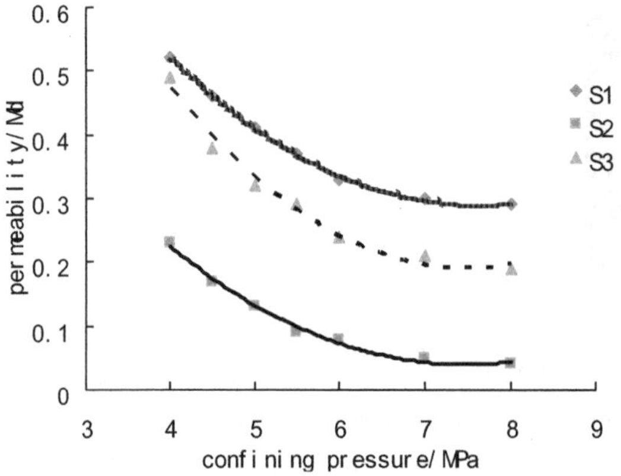


Figure 4. Fitted curve of relationship between permeability and confining pressure

From Table 2 we know the polynomials contain three parts. Quadratic term, the value of its coefficient is directly related with the permeability. One degree term, with the value of quadratic coefficient, the value of its coefficient effect the slope of the curves, whose size reflects the changing rate of the coal samples' permeability as confining pressure changing, also reflects stress sensitiveness of the coal samples, and is related with the coal samples' physical properties such as its strength, pole and fracture characteristics and so on. Constant term, whose size don't change the shape but the location of the curves and have no business with confining pressure, reflects the inherent essential attribute of the coal samples.

### 3.2. *Analysis of stress sensitiveness of coal samples' permeability*

Provided the particularity of the coal samples' structure and the interaction between stress field and seepage field in the process of seepage, the size of permeability is the result of synthetic action of many factors, for that it is hard to describe the process clearly. However, the key of coupling analysis between stress field and seepage field often depends on one's deep understanding to the variation law of permeability (Sun 2001). For this, define $C_k$ as the sensitivity coefficient of permeability to confining pressure, its mathematical expression shows as formula (1) below (Cheng 2008, Xiong 2002). Normalize the factors affecting the coal samples' permeability. Further examine the variation law of permeability as confining pressure changing $C_k$ can reflect the changing tend of permeability. The large the size of $C_k$ is, the higher the sensitivity of permeability presents as average effective stress changing, and vice versa. $k_0$ in the formula (1) represents the basis permeability of coal sample.

$$C_k = -\frac{1}{k_0}\frac{\partial K}{\partial \sigma} \tag{1}$$

Put the date in table 1 into formula (1) and calculate, we can get three mathematical expressions about the fitting relationship between sensitive coefficient $C_k$ and confining pressure (Table 3).

Table 3. Mathematical expressions about fitting relationship between sensitivity coefficient and confining pressure

| NO. | Equations of the fitted curve | Correlation coefficient |
|---|---|---|
| S1 | $C_k = 26.536\sigma^{-3.4392}$ | 0.9533 |
| S2 | $C_k = 26.78\sigma^{-3.3265}$ | 0.9706 |
| S3 | $C_k = 34.608\sigma^{-3.6435}$ | 0.961 |

Getting from Table 3, the relationship between $C_k$ and $\sigma$ presents as power function. That is:

$$C_k = a\sigma^{-b} \tag{2}$$

where: a and b are fitting parameters. If the efficient stress changes from $\sigma$ to $\sigma'$, we can get following by transforming formula (1).

$$K = K_0(1 - \int_{\sigma_0}^{\sigma'} C_K d\sigma) \tag{3}$$

Put formula (2) into (3), and take differential to $\sigma$ on the both sides of the equation, we can get formula between coal samples' permeability $K$ and coal samples' stress when stress changes from $\sigma$ to $\sigma'$.

$$K = K_0[1 + \tfrac{a}{1-b}(\sigma_0^{1-b} - \sigma'^{1-b})] \tag{4}$$

If we know the value of $K_0$, a, b, we can put them into formula (4) and obtain coal samples' permeability under different stresses.

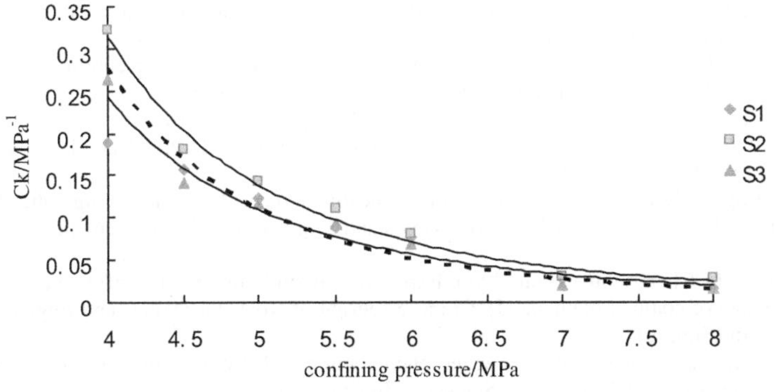


Figure 5. Fitted curve of sensitivity coefficient of permeability to stress

With the result of Table 3, we can get the sensitivity coefficient of permeability and its fitted curve of every group coal samples in the coordinate system, which is built by the sensitivity coefficient of permeability and confining pressure, and do further research on the variation laws of the sensitivity coefficient of permeability as confining changing. From Figure 5 we know the distribution of this three group coal samples' curve are very close. The coal samples' permeability are different form each other, while the difference of the sensitivity coefficient to different confining pressure of is not very big, which can be known by analyzing Table 1 and Figure 4. Take s2 for example, whose permeability is small but the sensitivity coefficient is large. All these indicate the sensitivity coefficient of permeability to stress is not completely related to the size of permeability and with the confining pressure increasing the coal samples' stress sensitivity gradually recede.

## 4. CONCLUSIONS

First, we did a test to three group coal samples' seepage properties under different confining pressure with the equipment of coal and rock permeability test system. By analyzing the results, we gained the conclusion that there were a relationship of quadratic polynomial with one variable between the coal samples' permeability and effective confining pressure.

Second, we defined $C_k$ as the coefficient of sensitiveness between permeability and stress. After normalizing the factors affecting the coal samples' permeability, we studied and found that the sensibility of permeability to stress of different coal samples are rather close and the sensitivity coefficient is not completely related to the size of permeability. Also we could calculate the value of permeability under different stress by ascertaining the sensitivity coefficient of permeability to stress.

Third, the conclusions we get from the study is basically identical with related reference and has certain directive significance to actual application of technology of coal seam water infusion. Meanwhile, it is validated that the test system is reliable.


## ACKNOWLEDGEMENT

This research is sponsored by fund of Key Scientific Project of Anhui Department of Education (2006KJ006A) and project of College Science and Technology Innovation Team.


## REFERENCES


Cheng Ming-jun, Xu Jiang, Tao Yun-qi, Wu Xin. 2008. Test and analysis on sensitiveness to gas permeability of coal seem under triaxial stresses [J]. Journal of Chongqing University 31(Special): 90-93. (In Chinese)

Huang Yuan-zhi, Wang En-zhi. 2007. Experimental study on coefficient of sensitiveness between percolation rate and effective pressure for low permeability rock [J]. Chinese Journal of Rock Mechanics and Engineering 26 (2): 410-414. (In Chinese)

Liu Jian-jun, Liu Xian-gui. 2001. The effect of effective pressure on porosity and permeability of low permeability porous media [J].Journal of Geomechanics 7(1): 41-44. (In Chinese)

Peng Yong-wei, Qi Qing-xin, Deng Zhi-gang, Li Hong-yan. 2008. Experimental research on sensibility of permeability of coal samples under confining pressure status based on scale effect [J]. Journal of China Coal Society 33 (5): 509-513. (In Chinese)

Sun Pei-deng. 2001. Testing study on the variation law of permeability of coal sample during solid deformation processes [J]. Chinese Journal of Rock Mechanics and Engineering (2) (supplement): 1801-1804. (In Chinese)

Tang Ju-peng, Pan Yi-shan, Li Cheng-chuan, etal. 2006. Experimental study on effect of effective stress on desorption and seepage of coalbed methane [J]. Chinese Journal of Rock Mechanics and Engineering 25 (8):1563-1567. (In Chinese)

Xiong Wei. 2002. Study on fluid-solid coupling seepage phenomenon [D]. Beijing: Institute of Porous Flow and Fluid Mechanics, Chinese Academy of Sciences. (In Chinese)

Zhao Yang-sheng. 1994. Mine rock fluid mechanics [M]. Beijing: Coal industry press. (In Chinese)

International Mining Forum 2010, Liu et al. (eds) © 2010 Taylor, Francis Group, London, UK. ISBN 978-0-415-59896-5

# Deformation properties and control models of surrounding rock in deep mines with complicated geological conditions

Wenhua Zha, Xinzhu Hua
*Key Laboratory of Coal Mine Safety and Efficient Exploitation of Ministry of Education, Huainan, Anhui Province, China*

ABSTRACT: The supporting technology of roadway in deep mine with complicated geological conditions is discussed. According to the complicated geological conditions in deep mine at Yuandian Mine, supporting effects of roadway based on existing supporting programs are monitored on-site. The deformation behavior of roadway is obtained and the causes of instability of surrounding rock are analyzed. Supporting principles and control model are proposed pertinently, providing reference for further decision-making of supporting programs in roadway.

KEYWORDS: Deep mine, complicated geological conditions, deformation properties of surrounding rock, control model

## 1. INTRODUCTION

As mining depth increases, a series of new problems such as increased ground pressure, roadway pressure, deformation of surrounding rock, overhaul of roadway and damage of supporting system come into being. Supporting effect has directly impact on safety and efficiency of mining (Wang 2010, Liu 2009, Wang 2008). Deep mining coal excavation started relatively late in China, lacking systematic, perfect pressure control system for deep mine. Nowadays most deep mine excavation still refer to experiences in excavating shallow mining. Engineering analogy method is blindly used for underground pressure controlling. In the practical application, the same underground pressure controlling method is applied in different roadway or the different sector of roadway, regardless of scientificalness and pertinence. As a result, the control effect of some roadway pressure is not good, which affect the regular production, and the pressure control of some roadway is too conservative to result in a waste of economical investment. All of these restricted the level enhancement of mine safety and highly effective. The designed productive capacity of Yuandian Mine of Huaibei Mining Group is 1.8 million t/a. Control model of surrounding rock is proposed according to the complicated geological conditions.

## 2. PROJECT OVERVIEW

102 transport rise of Yuandian Mine is near two faults of DF13 and DF64, Geological conditions in this area are complicated. Faults and fractures generate in roadway. Lithological character of roadway surrounding rock are mainly fine sandstone and fault mudstone. Burial depth of roadway is between -748 m and -470 m. The shape of fault surface is semi-circular arched, whose net width and net height are 3500 mm and 3350 mm respectively. Bolt-mesh-spurting supporting is the designed supporting program. Levorotatory rebar with the size of Φ20×2200 mm is used as bolt, resined anchors of 2×K2950 with the interval of 700×700 mm, round bar made of wire netting (Φ6 -100×100 mm) with the size of 1600×900 mm, steel strand with the

bolt sized 17.8×6300 mm are chosen and arranged with interval of 1500×1500 mm. Grouting bolts of 2200 mm are arranged with interval of 1500×1500 mm. Grouting slurry is normal portland cement of p.o 42.5 which lags the working surface about 120 m ~150 m. Spraying thickness is 150 mm and spraying concrete strength is C20.

## 3. SITE MONITORING OF SUPPORTING EFFECT

Surface displacement of roadway, deep displacement of surrounding rock and stress on anchor bolt/cable have been monitored. The layout of surface displacement and deep displacement are shown in Figure 1.

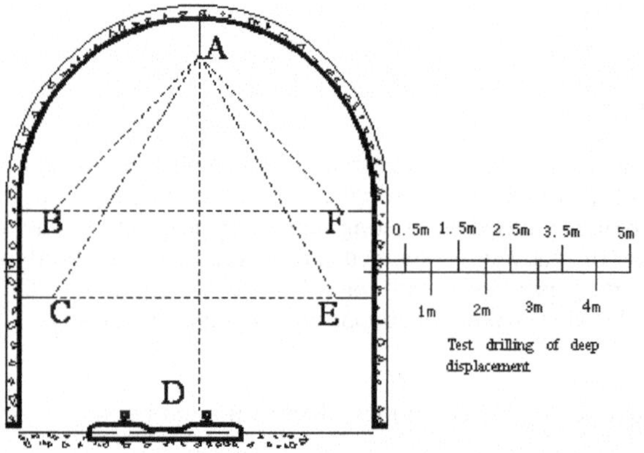


Figure 1. Surface displacement and deep displacement distribution diagram

### 3.1. *Stress analysis of Anchor bolt*

Figure 2 shows the anchor force change curve. Following conclusions can be drawn:

(1) The stress on anchor bolt gradually decreases, which indicates the anchor bolt did not reinforce surrounding rocks.

(2) The stress on anchor is merely 1.8 MPa eventually, which approximately equals the stress on bolt. Thus, the anchor is not able to reinforce surrounding rocks.

(3) The stress on both side bolt are not even, which are 1.8 MPa and 1 MPa respectively. Small stress on bolt shows the construction did not meet the design requirements. The construction management and supervision work should be strengthened.

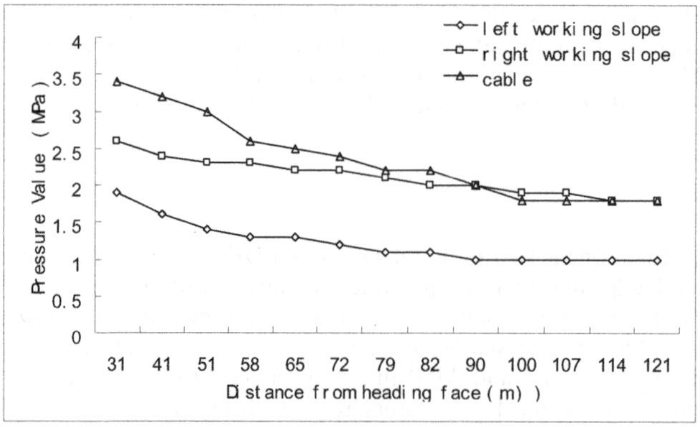


Figure 2. Anchor bolt force change curve

### 3.2. *Surface displacement of roadway*

Figure 3 shows the surface displacement change curve.

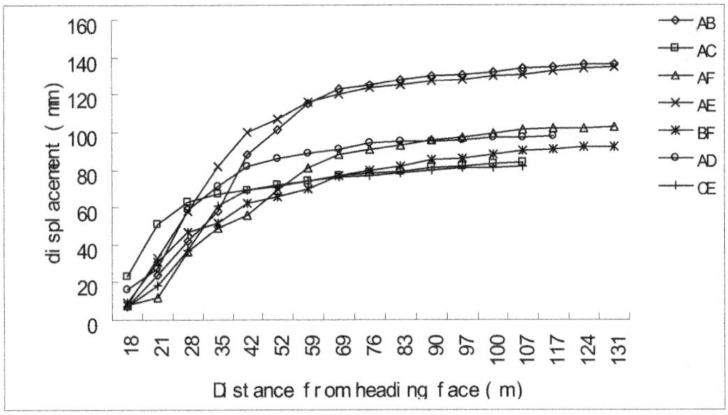


(a)The surface displacement change curve

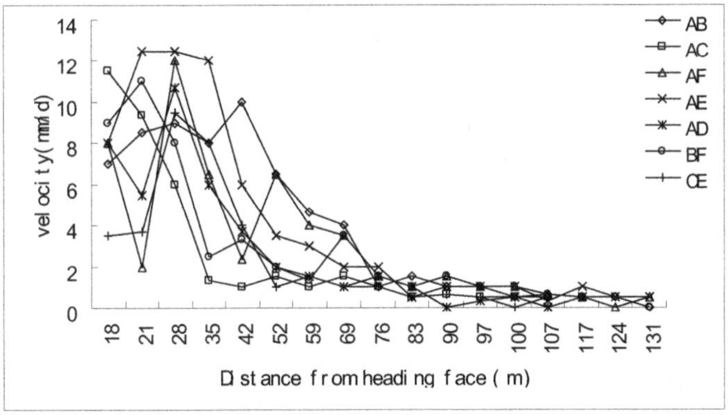


(b) velocity change curve

Figure 3. Surface displacement and velocity change curve

Following conclusions can be drawn:

(1) The surface displacement of roadway gradually increases with the advancing of heading face. It changes dramatically within 69 meters from the working face. The surface displacement changes slowly within the scope of 69~107 m and keeps stable 107 meters away from outside the working face.

(2) Deformation of BF is 92 mm and deformation of CE is 98 mm. Deformation of AD is 82 mm.

(3) Deformation of roadway is uneven. The difference between AB and AF is 32 mm and the difference between AC and AE is 53 mm.

The above analysis shows that with the process of roadway excavation, the deformation of roadway is chronic and uneven in each spot, leading to occurrence of differential deformation around the roadway. This differential deformation further reinforces assemble of stress in roadway and exacerbate deformation of roadway.

### 3.3. *Displacement of deep surrounding rock*

Figure 4 shows displacement of deep surrounding rock curve, from which following conclusions can be drawn:

(1) Displacement of deep surrounding rock changes dramatically within 127 meters from the working face and keeps stable 127 meters away from outside the working face.

(2) In the early process of excavation, displacement is not obvious. Once the working face move forward about 40 meters, no mutation of displacement happens 2.5 m within surrounding rock. However, within the scope between 2.5 m and 3.0 m, displacement increases to 11 mm, and take little change outside 3.0 m, Indicating the whole surrounding rock shift out in the range of 2.5 m and rock become loose between 2.5 m and 3.0 m, and range of surrounding rock loose circle is between 2.5 m and 3.0 m. According to on-site monitoring results, the bolts are too short to reinforce the surrounding rock. The proper lagged time for grouting is as soon as displacement occurs.Yet, it is postponed to 50~80 m form working face in order not to affect the construction.

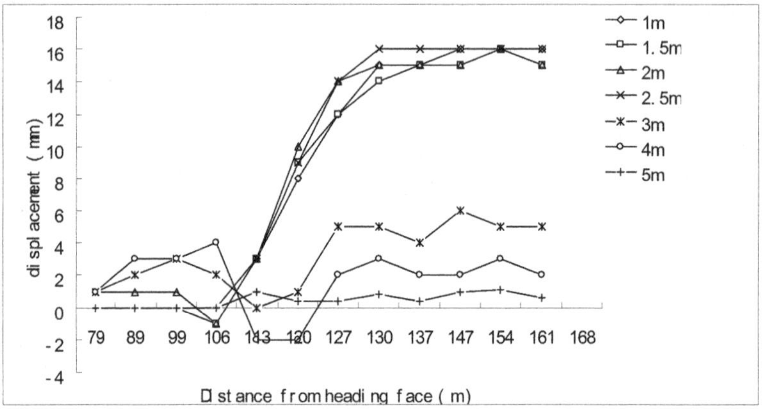

(a) displacement change curve

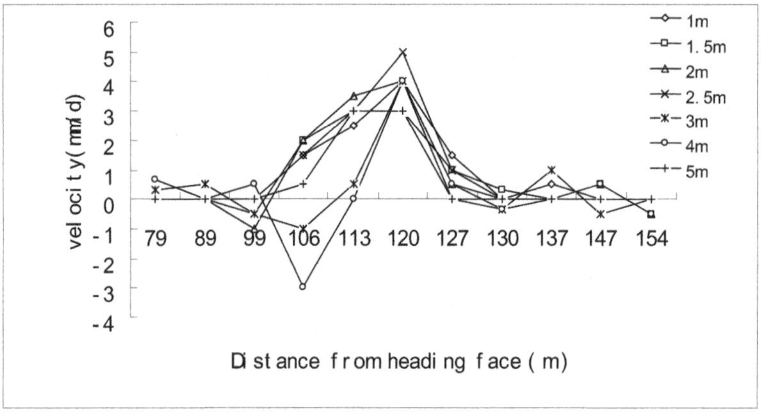

(b) velocity change curve

Figure 4. Deep displacement and velocity change curve

## 4. ANALYSIS OF INSTABILITY IN ROADWAY

(1) Poor integrity of roadway surrounding rock. The roadway construction has been completed by 500 m, and 700 m or so remain under construction. Geological conditions in this area are complicated. A number of associated and derived small faults will occur due to the faults in the construction. Lithological character is relatively fragmental in faults and fractured sections. Lithological character of roadway surrounding rock are mainly fine sandstone and sandy mudstone. The strength of rock block is relative taller than the rock strength because of the influence of internal structure. The integrity of rock is not good as well as overall stability. The broken rock zones of surrounding rock become larger and easier to deform under complex stress. The

broken rock zones are even larger after excavation. The deformation last longer and the floor heave is severe.

(2) Unreasonable supporting design. Using engineering analogy method, the original support design did not choose the right supporting shape and the right time according to the geological conditions and stress properties of surrounding rock. The bolt is designed too short and the intensity of supporting system is inadequate. Lacking of theoretical basis, the floor heave control is ineffective.

(3) Construction quality. Since this roadway is relatively high, procedures like roof controlling, drilling, installing anchor and bolt are difficult to deal with. Some leaky sections require high level of operational ability, construction quality and on-site management. According to on-site monitoring results, initial anchor-hold of the bolt is not adequate which weaken the load capacity of surrounding rock. So the deformation coordination requirements between supporting system and surrounding rock are hard to achieve.

## 5. SUPPORTING PRINCIPLES AND CONTROL MODEL OF SURROUNDING ROCK

### 5.1. *Supporting principles*

According to the depth and complex geological structure of this roadway, the following supporting principles are accepted.

(1) Active supporting principle. The rock is reinforced by high intensity cable anchor. Self load-bearing capacity of surrounding rock is fully exerted.

(2) Timely supporting principle. Once excavated, the roadway should be supported prior to the deformation of surrounding rock so as to ensure the integrity of surrounding rock and the formation of load bearing structure as soon as possible.

(3) Effective supporting principle. Ensure the quality of the project so that every bolt is intact and effective. Bolt should be strengthened when it fractured or became invalid to ensure the overall effectiveness of roadway supporting.

(4) Process control principle. In the early stages of roadway excavation, surrounding rocks are mainly destroyed by tensile-shear complex unloading. Two main reasons account for its intensity weakening: ① stress state converts from three dimensions to two dimensions. ② Excavation lead to surrounding stress concentration. Relatively large stress deviator is generated in the process of unloading pressure and main stress adjustment, resulting in shear fracture loops in the shallow surrounding rock gradually. In the following period, surrounding rock continue deformating in a relatively stable equilibrium. After a period of deformation accumulation, the balance of supporting surrounding rock structures tends to ultimate state. Affected by the environment, the intensity of surrounding rock further reduced. Stress perturbation generated at the same time the excavation works, direct impact of mining and other factors undermine the relative balance of the second phase, resulting in accelerating deformation of surrounding rock until the supporting become invalid. Thus, a good roadway supporting should be strengthened step by step according the cause of its damage and intensity weakening: ①Timely close the surrounding rock and compensate radial resistant force while excavating. ②Surrounding rock deformate severely once excavated. The destruction of the shallow rock is mainly shear stress produced during stress adjustment. In this dynamic adjustment process, provide shearing resistance across structure face to prevent damage from occurring and timely improve the shear intensity of surrounding rock. ③After the formation of fracture loops, surrounding rock stay in the fracturing state. A large number of macro-fractures appear and the intensity of rock decreases comprehensively. At this stage, the intensity of surrounding rock needs enhancing to improve the mechanical properties of rupture surface. Key damaged place should be reinforced to enhance the stability of its load bearing structure over a long period of time.

### 5.2. *Control model of surrounding rock*

According to the surrounding rock conditions and underground pressure monitored, an ideal supporting program for this roadway is stepwise unit reinforcement supporting including bolt-mesh- spurting first, then cable bolting and grouting. In the fracture zone of roadway, the supporting program includes erecting shed and cable bolting and grouting, as is shown in Figure 5.

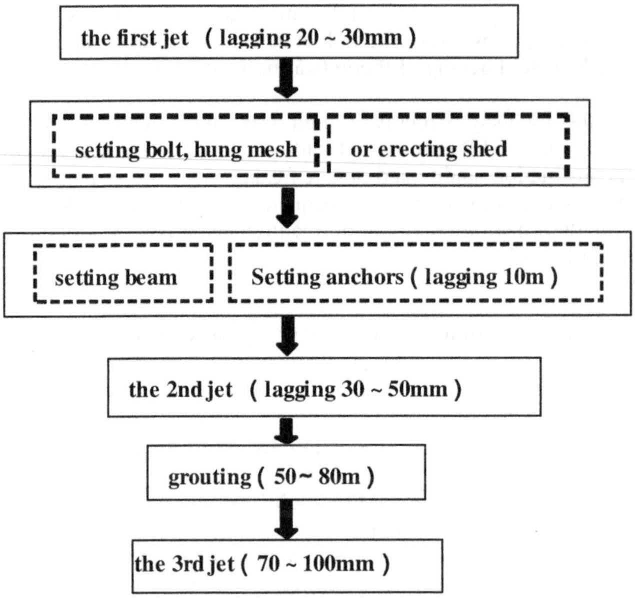


Figure 5. Stepwise unit reinforcement supporting flow chart

## 6. CONCLUSIONS

1) Affected by the fractures of faults and complicated geological conditions, the intensity of surrounding rock is weakened and its deformation becomes severe. This deformation lasts long and uneven.
2) Stress on anchor bolt is small, indicating that bolt did not reinforce the surrounding rock. Therefore, construction management and supervision should be reinforced to ensure construction quality and supporting effect.
3) By dynamic monitoring the supporting effect of roadway, reasonable delay distance for grouting is determined. A control model of surrounding rock is proposed and support parameters need to be further optimization.

## REFERENCES


Liu Wen-bao, Wu Long-quan, Liu Zeng-ping et al. 2009. Design and Technology Optimized Design of Mine Roadway Support in Deep Mine under Complicated Conditions [J]. Coal Engineering (3): 5-7. (In Chinese)

Wang Jian-tuan, Liu Zong-chao. 2010. The Supporting Practice on the Soft Rock Roadway with High Stress in Deep Underground Mines [J]. Energy Technology and Management (2): 23-24. (In Chinese)

Wang Qi-sheng, Li Xi-bing, Li Di-yuan. 2008. Surrounding Rock Deformation Properties and Determination of Support Parameters of Soft Rock Roadway in Deep Mine [J]. Journal of China Coal Society 33(4): 364-367. (In Chinese)

International Mining Forum 2010, Liu et al. (eds) © 2010 Taylor & Francis Group, London, UK. ISBN 978-0-415-59896-5

# Analysis of the dynamic stability of hard and thick strata of overlying multilayer spatial structures in deep coal mines and its application

Hong Shi
*Shandong Jiaotong University, Jinan, Shandong Province, China*

Huajun Wang
*Bijie University, Bijie, Guizhou Province, China*
*Shandong University of Science and Technology, Qingdao, Shandong Province, China*
*Yankuang Group, Zoucheng, Shandong Province, China*

ABSTRACT: According to the characteristics of sediment rock and the movement of hard and thick strata of overlying multilayer spatial structures in deep coal mine, by using the dynamic stability judgment rules controlled by the top coal's recovery ratio, the author worked out the judgment curves of dynamic stability of hard and thick strata and explained their application. This paper also analyzes the factors influencing the dynamic stability, and concludes that the thickness of key strata is the main factor influencing the stability of the whole roof deformation; the stability of the part contact of the main roof is influenced by the friction characteristic, the thickness and intensity of hard and thick strata; and the preternatural pressure at caving face is mainly caused by the instability of the part contact of the main roof. The suggested judgment rule and curve of dynamic stability have been tested by field experiments which achieved safe caving in isolated caving face by adjusting the top coal's recovery ratio. The theoretical basis of designing the safe top coal's recovery ratio can be provided according to different overlying strata, and the safe overlying strata controlling can be achieved in deep coal mining.

## 1. FOREWORD

With the increase of mining depth, there appear lots of mining dynamical disasters due to the instability of strata movement in deep coal mine. The strata concerning with the disasters consists of two sections: one is strata from the exterior region of overlying strata spatial structures to land surface, which causes high stress and determines whether stress level reaches the level where mining dynamical disasters happen. The other one is strata in overlying strata spatial structures, which is the origin of safety accidents in deep coal mine because its poor stress level generated by instability of strata movement in this section. The movement instability in overlying strata spatial structures results in dynamic pressure, whose intensity differs from the motion ranges of the overlying strata spatial structures. For in top coal caving face the top coal from caving to accomplish is a dynamic processing in which the ranges and internal structure of overlying strata spatial structures is changing and the transformation is determined by the top coal's recovery ratio. Therefore, it is a theory problem that has practical significance to study on the relationship between the strata movement stability in overlying strata spatial structures and the top coal's recovery ratio, setting the rational top coal's recovery ratio in different mine stope to accomplish safe control of the stope roof.

## 2. THE DYNAMIC STABILITY OF HARD STRATA MOVEMENT IN THICK LAYER OF MINE STOPE OVERLYING STRATA SPATIAL STRUCTURES

### 2.1. *The stability of hard-and-thick strata overlying strata spatial structures in the first weighting period*

The top coal's recovery ratio of fully-mechanized sublevel caving face determines the formation, development and stability of overlying strata spatial structures. The instability of rock stratum in overlying strata spatial structures includes two parts: One is the global deformation instability because the increase of top coal's recovery ratio causes great deformation of roof structure. The other is sliding instability in overlying strata spatial structures along the interface caused by the enhancement of working face's advancing distance, or the fault structure in overlying strata.

In the first weighting period, the limited convergence to maintain the global deformation stable in hard-and-thick strata overlying strata spatial structures is (Shi, Jiang 2005):

$$SA_0^{'} < h - \sqrt{(l^4(h^2+l^2))^{1/3} - l^2}$$

The permissible squat to prevent the hard-and-thick strata from sliding along the partial pin's interface and keep it stable is $SA_0^{''} \leq h - 2\rho g h l / \eta_0 \sigma_0 f$; where $l$ stands for the length of the rock block; $h$ stands for the thickness of the rock block; $\rho g$ is of force per volume of the rock block; $\eta_0$ stands for the coefficients of rock blocks' inserting of block corner; $f$ is the friction coefficient between the rock block and coal wall; $\sigma_0$ stands for the rock block's uniaxial compressive strength. In practice, the overlying strata's thickness is generally given information in fully-mechanized sublevel caving face, so to apply the above criterion to the site operation effectively, we can judge strata's stability from its thickness and let the strata's length represent the strata's thickness. The intermediate fault of the rock beam after the tensile failure generally makes the hard-and-thick strata's failure mode. Therefore the length of the fractured strata can be drawn, $l \leq \sqrt{\dfrac{h\left([\sigma_t] + \dfrac{\rho g h}{5}\right)}{3\rho g}}$,

where: $\sigma_t$ stands for the smith hammer strength. Put Equation 1 into the stable condition of strata movement, the stable condition of the global deformation and partial contact of the main roof can be concluded respectively with the strata's thickness being variables:

$$SA_0^{'} \leq h - \sqrt{\frac{(Q^4(h^2+0.5Q^2))^{\frac{1}{3}}}{1.59} - 0.5Q^2} \tag{1}$$

$$SA_0^{''} \leq h - \frac{1.41 h \rho g Q}{\eta_0 \sigma_0 f} \tag{2}$$

where: $Q = \sqrt{\dfrac{\sigma_t h^2}{(h+t)\rho g}}$

### 2.2. *The stability of hard-and-thick strata overlying strata spatial structures in the cyclic period*

In the cyclic period, the hard-and-thick strata, influenced by the strata's thickness, periodic weighting length and the position of contact points in mining gob, can shape two rock block in suspension: one is the triangle block structure in the contract and the permissible squat to maintain the global deformation stable in hard-and-thick strata overlying strata spatial structures is $SA_3^{'} \leq \dfrac{5(h - \sqrt{h^2 + l^2 - (h^4 l^2 + 2h^2 l^4 + l^6)^{\frac{1}{3}}})}{4}$, and the limited convergence to keep stability of partial contact of the main roof is $SA_3^{''} \leq \dfrac{5}{4}\left(\dfrac{7fh\eta_0\sigma_0}{6f\eta_0\sigma_0} + \dfrac{-\sqrt{12l \cdot w^2 - 10fhw\eta_0\sigma_0 + f^2h^2\eta_0^2\sigma_0^2} - 11w}{6f\eta_0\sigma_0}\right)$, where $w = \rho g h l$.

To apply the above criterion to the site operation effectively, we displace roof's length of periodic weighting of the strata's length in stability criterion, the stable condition of the global deformation and partial pin can be concluded respectively with the strata's thickness being variables:

$$SA_3^{'} \leq 1.25(h - \sqrt{h^2 + 0.33M - (0.04M^3 + 0.22M^2h^2 + 0.33Mh^4)^{\frac{1}{3}}}) \tag{3}$$

$$SA_3^{''} \leq \frac{35fh\eta_0\sigma_0 - 31.76h\sqrt{M}\rho g - 5N}{24f\eta_0\sigma_0} \tag{4}$$

where: $M = \dfrac{\sigma_t h}{\rho g}$

$$N = \sqrt{2.31h^2M^{3/2}(\rho g)^2 - 5.77fh^2M\rho g\eta_0\sigma_0 + (fh\eta_0\sigma_0)^2}$$

Therefore, the unified stable criterion to assure both global deformation and partial pin stable in the first weighting and cyclic period can be drawn respectively that squat in geometry state is less than the stable permissible squat during exercise, which are:

$SA \leq Min(SA_0^{'}, SA_0^{''})$, $SA \leq Min(SA_3^{'}, SA_3^{''})$.

## 3. THE APPLICATION OF STABILITY CRITERION OF HARD STRATA MOVEMENT IN THICK LAYER OF MINE STOPE OVERLYING STRATA SPATIAL STRUCTURES

### 3.1. *The judgment curves of dynamic stability in hard strata movement in thick layer of mine stope overlying strata spatial structures*

#### 3.1.1. *Hard strata's squat in kinematics geometry state*

For in top coal caving face the top coal from caving to accomplish is a dynamic processing in which the ranges and internal structure of overlying strata spatial structures is changing and the transformation is determined by the top coal's recovery ratio. Assume that H stands for mining height; $\eta$ stands for the top coal's recovery ratio; T stands for the thickness of top coal, so the roof squat in kinematics geometry state $SA = (0.15\sim0.25)(H+\eta T)$. In general conditions, the thickness of the coal seam is varies from 4 to 15 m, so by using the above formula the relationship curve between the squat in kinematics geometry state and the top coal's recovery ratio at different coal seam can be got as Figure 1.

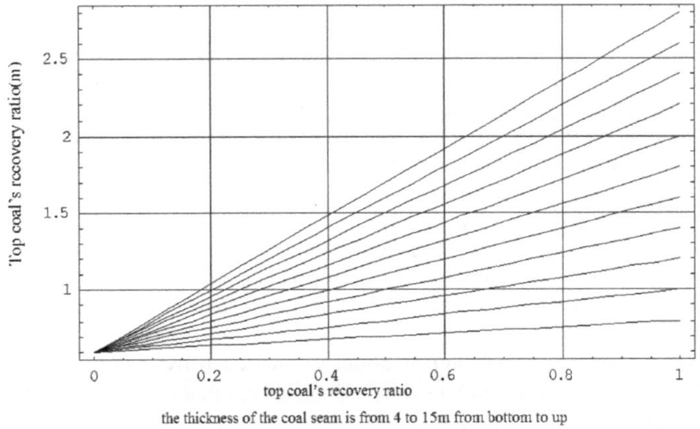


Figure 1. Relationship between the subsidence of roof structure and the top coal's recovery ratio

### 3.1.2. *The judgment curves of dynamic stability in hard strata movement in thick layer of mine stope overlying strata spatial structures*

To apply the strata's dynamic stability criterion into site operation effectively, from the given mine's histogram the rock stratum's stability can be Figured out according to its thickness. We can use a chart to illustrate the relationship between dynamic stability criterion of hard strata in formula (1), (2), (3), (4) and the strata's thickness. Then the relationship curve between the strata's thickness and strata movement stability can be concluded. Here take fine sandstone and conglomerate for examples as Figure 2 and Figure 3.

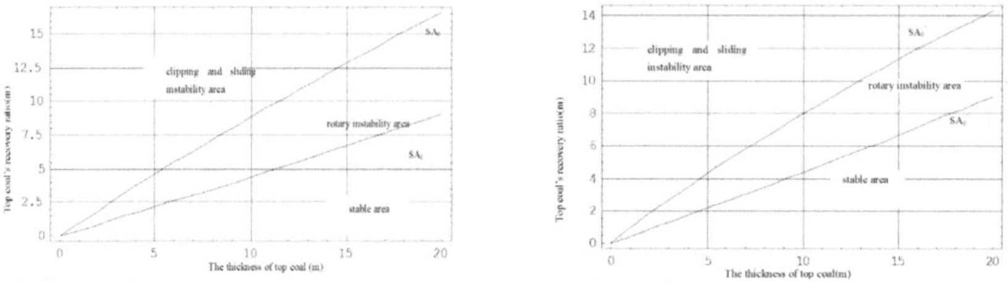


(a) fine sandstone (b) conglomerate

Figure 2. Relationship curves of roof subsidence and roof's thickness at first weight period

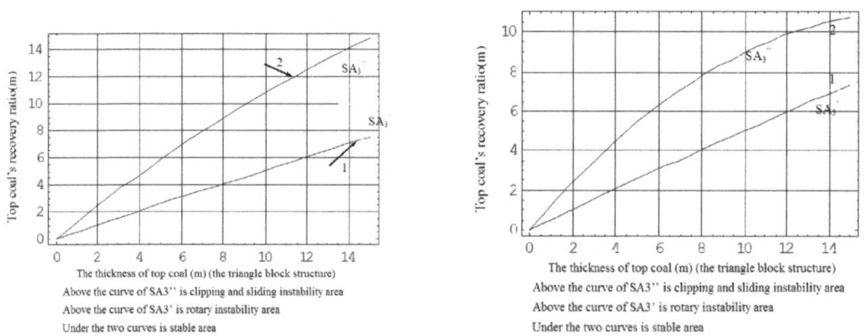


(a) fine sandstone (b) conglomerate

Figure 3. Relationship curves of roof subsidence and roof's thickness at cyclic period

### 3.2. *The application of dynamic stability judgment curve in hard strata movement in thick layer of mine stope overlying strata spatial structures*

In application, in which layer the engineering geology is can be judged by the histogram and immediate roof subsidence SA in roof track geometry and its mining condition can be Figured out according to the top coals' recovery ratio (Fig. 1). On the basis of the engineering geology's lithological character, we can determine which judgment curves to use in Figure 2 and 3. Then draw a plumb-line upward in the horizontal axis according to the engineering geology's thickness and an engineering geology to left in the vertical axis from where the squat places SA. Thus the strata's stability can be worked out by placing the intersection point of the plumb-line and horizontal line. If the structure is stable, the intersection point will be located in stable zone. If not, it will be in unstable zone and how the structure performs depends on in which form the intersection point is. For example, if the thickness of strata and coal seam is 8 and 7 m respective; the top coal's drawing ratio is 85%.

Seen from Figure 1, SA (squat) can be got and is about 1.3 m. Then by looking up Figure 2(b) and 3(b), the intersection point is in the stable zone, which shows that the strata movement is stable in the first weighting and cyclic period. However, the top's drawing ratio scope to maintain the strata movement stable can be also decided according to the strata's thickness and the squat.

What's more, when using this judgment curve, we can take advantage of the formula (1), (2), (3) and (4) according to the formation of the hard-and-thick strata in mining. By using the above method, we can draw the specific judgment curve of strata movement.

## 4. THE OVERLYING MULTILAYER SPATIAL STRUCTURE: THE ENGINEERING APPLICATION OF THE DYNAMIC STABILITY OF THE HARD AND THICK STRATA MOVEMENT

### 4.1. *Roof's thickness*

According to Figure 2 and Figure 3, the squat which ensures the stability of the whole roof deformation has a linear relationship with the thickness of strata. This is mostly because the thickness of strata makes it easier for the broken rock blocks to occlude on the articulated plane during the process of rotation and subsidence. The limited convergence which ensures stability increases as the strata become thicker. The squat which ensures the stability of the part contact of the main roof has a nonlinear relationship with the thickness of strata. However, the squat which ensures the stability of the part contact of the main roof increases as the strata become thicker. This is mainly because the main elements influencing the stability of the part contact of the main roof are the tangential force and the friction characteristics on the articulated plane. If the strata become thicker, then the area of the interface between the articulated rock blocks increases which will provide greater frictional resistance. But at the same time, the dead weight of the increases structure will also increase. Besides, when the strata become thinner, ensuring the stability of the whole roof deformation becomes more important. But when the strata become thicker, ensuring the stability of the part contact of the main roof is more essential.

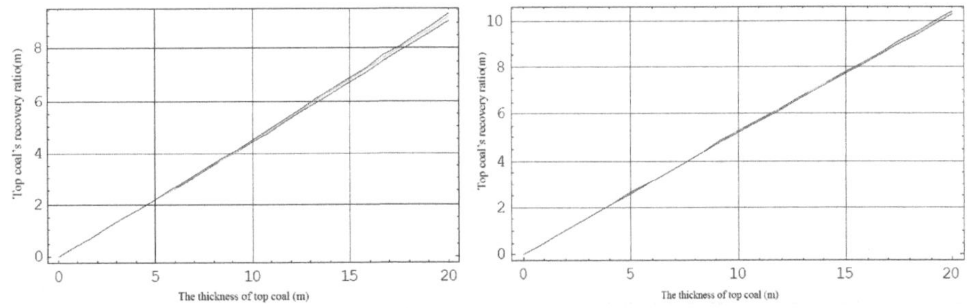


Figure 4. Stability curves of part contact of the main roof of different rock

### 4.2. *Material strength of strata*

According to the conditions of the stability of whole roof deformation, the relationship curves are drawn between the squat and the thickness of strata in fine sandstone, medium sandstone and grit stone strata, shown in Figure 4 (a) and (b). When the strata are thin, the stability curves of the whole roof deformation of different rock almost coincide. When the strata become thicker, the curves no loner coincide, but only with few differences. Therefore, it's not the nature of the strata that influence the stability of the whole roof deformation so much, but the thickness of strata.

Judging from the stability of the part contact of the main roof, the relation curves between the squat which ensures the stability of the part contact of the main roof and the thickness of strata are drawn. Take the fine sandstone, medium sandstone and grit stone for example, shown in Figure 5(a) and 5(b). When the thickness of strata is fixed, the more intense the material of rock, the more stable the par contact of main roof. Because the squat which ensures the stability of the part contact of the main roof has a nonlinear relationship with the thickness of strata, the influence of the material strength on the stability of the part contact of the main roof weakens. Therefore, the stability of the part contact of the main roof is influenced by both the thickness and the intensity of strata.

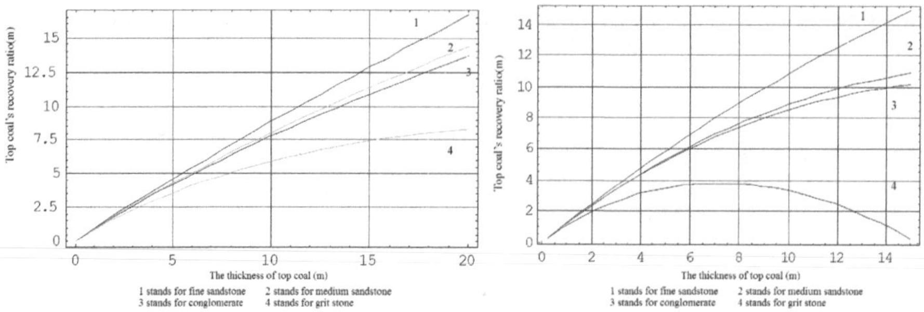


Figure 5. Stability curves of part contact of the main roof of different rock

### 4.3. *The coefficient of the frictional resistance on the articulated plane of strata*

The articulated plane caused by broken strata is irregularly undulating for the waviness and roughness of the articulated plane. The coefficient of friction resistance is the comprehensive reflection of all the factors influencing clipping frictional characteristics on the articulated plane. When the structure of the main roof is stable, the limited convergence has a nonlinear relationship with the coefficient of friction resistance when the other parameters are constant, the limited convergence ensuring the stability of the main roof decreases with the diminishing of the coefficient of friction resistance, so as to lead to the instability of the structure when there exists fault, phase transformation and water erosions. The first weighting period causes the full-depth pressing and clipping and sliding instability of the roof. The cyclic weighting period causes the sliding instability of the roof and the unusual pressure of the working face. The main reason for these is the decreasing of the coefficient of friction resistance.

## 5. THE OVERLYING MULTILAYER SPATIAL STRUCTURE: THE ENGINEERING APPLICATION OF THE DYNAMIC STABILITY OF THE HARD AND THICK STRATA MOVEMENT

Certain 21072 working face is an isolated caving face. The gigantic conglomerate forms the huge arch-bridge structure on the two sides of 21072 working face and puts the coal pillar under great abutment pressure. Within the overlying multilayer spatial structure of the 560-meter 21072 working face, the thicknesses of every set of key rock beams are: 14-meter coal 2-3; first set, 10.05-meter grit stone and the 6.76-meter medium sandstone; second set, 26.0-meter sandstone inter-bed; third set, 42-meter conglomerate. And above all the above mentioned rock beams is the 221-meter conglomerate. With the progress of the working face, the size of the coal pillar gradually becomes smaller. Thus the movement of the overlying multilayer spatial structures should be lessened. The movement of the strata is controlled by the top coal's recovery ratio. Therefore, during the progress of the working face we should limit the movement of the overlying multilayer spatial structures by 85%, 85-34.5% and 0%, respectively, during different phases, in order to ensure the stability of the roof structure.

The stability of every set of rock beam in the overlying spatial structure can be judged by 2.1. According to Figure 1, when the recovery ratio reaches 85%, 34.5% or 0, the squat SA determined by the geometrical condition are 1.68 m, 0.98 m and 0.5 m respectively. When the top coal's recovery ratio reaches 85%, the 6.76-meter medium sandstone and the 10.05-meter grit stone will fall according to the thickness of the roof by calculation. As to the 7.4-meter grit stone and the 42-meter grit stone, Figure 3 shows that the 1.68-meter squat point is within the stable area, which promises a stable structure. Under this condition, the overlying multilayer spatial structure is the 7.4-meter grit stone and the strata above. With the same principle, when the top coal's recovery ratio is 34.5% or 0, the overlying multilayer spatial structures are the 6.76-meter medium sandstone and the 10.05-meter grit stone and the strata above. And they are stable in structure.

## 6. CONCLUSIONS

Using the criteria and the judgment curves of the dynamic stability of the hard and thick strata in the overlying multilayer spatial structures, we can easily Figure out the stability of the key strata according to the thickness of key strata and top coal's recovery ratio. Also, with the stability of the roof, we can design the reasonable top coal's recovery ratio according to different overlying strata.

The thickness of key strata is an important factor which influences the stability of the whole roof deformation in the overlying multilayer spatial structures. The stability of the part contact of the main roof is influenced by the thickness and intensity of strata. When the overlying strata exist side fractures, transformation and water erosions, and the friction characteristic is another important factor which influences the stability of the part contact of the main roof.

The suggested judgment model and curve of dynamic stability of hard and thick strata have been tested by field experiments which achieved safe caving in isolated caving face by adjusting the top coal's recovery ratio. The theoretical basis of designing the safe coal's recovery ratio can be provided according to different overlying strata, and the safe overlying strata controlling can be achieved in deep coal mine.

## ACKNOWLEDGEMENTS

The research is supported by Doctor Foundation of Shandong Province (2008BS08003) and the Science Foundation of Shandong Provincial Education Department (J08LD52).

## REFERENCES:

Cheng Yun-hai, Jiang Fu-xing, Zhang Xing-min, et al. 2007. C-shaped strata spatial structure and stress field in long wall face monitored by microseismic monitoring. Chinese Journal of Rock Mechanics and Engineering 26(1): 102-107. (In Chinese)

Jiang Fu-xing, Zhang Xing-min, Yang Shu-hua, et al. 2006. Research On Overlying Strata Spatial Structures of Longwall In Coal Mine. Chinese Journal of Rock Mechanics and Engineering (5): 979-984. (In Chinese)

Pu Hai, Miao Xie-xing. 2002. Effect of the Key Strata Movement in the Mining-Induced Overlying Strata on the Abutment Pressure Distribution in Surrounding Rock. Chinese Journal of Rock Mechanics and Engineering 21(Z2): 2366-2369. (In Chinese)

Qian Ming-gao, Miao Xie-xing. 2000. Theory of Key Stratum in Ground Control. China University of Mining And Technology Press: 30-86. (In Chinese)

Shi Hong, Jiang Fu-xing. 2005. Study on relationship between roof stability and top coal's recovery ratio at first weight period in fully-mechanized sublevel caving face. Chinese Journal of Geotechnical Engineering 27(4):414-417. (In Chinese)

Shi Hong, Jiang Fu-xing. 2005. Study on Relationship between Roof Stability and Top Coal's Recovery Ratio during Cyclic Weighting in Fully-Mechanized Sublevel Caving Face. Chinese Journal of Rock Mechanics and Engineering. 24(23): 4233-4238. (In Chinese)

Shi Hong, Jiang Fu-xing. 2006. Application and analysis on rupture regularity of hard and massive overlying strata in fully-mechanized sublevel caving face. Chinese Journal of Geotechnical Engineering 28(4): 525-528. (In Chinese)

Shi Hong, Jiang Fu-xing. 2008. Study on Abutment Pressure Rule of Overlying Strata Spatial Structures Based on Microseismic Monitoring. Chinese Journal of Rock Mechanics and Engineering 27(supp.1): 3274-3280. (In Chinese)

Song Zhen-qi, Lu Guo-zhi, Xia Hong-chun. 2006. A New Algorithm for Calculating the Distribution of Face Abutment Pressure. Journal of Shandong University of Science and Technology 25(3): 1-4

Song Zhen-qi. 1988. Practical Ground Pressure. Xuzhou: China University of Mining and Technology Press: 41-78. (In Chinese)

International Mining Forum 2010, Liu et al. (eds) © 2010 Taylor & Francis Group, London, UK. ISBN 978-0-415-59896-5

# Study on dust collection and removal systems in heading face based on air curtain technology

Yucheng Li, Jian Liu, Bo Liu
*Liaoning Technical University, Fuxin, Liaoning Province, China*

ABSTRACT: This paper analyzes the present situation of research and application on controlling of coal-dust and air curtain technology at heading face domestic and abroad, analyzes the generation, distribution and migration rule of dust at heading face. without changing the original long-exhausted and short-forced ventilation system, it take the dry-type dust collection as starting point, seed a methods for dust control by using air curtain to collect dust and using windsocks to draw out dust, which can help to achieve the purpose of dust control at heading face. According to the theory of narrow-channel jet, the author devises a jet machine which could form an air curtain, analyzes the airflow field of pressure-in and draw-out style, demonstrates the mechanism of dust collection by air curtain, and forms the dust collection and control system by air curtain. By means such as theoretical analysis, numerical simulation, testing in laboratory, field testing and practical application in the pit etc., the research demonstrates the effect of U-shaped air curtain on dust control, especially on controlling of respirable dust, which has an obviously effect on reducing the dust concentration at heading face, in particular on reducing the diffusion of respirable dust to the driver. More over, the effect of air-curtain dust-collection and control system depends on design parameters of it, especially the width of the exit of air curtain jet-equipment and outlet air velocity of the jet. According to the theoretical and experimental analysis, this paper come to a conclusion that air curtain of the jet-equipment can completely play the role of dust collection, when the outlet air velocity of the jet lies between 10 to 20m/s and the width of the jet exit lies between 6 to 20 mm. This conclusion can provide a theoretical basis for field application of the jet machine.

KEYWORDS: Dust control, air curtain, dust collection and removal system, narrow-channel jet

## 1. INTRODUCTION

In the recent years, with the development of high-yield and high-efficiency mines, the concentration of dust is also growing in the comprehensive mechanized excavation face. So the dust pollution is more and more serious, and has become the largest pollution source. When there have no measure of dust control, the dust concentration of comprehensive mechanized excavation face is as high as 6000 mg/m$^3$, more over the respirable dust accounts for 20% percentage of them (Zhao 1988); therefore, we must take effective measures to control it. At this stage, domestic comprehensive mechanized excavation face usually adopts sprayed dust-falling technique inside and outside the set and "pumping and press-in" hybrid ventilation of dust removal technology. Although there have a lot of dust removal technology (Ge 2003, Rodi 1998), the existing technologies and management action of dust prevention and removal have a lot of limitations and little effect. The problems primarily are shown as the following:① The effect of dust removal is limited. Although there have used corresponding dust removal technology, the dust concentrations at driver's work area was still over the rule of safety regulations, especially the concentration of respirable dust; ② The respirable dust is a serious harm, and the control of it is difficult. Other than deteriorating the working face and spending a lot of water, conventional dust removal technology such as water spray and sprinkling couldn't control the respirable dust, and the effect of the poor flooding coal would not be good, for the water-deficient mine is more difficult. ③ The cost of dust control is high. Some new dust removal technology is of large investments, high cost and long cycle. The consequences of this situation is that drivers of tunneling machine making long-term presence in the harsh operating environment which contains high concentrations of dust, this seriously jeopardizes the health of drivers. This paper take the dry-type dust collection as starting point, uses an air curtain dust collection and removal system to separated dust sources from the working environment, and uses windsocks to draw out dust for purifying air.

## 2. THE MECHANISM OF DUST COLLECTION BY AIR CURTAIN

The dust removal system of comprehensive mechanized excavation face must be unified efficient dust collection and removal, moreover the first thing to address is the dust collection efficiency, and how to effectively control the dust sources of flow to prevent the spread of dust, then removal the dust. Through the analysis of a large number of measured data, in general, wind speed of airflow within the tunnel of excavation face is not high, and the law of dust distribution is (Mih 1967): With the movement of airflow farther from the termination of working face, the concentration of whole dust or respirable dust are rapidly reduced, then all kinds of size fraction deposit one by one far from dust source, and the deposition speed becomes slow and the concentration tends to stabilized when they are 20-30 m away from the working face. The concentration of fine particles and respirable dust has little change, but the high concentrations dust converge in the context of 3–4 m away from the work face, within this context relatively large particles or large particles deposit rapidly, and the remaining dust come close to the floating dust which almost move with the wind, so generally the dust can be removed when the wind speed is above 0.15 m/s. Therefore, as long as the airflow field of dust source under control, if the wind speed reaches 0.3 m/s general particulates will be discharged by driven or mixed disturbance.

In view of the advantages and disadvantages of application of traditional dust control technology in excavation face, domestic and foreign experts researched the dust collection and removal technology of air curtain (Bardbury 1965), in order to achieve a better effect of dust removal. Under the theory of narrow-channel jet, air curtain is like an cut-off gas screen formed by air injection from the outlet of the gap at a certain speed, which could separates the dust and ambient air, and the air within a certain range would be purified by the dust removal equipment, then cleaner air flows into the tunnel in order to ensure sanitary conditions in work area. The air curtain is installed on the body of tunneling machine. Its role is to turn back upwards the dust diffusing to the driver when the machine is working, and bring the surrounding air rushed into the roof, then the air spread simultaneously on both sides, coal dust were separated in the side of coal wall, meanwhile, the dust in the side of coal wall was cleaning away by the way that use the windsocks of dust removal fan to draw out them. The air curtain acting on excavation face would form an "invisible barrier" which can separate drivers from coal wall, with the purpose of preventing the dust from spreading out the heading face, especially preventing the respirable dust whose diameter is less than 5μm from diffusing to the driver. The separation effect of air curtain is not the same as the solid wall blocking the penetration of dust, but rather by the continuous entrainment of dusty air in the side of the coal wall and continuous dilution and removal of the dusty air coming in, which makes the dust particles can not penetrate the air curtain. Only a small number of dust particles may enter the center of air curtain by the transverse pulse of airflow, and part of dust particles penetrating the air curtain enter the other entrainment flow of air curtain on driver side, and then ascend and spread along with the airflow of curtain air to the non-breathing zone of drivers, thereby this ensure that the air in breathing zone of drivers is clean for the objective of dust separation. To some extent, exhausted ventilation has solved the dust-removal and discharge problems in excavation roadway, and with the use of dust - collection device by air curtain the efficiency of dust-removal and discharge is greatly improved and the dust concentration is greatly reduced at excavation face. As the superposition of potential flow, the attaching effect and the entrainment affect of jet, under the affect of negative pressure formed by the windsocks, the jet plane of air curtain device curves toward the direction of working face, forming a cambered air curtain at the exit of the air curtain device. This effectively controls most of the dust between the air curtain and working face, prevents the dust from spilling.

## 3. THE COMPONENTS OF AIR-CURTAIN DUST-COLLECTION AND REMOVAL SYSTEM

In this paper, the dust collection and removal system is that: On the basis of original ventilation and dust removal system of comprehensive mechanized excavation face, it use a reasonable air curtain to seal the excavation face, prevent the dust from spreading to the roadway by setting rational wind speed and air quantity, and does not affect the staff's operations, access and observation. After the dust source has been sealed by air curtain, then pumping the dusty air into the dust removal equipment, completing the purification process of dusty air which can effectively control the coal dust. In the traditional long-forced and short-exhausted ventilation system, the exit of forced-in windsocks is generally 25 m away from the excavation face, and according to the rules the exhausted windsocks should be away from the excavation face no more than 3 m. The air-curtain dust-collection and removal system adds an air curtain jet-equipment on the basis of original forced and exhausted ventilation system, as shown in Figure 1. The air curtain jet-equipment is composed by a high-pressure fan and forced-air pipe, as well as jet box and jet cavity which should be installed in comprehensive mechanized tunneling machine. And the exit of jet should be installed in front of the driver, so it can be done without affecting the operations of drivers and normal production, the effect of air curtain formed by jet-equipment and tunneling machine is shown in Figure 2.

In the system, about 20 m forced-air pipe is hanged on the belt conveyor, and the high-pressure fan is placed on a special trailer which is moving with the excavation work. And the exit of the traditional press-in windsock should be moved to the high-pressure fan of jet-equipment about 25 m away from the excavation face, to ensure the tunnel ventilation and in the meanwhile provide fresh air for the air curtain jet-equipment. The forced-air pipe chooses appropriate diameter of the mine ventilation windsock. The jet box is designed into the downward triangle, and of same body with the jet cavity and with a windshield in it. The exit of jet is a narrow-channel shaped downward U, and the width of narrow-channel is adjustable, the regulation of here could adjust the wind speed to the value of actual needs. When the tunneling machine is at work, starting the air-curtain dust-collection and removal system can form an air curtain in front of the driver, using the dust removal system to pump and purify the dusty air has an obviously effect on controlling the dust pollution at heading face, in particular on reducing the diffusion of respirable dust to the driver, which played a role of reducing dust pollution and protecting the safety production of coal mine. The main characteristics of this equipment are that: The jet cavity and jet box is of same body, which occupies the smallest space; it is placed properly, which does not affect the driver's operation and the three-dimensional movement of the tunneling machine; the jet box shaped downward-triangle not only has a smaller resistance, but also achieves a multi-level and multi-stage adjustment and controls the direction and wind speed of the jet.

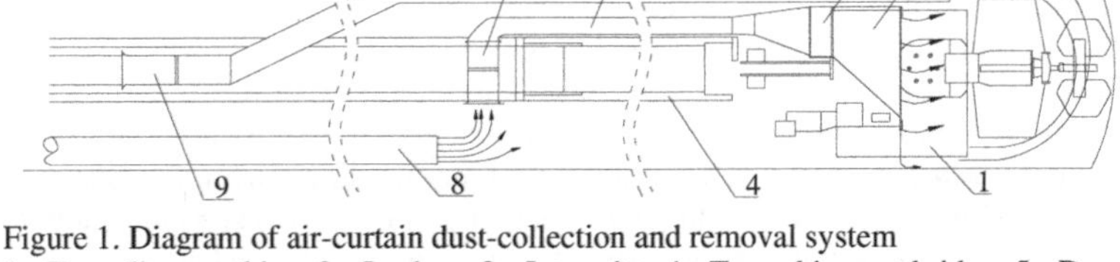

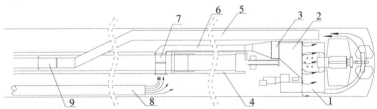

Figure 1. Diagram of air-curtain dust-collection and removal system
1 - Tunneling machine; 2 - Jex box; 3 - Jet cavity; 4 - Transshipment bridge; 5 - Draw-out windsocks; 6 - Press-in windsock of the air curtain; 7 - Press-in fan of the air curtain; 8 - Press-in fan of ventilation; 9 - Dust removal fan

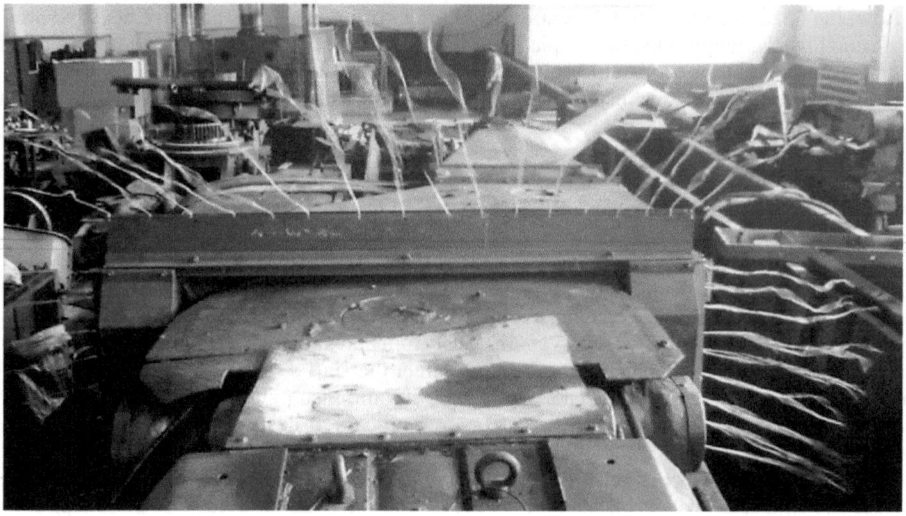

Figure 2. The effect of air curtain formed by jet-equipment and tunneling machine

## 4. PARAMETERS DESIGN OF THE JET EQUIPMENT

There are two main factors to impact the effect of dust separation of air curtain: One is the capacity of air curtain on entraining bilateral air, the other is the anti-diffusion capacity of air curtain. In order to improve the effect of dust separation of air curtain, it is hoped that the entrainment capacity of air curtain is the smaller the better. The less air volume of the lower air curtain entrained, the less dusty air volume would get into the dust-free area from the upper air curtain, which can reduce the dust concentration of dust-free area and improve the efficiency of dust separation of air curtain.

On the other hand, it is hoped that the anti-diffusion capacity of air curtain is the more the better, which can help stop the spread of dust from the coal wall to the driver side. However, the role of the two is restraining each other in practice, if the anti-diffusion capacity of air curtain is wanted to be enhanced, the thickness of air curtain must be wider and the wind speed should be larger. The wider the exit of air curtain is, the higher the jet velocity is, and its anti-diffusion capacity also stronger, but the entrainment capacity of air curtain become greater.

The key of experimental study is how to choose reasonable parameters in the design of the air curtain, which ensure that the air curtain can both effectively prevents the dust from spreading to dust-free area, and entrains less air volume by air curtain. In order to successfully apply the air-curtain dust collection and removal system to dust control of comprehensive mechanized excavation face, after the laboratory tests, the ground installation and commissioning of this equipment, it is proved that the wind speed attenuation law of narrow-channel jet is basically conform to the theory of wind speed attenuation law.

The measuring-point layout of wind speed at the exit of jet equipment is shown in Figure 3, test parameters with different width of narrow-channel of the jet equipment is shown in Table 1. According to the theory of narrow-channel jet and the test data of field experiment, for different width of narrow-channel this paper studies the relation between wind speed of air curtain ($u_{max}$) and wind speed of exit of jet ($u_0$) and distance from the exit of air curtain ($s$), as well as the relation between wind speed of air curtain ($u_{max}$) and wind speed of exit of jet ($u_0$) and half-width of the narrow-channel ($b_0$). If the width of narrow-channel is 10 mm, relation schema of $u_{max} = f(u_{0,} s)$ is shown in Figure 4, and if the width of narrow-channel is 1 m, relation schema of $u_{max} = f(u_{0,} b)$ is shown in Figure 5.

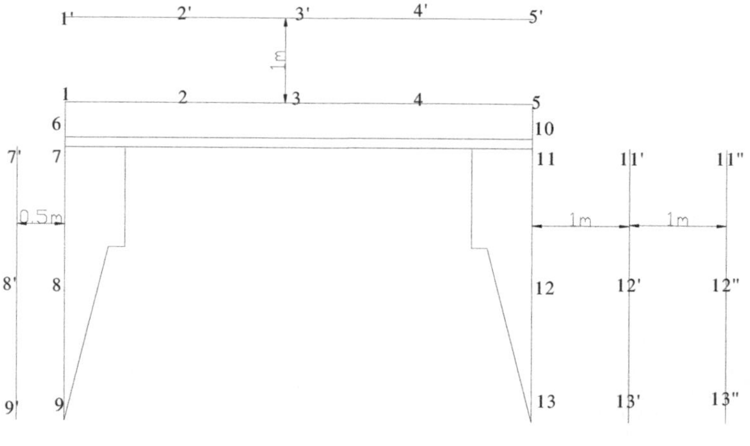


Figure 3. Measuring-point layout of wind speed at the exit of jet equipment

Table1. Wind speed test with different width of narrow-channel

| Test Project | | The first group | The second group | The third group |
|---|---|---|---|---|
| Width of exit of narrow-channel (mm) | Left side | 8 | 15 | 12 |
| | Upside | 12 | 20 | 15 |
| | Right side | 9 | 15 | 12 |
| The average wind speed of exit of jet (m/s) | Left side | 10.55 | 7.35 | 11.63 |
| | Upside | 18.2 | 18 | 20.6 |
| | Right side | 14.65 | 10.4 | 15 |
| The average wind speed 1 m away from the exit of jet (m/s) | Upside | 7.13 | 6.43 | 4.28 |
| | Right side | 5.6 | 3.3 | 4.73 |

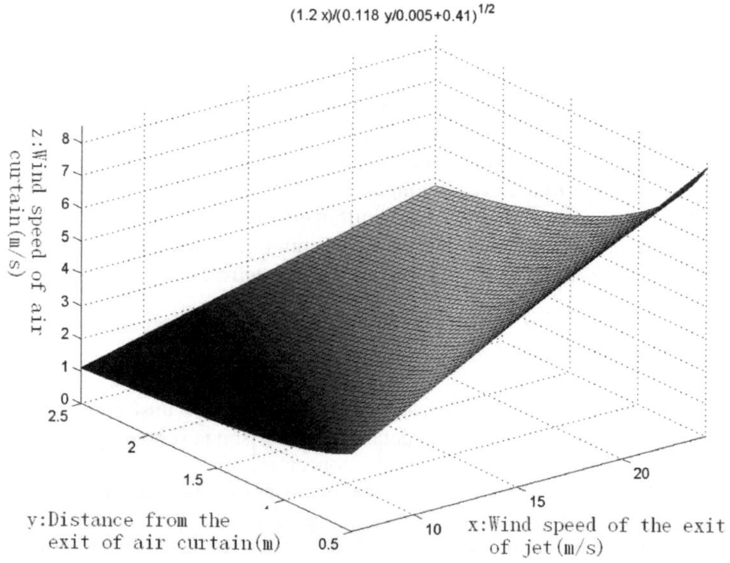


Figure 4. Relation schema of $u_{\max} = f(u_0, s)$ when the width of narrow-channel is 10 mm

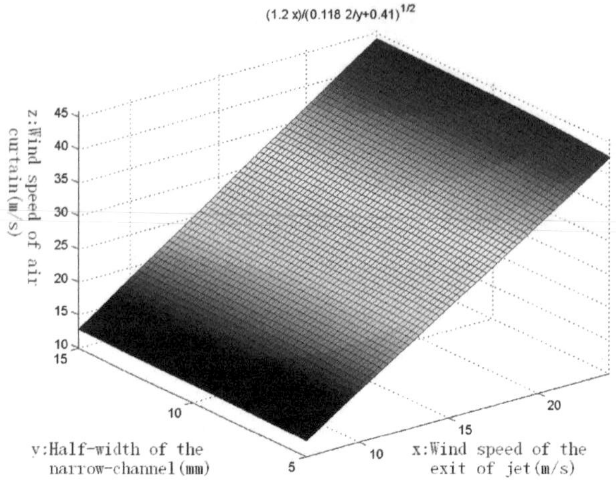


Figure 5. Relation schema of $u_{max} = f(u_0,b)$ when the width of narrow-channel is 1 m

As can be seen from Figure 4, the greater the wind speed of the exit of jet is, with increasing distance away from the exit, the greater the variation of wind speed is. From beginning of the main part of jet to the place 1m away from the exit of jet, axial wind speed decays quickly; when it is more than 2 m away from the exit of jet, the attenuating range of wind speed greatly reduces, and tends to steady state; the axial wind speed is basically above 2 m/s 2.5 m away from the exit of jet, but while the wind speed of the exit is below 10 m/s, the axial wind speed is basically below 2 m/s, 2.5 m away from the exit of jet.

As can be seen from Figure 5, while the wind speed of the exit of jet is minor, with the width variation of narrow-channel, the variation of wind speed is not obvious. But while the wind speed of the exit of jet exceeds 20 m/s and the half-width of the narrow-channel is greater than 5 mm, the variation of wind speed greatly increases. The wider the width of the exit of narrow-channel is, with wind speed of the exit of jet increasing, the better the air curtain be formed. However in practice, the power of fan is same, and the variation of wind speed would change with the width variation of narrow-channel, this can not ensure that both of them increase without limit.

In order to ensure that the wind speed in the end of air curtain attains the requirements of dust control, and does not lead to the secondary pollution which is due to the dust by the large wind speed blowing sidewall and the top, so the wind speed and the narrow-channel width of the exit of air curtain should be controlled within a certain range. After the simulation and field experiments, the wind speed of the exit of air curtain should be ultimately determined between 10 m/s to 20 m/s, and the narrow-channel width should be between 6 mm to 20 mm, so as to make the wind speed in the end of air curtain attain the requirements of dust control.

## 5. THE EFFECT OF DUST CONTROL

After the dust collection and removal equipment used in heading face, the dust concentration was compared before and after the use of air curtain, detailed comparison is shown in Table 2.

As can be seen from Table 2, after using air curtain the concentration of whole dust declines from initial 150 mg/m$^3$ to 21 mg/m$^3$, and the concentration of respirable dust declines from 50 mg/m$^3$ to 5 mg/m$^3$, The effect of dust collection is very obvious. Therefore we can say that the design of the system mainly make use of the advantage of the air curtain dust-collection, it can not only control the whole dust in the working space, but can control the respirable dust effectively. Various size of dust scattering in the working space are difficult to screen out the air curtain, especially the respirable dust would flow with the wind of air curtain. Therefore, using the air-curtain dust-collection and removal system can completely control the dust dissipation of heading face.

Table 2. Table of dust concentration right after the installation of air curtain

| No. | The distance away from the working face | The concentration of respirable dust (mg/m$^3$) | | The concentration of whole dust (mg/m$^3$) | |
|---|---|---|---|---|---|
| | | Before using air curtain | After using air curtain | Before using air curtain | After using air curtain |
| 1. | Driver Position | 50 | 5 | 150 | 21 |
| 2. | 5 m away from the tunneling place | 39 | 4 | 100 | 18 |
| 3. | 20 m away from the tunneling place | 33 | 4 | 92 | 14 |

## 6. CONCLUSIONS

According to the principle of air curtain dust control, without changing the original long-exhausted and short-forced ventilation system, it take the dry-type dust collection as starting point, seeds a methods for dust control by using air curtain to collect dust and using windsocks to draw out them, which can help to achieve the purpose of dust control at heading face. The author also develops air curtain dust-collect equipment, and the practice shows that the dust jet-equipment is reasonable. This is because the U-shaped jet cavity and jet box is of same body, which occupies the smallest space; it is placed properly, which does not affect the driver's operation and the three-dimensional movement of the tunneling machine; and the jet box shaped downward-triangle not only has a smaller resistance, but also achieves a multi-level and multi-stage adjustment and controls the direction and wind speed of the jet. So the equipment for dust control may be suitable for different section of the comprehensive mechanized excavation face. The effect of the dust separation of air curtain depends on parameters in the design of the air curtain, especially the width of the exit of air curtain jet-equipment and the wind speed of the exit of jet. According to the experimental analysis, this paper come to a conclusion that air curtain of the jet-equipment can completely play the role of dust collection, while the wind speed of the exit of jet lies between 10 to 20 m/s and the width of the exit of jet lies between 6 to 20 mm.

## REFERENCES


Bardbury J. 1965. The structure of a self-preserving turbulent plane jet. Fluid Mech 25(6): 168-172

Ge Shao-cheng, Jia Bao-shan, Liu Ya-jun. 2003. Research on function and mechanism of flow field of wind-curtain dust catching fan. Journal of Liaoning Technical University 22(4): 439-441

Mih, W.C., Cominmgs E. 1967. The pressure distribution in the turbulent jet. Fluid Mech 33(3): 154-160

Rodi W. 1998. Recent development in turbulence modeling (A). Proceedings of the third international symposium on refined flow modeling and turbulence measurements. Tokyo: University Academy Press: 97-106

Zhao Shu-tian, Chao Zhong-chun, Mao Tie-lin. 1988. Research of respirable dust of Shuangyashan Coalfield. Journal of China Coal Society 13(1): 1-10

*International Mining Forum 2010, Liu et al. (eds) © 2010 Taylor & Francis Group, London, UK. ISBN 978-0-415-59896-5*

# Study on coal samples' acoustic emission law in triaxial unloading compression with confining pressure

Hua Nan
*School of Energy Science and Engineering, Henan Polytechnic University, Jaozuo, Henan Province, China*

Yingming Wen
*Longshan Coal Mine, Anyang Xinlong Coal Group, Anyang, Henan Province, China*

ABSTRACT: In order to find out the acoustic emission law in triaxial unloading compression with confining pressure experiments of coal samples from Qian Qiu Coal Mine are made by using of RMT-150B rock mechanics experiment machine. In experiments of triaxial unloading compression with confining pressure, almost no acoustic emission appears during compaction state, and there is a little acoustic emission before unloading confining pressure and the coal samples' acoustic emission frequency and energy is ratio to time, and there is a sharp increasing of acoustic emission frequency during unloading confining pressure. However, there is an obvious increasing of acoustic emission frequency and energy near their damaged point and it get the maximum value when they are damaged.

KEYWORDS: Coal samples, acoustic emission, triaxial unloading compression, confining pressure

## 1. GENERAL INSTRUCTIONS

As a reflection of the change in internal state of rock tools, acoustic emission (AE) can detect crack formation or expansion during rock's loading process. The principle is that crack formation or expansion can result to stress relaxation and production energy stored in the form of stress waves which is released in the form of sound emission signal. It's well known that the sound emission signals information can reflect the activity of rock micro-damage. So based on rock acoustic emission signals analysis and research, behavior change within rock can be predictable. According to this theory, rock acoustic emission characteristics were studied by some scholars, and a large number of results was published in recent years (Fu 2005, Li et al. 2004, Li et al 2004, Liu et al. 2007, Yu et al. 2007). As sedimentary rocks, coal has a large number of pores, fissures, bedding and many other types of defects within its body. So the processing of coal sample is more difficult than any other rocks. As a result, few papers were published on coal sample acoustic emission characteristics.

Excavation of coal in coal mine is the process in which loaded stress on coal mass around working face gradually becomes smaller and smaller and the stress state change into two-way or one-way loading state from three-way loading state, so as a corresponding change coal mass's bearing capacity also becomes smaller and smaller. It's very likely to produce coal burst in the excavating process which can result to lots of loss and serious threat to coal mine worker. However, the study of coal sample acoustic emission characteristics can give great benefit to the prevention of coal burst.

Tested coal samples were collected from No.2 coal seam, Qianqiu Coal Mines of Henan Yima Coal Industry Group. The average thickness of this coal seam is 13.2 m. All the details of making coal samples are based on the relevant norms and rules.

## 2. TEST EQUIPMENTS AND METHODS

In the test the loading equipment is RMT-150B rock mechanics test system made by Wuhan Institute of Rock and Soil Mechanics of China Academy of Sciences. Acoustic emission testing system is CDAE-1 acoustic emission detection and analysis instrument system made by Beijing Kehai Hengsheng Technology Co. Ltd. Acoustic emission test system can be shown by Figure 1.

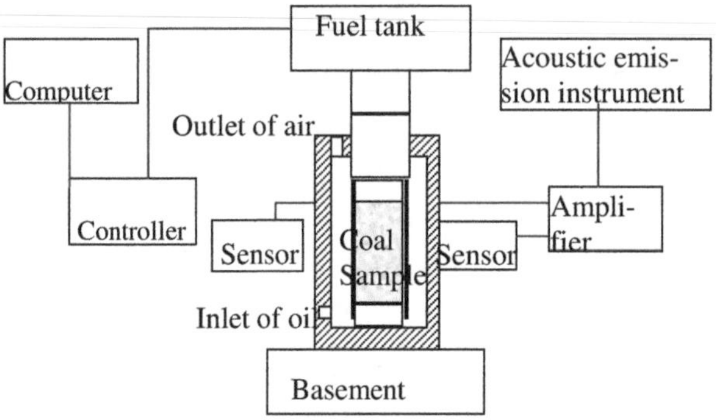


Figure 1. Acoustic emission test system

In the experiments, the confining pressure were selected at the level of 5 MPa, 10 MPa, 20 MPa, and loading rate of confining pressure is 0.5 MPa per second, and axial loading rate is 1 kN Per second. Firstly, make axial pressure ($\sigma_1$) equal to confining pressure ($\sigma_3$), then keep the confining pressure and increase axial pressure to sixty or seventy percent of expected bearing capacity, thirdly decrease confining pressure until coal samples destructed. Acoustic emission testing is made in accordance to axial loading. Three coal samples named A1, A2 and A3 were measured in these experiments.

## 3. DEFORMATION AND STRENGTH CHARACTERISTICS OF COAL SAMPLES

Three triaxial unloading confining pressure compression coal samples' stress-strain curve can be seen from Figure 2. The number upon the curve indicates the destruction confining pressure.

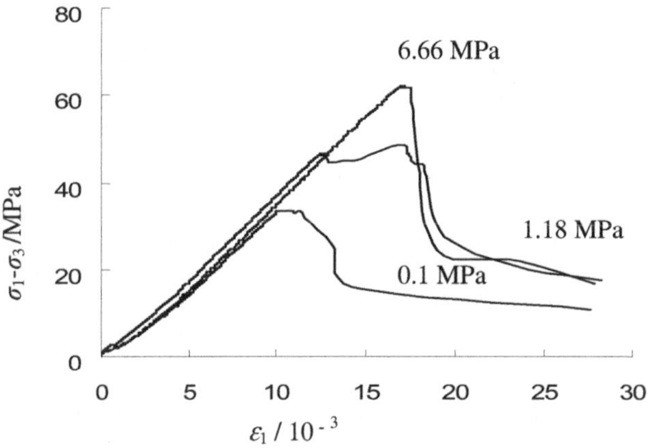


Figure 2. Coal samples' stress - strain curves in confining pressure triaxial tests

From Figure 2 it can be seen that the situation is very likely with the conventional triaxial compression test in front of the maximum deformation point. That is, the curve has similar shape, and the stress and strain has a linear relationship. However, the situation is completely different with the conventional triaxial compression test after maximum deformation point. For example, there is no obvious peak point and so there is a yield platform. This deference comes from deference of loading model. In the triaxial unloading confining pressure compression experiments, maintain the same value $\sigma_1 - \sigma_3$ when decreasing confining pressure until coal samples destructed. It indicates that when unloading confining pressure the friction of the normal stress decreases continuously, so there is obvious movement within coal sample until coal samples fully destructed. Also it can be seen from figure 2 that bearing capacity and elastic modulus grow with the increasing of confining pressure, and that after the maximum deformation point deformation drops very slow, as a result, there is a higher survival stress.

## 4. ACOUSTIC EMISSION CHARACTERISTICS OF COAL SAMPLES

It is well known that, deformation speed is not the same during the whole coal sample deformation process. So acoustic emission for coal sample monitoring is mainly based on acoustic emission parameters change in per time unit. In this paper, acoustic emission counts (N), total count (ΣN), acoustic emission energy (E), total energy (ΣE) are used to analyze coal samples' acoustic emission characters during triaxial unloading confining pressure compression process.

Acoustic emission situation of coal sample named A1 during triaxial unloading confining pressure compression process can be seen from Figure 3, Figure 4, Figure 5, and Figure 6. The initial confining pressure is 20 MPa.

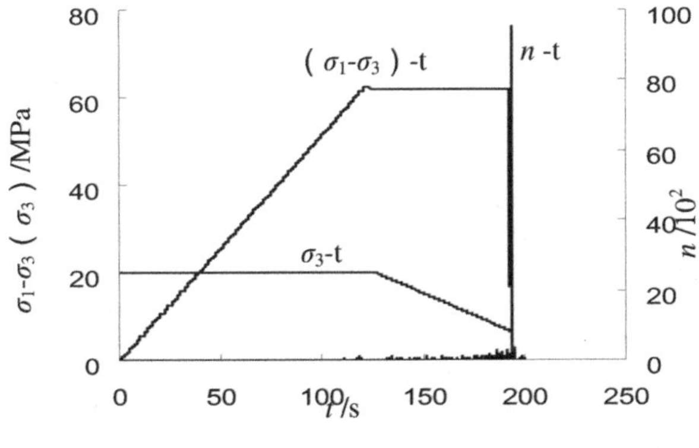


Figure 3. Relationship among stress, acoustic emission counts (N) and time

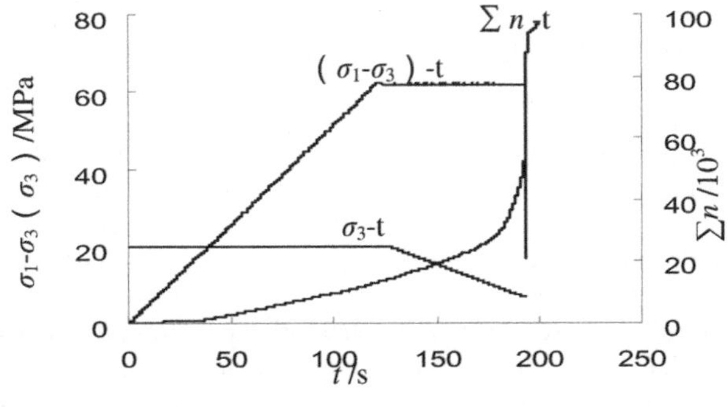


Figure 4. Relationship among stress, total count (ΣN) and time

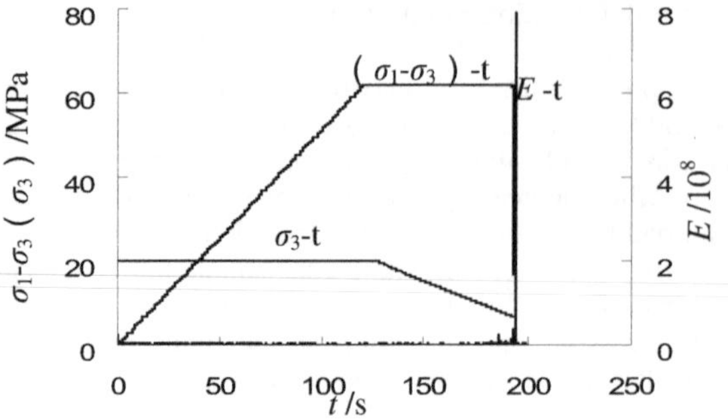


Figure 5. Relationship among stress, acoustic emission energy (E) and time

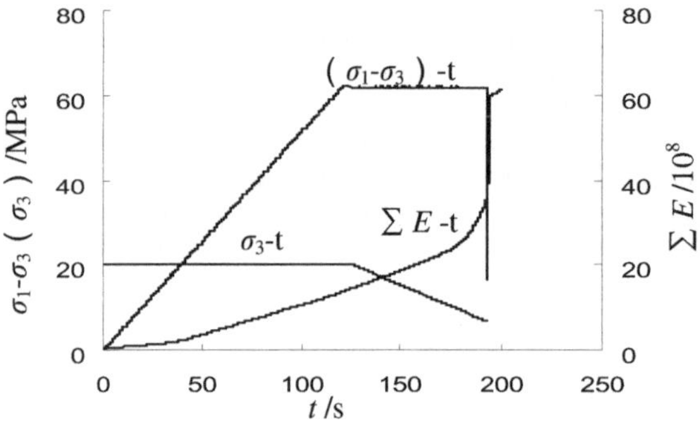


Figure 6. Relationship among stress, total energy (ΣE) and time

Acoustic emission characters can be summed as following:

(1) Under 20 MPa confining pressure which already exceeds coal sample uniaxial compression bearing strength 16.06 MPa (gotten from another test), most of raw cracks in coal sample has been closed. In the first twenty second, when the principal stress difference ($\sigma_1$ - $\sigma_3$) is under 10 MPa, nearly no acoustic emission signal is detected, very few AE counts and energy. Total count (ΣN) gave out in this stage only accounts for 0.4 percent of total test, and total energy (ΣE) gave out in this stage accounts for 1.7 precent.

(2) When time is from 20 seconds to 125 seconds and confining pressure is from 10 MPa to 61.8 MPa, which already nearly amount to coal sample triaxial compression bearing strength sixty percent (gotten from another test), the stress and time has a good linear relationship. In this stage, a small amount of acoustic emission signal is detected. Total count (ΣN) gave out in this stage only accounts for 13.8 percent of total test, and total energy (ΣE) gave out in this stage accounts for 21.5 percent.

(3) When time is from 126 seconds to 193 seconds and the principal stress difference ($\sigma_1$ - $\sigma_3$) remains the same, and confining pressure is unloaded slowly, both acoustic emission counts (N) and acoustic emission energy (E) has a sharp increase. Total count (ΣN) gave out in this stage accounts for 48.6 percent of total test, and total energy (ΣE) gave out in this stage accounts for 39.8 percent. A yield platform in the stress-strain curve figure of triaxial unloading confining pressure compression is produced in this stage. It indicates that in the process of unloading confining pressure the friction stress is constantly reduced, and some material within coal sample already started to slip.

(4) With continuous decrease confining pressure, when the same principal stress difference ($\sigma_1$ - $\sigma_3$) amounts to the maximum value 61.85 MPa, and confining pressure ($\sigma_3$) amounts to 6.66 MPa, the coal sample quickly rupture along the master instantaneous macro-slip surface and it bearing capacity nearly decrease to zero, and the whole time is only 0.2 second. In this

stage, a large amount of acoustic emission signal is detected. Total count ($\Sigma N$) gave out in this stage accounts for 37.2 percent of total test, and total energy ($\Sigma E$) gave out in this stage accounts for 37 percent. What's important, the increasing velocity of acoustic emission counts (N) and acoustic emission energy (E) amount to the maximum value.

## 5. CONCLUSIONS

Based on triaxial unloading confining pressure compression acoustic emission experiment, deformation and strength characteristics and acoustic emission characteristics of coal samples are studied in detail. Conclusions can be drawn as following:

1) The stress-strain curve of triaxial unloading confining pressure compression is different from the conventional triaxial compression test after the maximum deformation pint. This deference comes from deference of loading model. There is a yield platform in the stress- strain curve figure of triaxial unloading confining pressure compression.
2) There is little acoustic emission in compaction stage under high confining pressure, and there is a little acoustic emission in elastic stage under high confining pressure. There is a roughly linear relationship between acoustic emission counts (N), acoustic emission energy (E) and time.
3) There is a sharp increase both in acoustic emission counts (N), acoustic emission energy (E) during unloading confining pressure stage. And what's important, quantity and energy of acoustic emission gets the maximum value at the same time.

## REFERENCES

Fu X. M. 2005. Experimental study on uniaxial compression deformation and acoustic emission property of typical rocks. Journal of Chengdu University of Technology(Science and Technology) 17-21. (In Chinese)

Li J. P., Zhou C. B. 2004. Experimental research on acoustic emission characteristics of rock mass. Rock and Soil Mechanics 374-378. (In Chinese)

Li S. L., Yin X. G., Wang Y. J., et al 2004. Studies on acoustic emission characteristics of uniaxial compressive rock failure. Chinese Journal of Rock Mechanics and Engineering 2499-2503. (In Chinese)

Liu B. X., Zhao B. Y., Jiang Y. D. 2007. Study of deformation damage and acoustic emission character of coal under uniaxial compression. Chinese Journal of Underground Space and Engineering 647-650. (In Chinese)

Yu X. B., Xie Q., Li X. Y., et al 2007. Acoustic emission of rocks under direct tension brazilian and uniaxial compression. Chinese Journal of Rock Mechanics and Engineering 137-142. (In Chinese)

International Mining Forum 2010, Liu et al. (eds) © 2010 Taylor & Francis Group, London, UK. ISBN 978-0-415-59896-5

# Study on pre-evaluation model of support effect on mining roadways based on CBR

Xiangqian Wang, Xiangrui Meng, Zhaoning Gao, Huizong Li
*Anhui University of Science and Technology, Huainan, Anhui Province, China*

Feng Zhong
*Zhejiang International Studies University, Hangzhou,Zhejiang Province, China*

ABSTRACT: In view of the requirements of pre-evaluation on support effect for mining roadways and the working theory of case-based reasoning (CBR for short), a pre-evaluation model is established based on CBR, in which case representation, case retrieval, case revise and other key techniques of CBR have been discussed. The model offers a new and feasible way to pre-evaluation of support effects on mining roadway.

KEYWORDS: CBR, pre-evaluation model, support effect

## 1. INSTRUCTIONS

In mining shafts, the mining roadways take up over 60% of the whole shafts in length (Zhang 2009). Support effect of mining roadways directly influences the transference of working face and equipment in the gateways and the normal advancement of the working face (Yang et al. 2009). Generally, it is difficult to support and maintain the mining roadways for they are easily influenced by mining. Especially in mining enterprises' practices, the selection of support parameters is mostly referred to similar mining experience, without detailed and scientific analyzing and reasoning. If we pre-evaluate the support effect and thus reflect on its rationality after the selection of certain support parameters of roadways, a direct and effective approach would be provided to select and adjust the support parameters of mining roadways.

## 2. THE BRIEF INTRODUCTION OF CBR

CBR is a new reasoning way, growing up with the development of cognitive psychology in 80s in $20^{th}$ century (Chen et al. 2003). It is a way of analogism, with its main points extracting from the fact that when solving the problems, people always strongly rely on those similar cases, experience and acquired knowledge in the past and adjust according to the differences between the new case and some of past cases and thus find solution and form another new case. As a new reasoning model, CBR reflects human thinking habit in problem-solving (Slade 1991). Compared with traditional reasoning ways, it is endowed with many advantages (Gavin et al. 2003, David 2002.): people do not necessarily own clear field knowledge model and then avoid knowledge acquisition bottleneck; the system is easy to maintain and to some extent avoids the problem of co-relation and consistency with knowledge base with the increase of knowledge; coverage will expand, the reasoning speed will be faster and system better with the enlargement of case base. The working principle of CBR: to describe the successfully solved problems as the cases composed with problems features and their solutions and establish the case base according to the certain framework; when new similar problems need to be solved, the system will provide reference solutions by searching and matching the most similar case (or cases) and then adopt directly or modify the reference solutions as the solutions to the new problem after assessment, meanwhile, preserve the solutions to automatically upgrade the case base.

## 3. CBR MODEL OF SUPPORT EFFECT OF MINING ROADWAYS

### 3.1. *Case representation*

In case-based reasoning, case is the unit of knowledge and the basis and premise of reasoning. Case representation is actually the description of knowledge, that is, to encode the related knowledge as a group of data structure with some conventional signs which can be decoded by computers.(Liu 2004) There are many method of case representation including object-oriented method, framework, nerve-network, semantic network, memory network (Shi 2006) and so on. In view of the features of support of mining roadways and the factors influencing the support effect of mining roadways, this research adopts object-oriented method, dividing a case into six parts in description: case identity, case classification, case condition property, case assistant property, case result property and case explanation. As is shown below:

(1) Case identity: to mark the case with number.

(2) Case classification: to classify the cases with the guide of certain properties in order to organize, index and search the cases in the case base. For example, bolt supporting, bolt-net supporting, bolt and cable with steel-belt supporting, U-steel arch yieldable supporting, the ladder-shaper steel rib supporting and so on.

(3) Case properties: to describe the characteristic information of the cases and refer to those properties which play important roles in reasoning and have direct causal relationship with decision-making. Here cases properties mainly refer to the geologic parameters (density of surrounding rock, depth of mining roadways, integrality of rock, mining-induced effect, pillar influence coefficient, the ratio of direct roof to thickness of coal seam), roadway section parameters (section acreage, section shape), support manner and support parameters

(4) Cases assistant property: to refer to those parameters which are adopted to mark and help to describe the cases, such as the source of cases, the time of cases and so on.

(5) Case result property: it is the description of case solution, that is, the description of objective in reasoning. In this study, it refers to the assessment index of support effect of mining roadways (roof-to-floor convergence ratio, two sides convergence ratio, disrepair ratio of roadway, cost of supporting, speed of roadway excavation);

(6) Case explanation: to necessarily explain something to the users or to adjust the cases.

We adopt the following hexahydric groups to demonstrate the cases:

Case = (CID, CT, CBP, CAP, CRP, CE). CID (case identity)is the number of cases, CT (case type) is the type of case, CBP (case basic property) is the collection of conditions of cases, CAP(case assistant property) is the collection if assistant property, CRP is case result property, CE is case explanation.

### 3.2. *Case retrieve*

#### 3.2.1. *Case indexing*

With the increase of cases, it is necessary to adopt suitable case index strategies to improve the efficiency of case retrieval. Generally we can analyze the cases by clustering calculating method and then establish index according to the classifying. Because support type of mining roadway has the biggest influence on the roadway support effect, it is feasible to retrieve the cases according to the classification of roadway support type.

#### 3.2.2. *The determination of cast property weight*

The weight of case property refers to the degree of effect of the properties on the case matching. Only when reasonable weights are attached to different properties, can the matching result of cases objectively reflect the similarity among the cases. In multi-property cases reasoning, reasonable determination of property weight is crucial to improve the accuracy of case retrieval.

A normal way to determine cases property weight is Delphi method, namely, experts consulting method. This method involves the participation of a large number of experts, but the result is easily influenced by the experts' subjective factors (Wang et al. 2009). In information theory, Shannon proposes the concept of information entropy, which reflects the disorder of information. In order to determine the case property weight, we calculate information entropy of the cases, that is, the information amount. The smaller the information entropy of a certain property is, the smaller the disorder of this property in the system is, the smaller the discreteness degree of this property in the case is, and the smaller the influence of this property in differentiating the cases is.

To determine the weights of properties of the given case by calculating the information entropy of the property, the procedures are shown as follows:

1) Construct Matrix A of "n" properties of "m" cases;

$$A = a_{ij}, (i = 1,2,\dots, m; j = 1,2,\dots, n;) \tag{1}$$

where: $a_{ij}$ is the value of property j of case No.i.

2) Standardize the data by mean square error. Define the property values between 0 and 1 by standardizing and obtain Matrix B;

$$B = b_{ij}, (i = 1,2,\dots, m; j = 1,2,\dots, n;)$$

$$b_{ij} = \frac{a_{ij} - a_{j\min}}{a_{j\max} - a_{j\min}} \tag{2}$$

where: $a_{j\min}$ is the minimum of Property No. j of sample case, $a_{j\max}$ the maximum.

3) Define the information entropy of property No. j in the case as:

$$H_j = -k\sum_{i=1}^{m} f_{ij} \ln f_{ij} \; (j = 1,2,\dots, n;) \tag{3}$$

where: $f_{ij} = b_{ij} / \sum_{i=1}^{m} b_{ij}\ ; k = \frac{1}{\ln m}$.

4) Calculate the entropy weight of all properties in the case:

$$W_j = (1 - H_j)/(n - \sum_{j=1}^{n} H_j) \tag{4}$$

where: $W_j$ is the weight of Property No. j in the case.

### 3.2.3. *Value-obtaining method of case properties*

Owing to the difference in value types of properties in the case, the calculation of case similarity is directly influenced. In order to reflect the matching degree among the cases more vividly, we adopt different value-obtaining methods according to different case properties.

(1) Numerical Value Type

To numerical properties such as strength of surrounding rock, depth of mining roadways, section acreage of roadway, the ratio between two property values is adopted as the similarity:

$$\mathrm{sim}(a_i, a_j) = \frac{\max(a_i, a_j)}{\min(a_i, a_j)} \tag{5}$$

(2) Disordered Enumerating Type

This type is used when the properties belong to only one fixed type (e.g: support manner, the shape of roadway). The value of similarity in this type is whether 1 or 0, as is shown in the following formula:

$$\mathrm{sim}(a_i, a_j) = \begin{cases} 1, a_i \text{ is the same as } a_j \\ 0, a_i \text{ is not the same as } a_j \end{cases} \tag{6}$$

(3) Ordered Enumerating Type

This type is used when the values of the property are hierarchical and ordered (e.g. integrality of rock, mining-induced effect). In this case, value obtaining is determined according to the reality and the experts' experience. For example, the integrality of rock is often divided equally in values into five scales of "integrated" "fairly integrated" "average" "fairly cracked" "cracked", with value 1.0, 0.8, 0.6, 0.4, 0.2 respectively. And then calculate on the basis of the following formula:

$$\mathrm{sim}(a_i, a_j) = \frac{d_{ij}}{\max(r) - \min(r)} \tag{7}$$

where: r is the collection of enumerating values, max(r) is the maximum in collection r, min(r) is the minimum. $d_{ij}=|a_i-a_j|$ indicates the difference in the same property between the old case and the objective case.

3.2.4. *Calculation of similarity of the cases*

(1) Compare the new case with those "n" cases in case base one by one (focusing on "m" condition properties), and save the comparing results with the form of matrix *S*:

$$S = (a_{ij})_{m\times n} \tag{8}$$

$i = 1,2,..., n$; $j = 1,2,..., m$; *i* is Case No. *i*, *j* is Condition Property No. *j*, thus $S_{ij}$ is the local similarity degree of the new case on Condition Property No. *j*.

(2) Multiply Matrix *S* with weight vector $W= (w_1, w_2,..., w_m)$($w_i$ is the weight value of Condition Property No. *i*), we obtain the result vector $R=(r_1, r_2, ..., r_n)$, that is the whole similarity degree of the new problem to the "n" cases respectively. See the following formula:

$$R = (r_1, r_2, \dots, r_n) = \begin{bmatrix} s_{11} & s_{12} & \cdots & s_{1m} \\ s_{21} & s_{22} & \cdots & s_{2m} \\ \cdots & \cdots & \cdots & \cdots \\ s_{n1} & s_{n2} & \cdots & s_{nm} \end{bmatrix} \times \begin{bmatrix} w_1 \\ w_2 \\ \cdots \\ w_3 \end{bmatrix} \tag{9}$$

(3) Set a reasonable threshold value of similarity degree and select one or more case with comparatively bigger similarity degree as the retrieval result.

3.3. *Revision and reuse of the cases*

The revision and reuse of the cases are one of the difficulties in case reasoning. In most cases, the solution retrieved from the cases bank can be the solution to the new problem after it is revised. Many ways can be used to revise the case solution, such as manual revising method, weighted average method (Han et al. 2009), differentia-driven case revise strategy (Zhang et al. 2005), automatic revising strategy based on fuzzy logic (Fernández et al. 2003) and the revising model based on generalized operator (Duan et al. 2006) and so forth. This study adopts weighted average case revising method. Suppose "n" cases are matched according to the similarity degree calculation, among which the similarity degree of Case No. i to the objective case is $s_i$, Result Property No. j of Case No. i is $r_{ij}$, Result Property No.j of the objective case is $f_j$, then:

$$f_j = \frac{\sum_{j=1}^{n} s_j r_{ij}}{\sum_{j=1}^{n} s_j} \tag{10}$$

The result property values of the objective case can be obtained in turn, and also the vector of the solution to the objective case $T(f_1,f_2,f_3 \dots f_j)$,which is the pre-evaluation result of the support effect in the given mining roadway support case.

### 3.4. *The reasoning process of the pre-evaluation model of support effect on mining roadway based on CBR*

This paper establishes the pre-evaluation model of support effect based on CBR, and helps to pre-evaluate the support effect by inputting geological parameters and support parameter, that is, via the feedback information of support effect, to realize the pre-evaluation of the support effect under the given geological condition. Meanwhile, we can adjust the support parameters dynamically and match them repeatedly to obtain better support effect. The reasoning process of this model is shown as follows:

(1) Firstly, get corresponding geological parameters and support parameters from the objective case.

(2) According to the obtained parameters, retrieve the cases from the cases bank and calculate the similarity degree of them to the objective case, select those cases with bigger similarity degrees compared with the pre-set threshold value.

(3) Regard the similarity degree as the weight and process the result property values of the retrieved cases with weighted average method and obtain the pre-evaluation parameters of support effect of the objective case.

(4) Retain the cases which are in accordance with the practice in the case base after the assessment of the pre-evaluation result.

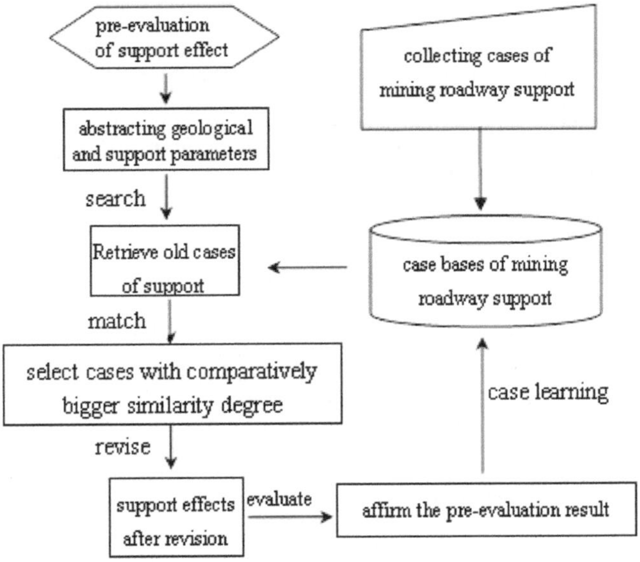


Figure 1. Reasoning process of pre-evaluation model based on CBR

## 4. CONCLUSIONS

The pre-evaluation model of support effect on mining roadway based on CBR is the application of case reasoning technology to the pre-evaluation of support effect of mining roadways. By using a great deal of existing slope cases to build case bases,

This paper initially applies CBR technology to the pre-evaluation of support effect on mining roadways. Based upon the analysis and the classification of the abundant support cases collected from coal mines, the geological parameters, support parameters and assessment parameters of support effect are represented via object-oriented method, and thus case base is established. After matching the support parameter and geological parameters in the new case with those in the old cases, the prediction of correspondent support effect is obtained as an important reference to the selection of support parameters in mining roadways. To testify the validity of the pre-evaluation model, some 30 cases of anchor-net support in Huainan and Huaibei were collected and put in the case base to test, the result of the test approximately being consistent with the truth. However, due to the inadequate cases at hand, the results of pre-evaluation of support effect have not been testify in the condition that there are many cases of various support types.

## ACKNOWLEDGEMENT

The research is supported by grant of Higher Specialized Research Fund Project for the Doctoral Program (200803610001), Higher Specialized Research Fund Project (new teacher) for the Doctoral Program (20093415120001), the Research Funds for Excellent Youngsters of Anhui Province ( No.10040606Y31 ) , Key Project of Science and Technology of Anhui Educational Committee (KJ2010 A081) and Project of Huainan City(2009A05015).

## REFERENCE

Chen Hong, Lin Li-min. 2003. Multi-leveled similarity measure model and its application to retrieval of management cases. Journal of Systems Engineering (1): 31-32. (In Chinese)

David, W. 2002. Efficient Similarity Determination and Case Construction Techniques for Case-Based Reasoning. ECCBR: 292-305. (In Chinese)

Duan Jun, Dai Jufeng. 2006. Research of Case Adaptation Method. Computer Engineering (6): 1-3. (In Chinese)

Fernández, R. F., Corchado, J. M. 2003. Employing TSK fuzzy models to automate the revision stage of a CBR system. San Sebastian, Spain: 10th Conference of the Spanish Association for Artificial Intelligence

Gavin, Finnie, Zhao haosun. 2003. R5 model for Case-Based Reasoning. Knowledge Based Systems.

Han Jianghong, Liu Xiaoping. 2009. Case-based Reasoning Model for Quality Prediction of Spinning Yarn. Journal of System Simulation 21(5): 1348. (In Chinese)

Liu Fang. 2004. Research on the Representation of Semantic Web Based Case and CBR System Architecture.Computer Applications 24(1): 17219. (In Chinese)

Slade, Stephen. 1991. Case-based reasoning: a research paradigm. AI Magazine 31(1): 42-55.

Shi Zhongzhi. 2006. Advanced Artificial Intelligence. Beijing: Scientific and Technical Publishers. (In Chinese)

Wang Xiangqian, Meng Xiangrui. 2009. Design on Strata Behaviors Observation Management and Prediction System Based on Web. Coal Science and Technology. (In Chinese)

Yang Jiachang, Li Qiang. 2009. Mechanism of Deformation and Failure in Mining Roadway Support and Its Practice. Science and Technology of West China 8(1): 28. (In Chinese)

Zhang Jinglong. 2009. Study on mining roadway support design of coal mine. Modern Science. (In Chinese)

Zhang Guangqian, Deng Guishi. 2005. Differentia-driven Case Revise Strategy in case-based reasoning. Computer Applications 25(7): 1658-1660. (In Chinese)

International Mining Forum 2010, Liu et al. (eds) © 2010 Taylor & Francis Group, London, UK. ISBN 978-0-415-59896-5

# Future perspectives of the coal mining sector in Poland

Zbigniew Rak, Jerzy Stasica, Zbigniew Burtan
*AGH University of Science and Technology, Krakow, Poland*

ABSTRACT: The foresight project "Scenarios of technological developments in the coal mining sector" allows for defining the conditions and scenarios of underground mining in Poland. This study briefly summarises the research work involved in the project whose main purpose is to define the directions for further development of the coal mining sector in Poland. The conditions for the development of various mining systems are explored and applicability of selected systems is investigated in the context of the above criteria.

## 1. INTRODUCTION

The performance of the coal mining sector in Poland in recent years viewed in relation to major world producers of coal prompts a thorough analysis of the current condition of the mining sector and its future perspectives. We observe the growing role of imported coal which is used to cater for the needs of the power generation sector and individual users. Once a major European coal producer and exporter, Poland is now becoming a consumer, aggravated by its own problems with the mining sector and the directions of coal imports. Still at a crossroads, now the time has come to conduct a thorough and reliable analysis of the current condition of the coal mining sector in order to design and plan its sustainable development, which should guarantee the longest possible period of self-sufficiency of the domestic power generating sector, at the same time, this sector is expected to act commercially in the market environment. No reaction to further loss of competitiveness of the domestic coal might lead to the uncontrolled productivity cutbacks and hence to irreversible changes of potential utilisation of the domestic coal reserves. That was one of the objectives of the foresight project "Scenarios of technological developments in the coal mining sector" completed in 2008. This study attempts to summarise the part of the research project exploring the future perspectives of the coal mining sector in Poland. Mining systems that are or can be adopted in Polish mines in the future are identified and evaluated.

## 2. PERSPECTIVES OF DEVELOPMENT OF THE COAL MINING SYSTEMS IN POLAND

In the context of the principle generally adopted in the Polish mines, which has it that "in favorable geological and mining conditions, the longwall mining systems with the caving-in strategy are recommendable wherever possible and when this method is justified in terms of technical and economic factors" the geological and mining conditions can be broadly categorized as favorable and unfavorable for the use of longwall mining systems. This, somewhat arbitrary, division of coal beds is shown in Figure 1.

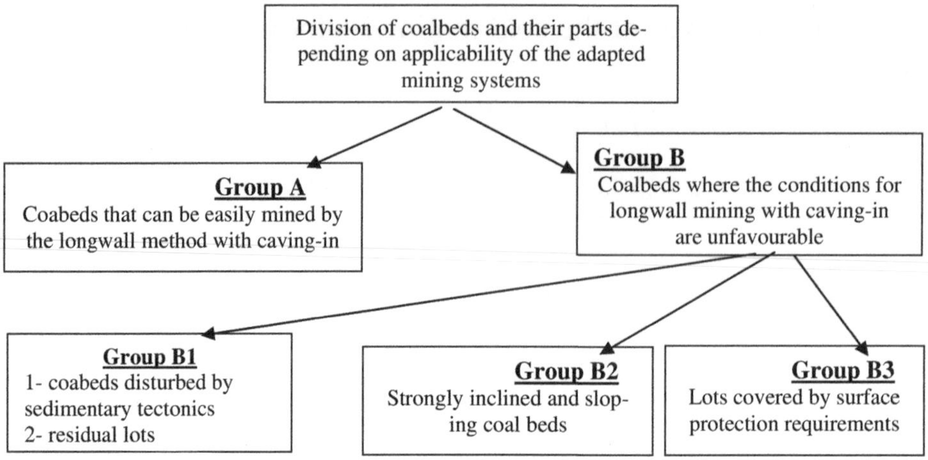


Figure 1. Division of coalbeds depending on the practicable mining systems

2.1.1. *Group B1*

2.1.2. *Longwall systems with caving-in in coalbeds of medium thickness*

In favourable geological and mining conditions the predominant coal extraction system involves the long face mining with caving-in. Expertise gathered to date shows that longwall systems with caving-in offer several advantages, such as:

- high safety level even in difficult geological and mining conditions;
- possibility of monitoring and restricting the scale of natural hazards;
- high productivity and relatively high level of concentration of coal production;
- good level of deposit utilisation and relatively low quality loss;
- small scale of development works;
- potentials of mechanisation and automation of all involved operations;

In order to mine coal by the longwall systems, it is required that coal panels be first developped, their range designed parallel to their strike (Fig. 2).

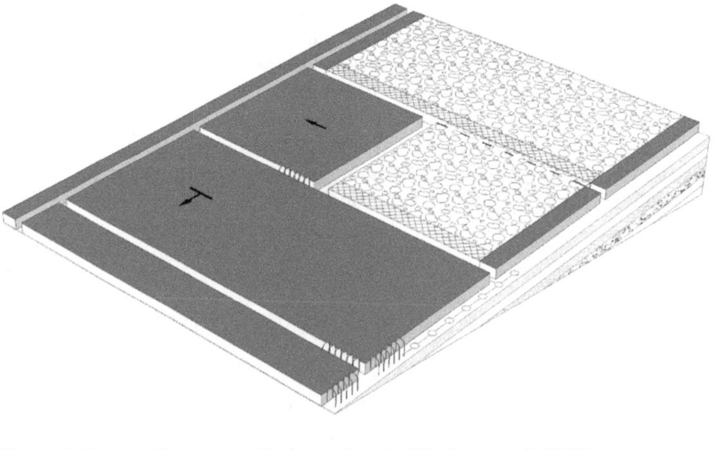

Figure 2. Longwall system with the caving-in (Piechota et al. 2009)

Assuredly, this system ensures the highest cost efficiency of coal productivity levels whilst the mining, hauling and materials handling operations as well as bolting and supporting are now fully mechanised. Longwall systems are relatively easy to ventilate and drain and conditions for systematic caving – in are encouraging. Hauling of the mined material to the gateway is easier in inclined seams, besides it can be easily handled onto the main conveyor.

The face range is actually restricted by faults and disturbances in the coal seam. The performance of the longwall system is unsatisfactory when the face range is shorter than 700–800 m, which is associated with work interruptions required in order to install the supports and move the machines from one site to the new development drivage. Depending on the regularity of the seam deposition and the type of employed mechanisation system, the face length ranges from 100 to 300 m. Long face systems help towards lowering the costs of development works and drifting gate roads necessary to separate the panels.

With the standard equipment, the longwall mining system with the caving-in can be employed to mine coal from horizontal coal beds and inclined at the angle up to 35°. Mechanical behaviour of the roof strata should enable the self-induced caving-in, right behind the working space. The only factor precluding the use of longwall systems is complicated tectonics in the vicinity of the coal deposit. Panel dimensions are restricted chiefly by faults intersecting the coal bed. Folded seams are rather rare and do not preclude the use of longwall systems.

It is reasonable to predict that conditions for using the longwall system with the caving-in in Polish mines are favourable and will continue to be favourable in underlying seams, which will be accessible when the currently mined panels are depleted.

In the nearest future the longwall mining system with the caving-in will continue to be the most popular extraction method because the involved operations have now been safely and effectively mechanised. Further development of the longwall systems will involve mainly the improvements of mechanised and automated systems and widespread use of remote-control devices. The modernised technologies used to mine the coal body will be energy-saving, will ensure the better yield of high-grade coal and effectively make use of the pressure in the seam ahead of the face front. Unmanned continuous miner systems will probably appear in the nearest future and the roof supports will be moved directly behind the continuous miner. The roof bolting and support strategies will be modernised, too. Smart support control systems will be introduced to reduce the risk of rock falling ahead of the face front and to enable pressure control in the seam ahead of the face wall. One of the most challenging operations is handling the crossings of the worked face region with the gateway and main gates. Currently used roof bolting and supporting procedures require the presence of miners to disassemble the arching when the gate conveyor is to be moved and the damaged doors behind the wall have to be repaired. These operations are extremely difficult to mechanise and that is why the number of people responsible for handling the crossing is even higher than the total number of miners engaged in face mining.

When the gateway is to be maintained for further use, the roof support has to be provided behind the crossing. Typically, the support uses backfilling layers made of fast-binding materials, concrete columns or timber piles. Such supports take long to construct and might block the face advance. That is why it is now recommended to change the system of the gateway supports to the bolting or shell systems allowing for longwall mining over the gate cross-section and the removal of the gateways behind the wall.

Removal of gateway requires the development drivages and drifting of the system of gates, comprising at least two and preferably three gates. This system of development drivages requires the exchange of the roof support into the bolting or shell systems and different types of roadheading machines have to be employed. The main advantage of such development procedure is high production levels from development works, comparable to that achieved in regular mining operations. The drawback of such approach is that pillars have to be left between particular workings, which enhance the risk of fires or rock bursts, to say nothing of the productivity loss. Hence new solutions are sought in this field.

The cross-worked longwall up system with caving –in is less frequent. It is employed chiefly when mining faulted seams which preclude the use of long face systems and when doing completion works, often in residues left after the long face mining in faulted regions or in old goafs. The cross-worked longwall mining systems with caving-in are not practicable because the roof strata over the supports tend to be displaced towards caving-in section, and when the arrangement of cleavage planes is less favourable this can lead to rock slide near the path front and to problems with proper positioning of the powered supports. Longwall up systems are also threatened by coal blocks falling from the wall front. Development drivages and gateway inclines are difficult to make in such conditions. Mechanisation of cross-working systems follows the same pattern as in long face systems.

Since the cross-worked wall systems with caving-in ensures the most effective re-consolidation of goafs, it is recommended when caved in goafs are to be backfilled with water-ash mixtures for the purpose of consolidation and to reduce the risk of endogenous fires in caved-in goafs (Zorychta, Burtan 2008). Hence the widespread application of this system in longwall mining of thick seams with cave-in wherever the roofs in particular rock strata need to be patched.

### 2.1.3. *Longwall systems with caving-in used in thin seams*

When longwall systems are employed to mine coal seams less than 1.5 m in thickness, the use of coal ploughs is recommended. Two long faces in the collieries "Bogdanka" and "Zofiówka" are now worked by coal ploughs. Coal ploughs will have to be increasingly used in Polish collieries as the proportion of coal deposited in thin seams is now growing. In the absence of methane hazard, the productivity levels from the coal plough operation should not be less than from a long face worked by a continuos miner. After the start-up operations, the daily advance rate in the colliery "Bogdanka" approaches 10 m, cutting in the seam of the average thickness 1.5 m.

A relatively small cross-section of the heading, when mining seams of high methane bearing capacity, necessitates certain production cutbacks. Another limitation which might preclude the use of coal ploughs is too small cross-section of the gallery. In order to reach the required production levels, the coal ploughs have to be operated in the gallery. To ensure the sufficient space required for operations of a coal plough with standard equipment, the development headings have to be driven whose cross-section must not be smaller that that of the shield support LP 11 or 12. Given that coal is now mined at increased depth, maintaining headings of such cross-section is not practicable. Deep mining operations and the associated levels of primary stress do not help towards maintaining such large headings. The coal working method by ploughs will have to be modified, in particular, new solutions are sought to reduce the dimensions of the driving stations in the discharge and return section in conveyors and coal ploughs.

An alternative solution to coal ploughs is offered by the combined cutter loader (Joy miner) systems. Major manufactures of cutter loader systems – JOY, EICOFF, FAMUR companies- are engaged in extensive research work aimed to design combined cutter loaders systems to mine thin coal seams (less than 1.5 m in thickness). Though it is now too early to predict the results, it is reasonable to expect that prototypes of the new machines will soon be used in Polish collieries. Conventional cutter loader systems used for working faces allow for driving and maintaining small-size headings and will help towards reducing the costs of driving headings in thin seams including coal and rock strata.

### 2.1.4. *Longwall system with caving-in and underhole systems in thick seams*

When cutting coal seams 5 m thick or more and when no constraints are imposed by surface protection requirements, two alternative mining systems can be selected:

- conventional long face systems with the seam being sub-divided into strata;
- longwall cutting system (benching).

In multi-strata mining systems, the seam is mined by top benching. Each next bench is found underneath the caved-in rocks from previously mined strata. For technical reasons, the systems whereby long faces in the strata are worked at the same time, with small delays only (30-40 m) has now been abandoned. Besides, the artificial roof technique has been abandoned too. This strategy typically used a steel fencing grid placed beneath the caved in rock from the roof strata or of a floor made of pillar fragments behind the chock shields in each bench (except the last one) to protect the top wall below. In today's mining practice benches are worked separately. After working the top bench over the entire panel, the working of the bench below will begin after driving the gateways and cross-cuts required for start-up operations because the main headings (haulage and ventilation headings) are typically common for all worked benches. The lower bench is worked directly beneath the caved-in bench (particularly in conditions of partial re-consolidation of caved-in goafs) and sometimes after providing a coal patch whose thickness should exceed 0.5 m and is associated with mechanical behaviour of coal in the mined seam. As coal left in the top wall tends to cave in, that leads to fire hazards alongside the productivity loss. That is why the cross-worked or diagonal face systems are employed to provide for effective stowing of goafs. Application of the stowing materials (also referred to as backfilling) is also a preventive measure reducing the fire hazard and helping in re-consolidation of goafs, which improves the roof stability in the underlying strata (Piechota et al. 2009). Backfilling of goafs is also possible in long face systems (Fig. 3), that is why the mining practice uses long face, cross-worked and diagonal systems with caving-in.

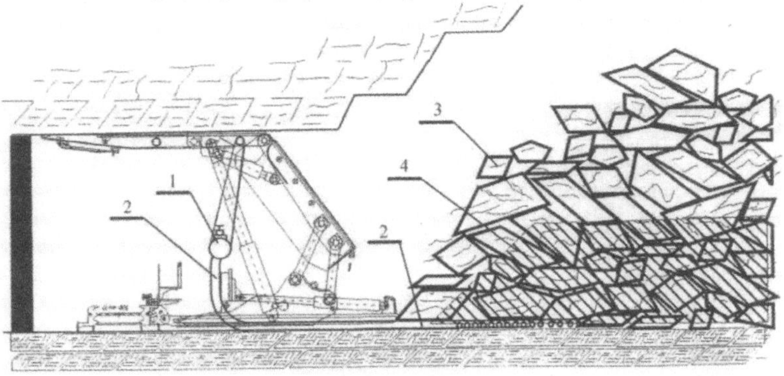


Figure 3. Backfilling of caved-in goafs in the long face with the binding material via injection pipes behind the sections (Piechota et al. 2009). 1-main pipe supplying the backfilling material; 2- outlet of the injection pipe; 3-caving-in; 4- backfilled caved-in section

Underhole (benching) systems, popular in Poland and in many other countries, are used to mine thick seams. Underhole/benching involves cutting coal stripe of about 3.5 in thickness from the lower bench. The remaining coal shelf 4-10 m in thickness is caved in. Depending on the applied technology, caved-in coal is gravity-loaded onto an additional chain conveyor located under a protecting cover of special design or to a standard gate conveyor located under the wall side.

The first method, widely applied, uses two conveyors in the face area (Fig. 4), one standard conveyor on which the continuous miner machine is positioned. The other conveyor dragged behind the roof support sections (under the cave-in protection) is used to collect caved-in coal. This technique is widely used in Chinese mines and in Slovakia, in Europe. The production levels in China reached 0.5 to 7 million Mg/year from a face, the proportion of coal lost in mining not exceeding 15%. Similar coal recovery rates are reported in Slovak mines.

The other underhole (benching) system uses powered support sections with an opening in the cave-in protection cover through which the mined material is discharged to the gate conveyor via a short chute (Fig. 5). In older systems the opening was also provided in the roof bar. This system is still used with great success in Slovakia. The extraction rate is about 85%.

The length of the face in both variants typically exceeds one hundred metres, the faces worked in Chinese mines are sometimes even 300 m long. The main advantage of this system is vast reduction of coal production costs in relation to conventional multi-level systems with caving in. The improved cost efficiency is the consequence of:

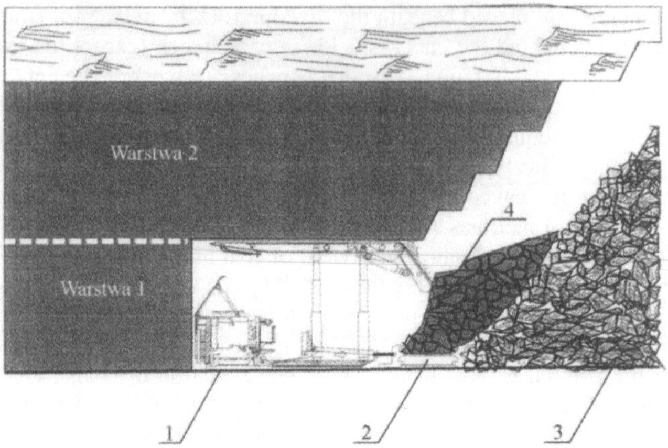


Figure 4. Schematic diagram of the underhole (benching) system (Piechota et al. 2009)
1 - haulage conveyor; 2 - conveyor to transport caved-in coal; 3 - goafs;
4 - mined material from the next bench

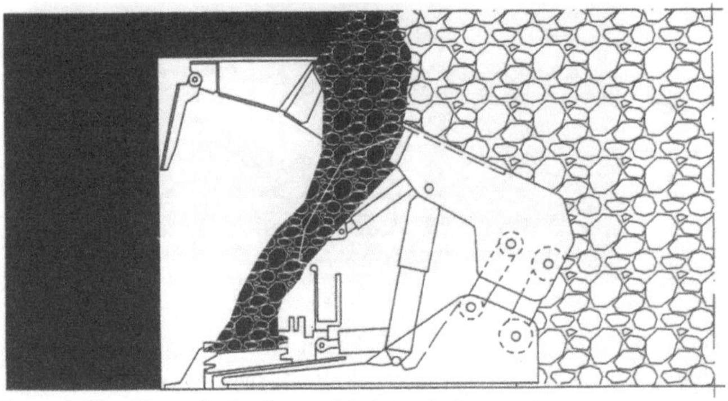

Figure 5. Handling of mined material through the opening
in the roof support - schematic diagram

- small number of development drivages – contour workings are driven only in the floor strata;
- few faces that have to be set out- faces set out in the bottom bench are worked to mine coal from higher levels;
- lower energy consumption during mining operations – only the floor levels are mined by electric-powered continuous mining machines.

However, the system has certain drawbacks too. The proportion of coal lost in mining accounts for 15%, and that is mostly coal left in goafs, which creates the risk of an endogenous fire. Since the mined material has to be brought down from the top wall, the recovery rates from faces are lower, which further increases that risk. While comparing the multi-level system to the underhole (benching) systems, we have to bear in mind that providing the coal patches on particular levels generates the production loss (comparable or higher than those reported in underhole systems) and enhances the risk of a fire in goafs. Both system, therefore, require that fire prevention measures should be put in place.

Certain difficulties may periodically arise in connection with the mined material in the cave-in zone that has to be crushed to the right size. Some coals display high mechanical strength and producing the cave-in might be difficult locally, particularly at points where the amplitude of operating pressure is too distant from the face region. In such situations blasting methods have to be employed.

The underhole system using two conveyors presents certain difficulties with flitting and maintaining the right conditions at the gateway - long face crossing. Two independent large-size and heavy driving units require an additional flitting device. Furthermore, opening of the crossing in such conditions may be considerable - up to 8–9 m. Keeping the roof support ŁP so wide open on the crossing requires the appropriate mining and geological conditions and engineering techniques. For that reason, the mining sector in Slovakia is now resorting to underhole systems using one gate conveyor, in this case the powered support has to act as the protecting support, having a relatively short roof bar (and the mined material is discharged via an extended cave protection cap). With weak coal roofs, this may periodically lead to support damage over the sections and in consequence to local rock slide in walls, making the face equipment a difficult task.

Nowadays this method is not used in Polish mines, mainly because of fire hazard and because coal has to be mined from thick seams located under surface – protected areas. In the case of thick anticline seams (No 501 and 510) mined beneath rural areas, this system might become a promising alternative to multi-level longwall mining. It seems that the underhole systems should be widely employed to mine deep anticline coal seams in the southern part of the Upper Silesia Coalfield, in the collieries managed by the Jastrzębie Coal Corporation. Cost-efficiency of the underhole system should be a sufficient reason to put it in practice in the remaining collieries working the coalbeds No 501 and 510. This technique should be adopted wherever the environmental impact assessment shows that this technology should produce no surface damage.

### 2.2. *Group B2 of coal seams*

This group includes 3 sub-groups of coal seams where the mining conditions preclude the use of the longwall system with cave-in. As regards the group B1 of seams, the main drawbacks include their small size and irregular shape due to the residual nature of mining lots and the dense grid of tectonic disturbances. The Group B has inclined coalbeds and B3 includes the seams where the mining operations are restricted by surface protection requirements.

#### 2.2.1. *Coal seams B1*

Most complicated tectonics of coal deposits in Poland is responsible for generation of residues where the long face mining is very difficult and sometimes impracticable. The face region may have a variable width, which means that the face to be mined has to be made longer or shorter, which occurs quite frequently in Polish collieries. In some cases, however, the cost-efficiency or safety considerations further add to the complications so instead of varying the face length, its shape is made regular at the price of leaving some residues. Residue plots sometimes emerge as a result of sedimentation tectonics disturbances, such as local compression or anticline features dividing the plot into smaller sectors, but sometimes forcing the miners to abandon them. For technical and engineering reasons, protecting pillars have to be left, too. It often happens that because of their dimensions and shapes these pillars cannot be mined or removed even though they cease to be used for surface support.

Protecting pillars are typically designed for protection of underground or surface structures, whilst barrier pillars are left in form of stripes left along the development headings. When mined by the long face method with caving-in, these pillars may lose stability and cease to perform their function. Pit pillars are a specific type of protecting pillars. Because of their size, they often contain large amounts of coal. In conclusion, residual and pillar plots are characterised by most specific geological and mining conditions, such as:
- small size;
- irregular shape;
- surrounded by goafs;
- surrounded by sedimentary or tectonic features;
- located in the proximity of existing gate roads.

As regards the mining technique, face equipment and organisation of mining work, the shortwall system is similar to the long face mining and can be employed to mine such plots (Fig. 6).

This system is also referred to as "open-end pillaring" system since coal is mined in the pillar lot, of up to 50 m in width.

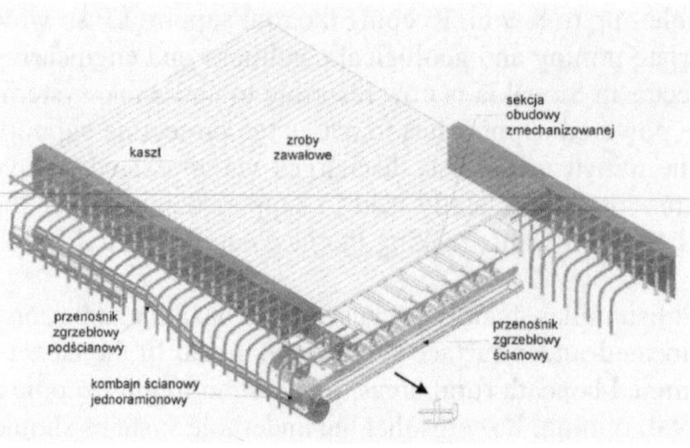


Figure 6. Shortwall system with caving-in (Tor 2007)

The equipment used in conventional openend shortwall systems need not differ from that used in long face mining. The face equipment includes a continuous miner (or a coal plough), a gate conveyor and power support sections. Because of the small face length (10–50 m), the daily output ranges from several hundred to about 1500 Mg. That is why the use of conventional high-efficiency mining machines does not seem justified, so shortwall system typically use single-arm continuous miners, such as KGU-130 manufactured by FAMUR company (no longer produced) or BESA-L, ESA-150-L (EICCOFF).

Costs involved in development of shortwall systems include the costs of driving two headings to contour the mining plot. The solution here might be to work the face along a single haulage and ventilation heading (Fig. 7).

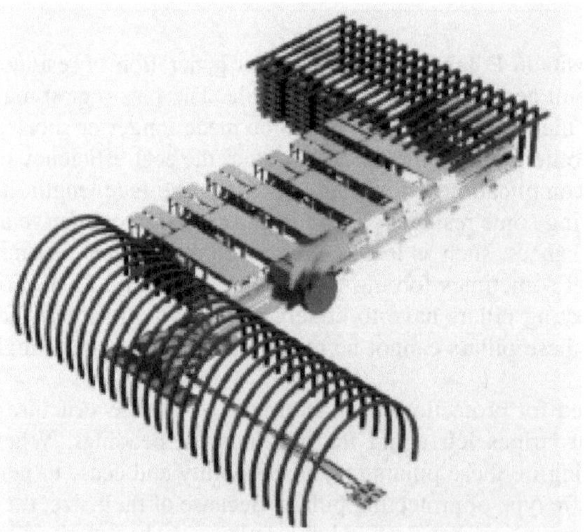

Figure 7. Shortwall system with caving-in, with the fenced ventilation heading behind the face (advertising materials supplied by Becker Warkop company)

However, that configuration requires a separate ventilation system in the face region, at the same time vastly reducing the costs of development of an open-end mining plot. Whenever the separate ventilation system proves to be too risky (because of methane hazard) or impossible to provide, the ventilation heading can be fenced off behind the face front. Such heading is drifted by the continuous miner alongside the face advance, and the roof support is set directly behind the gate conveyor. This solution is decidedly cheaper than driving a new heading before the coal extraction begins by traditional mining methods. This approach has been recently used in the colliery "Borynia" with great success (Tor 2007).

A part of coal reserves categorised as Group B are deposited in thick seams. Open-end shortwall systems seem a promising solution. The underlying principle is the same as of shortwall face systems, so the extraction works are done in the floor strata whilst the remaining portion of the output is hauled by one of the methods outlined previously. The costs of development works are reduced as they consist mainly in fencing off one of the tail gates in small plots or using a blind drift (particularly at low levels of methane risk).

Another solution is known as the cut-and-fill shortwall system, where the face is worked along a single pilot heading connecting a carrying gate and haulage gate. A mechanised version of such system incorporating the light frame support LOP, an uphill conveyor and the continuous miner ESA is operated in the colliery "Niwka-Modrzejów" (Fig. 8). The major draw-back, however, reducing the economic viability of the system is the need to use the hydraulic filling system. Besides, the use of conventional water-sand backfill produces huge amounts of water in the pilot heading and in transport and haulage headings.

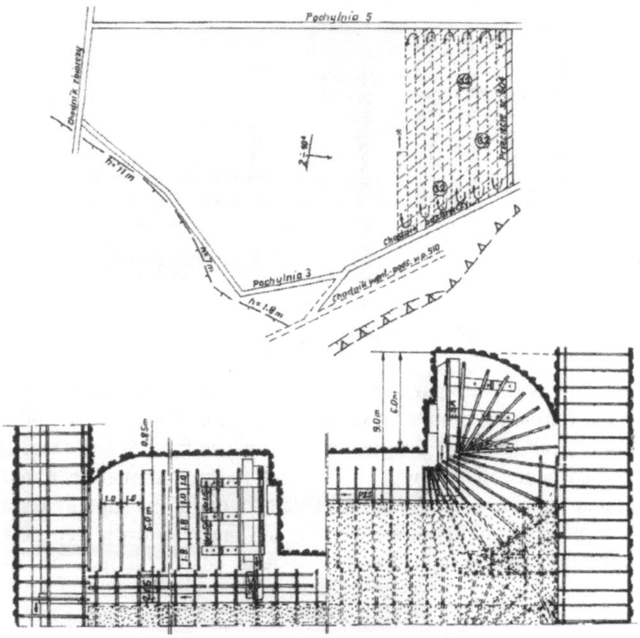


Figure 8. Cut-and-fill shortwall system in the colliery "Niwka Modrzejów" (Strzemiński 1999)

To a certain extent, this problem can be solved by using self-hardening or self-binding stowing materials made from various waste types. That would reduce the amounts of water present in the headings and help towards reducing the costs of the stowing operation. The widespread use of such system in the future does not seem very likely though and will be probably restricted to mining plots where surface protection requirements have to be satisfied.

Another group of systems that might be utilised to mine coal from pillars and residues are drift headings. Nowadays these systems are rarely used, recently only in the collieries "Staszic", "Siltech" and "Marcel". Historically, these belong to the oldest coal mining systems in deep-mines.

The general principle of mining operations is relatively simple: coal is mined selectively along the drifts (or along the dips or board gates) between the contour headings. Mining operations can be pursued also in headings contoured on side only and the extraction gallery will remain as a blind drift. Typically the extraction gate is 4–6 m wide and is supported by light props or bolts. Protecting pillars are left between the drifts, whose width is comparable to that of the drifts. Theoretically, coal can be subsequently mined from the pillars, too. Mechanisation processes are similar to those employed in road heading. The goafs are typically backfilled, using the hydraulic stowage or dry stowing systems.

A modified version of this system is the system of drifts and necks applied in the colliery "Staszic" (Fig. 9) (Juzek, Rojek 2007; Tajduś et al. 1999). In order to improve the face performance and to reduce material costs of roof supports, a system is designed whereby a pillar (8 m in width) between the extraction drifts was partially mined by necking, without any support. The drift extraction process involves two stages.

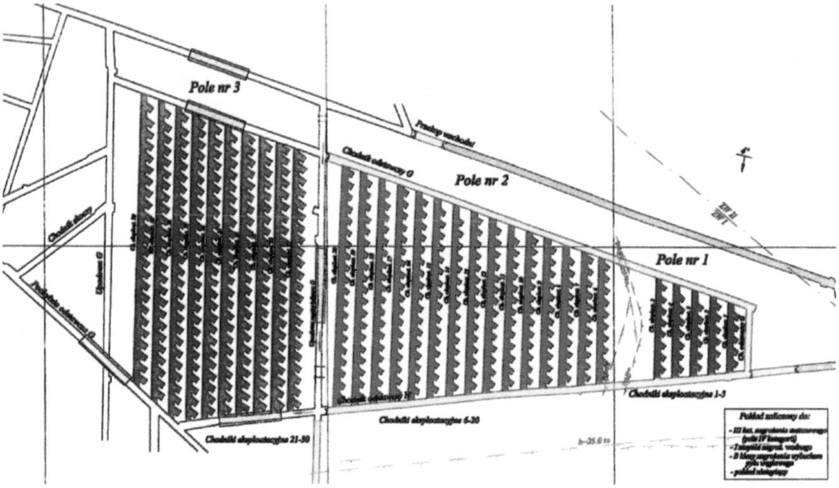


Figure 9. Panel extracted by the drift system with necking in the colliery "Staszic'

In the first stage an extraction drift is driven and supported by light props between the contour drifts. In the second stage the retreating mining machine makes sloped slots in the pillar. Thus cut necks are about 4.5 m in width (Fig. 10, dimension "a"), spaced by 2.5–3 m (Fig. 10, dimension "c") and left unsupported hence their length corresponds to the maximal opening of the continuous miner, assuming that the machine operator must remain under the supported roof section. After the last neck, the continuous miner makes a cut to drive another drift. Depending on the geological and mining conditions in the mine, the drift systems ensure the recovery rate of 50–70%.

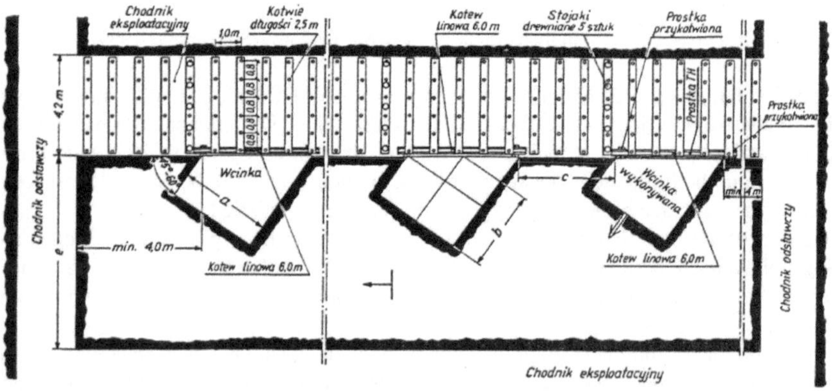


Figure 10. Drift system with necks in the colliery "Staszic" (Tajduś et al. 1999)

Because of the backfilling requirements and relatively low face productivity levels, these systems will be probably employed only to mine residual plots subjected to surface protection restrictions. The development of drift systems would be prompted if mechanised face systems became available that were based on continuous miners with integrated bolting units. Improving the face performance is the key determinant of cost-efficiency of the entire system. Alongside its several drawbacks, the drift system has numerous advantages, including:

- the shape and size of the mining plot play no important role;
- the mined output is relatively clean;
- technology based on standard solutions;
- good work safety features.

### 2.2.2. *Group B2 of coal seams*

Strongly inclined and sloping coal beds (the inclination angle in excess of 30°) were predominant in the Lower Silesia Coal Basin, no longer existing. In the regions of the Upper Silesia and Lubelskie Coalfields, slightly inclined coalbeds still tend to dominate. Sloping coal beds inclined at less than 30° have always been mined by longwall systems. In order to reduce the face sloping in the longitudinal direction, the faces are worked diagonally. Furthermore, to mitigate for the drag-down effect, the operations at the top gate crossing are delayed with respect to the gateway. Both operations help reduce the face inclination in the longitudinal direction, so there should be no obstacles to the longwall mining system. For coalbeds inclined at higher angles, alternative mining systems have to be adopted.

Leaving aside the ineffective semi-manual systems, the underhole systems can be chosen to mine seams of more than 2 m in thickness or, alternatively, open-end shortwall systems to mine thick seams. The classical example of the underhole system is that adopted in the colliery "Kazimierz-Juliusz" where an inclined and thick seam is mined by discharging the coal from a dip-worked face. The entire seam is divided into blocks and an extraction drift is excavated in the bottom of each block. The extraction drift is localised near the floor and is worked in the cross pitch direction (Fig. 11).

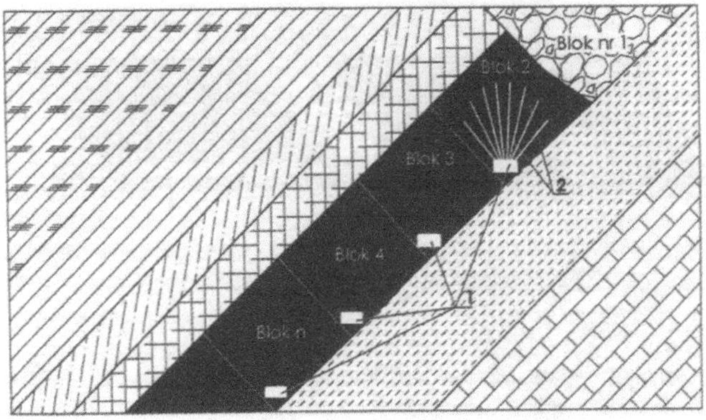


Figure 11. Coalbed sub-divided into extraction blocks (Tajduś et al. 1999)
1-extraction gallery; 2- blast holes

When the extraction drift is made, it is provided with two sections of specially designed powered support and a chain conveyor whose turning station is pushed behind the power support sections by about 5 m. Extraction involves blasting work above and behind the support sections, in the consequence the extracted coal falls onto the chain conveyor and is hauled from the face to other handling installations (Fig. 12). Once the mined material is hauled away, the conveyor operation is stopped and the face equipment is moved by about 2–3 m. The cycle is then repeated: blast hole drilling – charging – MW blasting – hauling the mined material – shifting the face equipment.

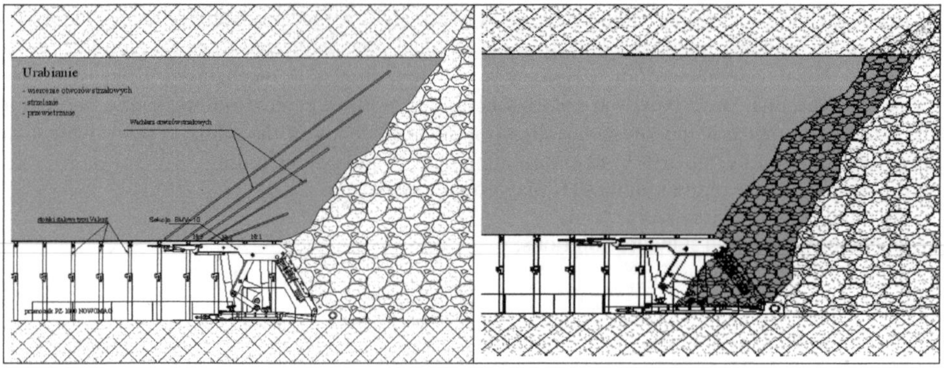


Figure 12. Schematic diagram of MW mining and haulage in the open-end shortwall system (Piechota et al. 2009)

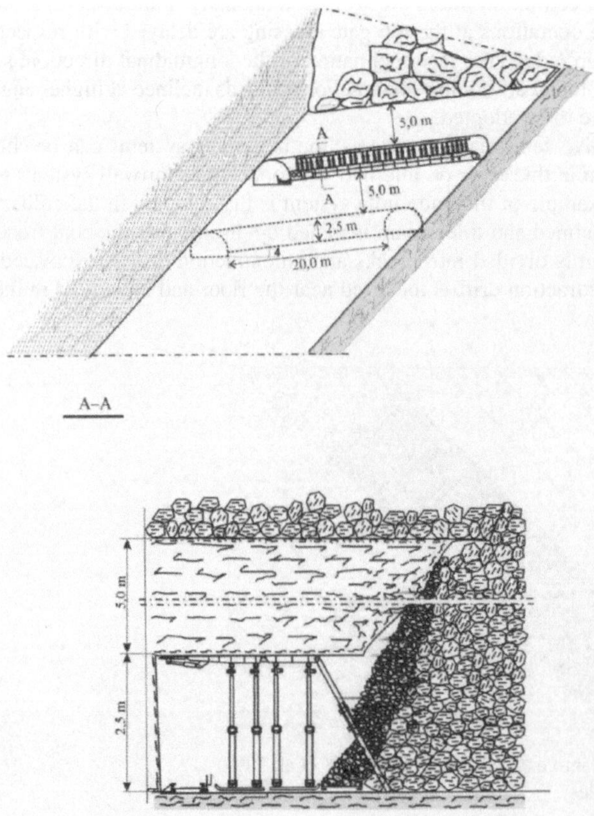


Figure 13. Cut-and-fill shortwall system using a light mobile support (Piechota et al. 2009)

The expertise gathered to date suggests that this technology might prove satisfactory when mining coal beds at least 2 m thick and sloping at more than 30°. In certain conditions this system can become a promising alternative to shearing of small and irregular residues of thick seams, even when they are inclined only slightly. However, further analyses are required, supported by testing done on mathematical and physical models.

Alternative methods of mining thick and inclined seams are cut-and-fill systems described in previous sections. Short faces are equipped with a support enabling the mined material to be discharged via a roof cap onto the face conveyor or an additional conveyor provided in the face region. In this case the thicker and more inclined the seams, the better system performance. In coal seams of variable thickness, the lighter face equipment can be used, that refers particularly to powered supports. Light mobile supports of relatively low mass allow the face to be lengthened or shortened, as necessary (Fig. 13). Short faces require lighter single-armed continuous miner machines. Operating costs may be reduced by fencing off the ventilation heading behind the face or the face can be worked with a separate ventilation system. Cut-and-fill mining of thick and inclined seams with the light frame support has been used with great success in a Spanish mine Hullera Vasco-Leonesa.

It seems now that application of such mining systems will render the mining of inclined coal seams a cost-effective undertaking. The shortwall working and drift mining systems should not be forgotten, either, as their cost-effectiveness is strongly related to the economic conditions and current coal prices on the market. The novel shortwall method was recently employed in the colliery "Kazimierz-Juliusz" in Poland to mine an inclined and thick coal bed No. 510. This mining method is referred to as sloped cutting. The coalbed is cut into a number of drifts under top wall and is worked downward, by short, inclined cuts (Fig. 14).

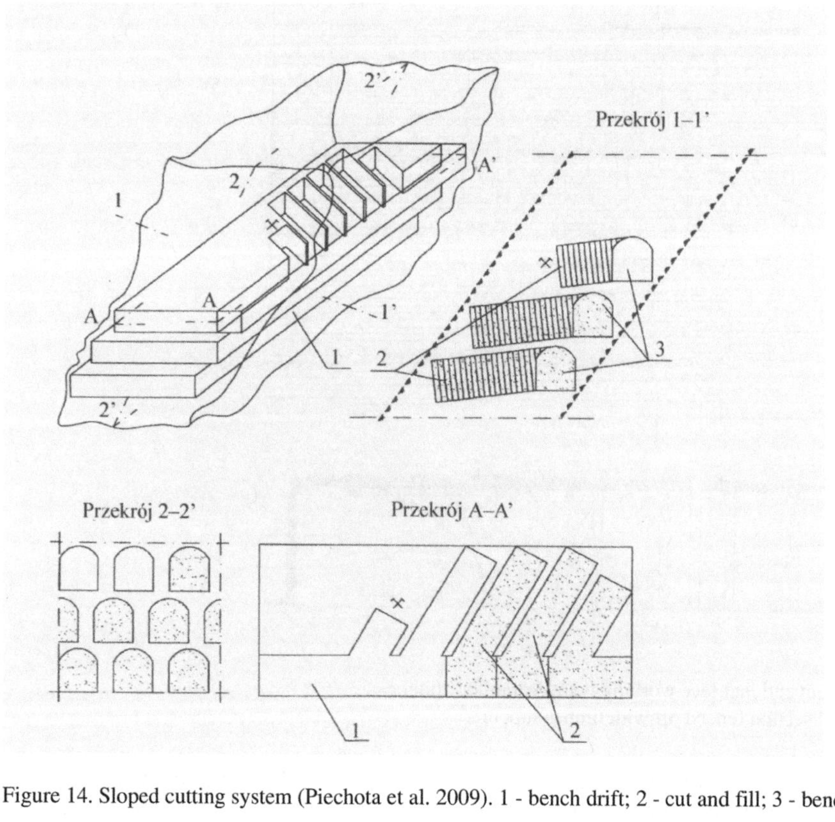


Figure 14. Sloped cutting system (Piechota et al. 2009). 1 - bench drift; 2 - cut and fill; 3 - bench drifts

The backfilling uses the conventional hydraulic stowage systems and the face is worked upwards. The coal fence 1–2 m in thickness is left between the juds and a coal shelf 0.5 m in thickness is left between the layers. In order to reduce the production costs, the juds are drawn between the spaced sections of the support ŁP using a continuous miner AM 50. The system, however, was abandoned when the coal price fell below the level that would guarantee breaking even.

2.2.3. *Group B3 of coal beds*

Recent analyses reveal that the approximated industrial reserves of coal located in protecting pillars account for nearly 20% of coal reserves. The only method of extraction in pillars is the cut-and fill system. Nowadays the dry stowage systems are not used in Poland, conventional hydraulic filling methods are in widespread use as they allow for reaching relatively high productivity levels – up to 3000 tons per day. Obviously, collieries employing only long face mining methods with hydraulic filling are not able to compete on the market with coal producers using the longwall systems with caving-in. In certain conditions application of both systems can prolong the life of a mine, at the same time making it possible to extract at least some part of coal reserves still remaining in protecting pillars.

Despite their long history and attempts at modifications, the cross face and diagonal face systems with hydraulic filling still have major drawbacks and no effective solution is available to install a self-moving stowage dam. Presently the backfilled section is fenced off manually with canvass. The front dam is based on timber support structured directly behind the spaced sections of the roof support (Fig. 15). The timber support in the backfilled section should protect from rock slide in unsupported parts. However, breaking the continuity of top wall layers will lead to caving in and spreading of the backfilling material to the working space in the face region.

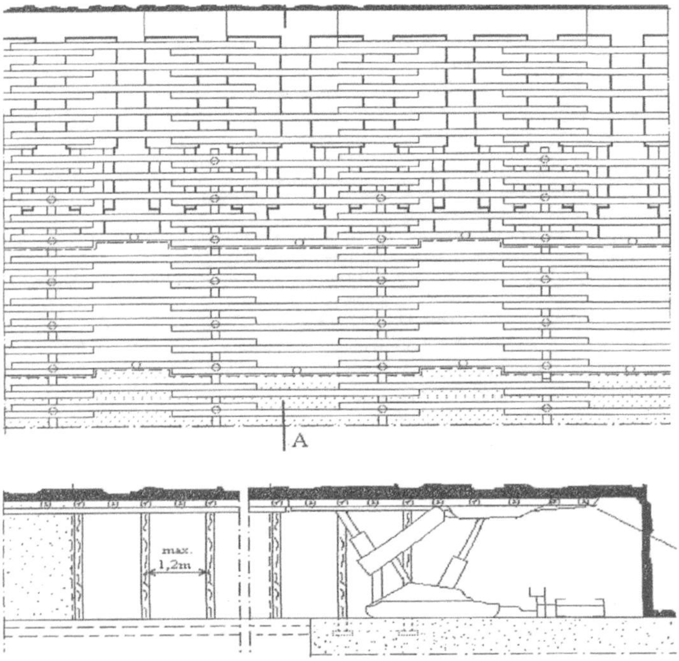


Figure 15. Conventional face working with hydraulic filling,
the backfilled section fenced off with timber supports

Cut-and-fill systems will be in widespread use because the coal reserves in the areas not requiring surface protection are fast running out. It is reasonable to predict that the self-moving dam will be designed and used in the modified system. When self-moving dams are used in mine workings with the tendency to rock sliding, the prop or bolting support will be required in the backfilled section. To reduce compressibility of the backfilling material, drainage channels will have to be cut in the backfilled space, which will act as sedimentation reservoirs. The water management will have to be modified accordingly and a water intake should be provided to eliminate the risk of contaminating the haulage and transport headings.

As regards residual panels and sloping seams in the areas subjected to surface protection requirements, mining companies will be forced to use alternative extraction systems. The performance of cut-and-fill shortwall systems as well as open-end systems or shortwall systems might prove unsatisfactory (Fig. 16).

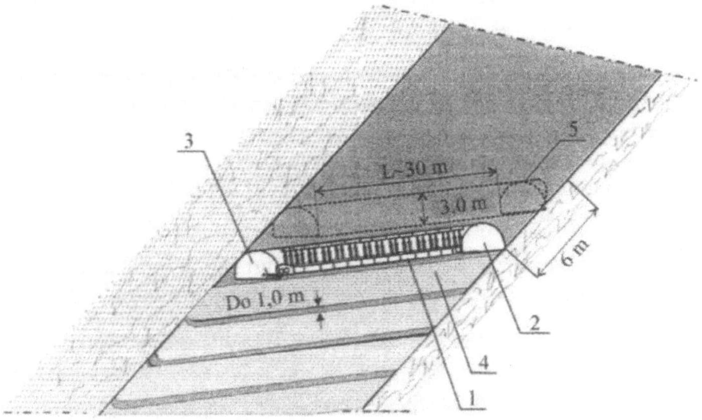


Figure 16. Open-end shortwall system with hydraulic filling (Tajduś et al. 1999)
1 - open-end; 2 - ventilation drift; 3 - haulage drift; 4 - coal shelf left in the bottom wall;
5- planned headings in the next bench

It is reasonable to expect that as coal reserves are shrinking, the open-end shortwall systems will be used alongside long face mining with caving in. To improve the efficiency of backfilling systems, the composition of the filling mixture will have to be modified to include mine sand, mine waste and to retain some part of water without compressibility losses. Using mine waste or waste from energy generation sector as ingredients of the backfilling mixture may help towards reducing the costs of filling systems and will produce positive environmental impacts.

## 3. CONCLUSIONS

Coal mining activities in the Upper Silesia Coalbeds continued for over 200 years have led to the depletion of the coal reserves. High levels of demand for coal coupled with strong competition between coal producers on the market encourage the discussions about the future of the Polish mining sector. One of the major issues to be addressed is the selection of mining techniques which, given the increased mining depth, pose a major challenge to mine operators and mining research teams. The analysis of potential developments in the coal mining sector in Poland leads us to the following conclusions (Burtan et al. 2008):

- longwall systems with caving in will still remain as the main extraction technique allowing the Polish producers to effectively compete on the domestic coal market;
- development of long face mining systems should be focused on further improvement of face productivity through the application of automated system and monitoring procedures;
- sheared walls and continuous mining of thin beds give an excellent opportunity of cost-effective and safe extraction of thin beds (up to 1.5 m in thickness);
- wherever surface protection requirements allow the cave-in of thick seams, mining companies should promptly adopt the shortwall systems and open-end shortwall systems to mine residues;
- shrinking coal reserves necessitate the extraction of coal from protecting pillars and residues;
- short face systems have become a most promising method of residue mining because the

operations can be fully mechanised, assuring high efficiency in comparison to alternative extraction systems;

- drift systems can be well used to mine irregular residual fields, as the operations are mechanised, the costs of development works are minimal and the roof bolting can be employed wherever the mining and geological conditions are appropriate;
- to improve the efficiency of drift systems using hydraulic filling, these systems should be considered as a major element in the management of mining waste, also utilising the waste from metallurgical and power generation plants;
- sloping seams in favourable mining conditions can be mined at relatively low costs from the extraction gallery by the benching system or by the open-end shortwall system;
- the presence of coal reserves in the areas subject to surface protection requirements prompts the search for more effective long face solutions using hydraulic filling;
- one of the ways to keep the cut-and-filled faces competitive is to improve the efficiency and reduce the coal production costs by application of mechanised face dams and using mine waste for backfilling purposes;
- intensive research is merited in the area of mechanisation, automation, production systems and work organisation in the extractive sector, to ensure the optimal utilisation of the domestic coal reserves;

## REFERENCES


Burtan Z., Rak Z., Stasica J. 2008. Priorytety rozwoju systemów wybierania złóż w polskim górnictwie węgla kamiennego. Gospodarka Surowcami Mineralnymi. Tom 24, Zeszyt 1/2. Wydawnictwo IGSMiE PAN. Kraków. (In Polish)

Piechota S., Stopyra M., Poborska-Młynarska K. 2009. Systemy Podziemnej Eksploatacji Złóż Węgla Kamiennego, Rud i Soli. Wydawnictwa AGH, Kraków. (In Polish)

Strzemiński J. 1999. Możliwości mechanizacji przy stosowaniu krótkofrontowych systemów wybierania węgla. Mechanizacja i Automatyzacja Górnictwa. (In Polish)

Juzek G., Rojek A. 2007. Wybieranie resztki pokładu 405 w KWK „Staszic". Bezpieczeństwo Pracy i Ochrona Środowiska w Górnictwie, nr 7. (In Polish)

Tajduś A., Kluka J., Rak Z., Stasica J. 1999. Prototypowy system wybierania węgla chodnikami w obudowie kotwiowej i wykonywanymi z nich wcinkami. Przegląd Górniczy, nr 3. (In Polish)

Tor A. 2007. Opracowanie innowacyjności stosowanych technologii eksploatacji węgla kamiennego w Jastrzębskiej Spółce Węglowej. Praca zrealizowana w ramach projektu „foresight węglowy" pt. „Scenariusze rozwoju technologicznego przemysłu wydobywczego węgla kamiennego". Katowice (unpublished). (In Polish)

Zorychta A., Burtan Z. 2008. Uwarunkowania i kierunki rozwoju technologii podziemnej eksploatacji złóż w polskim górnictwie węgla kamiennego. Gospodarka Surowcami Mineralnymi. Tom 24, Zeszyt 1/2. Wydawnictwo IGSMiE PAN. Kraków. (In Polish)

Zorychta A., Chojnacki J., Krzyżowski A., Chlebowski D. 2008. Ocena możliwości wybierania resztkowych partii pokładów w polskich kopalniach węgla kamiennego. Gospodarka Surowcami Mineralnymi. Tom 24, Zeszyt 1/2. Wydawnictwo IGSMiE PAN. Kraków. (In Polish)

International Mining Forum 2010, Liu et al. (eds) © 2010 Taylor & Francis Group, London, UK. ISBN 978-0-415-59896-5

# Injectory firming of a rock mass in the area of rebuilding an excavation as a way to prevent rocks from falling and from the results of these fallings

Tadeusz Rembielak, Dariusz Chlebowski
*AGH University of Science and Technology, Krakow, Poland*

ABSTRACT: A series of rebuilding of excavations with a corroded ŁP lining was performed in the KWK "Piast" mine, without a necessity of limiting their basic "motion" functions. One of the ways to increase safety during performing the rebuilding of crossroads and forks is applying injectory rock mass firming in their surrounding. An example of a rebuilding technology of the crossing 652 with by-pass to the shaft 3 on the level of 650 meters was presented in the paper.

## 1. INTRODUCTION

In active underground mines there is a permanent necessity to rebuild excavations, for of a number of reasons, such as for example considerable deformations of lining or corrosion of its substantial elements (Figure 1.). The work connected with firming of a rock mass, before starting a rebuilding, is usually limited to certain parts of an excavation and crossroads, as well as forks. It prevents from roof rocks embank and fall, and by this it also prevents from dangerous events with humans participation occurrence (Rembielak 2002, 2007).

Gliwicki Zakład Usług Górniczych Spółka z o.o. realised rebuilding of a fork cross-cut no 652 with a by-pass to a shaft 3 on 650 m level in KWK Piast, which was preceded by performance of necessary geotechnical interventions, by injectory firming of a rock mass and its surrounding (Fig. 1).

Before the rebuilding was started examinations of rock mass gaps was performed in 5 holes drilled in this fork's surrounding, and then based on the obtained results spatial distribution of injectory holes was worked out, their essential length was estimated, and a recipe of the firming-sealing agent was formulated.

Within the worked technology 48 injectory holes, at the length of 2 m and diameter of 42 mm each, were drilled in 8 sections. Through the injectory holes and through 6 holes for examining the gaps 0.708 $m^3$ of sealing-firming agent was pressed to the rock mass. After having performed injection, in the strengthened section of the cross-cut, 6 control holes, into which examinations on gaps were performed, were drilled in the roof. It was affirmed that the injectory work was performed correct, the technology proved to be efficient, and as a consequence of that it was possible to start rebuilding the fork.

## 2. GEOLOGICAL-MINE CONDITIONS IN THE AREA OF CROSS-CUT NO 652 FORK WITH A BY-PASS OF THE SHAFT 3 ON 650 M LEVEL

The cross-cut no 652 and a by-pass to the shaft 3 are stone excavations of a long-term usability, performed in medium-grained sandstone of the total thickness of about 68 m, located under the stratum 207 on the level of 650 m in the central part of the mine area in KWK Piast (KWK Piast

2010-2013) and fulfilling basic functions of transportation, haulage, ventilation and crew movement.

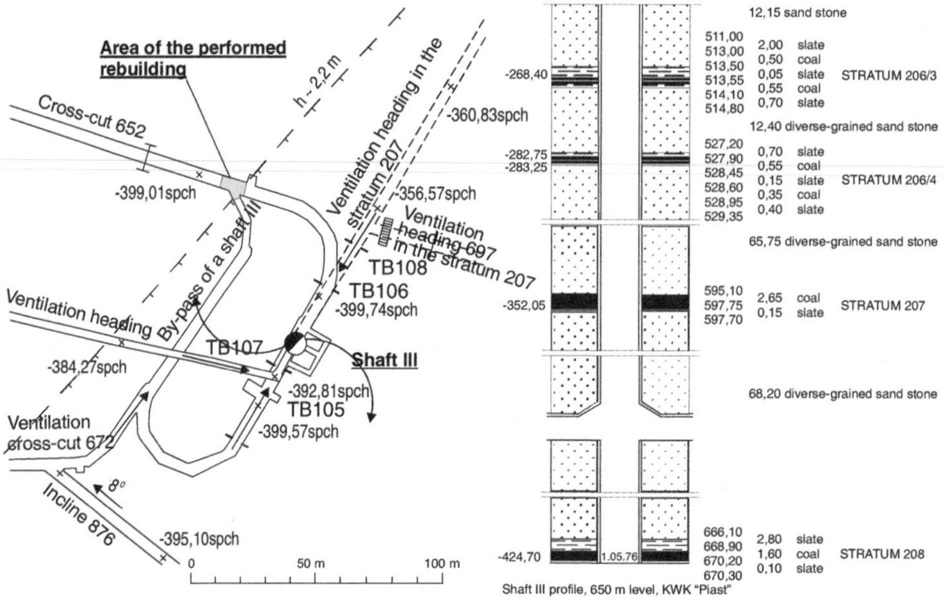


Figure 1. Map of excavations and a profile of shaft III on 650 m level

Figure 1 presents a draft of a map of excavations and a sector of lithological profile of the rocks, according to the typical data, as of its location at a vertical distance of around 60 m to the ventilation shaft 3 (eye ordinate + 258 m).

## 3. STRESS-STATE CREATION

Maintaining stability of an excavation in coal mines conditions is essentially important from the point of view of guarantee and its functionality. The process of losing stability manifested by lining destroying (not necessarily flexible) or an excessive movement of its elements as a result of increased weight from the part of the rock mass is amongst the other things a consequence of phenomena occurring in a surrounding rock centre as well as affecting chemically aggressive underground water and mine atmosphere.

A range and kind of excavation's influence on the neighboring rock mass area has a local character, however, depends on the size of lithostatic stress occurring at a given depth, geomechanical properties of the centre and its dimensions (width and height). Because, depending on the function, the size of cross-section of the excavations can change in a wide range, in practice the spheres of these influences are diversified. In certain geotechnical situations disturbance by performing a primary excavation the balance state of a rock mass and passing to the secondary balance state can be accompanied by creation of destruction area around the edges, as a result of exceeding the resistance to stretching, pressing or cutting. In other cases the created stress state causes solely deformations of springy character, provoking accumulation of a particular amount of energy of these deformations in the roof and the side wall, which can be released later after eventual achieving critical effort value. If for a number of reasons this phenomenon will not take place, the state of increased stress will stay until it relaxes as a result of relaxation processes (Goszcz 1999).

Taking into consideration the above, as well as a long-term, although difficult for a precise estimation, period of existing mentioned excavations, numeric simulations were performed, which were characterized by creating stress state in the nearest surrounding of the fork the time function, which passed since it was performed. It was assumed, in accordance with the results of in-site observation, that the rock wedge of a 0.4–2 m width between the cross-cut 652 and by-pass to the shaft 3 forms a cracked circle projecting onto a considerable range of the exposed roof.

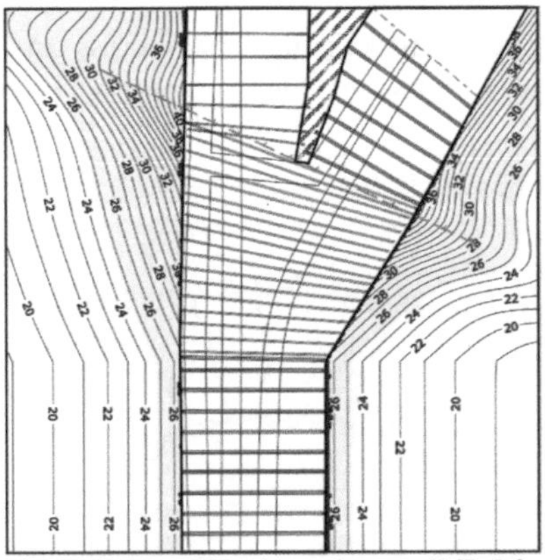
Figure 2. Map of horizontal stress [MPa] for t=0

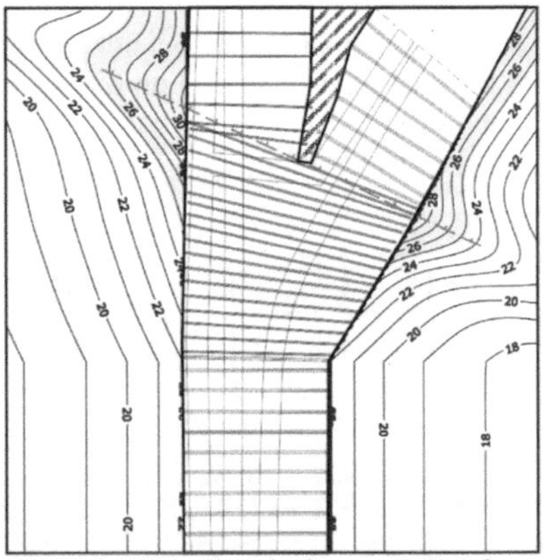
Figure 3. Map of vertical stress [MPa] for t=2

The results of the undertaken calculations are illustrated in contour-maps of the vertical variable pseudo-primary stress state at the level of analyzed excavations (fork) directly after having performed them (Fig. 2) and after two (Fig. 3), five (Fig. 4) and ten years (Fig. 5). On maps (Fig. 2 and 3) can be found areas highlighted in yellow colour, in which the value of vertical stress concentration factor value exceeds 1.5 (lack of these in Fig. 4 and 5), meaning the places, where the rocks pressure is 50% higher than at the primary state, corresponding to an undisturbed rock mass, which amounts to about 16,5 MPa for this depth.

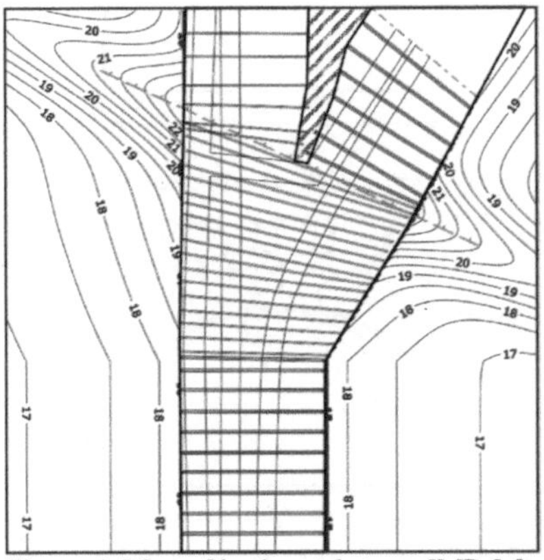
Figure 4. Map of horizontal stress [MPa] for t=5

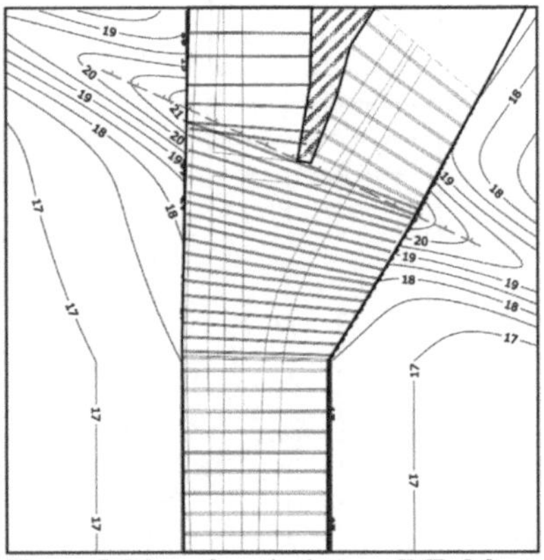
Figure 5. Map of vertical stress [MPa] for t=10

It can be concluded, from the presented distributions, that at the nearest area of the discussed fork occur raised local zones of compressive stress, which are determined by the width of opening the roof and a presence of fault thrust up to 2.2 m (Chlebowski 2009), a range and level of which successfully decreases with the time flow.

The analysis of obtained distributions in connection with the values of stress (in relation to uniaxial compression) of the surrounding rock circle shows a possibility of occurring local zones of cracking and destroying a primary structure of the rock mass in the closest neighborhood of excavations edges (taking into particular consideration the branched segment situated close to the fault).

## 4. CHARACTERISTICS OF GEOTECHNICAL INTERVENTIONS PRECEDING CROSS-CUT NO. 652 FORK REBUILDING WITH SHAFT III BY-PASS

### 4.1. *Examining gaps in a rock mass before injection*

Before starting formulating the technology for firming of roof rocks in the area of cross cut no 652 fork with the by pass shaft 3 on 650 m level a range of rock mass gaps examinations were planned and performed with applying peephole, which was essential for estimating the rock centre state behind the lining.

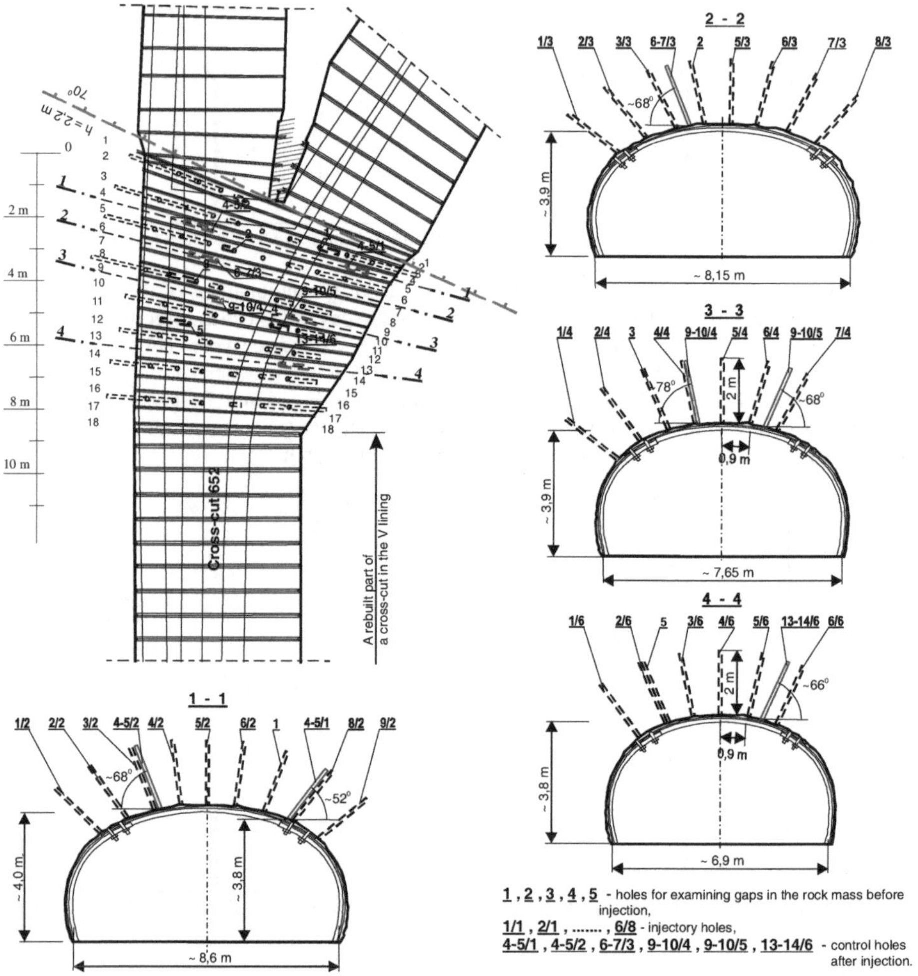


Figure 6. Diagram of injectory holes and control holes distribution for measuring gaps

In order to realise the established aim, 5 examination holes of 2 m length and 42 mm diameter each were drilled. These holes were performed in four sections perpendicular to the cross-cut axis and shaft 3 by-pass (Figure 6) within the area of the biggest exposure of the roof. A detailed scheme of drilling holes for examining gaps and firming of a rock mass in a surrounding of the subject fork was presented in Figure 6. Exemplary profiles of examination holes prepared on the basis of research obtained by a peephole before the injection are presented in Figure 7.

4.2. *A technology of rock mass firming in a surrounding of a fork*

In order to firm a rock mass in a surrounding of a fork cross-cut no. 652 with a shaft 3 by-pass on 650 m level it was decided to drill in the roof part of the fork a range of 48 injectory holes in 8 separate appropriately selected sections situated between the double timber (Figure 6).

In order to achieve the best possible interpenetration of the sealed rocks after having injected each hole, it was proposed to create their net in a shape of chessboard, however the inlets of the succeeding injectory holes at the edge of excavation were designed at the distance of 0.9 m. The 5 holes, which were drilled to examine gaps of the rock mass, were also used for the injection (Figure 6). As a consequence of the realised drilling work through all the 54 holes it was planned to press a sealing-firming medium, prepared on the basis of the Portland cement CEM I 32.5, mining plaster, and slaked lime, at a pressure not higher than 0.3 MPa.

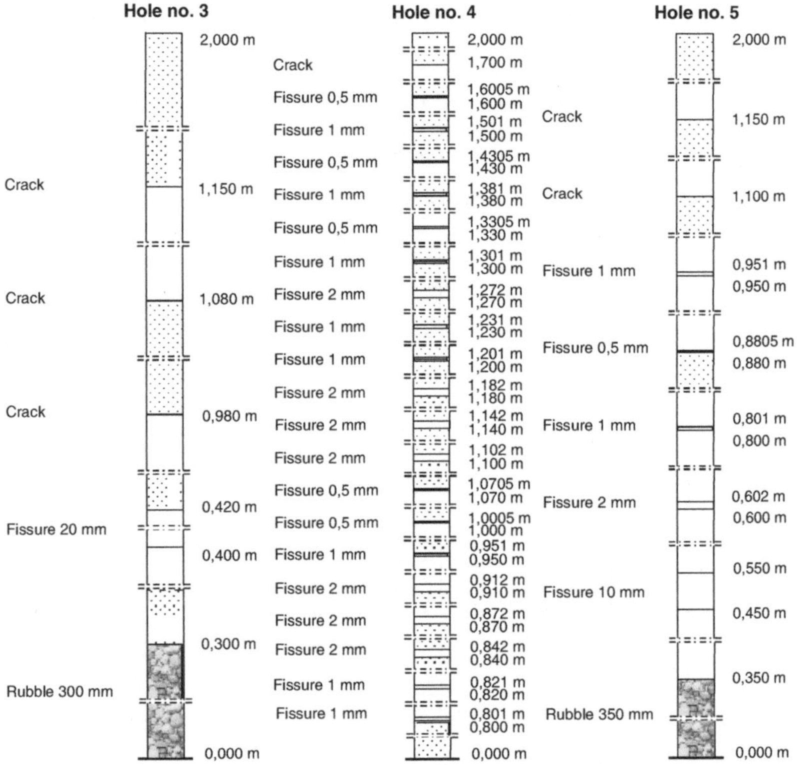


Figure 7. Exemplary holes profiles performed on the basis of results obtained by using a peep-hole before injection

During performing the injectory work in accordance with the formulated technology of firming the rock mass in a surrounding of cross-cut no. 652 fork with a shaft 3 by-pass on the basis of the drilled 48 injectory holes in 8 sections, and 5 examining holes to research the gaps, in total 0.708 $m^3$ of a sealing-firming agent was pressed to the rock mass. In the applied technology injectory installation was proposed, including a mono-helical pump Pdk type from Minova Ekochem S.A. Company and mechanically expanded sealing heads.

Table 1 presents amounts of sealing-firming agent pressed to the rock mass in a surrounding of the firmed fork in each section through the injectory holes. During pressing the firming agent through some of the injectory holes penetration of the pressed medium at a distance from 2 to 3 m from the presently injected hole was observed.

Table 1. Characteristics of pressing the inject in individual sections and holes

| No. | Section no. | Injectory hole no. | Amount of the pressed inject through the holes, $dm^3$ | Amount of the inject in section, $dm^3$ |
|---|---|---|---|---|
| 1. | 1 - 1 | 1/1, 2/1, 3/1 | 18, 18, 12, 0 | 48 |
| 2. | 2 - 2 | 1/2, 2/2, 3/2, 2, 5/2, 6/2, 1, 8/2, 9/2 | 18, 18, 12, 12, 0, 0, 6, 12, 12 | 90 |
| 3. | 3 - 3 | 1/3, 2/3, 3/3, 2, 5/3, 6/3, 7/3, 8/3 | 18, 24, 12, 18, 0, 12, 12, 12 | 24 |
| 4. | 4 - 4 | 1/4, 2/4, 3, 4/4, 5/4, 6/4, 7/4 | 18, 18, 12, 18, 0, 18, 24 | 108 |
| 5. | 5 - 5 | 1/5, 2/5, 3/5, 4/5, 4, 6/5 | 18, 12, 12, 6, 12, 18 | 78 |
| 6. | 6 - 6 | 1/6, 2/6, 5, 3/6, 4/6, 5/6, 6/6 | 18, 18, 18, 6, 6, 18, 24 | 90 |
| 7. | 7 - 7 | 1/7, 2/7, 3/7, 4/7, 5/7, 6/7 | 18, 18, 12, 0, 12, 18 | 78 |
| 8. | 8 - 8 | 1/8, 2/8, 3/8, 4/8, 5/8, 6/8 | 18, 18, 12, 12, 24, 24 | 108 |
| TOTAL | | | | 708 |

### 4.3. *Examining rock mass gaps after the injection*

After having realised the injectory firming of a roof fork cross-cut no. 652 with a shaft 3 by-pass between the injectory holes in the sections: 1 - 1, 2 - 2, 3 - 3 and 4 – 4, six additional control holes of a 2 m length and 42mm diameter were drilled to perform repeated gap examinations (Figure 7). Based on the subject examinations results the correctness of injectory work realisation was performed. Figure 9 presents exemplary profiles of control holes performed on the basis of research results obtained from optical peep-hole after the injection.

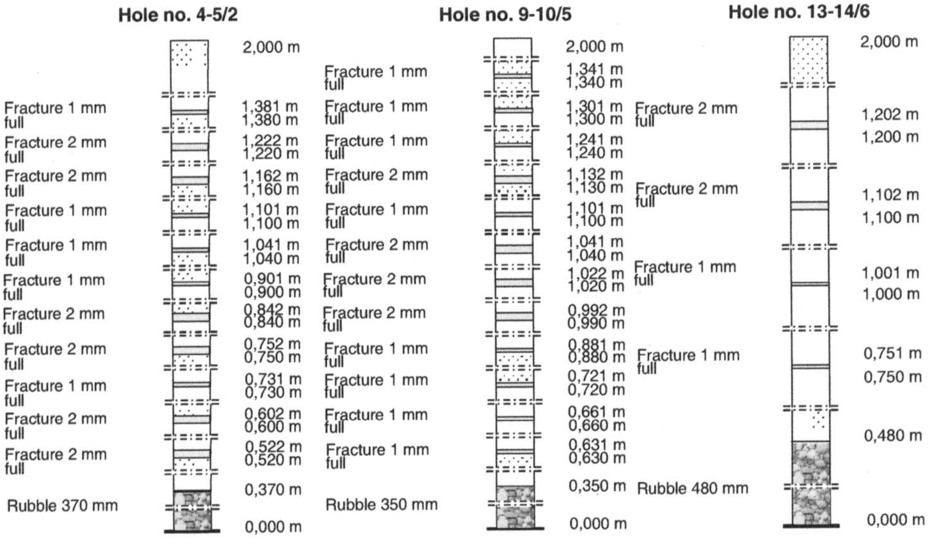


Figure 8. Control holes profiles performed on the basis of results obtained from optical peep-hole research after the injection

Having analysed all the profiles of control holes it was concluded that the injectory work realised within the formulated rock mass firming technology in a surrounding of a cross-cut no. 652 fork with a shaft 3 by-pass on 650 m level were performed correct. As a result of that it was possible to start rebuilding the fork. The presented mining work was realised without any problems and with maintaining an appropriate safety level of the crew.

## 5. CONCLUSIONS

During performing rebuilding of the excavations, their crossroads and forks, it is necessary to perform appropriate preceding geotechnical interventions, preventing from rocks fall and their consequences, such as dangerous occurrence with a participation of the employed crew.

Applying the injectory sealing and firming of a rock mass preceding the excavations' rebuilding, in particular with a corroded steel lining, proves to be an efficient way to limit a risk of such situations occurrence, and at the same time aiming at improving safety of the performed work.

The formulated and realized technology of rock mass sealing and firming in a surrounding of a cross-cut no. 652 fork with a shaft 3 by-pass on 650 m level in KWK Piast proved to be efficient, which enabled a safe rebuilding, and by this the aim of injectory work was fully achieved.

In order to estimate the stress and effort of the rock centre from the point of view of maintaining excavations stability in the area of designed renovation work, it can be helpful to acknowledge the results of analytical or numerical modeling, based on which it is possible to determine the static weight of a lining, character and range of the zones of damaging the structure, the probable level of stress relaxation, and a degree of rocks reconsolidation in the relaxed areas, taking into consideration a dose of a probability, depending on the quantity and quality of the data.

The achieved positive result of rock mass firming in the area of crossroads before rebuilding them, with a low engagement of force and means for realization of the project, and in particular decreasing a risk during the work realization from a small to permissible, justifies fully purposefulness of applying injectory sealing and firming of rocks as an effective way of solving problems of technical-motion nature, with maintaining an appropriate safety level, which should always, not only in the case of underground mining, be a priority.

## REFERENCES


KWK Piast 2010-2013. Plan Ruchu KW S.A. KWK Piast na lata 2010-2013 (unpublished materials). (In Polish)

Chlebowski D. 2009. Wpływ wybranych parametrów geomechanicznych na kształtowanie się przemieszczeń i naprężeń pionowych w sąsiedztwie uskoku. Miesięcznik SITG Przegląd Górniczy 65(1-2)(CV) (In Polish)

Goszcz A. 1999. Elementy mechaniki skał oraz tąpania w polskich kopalniach węgla kamiennego. Biblioteka Szkoły Eksploatacji Podziemnej. Kraków: Wydawnictwo Instytutu Gospodarki Surowcami Mineralnymi i Energią PAN. (In Polish)

Rembielak T. 2002. Zvýšovanie bezpečnosti práce v baníctve pri používaní injektovania hornín. Międzynarodowa Konferencja „Nerastné suroviny Slovenskej Republiky". Slovenská Banícka Společnosţ. Zbornik Prednasášok. Demänovska Dolina (In Slovakian)

Rembielak T. 2007. Injektory Rock Mass Firming During Rebuilding of Headings. BICTI Донецького Гірничого Інституту. Всеукраїнський науково-технічний журнал гірничого профілю (2), Донецьк. (In Russian)

International Mining Forum 2010, Liu et al. (eds) © 2010 Taylor & Francis Group, London, UK. ISBN 978-0-415-59896-5

# Technologies used to reinforce preparatory headings before longwall exploitation front in conditions of Polish hard coal mines

Jerzy Stasica
*AGH University of Science and Technology, Krakow, Poland*

ABSTRACT: This article presents technologies used to strengthen preparatory headings in the conditions of Polish hard coal mines. The researchers focused on the need to reinforce the linings of headings, especially those at longwalls, which are subject to frequent deformations in areas of exploitation pressures in front of and behind longwall exploitation fronts and at crossings with exploitation longwalls. The analysis conducted pointed out directions of development in the reinforcement technologies provided while driving the headings discussed, and while maintaining them in various mining situations.

## 1. INTRODUCTION

The exploitation of coal beds using longwall systems (Fig. 1) requires the construction of longwall headings, often with long runs reaching lengths of up to thousands of metres. In areas subjected to exploitation pressures generated by the mining process, the linings of longwall headings are subject to significant loads and deformations, which reduce the heading cross-sections. As a consequence, their functionality is diminished and they can be difficult to reuse (Zimonczyk et al. 2006).

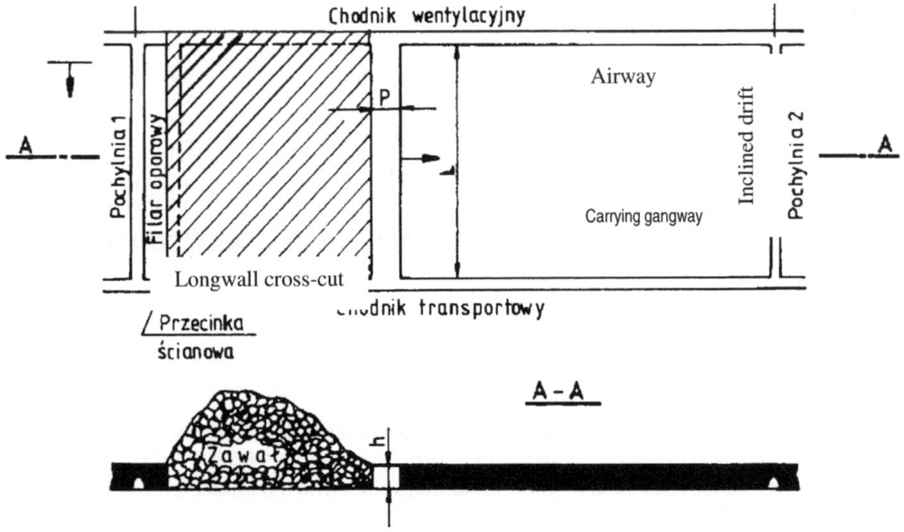


Figure 1. Longwall exploitation system (Piechota 2003)

Three-wall headings allow output haulage within the bed boundaries, delivery of equipment, machines, materials and power necessary during exploitation, comfortable and safe personnel traffic in the longwall area, supply of fresh air and removal of used air, and drainage of inflowing water. It is necessary to keep these headings in a condition that guarantees that they carry out the functions for which they were designed and driven, while not disturbing planned longwall progress. Certainly, this translates into financial results for the mine.

Difficult geological and mining conditions in which longwall headings are located can create the need for hard coal mines to use various methods to reinforce the linings of these headings in front of and behind longwall exploitation fronts.

Longwall headings are strengthened mainly in order to:

- improve their stability in front of the longwall front (in order to sustain them for the longwall currently under exploitation and at crossings with that longwall);
- improve their stability behind the longwall front (in order to sustain them for the next planned longwall).

## 2. CONDITION OF ROCK MASS EFFORT IN THE VICINITY OF LONGWALL HEADINGS

In the area of longwall exploitation, support linings in longwall headings are subject to intensified rock mass impact. Heading functionality may be lost if the lining stability is compromised. Potential hazards occur not only from the roof side, but also from the working side and the heading floor. Exploitation pressure is one of the most significant factors affecting stability of heading lining works near the exploitation front. The effect of the rock mass pressure known as exploitation pressure is felt in front of the extracted longwall front. Within the area of the exploitation zone, the gallery linings are subject to increasing loads. The effects of exploitation pressures in galleries at longwalls are manifested by increasing intensity of vertical and horizontal compression of the heading, deformation of the lining and floor uplift. Depending on depth and geological and mining conditions, these effects most frequently appear at the distance ranging from 40 to 100 m before the longwall front and intensify in the direction of the longwall front. Usually, highest stress concentration occurs at the longwall front crossings with galleries at the longwall. Figure 2 shows the stresses in the area of the longwall heading.

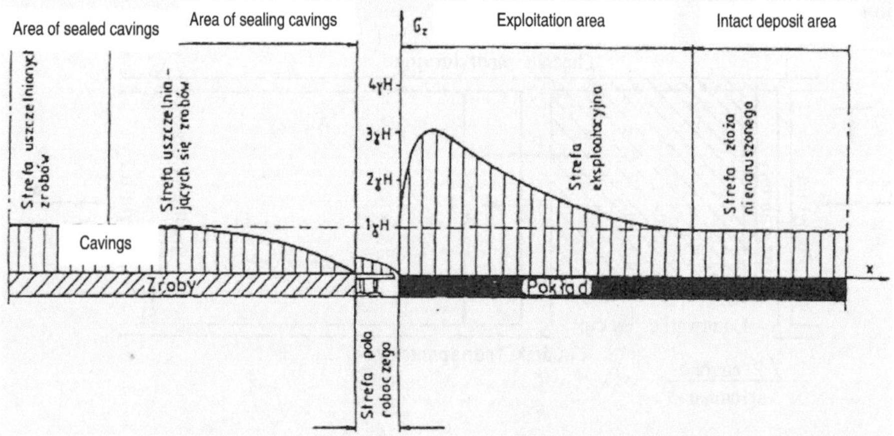


Figure 2. Diagram of stress distribution in the area of the longwall heading (Piechota 2003)

For characteristic deformations of galleries at longwalls, which include:

- settlement of roof rocks,
- floor uplift,
- shifting of side walls (horizontal compression),

it is possible to distinguish two time frames of occurrence of these rock mass movements:

- the period after the headings have been driven, when they are not exposed to exploitation impact,
- the period in which galleries at longwalls are in the area affected by exploitation pressures (Majcherczyk et al. 2008, Prusek 2008).

## 3. STAGES OF EXISTENCE OF LONGWALL HEADING

Three stages characterise the existence of a gallery at a longwall. They are determined by gallery position relative to exploitation front, and the consequently variable load conditions to which any gallery at a longwall is exposed during its service life.

Immediately after being driven, longwall heading is surrounded by the coal body on both sides. This is the first stage of its existence. It is then subjected to loading by layers, which are deposited in its roof and which are subjected to relieving. This loading condition persists until the longwall effect occurs. The front of that longwall moves along one of the side walls in the given heading. The heading then enters its second stage of existence, when disturbed rock mass becomes a strenuous rock mass, and the beginning of approaching longwall effect occurs (heading load increases). The third stage of existence of the longwall heading begins when the rock mass surrounding it becomes a relieved rock mass.

Additional factors (apart from longwall impact) affecting stability of longwall headings primarily include (Grzybek 1996, Prusek, Kostyk 2000):

- heading depth,
- exploitation edges,
- tectonic disturbances,
- location of heading in edge sections of protective pillars,
- longwall exploitation height,
- time existence of heading,
- strength of roof and floor rocks in the bed exploited,
- heading cross section shape, its overall dimensions, and lining type,
- lining function with rock mass.

## 4. REINFORCEMENT OF LONGWALL HEADINGS BEFORE THE LONGWALL EXPLOITATION FRONT

Longwall headings existing in unfavorable geological conditions, especially in areas of impact of the exploitation front, are subjected to various forms of deformation, followed by the effect of reduction of heading cross-section to dimensions that make its further function difficult. In order to prevent this effect, it is necessary to design and provide optimum reinforcement for the heading lining as early as is possible.

Delay in reinforcement of the lining may result in its deformation or in the worst case in the loss of its stability. The following are some of the most typical operations aimed at restoring a heading to its original functionality:

- floor taking,
- rebuilding of gallery headings using rock ripping,
- reinforcement of the heading lining structure.

Floor taking is conducted using manual methods, shotfiring works, or using machines for partial mining and loading. The organisation of floor taking must be adapted to the circumstances and the function of the heading. As a rule, floor taking needs to be carried out without causing interruption of heading functionality, especially output haulage. Therefore, heading floor taking is organisationally subordinate to other functions that are being carried out at the same time. Manual taking of squeezed bottoms is carried out using rippers and pickaxes, while the broken and loosened output is loaded by hand onto mining cars. These works are partially aided by shotfiring works carried out using small charges of explosive material, aimed at ensuring additionnal rock crushing and loosening. As a rule, these methods of floor taking are generally characterised by low productivity. Self-propelled machines, known as roadheading machines, working alongside the means of haulage employed, are used to increase productivity and reduce the length of floor taking operations.

Another way to restore the original heading shape and its functionality involves its rebuilding using roof rock ripping. These works need to be carried out very carefully, and as far as their organisation is considered are categorised as heading management operations. Rock may be ripped using power hammers – as a rule with the help of shotfiring with limited charges of explosive material. Lining adapted to new heading cross-section is provided after having finished required ripping. The following factors are crucial in making decisions concerning rebuilding of a heading: access to the heading interior, removal of the heading from service, carrying out of works "under operational conditions", geological conditions, and the mechanical condition of the old lining.

In order to avoid reconstruction and reduce floor rock taking, it is most convenient to reinforce the longwall heading before longwall exploitation front, still in the area that is surrounded on both sides by the coal body, which is in a zone free from exploitation pressure. Mining practice proves that greatest heading reinforcement effects are attained when it is carried out in advance relative to the longwall exploitation front line, at a distance no less than 60÷80 m. The best reinforcing effects are achieved when the heading lining is reinforced when it is driving or at a small distance behind the front of the driven heading face. Most frequently, these reinforcements are provided by means of:
- reduction of bearing lining spacing,
- use of closed lining,
- supporting the sets of bearing lining with individual wooden or steel props,
- applying horseheads,
- using horseheads supported by props,
- fixing lining elements with rigid, strand or string anchors (roof bolts),
- incorporation of ties in fields between bearing lining sets,
- fitting extra sets between the existing ones,
- incorporation of ground braces in the case of floor uplift.

The lining spacing is reduced at the stage of selection of a bearing lining for the designed longwall heading. The most frequently selected bearing lining is used in longwall headings is the three-piece flexible ŁP type. Its cross-section is shown in Figure 3. The lining spacing usually does not exceed 1.0 m. Experience acquired during previous exploitation proves valuable when selecting the lining spacing. Taking into account expected problems with maintaining a dog heading, based on experience acquired with the adjacent wall, it is possible to reduce the lining spacing, e.g. from 1.0 m to 0.75 m, or from 0.75 m to 0.5 m. It is also possible to use stronger ŁP profiles, e.g.: V32 or V36. This method has an critical disadvantage – it considerably increases the costs of building the gallery and slows its driving. Very often, higher heading cross-section is also planned due to compression. Flexible bearing linings dominate in Polish hard coal mines, with sets sized ŁP-9 and ŁP 10, made of sections of a minimum of V29.

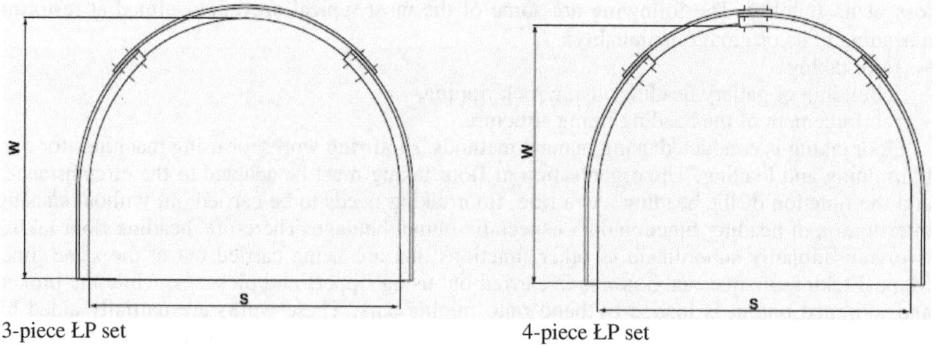

3-piece ŁP set 4-piece ŁP set

Figure 3. Flexible arched lining (ŁP): 3-piece and 4-piece (Łabędy 2009)

In the case of heading floor uplift, the most efficient lining strengthening method involves the use of closed lining. This procedure reduces lining deformations. As a rule, floor arches are fitted after a certain delay behind face of a driven heading. Larger overall heading dimensions frequently require a multi-piece lining to be used – most often four-piece (Fig. 4), or multi-piece including the use of a sectional ground brace (Figures 5 and 6). One disadvantage of the closed lining is higher labour requirements while making it, and the fact that ground brace often constitutes an obstacle during exploitation works, especially in case of further floor uplift, which frequently leads to its deformation. When it is necessary to perform another floor taking, it is necessary to remove the ground braces, which slows the process considerably. Mining practice proves that in case of high floor uplifts it is advantageous to detach ground braces before the front of the approaching wall, at a distance of 60 m from the front of the wall.

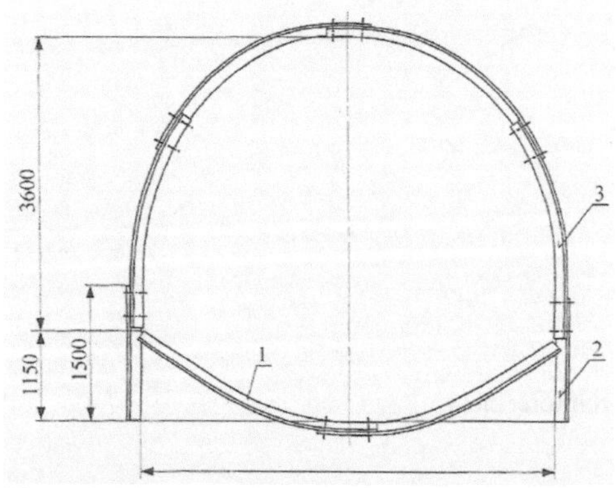


Figure 4. Diagram showing eight-piece closed lining with ground braces: 1 - combined ground brace, 2 - side wall arch elongation, 3 - original set (Piechota et al. 2009)

Figure 6 presents ground braces made of straight V connectors, overlapped using locks. In high-pressure conditions, ground braces may be additionally fastened with steel anchors, Figure 7. In these cases, it is also possible to introduce a floor lagging made of mesh, e.g. fence type.

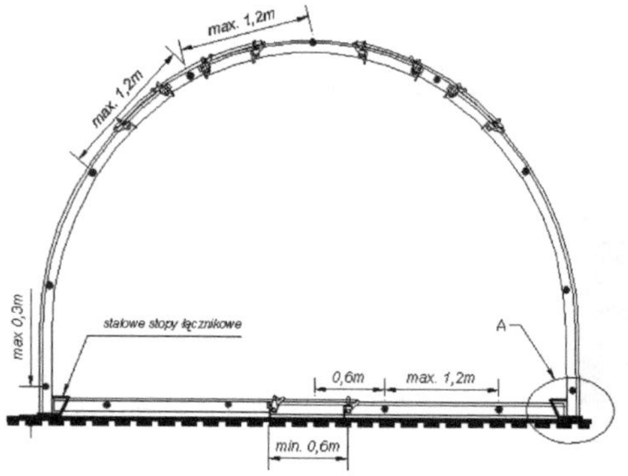


Figure 5. Flexible arched lining (ŁP) - closed

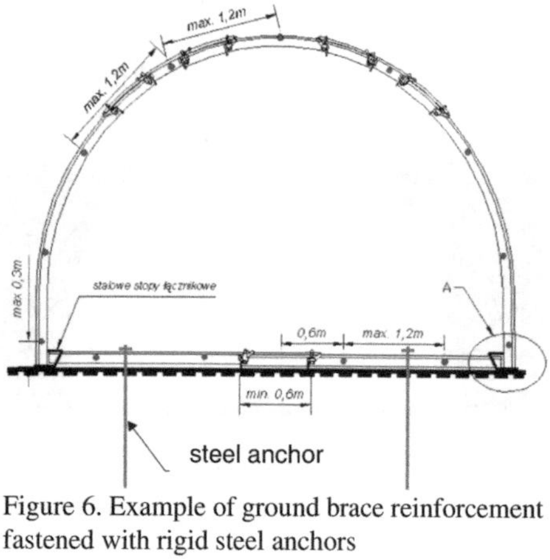


Figure 6. Example of ground brace reinforcement fastened with rigid steel anchors

Instead of ground braces, it is possible to use strand anchors encircling the heading floor. This requires so-called "floor sewing" carried out with lagging ropes and meshes (Fig. 7). This sort of strengthening may be acquired using old cables, e.g. withdrawn from cable transport. If further taking is necessary, or if its removal is necessary for technical reasons, this reinforcement method has the advantage of the possibility of quick disassembly of the whole floor reinforcement structure by unfastening or cutting the cables.

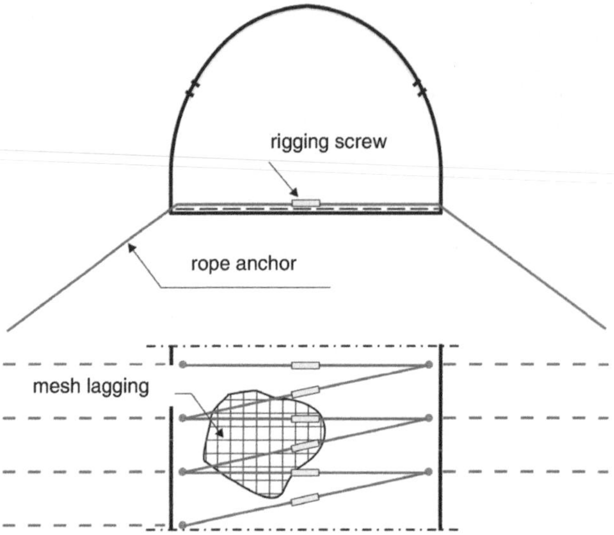


Figure 7. An example of a method for heading floor reinforcement with strand anchors and mesh

The simplest and most frequently used method of heading lining reinforcement before longwall exploitation front involves employing individual horseheads or horseheads supported by wooden or steel props along the axis of the gallery at the longwall. Props set under horseheads are most often located under each roof-bar arch of the bearing lining or, if the mining and geological conditions allow it, under every second or third roof-bar arch. If necessary, it is possible to concentrate the props. In case of occurrence of soft floors, props should be set e.g. on wooden sleepers (Fig. 8).

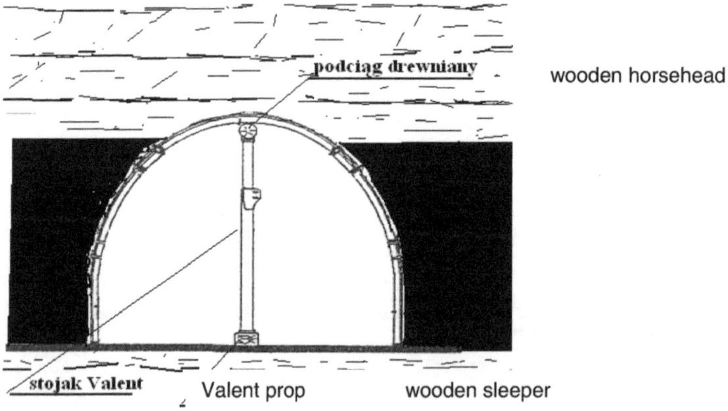


Figure 8. Reinforcement of gallery lining set with Valent props

SV props are very simple to make and inexpensive, while having considerable supporting properties. The materials used to make the SV prop may be acquired from old straight V connectors, e.g. during closedown of dog headings (Photo 1). Steel horseheads ensure more a uniform loading of the arched lining set, and thus its more efficient use. Steel horseheads made of V, KO and KS fraims or railway rails are used most often in the area of the wall – gallery crossings, or sometimes along the entire length of the gallery. The length of the sections used as steel horseheads ranges from 4 to 6 m. They are most often built on the axis of the gallery, but it does occur that steel horseheads are used to strengthen side walls. Straight V connectors or rails are usually coupled by of ŁP lining overlapping on two segments (Fig. 9).

Photo 1. Reinforcement of gallery lining set with SV-type steel props

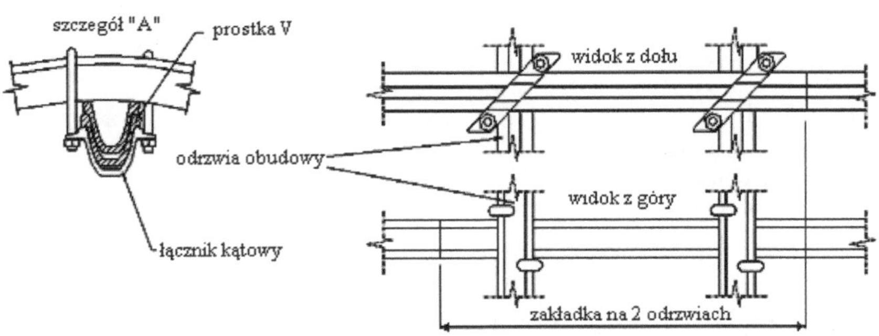


Figure 9. Method for incorporation of steel horseheads

One should remember to fasten any installed horsehead to each lining roof-bar arch using locks or special clamping rings. Then, they should be firmly tightened and fixed with wooden wedges. In the case of substantial heading height, each of the props should be tied to the horsehead using wire, steel cord or chain so as to secure it against falling (Photo 2).

Photo 2. Method for protecting a prop against falling using steel cord

The simplest way to increase stability of headings is to use anchors with varied rod length. Anchor lining has been commonly used in Polish coal mining industry for many years now.

The above-mentioned anchors may be installed in the fields between bearing lining sets (Fig. 10), or may support its elements with proper clamping rings or horseheads (Figs. 11 and 12).

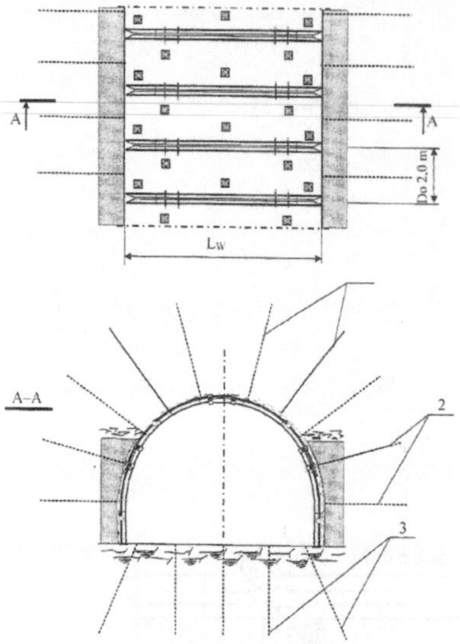


Figure 10. Rock mass strengthening with
1 - roof anchors, 2 - side wall anchors, and
3 - floor anchors. (Piechota et al. 2009)

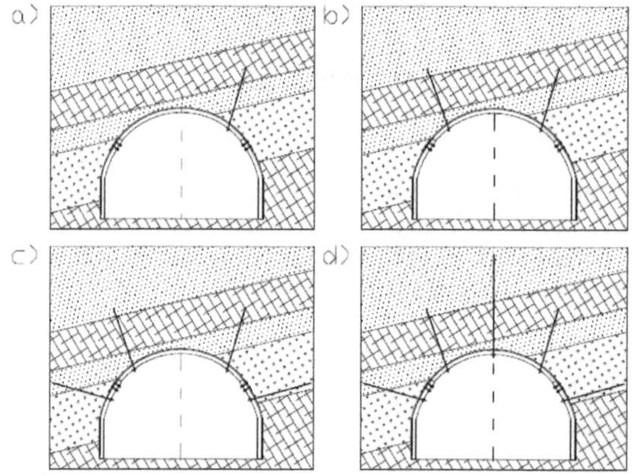


Figure 11. Methods used to reinforce bearing lining arches with anchors

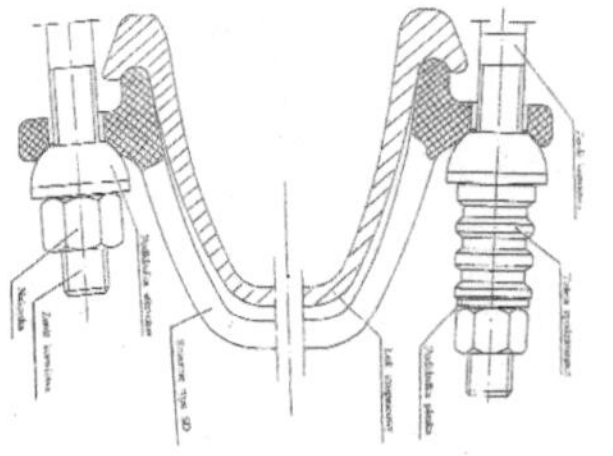

Figure 12. View of clamping ring carrying roof-bar arch
of bearing lining using a pair of installed anchors

The bearing lining of the longwall heading may be also reinforced by anchoring one or two horseheads made e.g. of straight V connectors (Figs. 13 and 14), or a so-called T-bar, Figure 15. In case of single horseheads (Fig. 13), each of string anchors usually possessing carrying power of min. 280 kN, supports roof-bar arches in heading axis. Reaction is transferred from the anchor to set using a suspended steel horsehead made of straight connector sections with holes, min. V25.

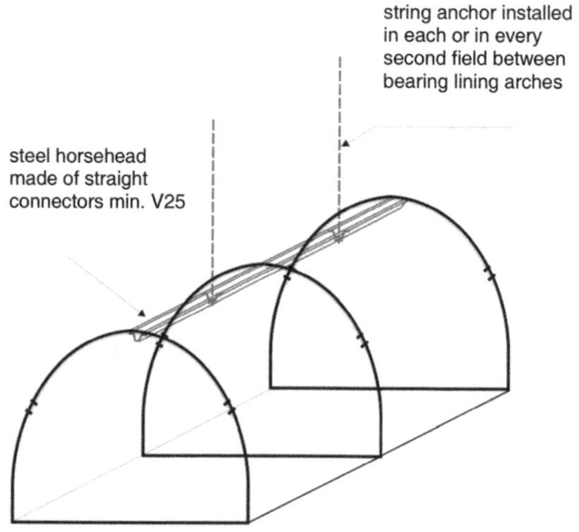


Figure 13. Diagram showing anchoring of the ŁP lining with horsehead and anchors

Double horseheads are usually installed symmetrically on both sides of heading axis (Fig. 14), just above bearing lining locks, and at a distance of approximately 0.5÷0.7 m from them. In case of an asymmetrical arrangement of the horseheads, one horsehead is installed along the heading axis, and the second above lining lock on the side of planned longwall exploitation. String anchors range from 3.5 to 6.0 m-long. Anchors are usually fastened using a resin adhesive loads at glue-in section of min. 1.2 m. Figure 14 shows string anchors used to suspend horse-heads in each field between the bearing lining sets.

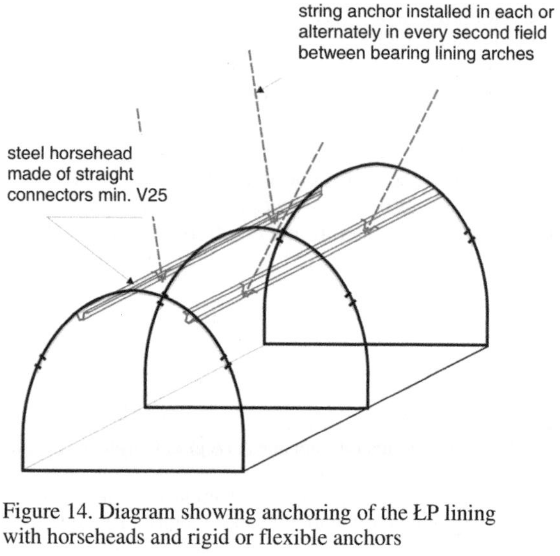


Figure 14. Diagram showing anchoring of the ŁP lining with horseheads and rigid or flexible anchors

In case of favourable roof conditions, these anchors may be installed in every second field between the bearing lining sets, and alternately on both sides of the heading axis.

Figure 15 presents example of a protection system for bearing lining of a gallery at longwall based on single T-bar sets. Each of the T-bars supports two adjacent bearing lining arches via string anchor installed in every second field between bearing lining sets. These T-bars may be also installed alternately (symmetrically or asymmetrically) on both sides of the heading axis.

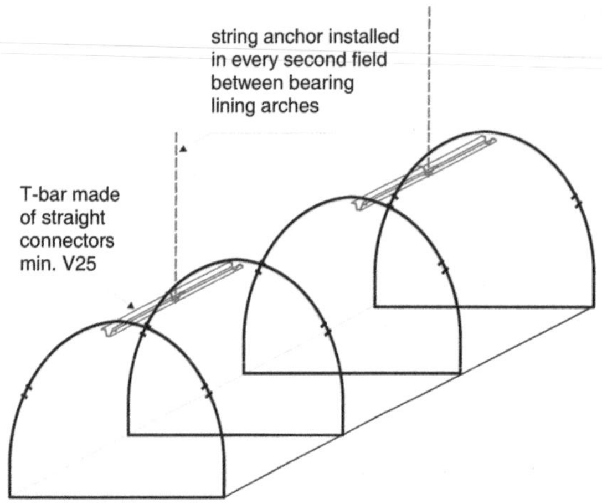


Figure 15. Diagram showing anchoring of the ŁP lining using horseheads in the form of T-bars and rigid or flexible anchors

## 5. SUMMARY

Longwall headings situated in unfavourable geological conditions, especially in areas affected by the exploitation front, are subject to different forms of deformation, followed by the effect of reduction of dimensions of the heading cross-section, making its continued functioning difficult. In order to prevent this effect it is necessary to design and implement optimum strengthening for heading lining as early as only possible. Properly designed and constructed linings for galleries at longwalls should ensure their stability in the area of exploitation pressures. Most frequently, the lining is strengthened with some delay relative to the front of driven galleries, or immediately before approaching the front of the exploitation works. The time over which exploitation pressures effect the lining is short, usually approximately 1 month (Piechota et al. 2009; Piechota 2008), thus it is sometimes considered rational to design heading lining without taking into account exploitation pressure value. Then, the lining is strengthened only in pressure impact area.

However, the delay in providing reinforcement for a heading may result in deformation of the lining, or in worst case, in loss of its stability. A three-wall heading, which has already been subjected to deformation combined with a reduction in its cross-section dimensions that do not guarantee its continued functionality, must be restored to operational condition, otherwise it will be necessary to build a new, parallel heading near the compressed one, which would considerably increase the cost of exploitation.

## REFERENCES

Zimonczyk J., Kiełkowski A., Schopp W. 2006. "Borynia" Mine experience in maintaining galleries at longwalls behind exploitation front, surrounded with cavings on one side. Przegląd Górniczy (Mining Review Monthly) (5), Katowice

Piechota Stanisław. 2003. Basic principles and technologies applied to extraction of solid minerals. Biblioteka Szkoły Eksploatacji Podziemnej (Underground Exploitation School Library), Krakow

Majcherczyk T., Olechowski S. 2008. Hazard areas exposed to lining deformations in galleries at longwalls. Przegląd Górniczy (Mining Review Monthly) (6), Katowice
Prusek S. 2008. Methods allowing to forecast deformations of galleries at longwalls in areas exposed to the impact of roof fall exploitation. Scientific Works of GIG (Central Mining Institute), No. 874, Katowice
Grzybek J. 1996. New look at the impact of mining and geological factors on maintaining dog headings in vicinity of roof fall cavings. Szkoła Eksploatacji Podziemnej (Underground Exploitation School), Krakow
Prusek St., Kostyk T. 2000. Gallery compression forecast (GIG), Katowice
Piechota St., Stopyra M., Poborska-Młynarska K. 2009. Systems of Underground Exploitation of Hard Coal Deposits in Poland. Published by AGH-UST, Krakow
Piechota St. 2008: Technology of Underground Exploitation of Deposits and Liquidation of Mines. UWND AGH-UST, Krakow
"Łabędy" S.A. 2009. Steelworks Product Catalogue

International Mining Forum 2010, Liu et al. (eds) © 2010 Taylor & Francis Group, London, UK. ISBN 978-0-415-59896-5

# Comparison of methods used in Poland for the evaluation of dewatering wells

Krzysztof Polak, Karolina Kaznowska
*AGH University of Science and Technology, Krakow, Poland*

ABSTRACT: Wells are the key element of the drainage system in lignite mining in Poland. Currently, more than thousand are in operation, pumping groundwater in quantities reaching 1000 $m^3$/min. Because of their use, mining wells work in extreme conditions. In contrast to public groundwater intakes, their task is to cause depression of groundwater, allowing safe lignite exploitation. Working conditions result in accelerated aging of wells' filters. This increases the hydraulic and energy losses. To reduce the losses, it is necessary to objectively assess the technical status, what allows to decide whether to renovate wells or drill new ones. Assessment of the wells is conducted basing on the macroscopic or parametric methods. The article presents a comparison of parametric methods. This comparison was based on the interpretation of pumping test results. The study was conducted in a selected deep well, belonging to AGH-University of Science and Technology.

## 1. INTRODUCTION

Wells are one of the basic elements of water supply for drinking and industrial purposes. They also perform a security role for deep-seated structures, allowing their dewatering. The collected underground water is usually characterized by favorable physico-chemical parameters and the intakes are characterized by resistance to contamination from surface pollution outbreaks. Underground water intakes, compared to surface water intakes, have higher operation cost. Total demand for energy needed to pump water from a deep well is a result of spending it for:

- the work of lifting water from the well,
- overcoming the hydraulic resistance,
- overcoming the resistance of the well's filter.

Over time the cost of deep-water intake's operation increases. It is related to the ongoing process of well's aging. This process is a natural phenomenon occurring in the hydrogeological holes, regardless of their use. However, it is particularly intense in the case of dewatering wells. Mining wells are characterized by extreme working conditions. Their purpose is to obtain a significant lowering of water level in a relatively short time. This allows safe acquisition of mineral deposits. Exceedance of the optimal working parameters results in an accelerated clogging of the well's filter. Clogging of filters increases the well entry loss and thus the hydraulic losses. The end result is an increased demand for electricity, which adversely affects the cost of the project.

Energy intensity of the hydraulic losses associated with wells' filters has its economic dimension. For example, it is estimated that, in the sector of lignite mining in Europe, energy consumed for the purpose of pumping water increases by about 20% as a result of clogging of filters. Losses associated with this fact are estimated at hundreds of thousands of euros per year. In

Poland energy sector is based on lignite in about 35%. More than one thousand deep wells are used for the purpose of drainage.

Clogging of filters in the wells occurs even if the intake is not operating or is operating with low efficiency. The current state of knowledge suggests that following types of clogging may appear in the zone surrounding the well filter (Batu 1998, Driscroll 1995, Kasenow 2001, Rubbert et al. 2008, Van Beek et al, 2006, DeZwart 2006):

- Chemical – caused by certain chemicals (e.g. calcium carbonate, iron compounds); chemical reactions may occur in the zone surrounding the drillhole due to the changes in redox potential of the aquatic environment (oxidation of mineral compounds present in the underground water), mixing water from different aquifers with different physicochemical characteristics and precipitation of secondary minerals formed.
- Biochemical – this type of clogging occurs in association with bacteria that cause the transformation of chemical reactions' products.
- Electrochemical – the process takes place in connection with the formation of an electrostatic potential difference during the flow of water on the surface of the filter. As a result, a process of catalysis occurs, resulting in the precipitation of certain minerals on the active surface of the filter.
- Mechanical – this kind of clogging results from a force of transportation related to the movement of water. The flow through the ground skeleton causes the lifting of ultrafine particles of rock (suffosion) towards the well. Particles get inside the well. When rock particles are retained on the filter surface, decrease of an active filter surface takes place, resulting in accelerated aging.

In practice, the simultaneous occurrence of several types of clogging is observed. The emergence of chemical clogging on the filter reduces the active surface of the filter. In such conditions, maintaining fixed well of yield increases the speed of water flow, resulting in the emergence of turbulence in the zone surrounding the well which leads to sweeping of ground particles and mechanical clogging.

The problem of eliminating energy losses is currently the subject of deeper studies in lignite mining sector in European countries. The works are moving towards the use of new filtering materials, which reduce chemical clogging and towards new methods of filter restoration (Houben et al. 1999; Houben, Treskatis 2003, 2007; Rubbert et al. 2008; Weihe, Houben 2004).

## 2. METHODS OF ASSESSING WELL TECHNICAL STATUS

Currently in surface mining due to the development of digital technology and the increasing availability of TV technology, inspection methods are applied more often. Inspection is carried out from inside the well during operational stoppage and with disassembled pipe fittings. The result of the evaluation is not parameterized and is a subjective view of the person conducting the inspection. TV inspection method allows assessing only the macroscopic state of the well filter. Only the effects of clogging are assessed. The issue of assessing the state of deep wells had already been taken in the 40 s of the last century. Jacob (1947) proposed a method for evaluating the technical status of wells basing on the interpretation of the results of pumping tests. The relationship between depression in the well and well of yield is presented in the form:

$$S = BQ + CQ^2 \tag{1}$$

where:

$S$ – total depression in the pumping well, [m];
$Q$ – yield of well reached when depression equals S, [$m^3/s$];
$B$ – coefficient of resistance of laminar flow in aquifer, [$s/m^2$];
$C$ – Jacob's coefficient of turbulent flow of the well, [$s^2/m^5$];
$BQ$ – actual depression in the well, [m];
$CQ^2$ – well entry loss, [m].

This issue was taken up and developed by many authors, including: Rorabough (1953), Walton (1962), Bierschenk, Kasenow (Kasenow 2001). In Poland, the knowledge regarding this subject has been disseminated in publications of such authors like: Siwek (1978, 1979, 1980) and later Sozański, who proposed an algebraic method of determining the coefficient of wells' resistance (1985). Implementation of methodology for conducting pumping tests and interpretation of results for dewatering wells was proposed by Klich (1998). The method consists of carrying out short-term, multi-stage pumping tests and a modified approach to the interpretation of research results. The issue of assessing the technical status of the hydrogeological drillholes has also been addressed in the work of Marciniak (1999), which is devoted to the methodology of identification of hydrogeological parameters, as well as in the work of Dabrowski (2005), documenting the resources of operating underground water intakes. Table 1 shows the characteristics of the parametric methods of assessing the well status.

Table 1. The characteristics of the parametric methods of assessing the well status

| Method | Characteristic of the pumping test | Method for calculating C coefficient | Comments |
|---|---|---|---|
| Jacob | $Q_1$, $Q_2=2Q_1$, $Q_3=3Q_1$ $t_1=t_2=t_3$ | analytical | turbulent flows within the well |
| Rorabaugh | $Q1\neq Q2\neq Q3$ $t_1=t_2=t_3$ | graphical | $S = BQ + CQ^n$ |
| Kasenow | $Q1\neq Q2\neq Q3$ $t_1=t_2=t_3$ | analytical | 3-stage pumping test / 3 independent pumping tests |
| Bierschenk | $Q_1\neq Q_2\neq Q_3$ $t_1=t_2=t_3$ | graphical | 3-stage pumping test / 3 independent pumping tests |
| Sozański | $Q_1$, $Q_2= 2Q_1$, $Q_3=3Q_1$ $t_1=t_2=t_3$ | analytical | the solution is a positive root of the quadratic equation |
| Klich | $Q=0\rightarrow Q_{max}$ | graphical/analytical | autoregulating character of the pumping test |
| Bełchatow | t = 1h | analytical | one step-drawdown test |
| Multiple pumping | $Q_1$, $Q_2= 2Q_1$, $Q_3=3Q_1$ t = 1 – 1.5h | analytical | 3 independent pumpings (after recovery) |
| Jacob – Thais | $t_{min}$= 1h | analytical/ graphical | one step-drawdown test |

Q – yield of well [$m^3/h$], t – pumping period [h], $n \neq 2$

Pumping test results allow to calculate the hydraulic parameters of the aquifer and the analyzed well: well's efficiency $\eta$, actual depression BQ, well entry loss $CQ^2$, coefficients of layer's resistance B and of the well C, which is used to control the degree of improvement of the intake, but also as a measure of acceptance by an investor.

Table 2 shows the classification used in Poland, based on Walton's criteria (1962).

Table 2. Boundary values of coefficient C

| | Well's technical status | | | | |
|---|---|---|---|---|---|
| Jacob coefficient | clean | barely contaminated | fairly contaminated | strongly contaminated | very contaminated |
| $s^2/m^5$ | <2000 | 2000-4000 | 4000-8000 | 8000-15500 | >15500 |
| $min^2/m^5$ | <0.6 | 0.6-1.1 | 1.1-2.2 | 2.2-4.2 | >4.2 |

## 3. DESCRIPTION OF THE RESEARCH EXPERIMENT

This paper presents a comparison of methods used for assessing the well status in terms of hydraulics. Parametric methods based on pumping tests were used. The results of this research led to the determination of hydraulic parameters. Comparison of the results allows to assess the applicability of the various parametric methods for evaluating technical status of dewatering wells.

Pumping tests were conducted in a deep well belonging to AGH-UST, located in Krakow. This 16-meter-deep well takes water from a neogen unconfined aquifer. Permeability coefficient is $k = 6 \times 10^{-4}$ m/s. Grains of diameter between 3 and 5 mm are sprinkled along the active parts of the filter and along the subfilter's pipe. The diameter of the filtered drillhole is 225 mm, depth 16 m. In the vicinity of the intake 4 observations well are situated. In order to carry out the calculations four pumping tests were conducted:

1) Three-stage pumping I

Pumping was carried out with a fixed well of yield for each stage, in keeping with its multiples compared to the previous steps. Pumping efficiencies were as follows: $Q_1$=9.815 m$^3$/h, $Q_2$=19.649 m$^3$/h, $Q_3$=30.004 m$^3$/h. Pumping lasted for 100 min. The resulting depressions were as follows: $S_1$=0.17 m, $S_2$=0.38 m, $S_3$=0.617 m. The results of pumping I were used to assess the hydraulic state of the intake using following methods: Jacob's, Jacob-Kasenow's, Sozanski's.

2) Three-stage pumping II

Pumping was carried out with a fixed well of yield for each stage. Pumping efficiencies were as follows: $Q_1$=21.99147 m$^3$/h, $Q_2$=27.40019 m$^3$/h, $Q_3$=35.23878 m$^3$/h. The depressions were as follows: $S_1$=0.4805 m, $S_2$=0.6396 m, $S_3$=0.8576 m. Pumping lasted for 180 min. were used to assess the hydraulic state of the intake using following methods: Rorabaugh's, Bierschenk's and Kasenow's.

3) Multistage pumping

The analysis of the technical status of the well was based on consecutive stages of the test pumping. Measurements were carried out from zero to a maximum flow rate $Q_{max}$=35.64 m$^3$/h. A 10-stage pumping, which lasted for 48 minutes, was conducted. Depressions $S_1$ to $S_{10}$ were read after stabilization of the water surface level in any given stage of the pumping. The results were used to calculate the coefficient C using the method proposed by Klich. The graph (Figure 1) presents the results of the multi-stage pumping test.

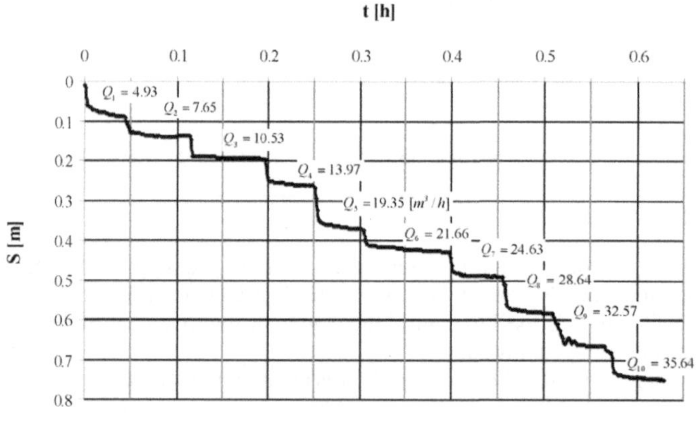


Figure 1. Multi-stage pumping test's results

## 4. MULTIPLE ONE-STAGE PUMPINGS

Pumping was conducted for the following yield of well: $Q_1$=35.815 m$^3$/h, $Q_2$=23.597 m$^3$/h and $Q_3$=11.901 m$^3$/h. Each pumping lasted for 60 min. The resulting depressions were respectively $S_1$=0.825 m, $S_2$=0.508 m and $S_3$=0.235 m. After the completion of each pumping the water level was stabilized. Total time of the assessment was about 7 h. The depression in observation wells was also observed. The results were used to calculate the hydraulic status of the well using following methods: multiple one-stage pumping, Jacob - Thais' (for Q1) and Belchatow method (for $Q_2$). The results obtained allowed to determine both the well's and the aquifer's hydraulic parameters. The calculation results are summarized in Table 3.

Table 3. Calculation results

| Method | B [s/m$^2$] | C [s$^2$/m$^5$] |
|---|---|---|
| Jacob | - | 1853.28 |
| Jacob-Kasenow | 58.896 | 1853.28 |
| Sozański | - | 332.069 |
| | - | 12754.318 |
| Rorabaugh* | 56.88 | 782.65 |
| Bierschenk | 64.8 | 2592 |
| Kasenow | 64.152 | 2592 |
| Klich | 64.8 | 1231.2 |
| Multiple pumping | 65.52 | 1684.8 |
| Jacob – Thais | 60.408 | 2332.8 |
| | 60.912 | 2203.2 |
| Belchatow | - | 1036.8 |

*n=1,55

## 5. DISCUSSION OF EXPERIMENT'S RESULTS

Conducted pumpings and calculations showed divergent results of the hydrogeological and hydraulic parameters. The results indicate that Rorabaugh's and Sozanski's methods do not allow for an objective assessment of the well's status.

The solution of Sozanski's algebraic method produced results which are inconsistent with the assumptions. One of the solutions of the expression should be negative.

Rorabaugh's method produced coefficient value diverging from the values of C calculated using other methods. According to Rorabaugh's assumptions, a flow different from laminar or turbulent may occur (exponent greater than 2), making it impossible to compare results.

According to Kasenow (2001), a well which is designed and constructed properly, should have coefficients B and C similar to results obtained by Bierschenk's graphical method. The obtained results of calculations of the layer resistance coefficient are almost identical, while the values of the coefficient of hydraulic resistance are exactly the same. This confirms the validity of Kasenow's assumptions and the credibility of this method of assessment.

The method used in Belchatow gave the lowest value of coefficient C. This is a method that takes the geological situation into account. Determination of individual elements of the model can be lead directly in the pumping well, without additional observation wells. In contrast to this method, Jacob - Thais' method is limited to only one stage of pumping and must be conducted in a pumping well-observation well system. The calculations were performed for two observation wells which were 10-15 meters away from the pumping well. The results of the coefficient C were relatively high. This was probably due to a very large resistance in the turbulent regime. The pumping was conducted at maximum efficiency, Q=35.815 m$^3$/h.

The method proposed in AGH-UST by Klich helped to define the parameters for the laminar and turbulent motion in the zone surrounding the well. The calculations made clear that the favorable parameters are reached at the yield of well: Q=19.37 m$^3$/h. This output is optimal in terms of energy intensity of the pumping. The method is characterized by short duration and very high accuracy of the obtained results.

In contrast to the hydraulic resistance coefficient C, which has a variable size depending on the actual status of the intake, the coefficient B is constant for a given layer. On the basis of the coefficient B values which were both calculated and read off the charts prepared for particular methods, one may notice that the results are similar. They oscillate in the range from 56.88 do 65.52 s/m$^2$ (neither Jacob's nor Sozański's nor Belchatow's method allow to calculate the coefficient B).

The results show that the test well is of good quality. The efficiency of the intake is ranging from 72 to 92%, noticing that the lowest value was obtained during the maximum efficiency Q=35.64 $m^3$/h. This is due to high resistance of turbulent flow. According to the results calculated using the methodology proposed by Klich (1998), outputs higher than optimal should be avoided. Pumping with well of yield exceeding this value will lead to reduced efficiency and will shorten the operation time of the intake. However, the conditions in the mining wells often require higher than optimum efficiency to lower the water level below the exploited deposits.

Parametric and macroscopic studies are mutually complementary methods – they allow to identify the problem, assess the technical status of the wells and to choose the restoration method.

## REFERENCES


Batu V. 1998. Aquifer Hydraulics: A Comprehensive Guide to Hydrogeologic Data Analysis. New York: John Wiley & Sons

Dąbrowski S., Przybyłek J. 2005. Metodyka próbnych pompowań w dokumentowaniu zasobów wód podziemnych. Warszawa: Bogucki Wydawnictwo Naukowe. (In Polish)

DeZwart B. R., Van Beek K., Houben G., Treskatis C. 2006. Mechanische Partikelfiltration als Ursache der Brunnenalterung, Abb vwgw Bonn: Teil 2. – bbr 9/2006: 32 – 37, 5. (In German)

Driscoll F.G. 1995. Groundwater and Wells. St. Paul. Minnesota: Johnson Screens. 1089 p.

Houben G., Treskatis C. 2003. Regenerierung und Sanierung von Bohrbrunnen. München: R. Oldenbourg Verlag. 268 p. (In German)

Houben G., Treskatis C. 2007. Water Well Rehabilitation and Reconstruction. New York: McGraw Hill. 391 p.

Houben G., Mertens S., Treskatis C. 1999. Entstehung, Aufbau und Alterung von Brunneninkrustationen. - in: bbr 10/99: 29 – 35, 2 Abb., 5 Tab.; Köln (R. Müller). (In German)

Kasenow M. 2001. Applied Ground-Water Hydrology and Well Hydraulics, LLC: Water Resource Publ. 2nd Edition

Klich J., Polak K., Sobczyński E. 1998. Opis metody oceny jakości wykonania i stanu studzien ujęciowych i odwadniających. Międzynarodowa Konferencja Naukowo-Techniczna pn. „Zaopatrzenie w wodę wsi i miast". Zeszyt 1-III. Poznań (In Polish)

Marciniak M. 1999. Identyfikacja parametrów hydrogeologicznych na podstawie skokowej zmiany potencjału hydraulicznego. Metoda PARAMEX. Poznań: Wydawnictwo Naukowe UAM. (In Polish)

Rorabaugh M. I. 1953. Graphical and theoretical analysis of step-drawdown test of artesian well. ASCE Proc. 79(362)

Rubbert T., Treskatis C., Benz P., Urban W. 2008. Brunnenalterung – Systematisierung eines individual problems. - bbr - Fachzeitschrift für Wasser und Leitungstiefbau, Heft 07+08/2008, wvgw, Bonn: 44 - 53. (In German)

Siwek Z. 1978, 1979. Amerykańskie mierniki usprawnienia ujęcia wód podziemnych, część I & II, Technika Poszukiwań Geologicznych (1). (In Polish)

Siwek Z. 1980. Techniki usprawniania ujęć wód podziemnych, część I & II. Technika Poszukiwań Geologicznych (4), (5). (In Polish)

Sozański J. 1985. Badanie sprawności studzien. Górnictwo Odkrywkowe (4) - (6). (In Polish)

Van Beek K., DeZwart B. R., Houben, G., Treskatis C. 2006. Mechanische Partikelfiltration als Ursache der Brunnenalterung. vwgw Teil 1. – bbr 7/8/2006: 42 – 49, 8 Abb.; Bonn. (In German)

Walton W.C. 1962. Selected analytical methods for well and aquifer evaluation. Illinois State Water Survey: Bull. 49

Weihe U., Houben G. 2004. Räumliche Verteilung von Inkrustationen in Brunnen. wvgw – bbr 08/2004. Bonn. (In German)

International Mining Forum 2010, Liu et al. (eds) © 2010 Taylor & Francis Group, London, UK. ISBN 978-0-415-59896-5

# Mechanised sublevel caving systems for winning thick and steep hard coal beds

Zbigniew Rak
*AGH University of Science and Technology, Krakow, Poland*

ABSTRACT: This article presents a process for mechanised exploitation of a steeply inclined, thick coal bed. It presents the mine face mechanisation method, organisation of works, and the scope of preparatory works. The system was developed and implemented in 2002 for the purpose of exploitation of a bed approximately 20 m thick inclined at 45° under the mining and geological conditions of the Kazimierz-Juliusz Sosnowiec Hard Coal mine in Poland. The production output and the system efficiency assessment have been illustrated on the basis of previous experience at that mine.

## 1. INTRODUCTION

This article is based on experience acquired during exploitation of a thick coal bed in one of the oldest mines in Upper Silesian Coal Basin (Górnośląskie Zagłębie Węglowe) in Poland – the Kazimierz-Juliusz Sp. z o.o. [Ltd.] Hard Coal Mine in Sosnowiec. Because resources stored in largely horizontal beds in this mine were depleted, an attempt was made to work out last part of the bed, a high-quality coal deposit in bed no. 510. The thickness of this bed in the field discussed here is approximately 20 m, and the deposit angle is greater than 40°. In cooperation with the mine engineering staff, a team of AGH-UST employees in Krakow attempted to develop an efficient, mechanised bed exploitation system based on a block system with frontal output dumping, known as "sublevel caving". The system employed in the mine was referred to as an extraction gallery. It has now been functioning for seven years in the Kazimierz-Juliusz mine, and the process has been subject to corrections with time in order to reach optimal production results. There is currently fairly broad interest in this system. The design team has already completed two conceptual analyses and one technical design analysis for implementation at two hard coal mines in Russia. The following discussion summarises the production experience and the research conducted to date.

## 2. MINING AND GEOLOGICAL CONDITIONS OF BED NO. 510

Bed No. 510 in the area where the sublevel caving system is currently in use has a thickness ranging from 16 to 20 m and is deposited at depth from about 300 to about 520 m at an incline of approximately 40°. There are many faults within the boundaries of the exploitation fields, mainly running north – south, with throw sizes of up to a several metres.

Bed No. 510 in this area consists of semi-shiny coal, or stratified shiny coal with carbonates, or in some places semigloss. Its basic strength parameters are as follows:

- compactness index f: from 0.91 to 1.13, that is easily mineable,
- immediate uniaxial compression strength Rc - from 13.4 to 35.7 MPa.

The geological structure of the roof and floor in bed No. 510 allows sandstones with variable grain-size distribution, mudstones, clay-stones and coals to be distinguished. The percent share

of individual rock types up to a height of 100 m above the roof and 30 m below the bed is as follows:
- sandstones – 39%
- mudstones – 23%
- clay-stones – 34%
- coals – 4%

In vertical profile, up to a height of 100 m above roof of bed No. 510 there are 3 sandstone layers each with a thickness exceeding 10 m, at distances of approx. 20 m, 47 m and 73 m above the roof bed. The average compression strength of the above-mentioned sandstones is about 50 MPa. In places, individual rock layers show distinct cracks. The average strength of the roof rock package (100 m) and floor rock package (30 m) is about 45.0 MPa. The rocks surrounding the bed should be ranked as being among rocks with considerable rockburst tendency.

The methane carrying capacity of bed no. 510 remains within the limits of the values of hazard category I.

In areas of the mine, including the M-3 lot, the coal from bed No. 510 is rated in self-ignition group V, that is, coals highly inclined to self-ignition. The self-ignition index for coal sampled from the M-3 lot was 179°C/min, and its activation energy: 37 kJ/mol. All of the headings made and under construction in the area of sublevel exploitation from the extraction gallery were subject to a category B coal dust explosion hazard.

The bed in the area of the mining works conducted was subject to a degree II water hazard. The hazard is posed by a water-bearing horizon located in sandstone above bed No. 510. This level may be insufficiently isolated by impermeable layers and one should take into account possibility of outdropping into the space where mining works are underway. Bed No. 510 in the area of field A and field B has:
- rockburst hazard degree I at deposit depth up to 400 m,
- rockburst hazard degree III at depth exceeding 400 m.

## 3. GENERAL CHARACTERISTICS OF THE SUBLEVEL CAVING SYSTEM SYSTEM KAZIMIERZ-JULIUSZ-TYPE

Using planes perpendicular to the roof and floor planes, the sublevel system the Kazimierz-Juliusz type divides the thick and steeply inclined bed into horizontal exploitation blocks with the same thickness as deposited bed thickness.

As it has been mentioned before, this system is based on a block system with frontal output dumping – "sublevel caving", which has been successfully used in ore mining for more than 100 years now and occasionally in hard coal mining as well.

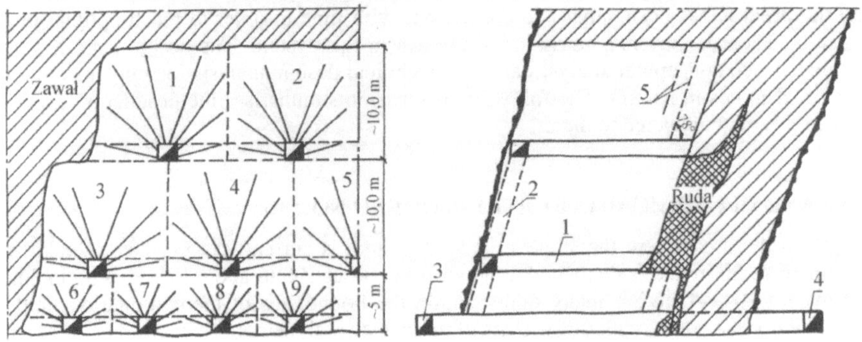


Figure 3.1. Diagram of the sublevel caving system

Bed exploitation in the Kazimierz-Juliusz mine involves the winning of individual panels, starting from that located at the highest level (Fig. 3.2).

The panels are exploited starting from field boundaries, by releasing the coal mined using explosives.

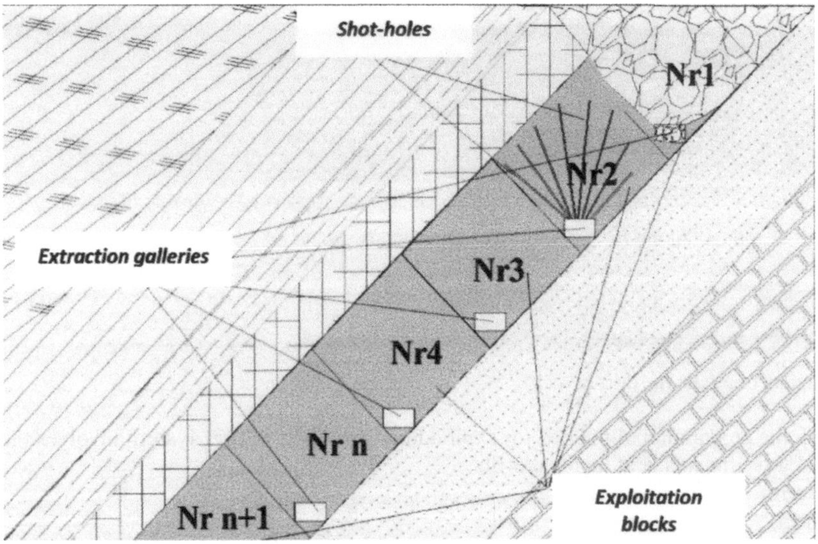


Figure 3.2. Division of bed into panels, and their order of exploitation

## 4. BED OPENING OUT

Opening out of beds slated for exploitation using the sublevel system does not differ significantly from the model of opening out beds intended for exploitation using other systems. In general, exploitation with the sublevel system applies to mining of thick beds, and thus the most advantageous opening out structure, especially in case of coal that is inclined to self-ignition, is the rock structure.

During exploitation using the sublevel system, the essential difference in opening out of individual bed sections is bed slope and the available wall system (up to approximately 35°) or block system (over 35°) connected with this. In certain cases, especially when bed is thin and steep, it is useful and sometimes necessary to divide an opened-out bed level into stages by making cross headings or inclined drifts.

In every case, the opening out method should be evaluated separately due to considerable differences in bed thickness and its inclination, as well as size (and thus resource available) of the part of the bed planned for sublevel exploitation, taking into account mining and economic factors.

Two exemplary bed opening-out methods including main preparatory headings, strongly inclined and steep coal beds, for the extraction gallery sublevel system (as in the Kazimierz-Juliusz Hard Coal Mine), are presented below.

In variant I, bed dissection is carried out by making the so-called stairway drift, built in stone, with inclination not exceeding 25°. This inclined drift cutting through the bed gives access to individual exploitation panels. The drift may be also built within the bed's boundaries, but this will result in higher deposit losses.

These inclined drifts will have the following functions: transport of people, machinery, equipment and materials, output haulage and ventilation. Drift inclination will allow the use of transport mechanisation (both for people and materials) and mechanical output haulage using armoured conveyors or special belt conveyors.

This position of the inclined drift relative to a steep bed makes it possible to carry out all functions required by the process of sublevel exploitation and also allows to fully mechanised driving of extraction galleries, in practice from its inlet.

This solution allows for two-way exploitation relative to the so-called stairway drift. A spatial diagram of exploitation headings with the essential element of the opening out structure in the form of a stairway drift is shown in Figure 4.1.

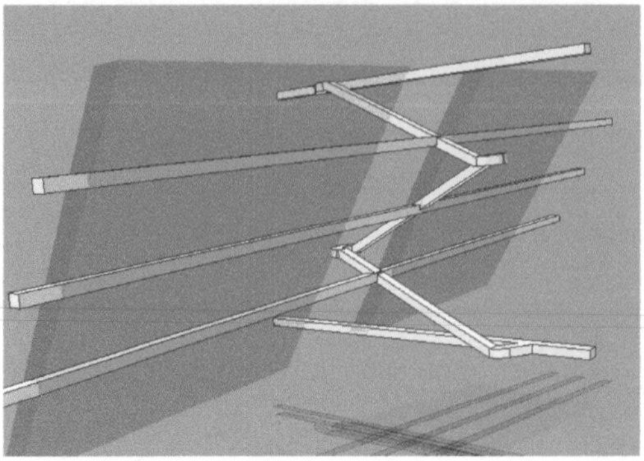

Figure 4.1. Diagram of a stairway drift in rock

In variant II, the preparation of the bed for exploitation with the sublevel system involves the construction of three independent inclined drifts:

- a transport drift
- a haulage drift
- a ventilation drift

The replaced drifts are located in bed parallel to the dip (that is at the angle of ca. 45°). A transport drift constructed under the bed roof plays the role of a heading for transport of materials, machines and equipment, and it will also form an access way to the mine faces for personnel.

A haulage drift, built along the bed floor, connects the extraction galleries with the bunker or other reception point for mine output and is used for gravitational haulage.

A ventilation drift connects the extraction galleries. This drift contains built-in ventilation pipes allowing to air exploitation faces and driven extraction galleries. In this variant, the above-mentioned drifts are driven using the conventional method, that is, using explosives. Figure no. 4.2. shows the location of the individual headings in the bed relative to the opening out levels.

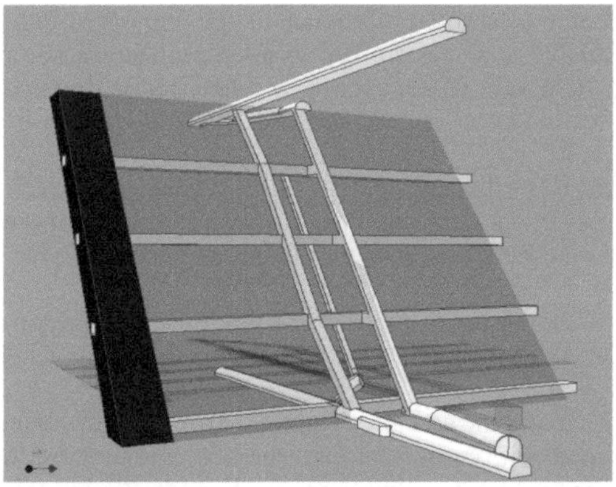

Figure 4.2. Diagram showing headings passing through the exploitation field in variant II

## 5. TECHNOLOGY USED TO CARRY OUT EXPLOITATION WORKS

In this the sublevel caving system, the galleries built along the bed floor in the lower boundary of the individual panels are exploitation headings. These galleries, called extraction galleries, are built in a rectangular steel lining of the following dimensions: width up to 4.5 m and height up to 3.0. Gallery lining diagram is shown in Figures 5.1 and 5.2.

The rectangular cross-section allows trouble-free use of the powered roof support in the gallery exploitation's end.

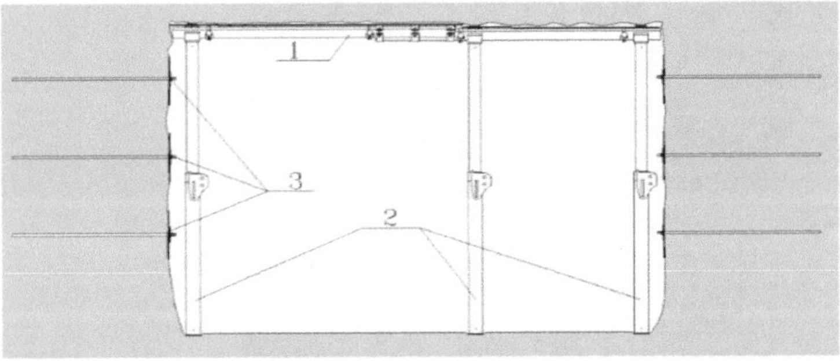


Figure 5.1. Extraction gallery lining – cross-section

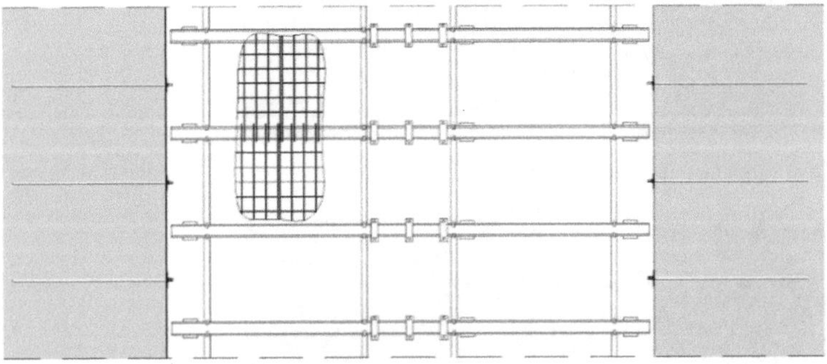

Figure 5.2. Extraction gallery lining – horizontal projection

The exploitation face equipment is brought to an extraction gallery made in this way to the end of its run. The basic elements of this equipment are: a powered roof supports which strengthens the gallery lining and protects the mine face, and an armoured conveyor used for output haulage (Fig. 5.3, Photos 5.1 and 5.2).

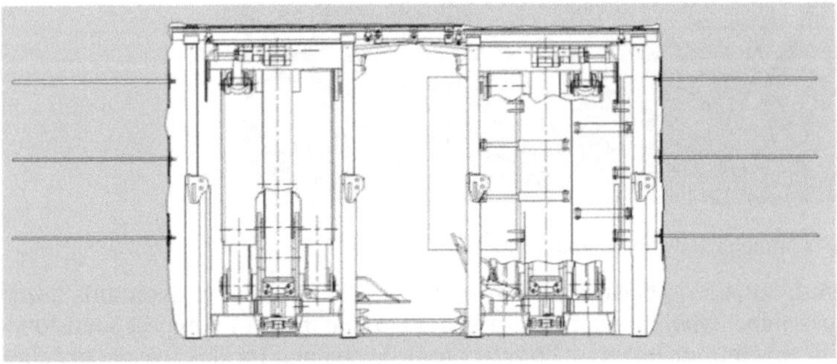

Figure 5.3. Mine face equipment – cross-section

Bed winning in a given panel involves execution of shotfiring works arranged in a fan, and then successive coal output release and cyclical moving of the powered roof supports and armoured conveyor inside the extraction gallery, followed by systematic removal of the gallery lining – Figure 5.4 and Figure 5.5. The moving mine face forms an exploitation front of the width of the panel along the entire width of the bed being won. The length of the extraction gallery is systematically reduced along with the exploitation progress.

Caving, as an integral element of mine face progress, should cyclically follow the progress of the sublevel front. In case of so-called roof "hang-up" there is a risk of cavern formation in the space after exploitation, in which primarily methane and shotfiring gases are accumulated. Uncontrolled caving in this space may result in abrupt ejection of a mixture of the abovementioned gases into working space of the exploitation face.

Photo 5.1. General view of mine face during output release

Photo 5.2. General view of mine face during output release

A system of so-called "torpedo" shotfiring has been introduced in order to avoid this hazard in the Kazimierz-Juliusz mine, where main roofs of bed no. 510 consist of compact sandstones. "Torpedo" shotfiring is aimed to disintegrate compact roof layers, and for this reason, excellent knowledge of the lithological structure of the roof deposited in the area of exploitation is very important. Besides conventional roof disintegration methods, other methods have been developed, such as "hydrofracturing".

## 6. PRINCIPLES OF THE ORGANISATION OF WORKS, SUBLEVEL FACE STAFFING AND FACE PRODUCTIVITY

A four-shift work system is used in the Kazimierz-Juliusz mine at its extraction gallery face. All shifts are production shifts, provided that organisational and technical capabilities allow repair works to be carried out during each shift while sublevel production is stopped. Effective work-time during a shift is about 6 hours.

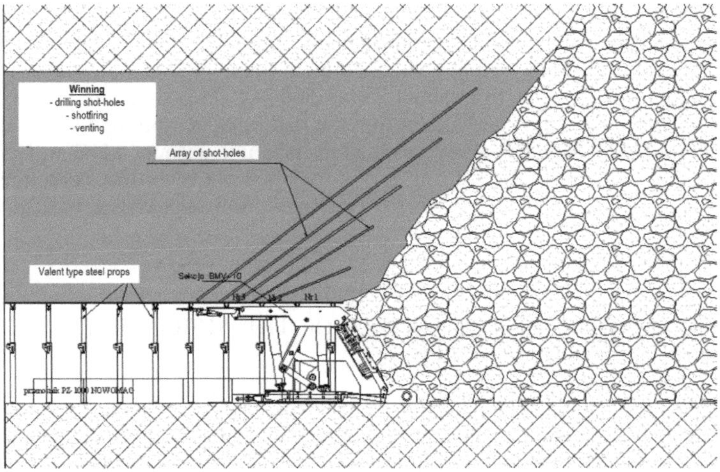


Figure 5.4. The method of bed winning in a panel

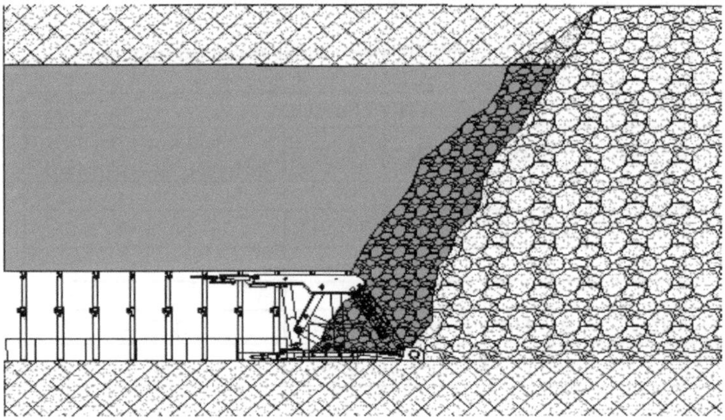

Figure 5.5. Output releasing method

The production cycle consists of the following basic working operations:

1 – drilling of shot-holes in mine face,

2 – shot-hole loading,

3 – firing of explosives and mine face ventilation,

4 – startup of mine face armoured conveyor and removal of side (internal) shields protecting against caving in order to release coal output (crushing oversized rounds using an output crushing set, pneumatic hammers or explosives when conveyor is halted),

5 – halting of mine face armoured conveyor and lifting of side (internal) shields protecting against caving as soon as coal output release is completed and rock appears in the output (in case of “suspension” of coal body – insufficient volume or no output – operations described in steps 1÷4 are additionally repeated),

6 – removal of gallery lining elements,

7 – moving the powered roof supports,

8 – moving of the armoured conveyor,

9 – startup of the mine face armoured conveyor and removal of side (internal) shields protecting against caving in order to release coal output (crushing oversized rounds using output crushing set, pneumatic hammers or explosives while conveyor is halted),

10 – stopping of the mine face armoured conveyor motion and lifting of side (internal) shields protecting against caving as soon as coal output release is completed and rock appears in the output,

11 – moving the armoured conveyor,
12 – start of another production cycle by drilling shot-holes in mine face.

In case of bed thicknesses of over 20 m, output volume during a single production cycle covering two successive working shifts ranges from about 250 to 500 Mg. Two complete production cycles are carried out during one day, and as a result, the daily output from one sublevel face ranges from about 500 to 1000 Mg. The above calculations take into account a bed utilisation index of minimum 75%. Assuming the staffing levels specified above in the extraction gallery (20 workers during a day), mine face productivity ranges from approximately 25 to 50 Mg per worker per day.

## 7. PRODUCTION AND ECONOMIC EFFECTS OF USE OF THE SUBLEVEL CAVING SYSTEM THE KAZIMIERZ-JULIUSZ TYPE

The following table presents the fundamental indicators, that is, the volume of output extracted from one face, and face and division costs in the area of exploitation using the sublevel system from the extraction gallery in the M-3 lot in the period starting from the beginning of exploitation until the end of 2006. In the period from June 2003 (that is from the beginning of exploitation) to June 2004, exploitation was carried out on an operational trial basis.

Table 7.1. Output volume and operating costs in the sublevel system during the period from June 2003 to the end of 2007 in the Kazimierz-Juliusz Hard Coal Mine in Sosnowiec

| Specification | EXPLOITATION PERIOD | | | | | |
|---|---|---|---|---|---|---|
| | 2003 | 2004 | 2005 | 2006 | 2007 | Cumulative from the beginning of exploitation |
| Output | | | | | | |
| Mg | 45,681 | 156,603 | 243,110 | 193,828 | 202,914 | 867,869 |
| Mg/d* | 365 | 619 | 965 | 775 | 807 | 769 |
| Cost in face | | | | | | |
| outlays in € | 266,542 | 721,827 | 915,666 | 589,174 | 682,570 | 3,177,440 |
| €/Mg | 5.8 | 5.4 | 4.2 | 3.4 | 3.6 | 4.2 |

* - daily output includes output during preparatory works

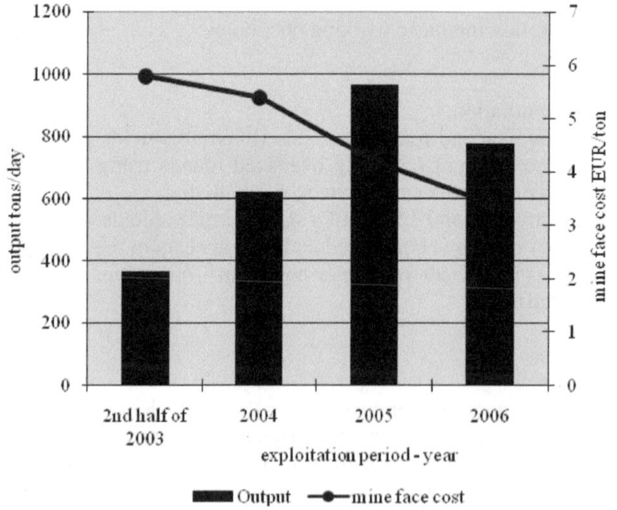


Figure 7.1. Sublevel caving system output and costs

## 9. SUMMARY

The assumptions specified at the design stage of the sublevel caving system for the extraction gallery in thick and highly inclined bed No. 510 in the conditions at the Kazimierz-Juliusz mine have generally proven to be correct. To date, exploitation from the extraction gallery using the sublevel system has allowed the development of more experience in the process of winning beds deposited in very difficult geological and technical mining conditions, especially as regards thickness and slope. Meanwhile, the technical and economic results attained are completely satisfactory for the mine, all the more so because coarse coals extracted from the sublevel face considerably exceeds 60%. The steel props retrieved from the exploitation face are used to drive other extraction galleries, which considerably reduces costs of running these galleries so that with field winning progress, these costs systematically drop.

Regarding the employment of the discussed exploitation system in other hard coal mines, the author of this paper wishes to point out at the following system elements:

- Strength parameters of beds are important from point of view of the process of winning the coal. Weaker coals will certainly require a small scope of shotfiring works, primarily as regards use of explosives and the number of shot-holes in production cycle. On the other hand, the exploitation system discussed here requires drilling of holes inclined at 60° to 70° from horizontal. Keeping holes that are inclined in this way clear at low strength parameters may give rise to some problems when loading the explosives. However, on the other hand, world mining experience proves that at these low strength parameters there is high probability of autogenous coal output dumping. Shotfiring be of a supporting character in this case. Finally, operational tests in each mine will allow the determination of the necessary scope of use of explosives in coal winning.
- Natural hazards occurring in different coal beds will have various effects on the safe exploitation carried out using the sublevel system. According to the author of this paper, the following two factors will have an essential impact on the safety of mining works: methane hazard and the risk related to endogenous fires.
  - Methane hazard – a relatively high methane carrying capacity of beds (up to 10 $m^3$ from 1 Mg of coal) will result in the need to carry out exploitation works using special safety precautions. During large-scale winning of coal using explosives (up to even 300 Mg in a single instance) one should expect high methane emissions into the blind heading of the extraction gallery. According to the author it may be necessary to provide an automatic methane detection system in headings extending from the extraction gallery face to grouped used air ducts. In the Kazimierz-Juliusz mine, the methane hazard practically does not occur, therefore it is difficult to speak from experience here;
  - Endogenous fire hazard – in using exploitation of this type, where there are wastes of crushed coal left in abandoned workings, it is necessary to monitor the self-heating effect. It is advisable to take air samples for analysis, and to determine the degree of endogenous fire hazard using an early fire detection system. In regard to this hazard, experience acquired in the Kazimierz-Juliusz mine requires a specific approach to analysis and assessment of results obtained from air samples being taken, since the presence of shotfiring gases may considerably obscure the actual degree of this hazard.
- Mechanisation of works is an important determining factor for the safety of workers employed directly in mine face, which has considerable impact on face productivity. The powered roof supports effectively protects the immediate face area and makes it easier to control inflow of the product from caving area. The armoured conveyor with its reinforced structure and properly selected power source guarantees continuous output reception with a relatively low breakdown frequency. A sliding station coupled with the above-mentioned elements allows relatively simple movement of the face equipment. Hydraulically powered drills used in the Kazimierz Juliusz mine (specially designed for the use in this exploitation system) allow trouble-free boring of shot-holes up to 50 m-long. Their relatively low weight and overall dimensions guarantee that they are easy to manoeuvre in mine face. However, using this sort of mechanisation (in particular mechanised sections and a reinforced armoured conveyor) requires that suitable transport routes covering the whole exploitation field be built.
- The relatively low operational costs of production presented above ~ 4 €, apply to the so-called face cost in Kazimierz-Juliusz mine. Therefore, they cover all operations connected

with winning process and haulages in a given extraction gallery. In order to obtain complete view of production costs it is also necessary to take into account the so-called "division" costs, which among other things include costs of transport, ventilation, drainage, methane removal, main haulage, energy and machine service, etc.

International Mining Forum 2010, Liu et al. (eds) © 2010 Taylor & Francis Group, London, UK. ISBN 978-0-415-59896-5

# Exploitation of thin hard coal beds in Poland – strategic decisions at the threshold of the $21^{st}$ century

Artur Dyczko, Jacek Jarosz
*Mineral and Energy Economy Research Institute of Polish Academy of Sciences, Krakow Poland*

ABSTRACT: The article discusses coal mining situation in Poland. It also defines hard coal deposits classification depending on their thickness, and presents the condition of recoverable and operative resources according to bed thickness, as well as presents their share in up to 1.5 m thick deposit production. The directions of thin beds exploitation technology development in national conditions have been signalized and a need to change the place of thin beds in the structure of hard coal mining in Poland has been shown.

KEYWORDS: mining, resources, hard coal, thin beds

## 1. INTRODUCTION

Primary energy consumption in 25 UN countries is on the level of 2.4 billion Mg of conventional fuel units, including 18% from hard and brown coal (Kicki, Sobczyk 2007). However, limited reserves of UN's energy carriers, such as petroleum, natural gas, and hard and brown coal, sustain hope that coal will remain its position on the UN energy market. It seems highly possible, since within last years petroleum and natural gas prices are high and it is expected that with the increasing global demand on oil and gas they will remain so.

At present, hard coal is produced in only six countries of the UN. Due to output quantity, they can be ordered as follows: Poland, Germany, Great Britain, the Czech Republic, Spain and Romania. Poland is the greatest producer of hard coal in the UN. Poland's coal production constitutes over 50% of Union's production, 59% of power coal and 39% coking coal. Poland is the second, after Germany, producer of coking coal in the UN.

In Poland we deal with hard coal production decrease for years. Polish mining had its finest period in the 70's and 80's of the $20^{th}$ century. Back then, hard coal production was around 200 million Mg/year. Maximum output of 201 million Mg was reached in 1979. It has turned out to be excessive quantity in relation to needs of the changed political and economic conditions of the 90's. Since then restructuring actions are taking place in Polish mining to adjust the sector to market economy conditions, as well as to increase Polish mining competitiveness on the world's markets (Paszcza 2010).

## 2. HARD COAL RESERVES IN POLAND

Recoverable resources of all managed national hard coal reserves according to the state from $31^{st}$ December 2008 were 16,002 million Mg. As far as economic resources are concerned, which are mainly recorded in active mines deposits and constitute the basis for mining planning and designing, they amounted to 4,898 million Mg.

Average complete operative reserves sufficiency to a scale of whole sector, at the rate of output established by coal companies and independent mines in 2009 is about 37.7 years.

Current state of hard coal reserves base is a consequence of changes in active mines deposits evaluation resulting from the rules of market economy implementation and consecutive restructuring processes.

## 3. THE ROLE AND PLACE OF THIN BEDS IN HARD COAL RESERVE BASE IN POLAND

Classification of hard coal and ores deposits in Poland according to their thickness is presented in the table 3.1. (Piechota 2004)

Table 3.1. Thickness classification of hard coal deposits

| Hard coal deposits classification | Thickness range, m |
|---|---|
| thin deposits | to 1.5 |
| medium deposits | 1.5 to 4.0 |
| thick deposits | over 4.0 |

One of the basic parameters influencing the choice of coal exploitation technology is deposits thickness. The minimal bed thickness in accordance with the valid criteria of balance should not be less than 1.0 m. Only beds of thickness greater than 1.2-1.5 m are qualified to industrial reserves in view of exploitation effectiveness, and consequently also the kind of mechanization used in mines.

Aiming at all costs to production effectiveness improvement in conditions of combined cutter loader technique, is forcing majority of mines to resign from exploitation of beds of thickness smaller than 1.5 m. The process of resignation from thin beds exploitation reflects in their constantly dropping share in reserves base, shown by Figure 3.1.

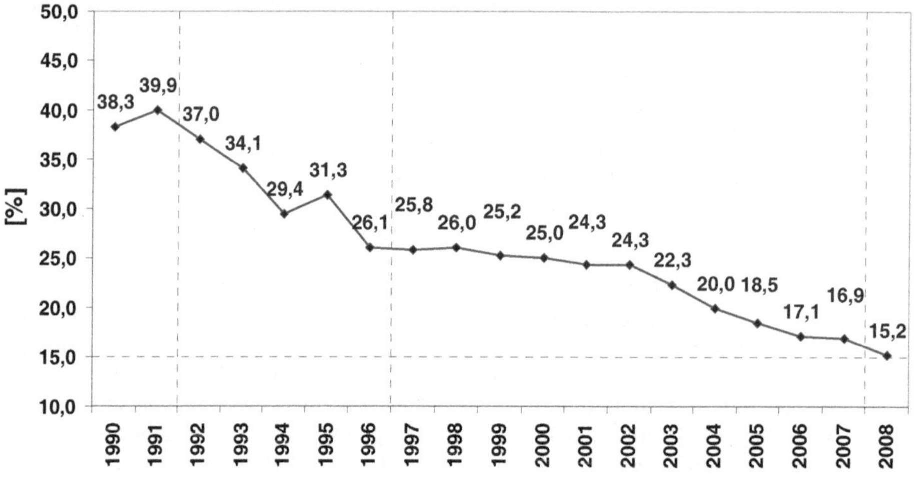


Figure 3.1. The economic resources share in beds of thickness smaller than 1.5

Detailed distribution of reserves base according to management degree and deposit accessibility is presented in the Figure 3.2.

Comparing output quantity form particular ranges of bed thickness, it can be stated that thin beds exploitation share is considerably lower than it results from their share in reserves and it shapes on the level of 1%, what is presented in Table 3.4.

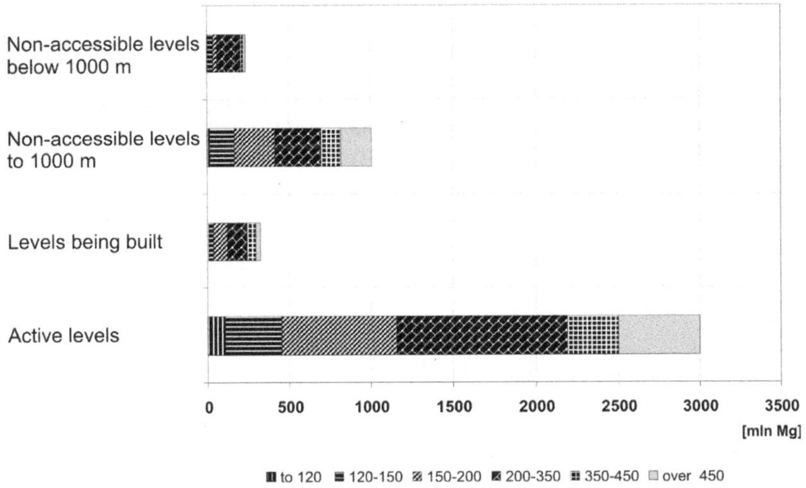


Figure 3.2. Economic resources according to bed thickness and the degree of accessibility (according to the state from 31.12.2008)

Table 3.4. Coal output according to bed thickness (Sikora 2007)

| Bed thickness | Percentage share in production | Percentage share of economic resources |
|---|---|---|
| [m] | [%] | [%] |
| to 1.5 | 1% | 17.1% |
| 1.5–2.0 | 11.4% | 23.7% |
| 2.0–3.5 | 74.6% | 33.5% |
| 3.5–4.5 | 5.4% | 8.8% |
| above 4.5 | 7.6% | 16.9% |

The data presented above clearly shows the direction coal mining industry in Poland heads. The best results within output concentration are gained at the range of walls height 2.0÷3.5 m. Moreover, almost 75% of the output comes from that range. The second range attractive for high concentration accessibility is the height range 3.5÷4.5 m. However, only 5.4% of output comes from this range. Some Polish experts claim that the chances of exploitation concentration increase should also be seen in thin beds exploitation conditions, on condition that the plough technique will be used.

Some mines, having significant reserves in thin beds, do not exploit them at all. It also concerns the best and the greatest mine in Poland LW Bogdanka S.A. (at present privatized) that owns 36% of industrial reserves in beds of thickness below 1.5 m and it does nor exploit them. Nowadays, mines exploit beds of thickness up to 1.5 m only when it is necessary, e.g. in case of danger (crump threat, spontaneous ignition), or when they do not own any other beds.

The above analysis of reserves in thin beds has shown that the share of thin beds (to 1.5 m) in Polish mining is significant, and in general industrial reserves amounts to 15.21%. Such situation confirms a very concerning tendency observed these days, that as thick beds become used up the role of thin beds will grow even more.

Form the engineering point of view, the main parameter determining the choice of exploitation technology is the thickness of chosen beds. At present two basic techniques to extract thin coal beds are used in world's mining: plough and continuous miner. The plough technique is used in thin and very thin beds. Continuous miner, on the other hand, is used in walls of height 1.4 m to 4.5 m (Krauze 2002, Sikora 1997, Tor et al. 2006)

To exploit thin coal beds in Poland in majority of cases coal combined cutter loaders were used, to a smaller degree coal ploughs. The number of ploughs in use in the first half of the 90's has been successively and quickly decreased to a complete elimination (Karbowik et al. 1995; Karbowik 2007).

Lately, in Polish mining two attempts of thin beds exploitation (thickness below 1.5 m) were taken. The first experiment was conducted in belonging to Jastrzębska Spółka Węglowa coal mine "Zofiówka" in August 2009. Now, plough wall exploitation has finished and conversions period is taking place. The other experiment is plough installation that is being activated at the moment (April 2010) in Polish best mine LW Bogdanka S.A. Both installations are more widely characterized below.

## 4. EXAMPLE S OF HARD COAL THIN BEDS EXPLOITATION IN POLAND AT THE THRESHOLD OF 21$^{ST}$ CENTURY

### 4.1. *The exploitation of hard coal thin beds in KWK Zofiówka conditions*

Jastrzębska Spółka Węglowa S.A. which owns the mine Zofiówka is the only in Poland and the greatest in Europe producer of coking coal type "hard" that is the main component of carbonic mixture used for producing blast-furnace coke that has high mechanical strength, as well as low reactiveness. The company was established 1.04.1993, joining seven independent mines which technological systems ensure production realization at the level of average annual daily output of 54,800 ÷ 56,300 Mg/day.

Zofiówka coal mine running since December 1969 is a producer of high quality ortho-coking coal type 35.2 used in blast-furnace coke production of finest quality. At present mine's daily output is on average 9,600 Mg/day (Tor et al. 2006)

During the years 2005–2008 in JSW actions were taken to run the first plough installation in KWK Zofiówka conditions. Chain of preparing actions allowed to design B-1 wall in mine's 406/1 bed as fully equipped in plough complex. During designing works it was assumed that plough complex should enable medium beds exploitation, too. It secures the continuity of work of the modern complex in case of unexpected geological or mining problems in the course of cutting of thin beds. As an effect of a conducted analysis technical parameters were established for plough complex ensuring its effective work in KWK Zofiówka. It was assumed that plough should get coal beds of thickness 1.0 – 2.2 m and has to have ability of fluid head height regulation at the minimal range 250 mm.

B-1 wall exploitation was done in the bed of thickness from 1.05 to 1.30 m, in the examined part of deposit at the depth of about 869 to 916 m. The length of exploited wall was 244 m and the coasting 800 m. Wall lengthwise inclination changed at the range 0÷8$^0$, and crosswise inclination about 4÷6$^0$. Average primary temperature of rocks is 37÷42$^o$C. In bed roof were respectively: layer of shale, dark-grey changeably carbon laminated with the thickness of about 0.10÷0.90 m; complex of shale, gritty mudstone and arenaceous shale with the thickness of about 10÷15 m. In the 406/1 bed floor were present: shale, dark-grey with single carbon laminates with thickness of about 0.0÷0.55 m, sandstone with thickness of 0.0÷0.70 m, layered arenaceous shale with sandstone laminas (thickness of about 1.5-2.5 m).

Technical equipment of B-1 wall – the first for 20 years plough wall in Poland – contained (Tor at al. 2006):

- GH 1600 type plough with the work range 0.98÷1.745 m and ability of fluid head height regulation at the range to 0.3 m and with maximum shortwall 0.2 m – moving on the wall conveyor with working speed 2.24 m/s;
- GH PF-4/1032 type wall carding conveyor with head dump;
- GLINIK-09/23-POzS type mechanized lining sections together with necessary equipment (including 9 extreme/terminal sections with changeable pitch with limits from 1.75 to 2.25 m) equipped with PMC-R type electro-hydraulic steering and sequence sprinkling of plough head;
- grot 1100 bottom gate carding conveyor 60 m long and with trough profile outside width of 1100 mm, equipped with KDBW 800-1200 dynamic crusher;

- UPP-2 device to wall conveyor main power separator with plough drive, and to bottom gate carding conveyor rebuild with turning station of belt conveyor;
- strandless device to wall conveyor assistance power separator together with UKP-1/N type of plough drive,
- 2 floor chargers DH-L800 type in bottom gate and top gate galleries at the area of plough drives;
- 5 belt conveyors with the belt width of 1,200 mm (2 x Pioma-1200 and 3 x Intermet-1200) and total length of 1,800 m.

B-1 wall starting in 406/1 bed in KWK Zofiówka took place on 17$^{th}$ August 2009. A half of September and the whole October 2009 were for the B-1 wall operating staff the months of intensive practical learning of work in the wall with activated fully automatic production processes.

As the result of conducted automatic exploitation, complex possibilities in the whole area of factors significantly influencing quantity of wall's output and staff employed in the wall safety, parameterization were tested, with positive outcome.

Leading of B-1 wall front in Zofiówka mine was empirically set within the limits of 12-16 m advance of main power by assistance power and 6-10 m advance in the wall's line middle showing chord between drives. Such wall setting minimized to a large degree movement resistance of plough chain, trough waste in working areas of conveyor channel and plough drive engines load – without observable changes in wall conveyor work characteristics and counteracts gravity forces causing complex "creeping" to bottom gate (Kubaczka et al. 2010).

A crucial element and condition to successful exploitation was the proper accomplishment of preparatory underground mining, which is presented in details in the Figure 4.1. The designed door-frame of roof support ŁPZof (Prusek 2009) was asymmetrical and composed of two identical roadside arches and two roof arches of different length.

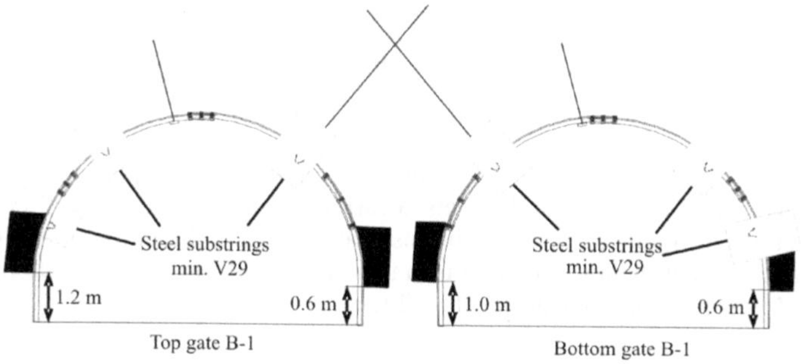


Figure 4.1. Asymmetrical construction of the longwall gates and the reinforcement

In the longwall gate they were built up with longer roof arch on the wall side. Identical arch curves were joined with regulated pleat what enabled door-frame adjustment to bed roof position.

In order to keep optimal output from plough wall, bed floor in bottom gate was at the average height of about 1.0 m above the floor, measured by side wall from the wall and 0.6 m from the non-wall side wall. In the top gate bed floor was above the floor, at the height of about 0.6 m on the wall side and 1.20 on the non-wall side wall as shown at Figure 4.2 and 4.3 (Kubaczka et al. 2010).

Constant monitoring of the wall work, as well as influence on many of its parameters enabled the mine to look at the exploitation process through a completely new and until now inexperienced organization prism.

Jastrzębska Spółka Węglowa, as the first in Poland (after over 20 years)that had implemented in Zofiówka Mine fully automatic low bed exploitation system, observes, with the flow of time and growing experience of its staff responsible for production quantity, constant increasing tendency in the area of efficiency possibilities of the complexes mentioned.

At the same time, it should be noticed that despite initial fears concerning miners work comfort at the wall discussed, in spite of small measurements of people's passage, not much larger than 0.4-0.6 m, it did not bring any negative opinions staff employed at the wall.

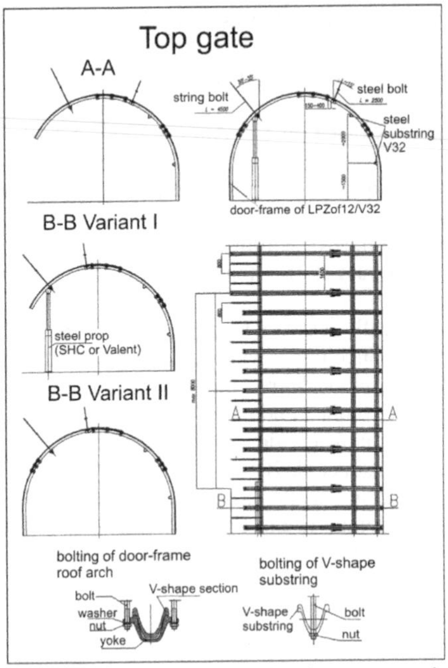


Figure 7. Combined steel set and anchor support at the junction of the top gate and B-1 plough longwall in 406/1 coal seam (Prusek 2009)

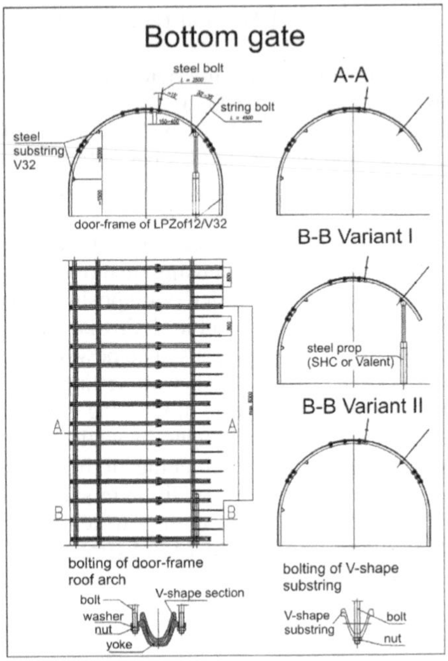


Figure 8. Combined steel set and anchor support at the junction of the bottom gate and B-1 plough longwall in 406/1 coal seam (Prusek 2009)

Economic results and output reached on the level of average combined wall output in Poland, i.e. 3,000 Mg/day in extreme conditions of natural threats occurrence (high methane threat, high degree of rock burst threat and difficult climate conditions) have justified the employment of plough technique. It is estimated that thanks to plough technique used in Zofiówka the exploitation of almost 113 million Mg is possible (Kubaczka et al. 2010)

### 4.2. *The exploitation of hard coal thin beds in LW Bogdanka S.A.*

LW Bogdanka S.A. is the only mine exploiting hard coal in Lublin's coal basin. The Bogdanka Mine reaches over 5 million Mg of output from two walls being exploited (and having one wall standing by) at the wall length of 300 m and coasting reaching 3,000 m, as well as average wall height 2.0 m.

LW Bogdanka is at present a mine of the highest technical and economic results in Polish mining. It concerns not only the output concentration indexes, work efficiency, but also economical outcomes, such as profit and sale for each one employed, common profit at the scale of whole mine.

Taking gaining the economic ability to exploit seams with thickness from 1.2 m as strategic for mine's development, in 2008 actions to start first plough wall in LW Bogdanka in the first quarter of 2010 were taken.

Main project assumptions concentrating on the improvement of net coal output ratio, as well as maintaining the hitherto output concentration were adjusted to technical and technological capacities of GH 42 plough device.

Consequently, in April 2010 starting of 1/VI wall in 385/2 poz. 910 bed localized in the eastern area of Bogdanka mine has begun. The wall lead in the direction from the South to the

North in the lot between the eastern gates 4/385 and 4/385; from the western field border (on the right field gate 5 in bed 385/2) characterizes the following mining-geological conditions:
- bed thickness 1.3–1.5 m,
- inclination 1°10'÷1°40' in the western direction,
- resistance on coal compression 8–19 MPa,
- gettability factor ap. 1,6.

In May 2010 – after the first phase of plough complex starting was finished, over a month after starting – the wall reached efficiency close to the one presumed on the level 10,000 Mg of coal a day. Totally, in April 2010 the wall 1/IV realized presumed progress of 170 m, reaching at the end of April 202 m. Average advance of the wall in the analyzed period of time was 8.1 m, and average daily output was 5,900 Mg.

At present, i.e. at the end of August 2010, so 5 months after the plough wall has been started in LW Bogdanka, G-1 Unit reaches at times maximum efficiency on the level of 17,000 Mg/day, trying to keep the assumed daily progress at the level of 12 m, what gives planned efficiency at the level of 10,000 Mg/day.

Performing pre-design analysis has shown that in reserves base of the mine we deal with considerable share of beds below 1.5 m, as in Polish mining in general.

In the LW Bogdanka conditions continuous miner technique allowed to gain daily output at the level of almost 20,000 Mg with the thickness of 2.0–2.5 m and to 15,000 Mg/day in walls 1,6–1,8 m.

Until April current year in Lublin's mine a question has been asked if it was possible to reach close production results in seams below 1.5 m. The data presented above indicate that plough technique in LW Bogdanka passes the exam and planned – after the investment program to build up mine output capacity is finished in 2012 – coal exploitation on the level of 9.3 million Mg net a year from four simultaneously active walls, i.e. 2 with continuous miner and 2 with plough is absolutely real.

## 5. SUMMARY AND CONCLUSIONS

Hard coal reserves base in Poland in the last few years has gone through large changes resulting from drastically implemented rules of market economy aiming at adjusting hard coal mining to new economic conditions.

Those new rules are the consequences of transformations in Middle-Eastern Europe in the last twenty years and Poland joining the EU in 2004. The last fact precisely will have a significant meaning for coal presence as the energy carrier in Poland, and consequently shaping reserves base in years to come. The significant share of hard coal in energy balance of Poland forces a need to exploit thin beds as well. Beds that have their important share in reserves base magnitude in Poland. The presented examples of their exploitation in Zofiówka and Bogdanka Coal Mines show that effective exploitation is possible due to development achieved in the area of plough technique in recent years.

The example of plough wall exploitation with thickness from 1.2 to 1.5 m in LW Bogdanka and reaching daily output at the level over 10 thousand Mg confirms large possibilities of the applied technique.

## REFERENCES:

Darski J., Kicki J., Sobczyk E. 2001. Raport o stanie gospodarki zasobami złóż węgla kamiennego. Studia i Rozprawy nr 85. Wyd. IGSMiE PAN, Kraków. (In Polish)

Dyczko A. 2007. Thin coal seam, their role in the reserve base of Poland. Proc. International Mining Forum 2007, str. 81-89, Taylor & Francis Group plc, London, UK.

Karbownik A., Woźnica E. 1995. Eksploatacja pokładów cienkich w polskim górnictwie węgla kamiennego. Wiadomości Górnicze nr 9. Katowice. (In Polish)

Karbownik A. 2007. Zasoby węgla kamiennego w Polsce. Wyd. Karbo, nr 2. (In Polish)

Kicki J., Sobczyk E. 2007. Węgiel kamienny jako podstawowy nośnik energetyczny w Polsce z perspektywy bazy zasobowej. Polityka Energetyczna, Tom 10 Zeszyt specjalny nr 1. Wyd. IGSMiE PAN, Kraków. (In Polish)
Kicki J., Sobczyk E., Jarosz J., Dyczko A., et al. 2007. Analiza stanu techniki strugowej na świecie i możliwości jej wdrożenia w LW Bogdanka S.A. (unpublished).
Kicki J., Sobczyk E. 2006. Restrukturyzacja górnictwa w Polsce a struktura i wystarczalność zasobów węgla kamiennego. Studia, Rozprawy, Monografie nr 134. Wyd. IGSMiE PAN, Kraków. (In Polish)
Krauze K. 2002. Ocena możliwości technicznych efektywnego wybierania cienkich pokładów. Proc. of Szkoła Eksploatacji Podziemnej. Wyd. IGSMiE PAN, Kraków. (In Polish)
Kubaczka C, Zabój K., Witamborski Z., 2010. Wdrożenie pierwszej, zautomatyzowanej ściany strugowej w Polsce. Strugowa eksploatacja węgla w KWK „Zofiówka". Proc of Szkoła Eksploatacji Podziemnej. Wyd. IGSMiE PAN, Kraków. (In Polish)
Paszcza H. 2005. Analiza zasobów węgla kamiennego w Polsce z uwzględnieniem ich ekonomicznego wykorzystania w świetle przepisów prawa unijnego. Gospodarka Surowcami Mineralnymi, Tom 21 Zeszyt specjalny. Wyd. IGSMiE PAN, Kraków. (In Polish)
Paszcza H., 2010. Procesy restrukturyzacyjne w polskim górnictwie węgla kamiennego w aspekcie zrealizowanych przemian i zmiany bazy zasobowej (unpublished).
Piechota S. 2004. Technika podziemnej eksploatacji złóż. Część I: Podstawowe zasady i technologie" Wydawnictwa Naukowo-Dydaktyczne AGH Kraków. (In Polish)
Prusek i inni 2009. Wykonanie projektu obudowy podporowo-kotwiowej skrzyżowania wyrobisk korytarzowych ze ścianą strugową B-1 w pokładzie 406/1 w KWK „Zofiówka". Dokumentacja pracy badawczo-usługowej opracowana przez Zakład Technologii Eksploatacji i Obudów Górniczych GIG w Katowicach (unpublished).
Sikora W. 1997. Pozytywne i negatywne strony technik urabiania: strugowej i kombajnowej w systemach ścianowych - tendencje rozwoju. Proc. of Szkoła Eksploatacji Podziemnej. Wyd. IGSMiE PAN, Kraków. (In Polish)
Sikora W. 2008. Innowacje w scenariuszach rozwoju mechanizacji procesów eksploatacji w górnictwie węgla kamiennego. Foresight. Gospodarka Surowcami Mineralnymi i Energią nr 2008. Wyd. IGSMiE PAN. (In Polish)
Stopa Z. 2008. Perspektywy eksploatacji cienkich pokładów węgla kamiennego w LW „Bogdanka" S.A. Proc. of Szkoła Eksploatacji Podziemnej, Wyd. IGSMiE PAN, Kraków. (In Polish)
Tor A., Wróbel A., Łukosz M., Olma R., Kapcia J. 2006. Nowoczesna technika strugowa jako sposób na zwiększenie racjonalnego wykorzystania złóż węgla w JSW S.A. KWK „Zofiówka" Proc. of Szkoła Eksploatacji Podziemnej, Wyd. IGSMiE PAN, Kraków. (In Polish)
Turek M. 2002. Analiza ekonomicznej efektywności wybierania cienkich pokładów węgla kamiennego. Proc. of Szkoła Eksploatacji Podziemnej. Wyd. IGSMiE PAN, Kraków. (In Polish)
Turek M. 2003. Rola i uwarunkowania ekonomicznej eksploatacji pokładów cienkich węgla kamiennego. Proc. of Szkoła Eksploatacji Podziemnej. Wyd. IGSMiE PAN, Kraków. (In Polish)

International Mining Forum 2010, Liu et al. (eds) © 2010 Taylor & Francis Group, London, UK. ISBN 978-0-415-59896-5

# Author indeks

T - #0061 - 071024 - C0 - 244/170/21 [23] - CB - 9780415598965 - Gloss Lamination